Enzymes in Cardiology

Diagnosis and Research

Enzymes in Cardiology

Diagnosis and Research

Edited by
David J. Hearse
Head of Myocardial Research and
Senior Lecturer in Chemical Pathology
St. Thomas's Hospital Medical School London

Joël de Leiris
Maitre Assistant
Laboratoire de Physiologie Comparée
Université de Paris Sud, Orsay

Associate Editor
Daniel Loisance
Chief Clinician
Hôpital Henri Mondor Créteil

A Wiley–Interscience Publication

RC685
I6
E58
1979

JOHN WILEY & SONS
CHICHESTER · NEW YORK · BRISBANE · TORONTO

Copyright © 1979, by John Wiley & Sons, Ltd.

All rights reserved

No part of this book may be reproduced by any means, nor transmitted, nor translated into a machine language without the written permission of the publisher.

Library of Congress Cataloging in Publication Data:

Main entry under title:

Enzymes in cardiology.

'A Wiley–Interscience publication'.
Includes index.
   1. Heart—Infarction—Diagnosis. 2. Clinical enzymology. 3. Enzymes—Analysis. I. Hearse, David J. II. Leiris, Joël de. III. Loisance, Daniel.
RC685.I6E58   616.1'23'075   78-13633
ISBN 0 471 99724 2

Filmset in Northern Ireland at the The Universities Press (Belfast) Ltd, and printed at the Pitman Press (Bath) Ltd.

Contributors

J. P. Bourdarias	Department of Cardiology, Ambroise-Paré Hospital, Boulogne, France.
E. Braunwald	Department of Medicine, Harvard Medical School and Peter Bent Brigham Hospital, Boston, Massachusetts, United States of America.
G. R. Bullock	The Research Centre, CIBA Laboratories, Horsham, West Sussex, United Kingdom.
R. Coleman	The Department of Biochemistry, The University, Birmingham, United Kingdom.
D. Feuvray	Department of Comparative Physiology, University of Paris, Orsay, France.
M. F. Groseth-Robertson	Department of Cardiology, Cedars-Sinai Medical Centre, Los Angeles, California, United States of America.
P. Guéret	Department of Cardiology, Ambroise-Paré Hospital, Boulogne, France.
P. Y. Hatt	Emile-Roux Hospital, Limeil Brévannes, France.
D. J. Hearse	Myocardial Metabolism Research Laboratories, The Rayne Institute, St. Thomas' Hospital, London S.E.1., United Kingdom.
W. Th. Hermens	Department of Biophysics, State University of Limburg, Maastricht, The Netherlands.
Å. Hjalmarson	Department of Medicine, Sahlgren's Hospital, Göteborg, Sweden.
S. M. Humphrey	Myocardial Metabolism Research Laboratories, The Rayne Institute, St. Thomas' Hospital, London S.E.1., United Kingdom.

R. B. Jennings	Department of Pathology, Duke University Medical Centre, Durham, North Carolina, United States of America.
J. C. Kahn	Department of Cardiology, Ambroise-Paré Hospital, Boulogne, France.
J. K. Kjekshus	The Riks Hospital, Oslo, Norway.
A. Van der Laarse	Department of Cardiobiochemistry, Academic Hospital, Leiden, The Netherlands.
J. de Leiris	Department of Comparative Physiology, University of Paris, Orsay, France.
D. Maclean	Department of Medicine Harvard Medical School and Peter Bent Brigham Hospital, Boston, Massachusetts, United States of America.
P. R. Maroko	Department of Medicine, Harvard Medical School and Peter Bent Brigham Hospital, Boston, Massachusetts, United States of America.
R. H. Michell	The Department of Biochemistry, The University, Birmingham, United Kingdom.
D. W. Moss	Department of Clinical Enzymology, Royal Postgraduate Medical School, Hammersmith Hospital, London W.12., United Kingdom.
W. G. Nayler	The Cardiothoracic Institute, 2 Beaumont Street, London W.1., United Kingdom.
H. Nordbeck	Institute of Physiology, University of Gottingen, W. Germany.
R. M. Norris	Green Lane Hospital, Auckland, New Zealand.
L. H. Opie	Medical Research Council Ischaemic Heart Disease Unit, Department of Medicine, Groote Schuur Hospital, University of Capetown, South Africa.
C. J. Preusse	Institute of Physiology, University of Gottingen, W. Germany.
K. A. Reimer	Department of Pathology, Duke University Medical Centre, Durham, North Carolina, United States of America.

CONTRIBUTORS

L. G. T. Ribeiro	Department of Medicine, Harvard Medical School and Peter Bent Brigham Hospital, Boston, Massachusetts, United States of America.
R. Roberts	Cardiovascular Division, Washington University School of Medicine, St. Louis, Missouri, United States of America
T. J. C. Ruigrok	Department of Cardiology, University Hospital, Utrecht, The Netherlands.
W. E. Shell	Department of Cardiology, Cedars-Sinai Medical Centre, Los Angeles, California, United States of America.
A. M. Slade	The Cardiothoracic Institute, 2 Beaumont Street, London W.1., United Kingdom.
A. F. Smith	Department of Clinical Chemistry, Royal Infirmary, Edinburgh, United Kingdom.
B. E. Sobel	Cardiovascular Division, Washington University School of Medicine, St. Louis, Missouri, United States of America.
P. G. Spieckermann	Institute of Physiology, University of Gottingen, W. Germany.
A. Waldenström	Department of Medicine, Sahlgren's Hospital, Göteborg, Sweden.
J. H. Wilkinson	Late of the Department of Chemical Pathology, Charing Cross Hospital Medical School, London W.6., United Kingdom.
S. A. G. J. Witteveen	Department of Cardiobiochemistry, Academic Hospital, Leiden, The Netherlands.
A N. E. Zimmerman	Department of Cardiology, University Hospital, Utrecht, The Netherlands.

Contents

Foreword xiii
    P. Y. Hatt

Preface xv

1 **Cellular Damage during Myocardial Ischaemia: Metabolic Changes Leading to Enzyme Leakage** 1
    D. J. Hearse

2 **Biology of Experimental, Acute Myocardial Ischaemia and Infarction** 21
    R. B. Jennings and K. A. Reimer

3 **Structure and Permeability of Normal and Damaged Membranes** 59
    R. H. Michell and R. Coleman

4 **From Heart to Plasma** 81
    P. G. Spieckermann, H. Nordbeck and C. J. Preusse

5 **The Distribution, Inactivation, and Clearance of Enzymes** . . . 97
    R. Roberts and B. E. Sobel

6 **Tissue Enzymes** 115
    A. F. Smith

7 **Tissue Isoenzymes** 133
    A. F. Smith and J. H. Wilkinson

8 **The Measurement of Enzymes** 145
    D. W. Moss

9 **Enzymes and Routine Diagnosis** 199
    A. F. Smith

10 **Radioimmunoassay of Creatine Kinase Isoenzyme** 247
   R. Roberts and B. E. Sobel

11 **Enzymatic Estimation of Infarct Size** 257
   B. E. Sobel, J. K. Kjekshus and R. Roberts

12 **The Prediction of Infarct Size** 291
   W. E. Shell and M. F. Groseth-Robertson

13 **Clinical Experience with Infarct Sizing and Its Value in the Prognosis of Myocardial Infarction** 319
   J. C. Kahn, P. Guéret and J. P. Bourdarias

14 **Enzymatic Infarct Sizing: Factors Influencing the Choice of the Marker Enzyme** 339
   W. Th. Hermens, A. Van der Laarse and S. A. G. J. Witteveen

15 **Infarct size quantification present and future** 355
   R. M. Norris

16 **Experimental Models for the Study of Myocardial Tissue Damage and Enzyme Release** 379
   A. Waldenström and Å. Hjalmarson

17 **The Effect of Calcium on Myocardial Tissue Damage and Enzyme Release** 399
   T. J. C. Ruigrok and A. N. E. Zimmerman

18 **Reoxygenation, Reperfusion and the Calcium Paradox: Studies of Cellular Damage and Enzyme Release** 417
   D. J. Hearse, S. M. Humphrey and G. R. Bullock

19 **Morphological Correlates of Myocardial Enzyme Leakage** . . . 445
   J. de Leiris and D. Feuvray

20 **Preservation of Myocardium for Ultrastructural and Enzymatic Studies** 461
   G. R. Bullock

21 **Metabolic Manipulations: Tissue Damage and Enzyme Leakage** 481
   L. H. Opie and J. de Leiris

22 **Pharmacological Protection of the Hypoxic Heart: Enzymatic, Biochemical, and Ultrastructural Studies in the Isolated Heart** 503
  W. G. Nayler and A. M. Slade

23 **Pharmacological Limitation of Infarct size: Enzymatic, Electrocardiographic, and Morphological Studies in the Experimental Animal and Man** 529
  P. R. Maroko, D. Maclean, L. G. T. Ribeiro and E. Braunwald

24 **Metabolism, Enzyme Release, and Cell Death: Possibilities for Future Investigation** 561
  L. H. Opie

Index 569

Foreword

P. Y. Hatt

Enzyme release and myocardial cell damage are both closely related to the deterioration of the myocardial cell membrane and its permeability characteristics. The cell membrane is an efficient diffusion barrier which ensures the maintenance of the cellular integrity by insulating the intracellular milieu from the noxious components of its environment and by preventing the outward diffusion of intracellular macromolecules such as enzymes. The loss of the semi-permeability properties of the cell membrane is one of the earliest manifestations of cellular damage and it is this myocardial lesion which will be extensively discussed in many chapters of this book.

The word 'lesion' is derived from the latin verb *'laedere'* meaning to wound or to do an injury. When applied to the cell this damage would be evidenced by changes in morphology, biochemistry, electrophysiology, etc. Until recently myocardial cell damage has been classified morphologically on the basis of structural changes revealed by light microscopy. These rather descriptive changes included 'coagulation necrosis', 'contraction bands', 'hyalinization', etc. The advent of electron microscopy has permitted far more specific ultrastructural changes to be described. Furthermore, it is now possible to correlate structural and functional changes within sub-cellular organelles such as mitochondria, myofibrils, tubular systems and sarcolemma, to differing facets of myocardial damage. Thus, for example, the observation of disappearing glycogen granules and increasing lipid deposits in the cell during ischaemia may be specifically related to ischaemia-induced changes in the metabolic pathways of glycolysis, glycogenolysis and lipolysis. Similarly the observation of cell swelling can be attributed to intracellular oedema resulting from the ischaemia-induced impairment of pumps which are located in the sarcolemma and which control water and ion movements.

Other ultrastructural characteristics of myocardial cell damage are less easily related to specific biochemical changes. Ischaemia-induced changes in mitochondria include for example: loss of dense intracristal granules, swelling, disruption of cristae and the appearance of dense calcium-rich deposits. The molecular mechanisms underlying these structural alterations are complex but may include changes in ATPase activity, changes in the chemoosmotic properties of various mitochondrial membranes and decreased availability of cellular high energy phosphates. Ischaemia-induced myofibrillar lesions, such as the lysis of the myofilaments, are also complex and inadequately understood but might be related to proteolysis secondary to lysosomal disruption or to inadequate supplies of energy necessary for the maintenance of highly organized macromolecular structures such as the myofibrils.

Clearly, much remains to be learned about myocardial cell damage, for example, the nature of reversible and irreversible damage and the critical point of transition between these states; the effects of transient ischaemia and reperfusion; the relationship between myocardial ischaemia and other lesions with closely related mechanisms such as acute heart overload. The investigation of these molecular mechanisms will undoubtedly be advanced by a better understanding of the phenomena of myocardial enzyme release and its relationship to myocardial cell injury.

Preface

The phenomenon of enzyme leakage has been recognized since the beginning of this century and for over 50 years it has been used for the detection and assessment of tissue injury. From its early empirical nature diagnostic enzymology has developed into a highly sophisticated and precise science which has been exploited for the benefit of mankind, particularly in the sphere of cardiac disease.

Despite its established history, diagnostic enzymology as a methodology has grown faster than our knowledge of the principles upon which it is based. Despite our ability to use enzyme leakage to pinpoint an exact disease process in a specific tissue or group of cells, to measure the mass of tissue involved, and maybe even predict the ultimate extent of injury we still remain ignorant about many of the fundamental cellular processes which lead to, and support, the phenomenon. It is the objective of this book to discuss and define the principles underlying these complex processes.

The book commences with a consideration of the biochemical changes which predispose to cellular injury and in particular the changes which occur during ischaemia and evolving myocardial infarction. The ischaemia-induced loss of intracellular enzymes is related to changes in cellular ultrastructure. The complex problems associated with the translocation of enzyme macromolecules from the cytoplasm to the extracellular space is discussed as is the movement of these molecules from the interstitial fluid to the circulating blood. These latter processes require a consideration of the respective roles of coronary and lymphatic drainage, the distribution of the enzymes in various fluid compartments and the mechanisms responsible for the clearance of the enzyme from the blood.

Having defined some of the basic principles governing enzyme leakage we have directed the emphasis of the book towards the concept of tissue specificity and the way in which the leakage of specific enzymes, groups of enzymes or isoenzymes can be used to detect and assess tissue damage. The meaningful measurement and interpretation of enzyme release profiles necessitates a full understanding of the complexities of the measurement of enzyme activity and the instrumentation required to accomplish it. These topics are covered in the early chapters of this book.

The cell contains many hundreds or thousands of enzymes and the practical considerations which determine which of these enzymes are best suited to the detection of myocardial damage are discussed as is the application of enzyme release to routine diagnosis.

From the relatively straightforward and well established use of enzyme leakage for the detection of damage has grown the concept that the phenomenon may be used to accurately quantitate and possibly even predict tissue damage. This important and controversial aspect of myocardial enzyme leakage is given major consideration in this volume.

While enzyme leakage is a proven and indispensible diagnostic tool its value is by no means limited to the practicing clinician and is utilized by many investigators as a research tool for probing the molecular complexities of disease processes and cellular damage. We have therefore attempted in the concluding chapters of this book to give an insight into enzyme leakage as an instrument of study and show how it may further extend our knowledge of the workings of the cell in both health and disease.

We would like to take this opportunity to thank the many eminent scientists who have contributed to this volume. We would also like to record with deep regret the death of Professor Henry Wilkinson who has contributed so enormously to the understanding of enzymes and isoenzymes. At the time of his death Professor Wilkinson had partly written Chapter 7 of this book and we are indebted to Dr. Alistair Smith for completing, at very short notice, this valuable contribution. Finally, we would like to thank those whose names are not appended to chapters: Mrs. Christine Boles for secretarial help and endless patience and Dr. Howard Jones of John Wiley for making the book possible.

DAVID J. HEARSE
JOËL DE LEIRIS

1978

CHAPTER 1

Cellular damage during myocardial ischaemia: metabolic changes leading to enzyme leakage

D. J. Hearse

INTRODUCTION	1
THE ISCHAEMIC PROCESS	3
THE LEAKAGE OF CYTOPLASMIC CONSTITUENTS	11
THE FATE OF LEAKED MATERIALS	13
CHOICE OF MARKERS OF ISCHAEMIC DAMAGE	15
CONCLUDING REMARKS	15
REFERENCES	16

INTRODUCTION

This book deals with the detection and assessment of cardiac injury through the measurement of myocardial enzyme leakage. Before we are able to exploit fully the diagnostic or research potential of enzyme leakage, it is necessary to understand the mechanisms controlling leakage and the origins of the leakage itself. While a number of pathological conditions result in the loss of myocardial enzymes, the one which will dominate this book is myocardial ischaemia. The aim of this first chapter is to describe in general terms the cellular conditions which characterize the process of myocardial ischaemia and which eventually lead to the leakage of cytoplasmic constituents from the intracellular to the extracellular space.

The broad overview presented in this introductory chapter has, by necessity, been assembled from a multitude of studies using different models, different species, and different conditions of oxygen deprivation. Caution must therefore be exercised in extrapolating many of the findings to the clinical situation. In particular, in this and subsequent chapters, it is most important to distinguish between ischaemia and anoxia or hypoxia.

Ischaemia, which is the more frequently encountered clinical condition is conventionally defined as a lack of blood in a particular tissue. However, a more broadly based definition which accounts for the dynamic nature of the condition and accounts for the removal as well as the supply of compounds to the myocardium is more appropriate. Essentially, ischaemia represents an imbalance between the myocardial demand for, and the vascular supply of, coronary blood. Not only does this create a deficit of oxygen, substrates, and energy in the tissue, but also, and of considerable importance, it results in an insufficient capacity for the removal of potentially toxic metabolites such as lactate, carbon dioxide, and protons. The total cessation of coronary flow is not a prerequisite of myocardial ischaemia, it rarely occurs clinically and even under experimental conditions with multiple coronary artery ligations, the collateral circulation may provide substantial perfusion in the ischaemic zone. Indeed, myocardial ischaemia could arise without a reduction in coronary flow under circumstances where there was an inadequate vascular response to an increased work or energy demand on the heart.

Anoxia and hypoxia are totally different[74] to ischaemia in both their origins and consequences and are conditions which are less frequently observed clinically. In anoxia or hypoxia the oxygen delivery to the myocardium is reduced by removing all or some of the oxygen in the coronary supply. Thus while the Po_2 is reduced, coronary flow may be normal or even elevated, and substrate delivery and metabolite removal may also be normal.

In considering ischaemic or hypoxic damage it is important to appreciate that they are not static conditions, but dynamic processes. Thus as discussed in Chapter 2, ischaemic tissue evolves[37,38] through reversible to irreversible damage, cell death and tissue necrosis; the whole process representing myocardial infarction. The rate at which these processes occur and the nature of the processes themselves are influenced by a number of factors including the severity of ischaemia or hypoxia, the age, sex, and species of the tissue under investigation, the hormonal, nutritional and metabolic status of the tissue and the coexistence of other disease processes.

A striking and important characteristic of ischaemia and to a lesser extent hypoxia, is its macroscopic and microscopic heterogeneity.[35,37] Varying conditions of work load and tissue perfusion may create a transient or patchy ischaemia. In the latter instance islands of severely ischaemic tissue may be interspersed with, or lie adjacent to, areas of normal tissue. Within a single ischaemic area, regional differences may exist with concentric zones of decreasing ischaemia radiating outwards from the core of the ischaemic area. This situation may create border zones[12,27,68] of marginally damaged tissue which separate severely ischaemic from normal tissue. Within these border zones gradients of metabolism, electrophysiology and flow may exist. Some investigators[1,9,54] believe these gradients to be very sharp such that within a single cell one may find two adjacent mitochondria one being aerobic and fully functional, the other being anaerobic and non-functional. Other investigators[27] believe that the gradients may be less abrupt with a gradual transition of damage creating a quantitatively significant border zone of intermediately damaged cells. Superimposed upon this heterogeneity there may be transmural gradients[67] with the endocardium exhibiting more advanced ischaemic damage than corresponding epicardial tissue.

The heterogeneity of damage makes it very difficult to develop adequate experimental models of ischaemia (see Chapter 16); a problem which is compounded by the as yet unclarified controversy over the suggestion that coronary arterial thrombosis need not be the primary event in the initiation of tissue ischaemia and myocardial infarction.[22]

A further point to consider when assessing the consequences of ischaemia or anoxia upon the myocardium is tissue subtypes within the organ itself. In general, the consequences of ischaemic damage have been applied to the contractile tissue of the heart. However, the heart is not composed solely of muscle cells and the effect of ischaemia upon the conducting tissue and the vascular tissue warrants equal consideration. There is for example evidence[49] for a differential susceptibility to damage between contractile and conducting tissue. Similarly, the susceptibility of myocardial vascular tissue and vascular responses to ischaemia is well known and is critical. Thus although the contractile tissue of the heart may be quantitatively the most significant component, the responses of that tissue to ischaemia may be considerably influenced by the responses of other tissue types.

THE ISCHAEMIC PROCESS

Despite the problems of the heterogeneity of ischaemia and the likelihood that at a single moment in time different cells within an ischaemic area will have developed different extents of damage,[38] a remarkably detailed picture[4,6,67,82,84,92] has emerged of the sequence of events thought to occur during both ischaemia and hypoxia. The following section attempts to give a very generalized and simplified overview of some of the deleterious changes which are initiated by myocardial ischaemia. Some of these changes, which affect cellular metabolism, electrical activity, contractile function, vascular responsiveness, and tissue ultrastructure are depicted in Figure 1. This figure is not intended to convey the impression that the events occur in a strict sequence or in the exact order listed. Some of the changes may occur simultaneously and the order of others may vary from condition to condition, indeed the sequence of the changes and the times ascribed to them can only be speculative, but are most likely representative of the situation prevailing following the onset of very severe ischaemia. The dynamic nature of the process and the individual changes should also be appreciated, thus although Figure 1 only indicates the onset of a change, the individual changes may continue for some time.

SECONDS

ONSET OF SEVERE ISCHAEMIA

Reduced oxygen availability
Disturbances of transmembrane ionic balance
Utilization of dissolved oxygen
Cyanosis
Reduction of mitochondrial activity and oxidative metabolism
Reduced ATP production
Reduction of creatine phosphate stores
Reduction of amplitude and duration of action potential
Leakage of potassium ⟶
ST segment changes
Accumulation of sodium and chloride ions
Catecholamine release
Stimulation of adenyl cyclase
Cyclic AMP mediated activation of phosphorylase
Stimulation of glycogenolysis
Net utilization of high energy phosphates
Accumulation of protons, carbon dioxide and inorganic phosphate
Stimulation of phosphofructokinase activity
Increase of glycolytic flux
Development of intracellular acidosis
Reduction or blockage of mitochondrial electron transport
Repression of fatty acid oxidation
Utilization of glycogen
Leakage of inorganic phosphate ⟶
Accumulation of NADH
Increased lactate dehydrogenase and α glycerophosphate dehydrogenase activity
Accumulation of lactate and α glycerophosphate
Leakage of lactate ⟶

MINUTES

Accumulation of fatty acyl CoA derivatives
Depletion of creatine phosphate
Leakage of adenosine, inosine and other metabolites ⟶
Vasodilation
Inhibition of adenine nucleotide transferase activity
Possible stimulation of triglyceride synthesis and degradation
Increasing cellular acidosis
Repression of phosphofructokinase and glyceraldehyde-3-phosphate dehydrogenase activity
Slowing of glycolytic flux
Increasing depletion of energy stores
Cell swelling
Increase in cytoplasmic ionized calcium content
Leakage of magnesium ions ⟶
Possible exhaustion of glycogen reserves
Development of mitochondrial damage
Inhibition of glycolysis
Severe reduction of ATP
Minor ultrastructural changes, e.g. mitochondrial swelling
Possible onset of contracture

?

ONSET OF IRREVERSIBLE DAMAGE?

Lysosomal changes and activation of hydrolases
Activation of lipoprotein lipases
Increasing cellular oedema
Loss of mitochondrial respiratory control
Non-specific electrocardiographic changes
Ultrastructural changes in mitochondria and myofibrils
Complete depletion of energy reserves
Metabolic disruption
Loss of mitochondrial components
Leakage of macromolecules to interstitial space and lymph ⟶
Severe ultrastructural damage and membrane deterioration
Cellular disruption
Extensive enzyme leakage ⟶
Disruption of mitochondria
Disintegration of myofibrils
Disruption of cell membranes
Cellular autolysis

HOURS

CELL DEATH AND TISSUE NECROSIS

Immediately following the onset of ischaemia (i.e. within a few seconds or a few beats in severe ischaemia) there is a precipitous decline of contractile activity.[35,44,84,86] This decline occurs at a time when excitability remains essentially normal.[41] It remains controversial whether this very rapid, conservative response can be attributed[29,42,44,87] to adenosine triphosphate (ATP) depletion in a small compartment, to a direct ischaemia-induced interference of calcium homeostasis, or to an effect of accumulating protons upon calcium binding within the contractile apparatus, or to some other, as yet unidentified process. During these first few seconds in *anoxic* tissue, or in *severely ischaemic* tissue with little or no collateral coronary flow, available oxygen dissolved in the cytoplasm will be utilized and anaerobic conditions will develop within the cell.[47,90] Associated with this will be a major reduction (or even a complete abolition) of oxidative metabolism, electron transport, and mitochondrial ATP production and only the much less efficient anaerobic pathways of metabolism remain for the production of ATP. In hypoxic tissue, or less severely ischaemic tissue with significant collateral flow, there will be a major reduction of oxygen availability and a corresponding reduction of oxidative ATP production; under these conditions ATP will be derived from both aerobic and anaerobic pathways.

In assessing the relative contribution to ATP production made by aerobic and anaerobic pathways, two factors must be considered. First, there is a large difference between the pathways in their efficiency of ATP production. Second, it appears[27,28,53,67] that substantial collateral flow exists even in the core of an ischaemic area. Taken together these two factors would suggest[11] that oxidative phosphorylation may often remain the dominant source of ATP despite a severe reduction in oxygen availability.

During the early seconds of ischaemia or hypoxia, sensitive cellular control mechanisms will trigger major changes[58,61,66,67] in substrate uptake and utilization patterns (see Chapter 21). Reduced mitochondrial metabolism will result in a rapid reduction in the flux through the beta oxidation pathway for fatty acids. Despite the reduction of fatty acid utilization, uptake may not be diminished and as a result of these two factors fatty acid acyl CoA derivatives may accumulate during ischaemia.[67] This accumulation may[32] be exacerbated by cyclic AMP-mediated lipolysis of endogenous triglyceride which itself may be triggered by the early ischaemia-induced release of catecholamines.[13,55,91]

Figure 1. Some of the cellular events thought to occur following the onset of severe myocardial ischaemia. This figure, which is *highly speculative*, is not intended to convey that the events occur in a strict sequence or in the exact order listed. The progression of changes is a dynamic process and only the onset of some changes is indicated. In general the early changes are reversible but, with increasing durations of ischaemia, progress to irreversible damage. The exact point of the transition is unknown

During the early moments of ischaemia while fatty acid utilization is repressed,[15] carbohydrate utilization is stimulated. In ischaemia (in contrast to anoxia), this stimulation does not result from increased glucose uptake, but from an increase in glycogenolysis and glycolysis.[67,74] Thus although glucose uptake falls, total glycolytic flux is increased (albeit only transiently) probably as a combined result of increased glycogenolysis (which most likely results from the cyclic AMP and/or calcium-mediated activation[75] of phosphorylase and an allosteric stimulation of phosphorylase b) and stimulation of glycolysis (by the activation of rate-limiting enzymes[15,67,74] such as phosphofructokinase).

The control of glycolysis and the mechanism for the stimulation of glycogenolysis serve to illustrate the metabolic differences which may occur between different models of oxygen deprivation. Thus it appears[15,67,74] that while glycolysis is controlled at the level of phosphofructokinase during anoxia and possibly during regional ischaemia, it is controlled by glyceraldehyde-3-phosphate dehydrogenase during global ischaemia. Control may however oscillate between these and other enzymes at different times in ischaemia.[14,70] It has also been suggested[55] that while catecholamines play a major role in the activation of glycogenolysis during ischaemia, they play a minor role during anoxia.

The stimulation of anaerobic glycolysis (the Pasteur effect) represents an attempt to maintain, through non-oxidative mechanisms of substrate-level phosphorylation, the declining myocardial ATP content. The stimulation of glycolysis in the face of reduced mitochondrial activity leads to the accumulation of glycolytic intermediates and reduced nicotinamide adenine dinucleotide phosphate (NADH). In an attempt to regenerate declining and limited reserves of NAD^+ for continued glycolytic activity, pyruvate is reduced to lactate, which accumulates in and leaks from, the cell.[48,63] Additional regeneration of NAD^+ may be achieved by promoting the activity of α-glycerophosphate dehydrogenase and thus the conversion of dihydroxy acetone phosphate to α-glycerophosphate.[67]

An early feature of myocardial ischaemia is the accumulation in the cytoplasm of protons and the progressive development of intracellular acidosis. These protons are derived from a number of sources[20,67] including the net degradation of adenine nucleotides, the accumulation of carbon dioxide (from residual oxidative metabolism and various decarboxylation reactions) leading to a 'respiratory' intracellular acidosis, and the continuing triglyceride synthesis and degradation which may possibly[67] occur. Although protons are also generated in anoxia, the presence of near-normal coronary flow will facilitate proton and carbon dioxide washout thus moderating the development of intracellular acidosis.[59] The cellular accumulation of protons, lactate, and NADH may[75] override the earlier activation (by decreased ATP and increased ADP, AMP, and inorganic phosphate) of glyceraldehyde-3-phosphate dehydrogenase and phosphofructokinase. The resulting inhibition of these enzymes will lead to the slowing of glycolytic flux.

In addition to their inhibitory effect upon glycolytic activity and their possible role in early contractile failure, protons may[67] contribute to the development of later stages of ischaemic damage. This may result from the activation of lysosomal hydrolases[7,72,88] and lipoprotein lipases, which together with proton-induced calcium shifts and proton-induced membrane conformational changes may contribute towards cell leakage.

As a result of the ischaemia-induced diminution of the control which can be exerted over transmembrane ionic gradients, the consequent cellular loss of potassium and the accumulation of sodium and chloride, osmotic changes lead to cellular water accumulation and cell swelling.[51] This swelling may contribute to cellular injury and in particular to cell membrane damage and the development of cell leakiness.

One aspect of early ischaemia and anoxia is the apparently paradoxical behaviour of certain aspects of fatty acid metabolism. For example, experimental studies[21] have revealed that anoxia stimulates the energy-requiring incorporation of acetate into triglycerides. Although this process utilizes much needed ATP, it is also a reductive process and as such may be of value by utilizing the increasing content of reduced nicotinamide adenine nucleotide cofactors. The quantitative importance of this process, if indeed it occurs in ischaemia, may however be limited by the very low levels of available acetyl CoA and the small amounts of acetate available as a substrate. Another apparently paradoxical aspect of fatty acid metabolism is the proposed energy-wasting[67] triglyceride–fatty acid cycle which is thought to occur in ischaemic tissue.[13,64,67] This process (which may be stimulated by the ischaemia-induced increase in α-glycerophosphate, by the possible cyclic AMP-mediated stimulation of lipase activity and by the possible stimulation of *de novo* fatty acid synthesis) would result in the net utilization of ATP and the net production of protons.

It is not only metabolic changes which occur during the early seconds and minutes of ischaemia and hypoxia, major ionic and electrophysiological changes also occur.[8,11,29] It has long been known[69] that ST segment changes (relative to TQ segment) occur as a result of coronary artery occlusion. These can be detected within 30 seconds of the onset of ischaemia and usually reach a maximum some 5 or 10 minutes later.[29] The electrophysiological basis of ST segment changes is complex and controversial.[5,17,29,73] The changes may result from inadequate polarization of the ischaemic region during diastole whilst normal tissue is well polarized or alternatively may result from inadequate depolarization of the ischaemic tissue during excitation.[29] Disturbances of membrane ion transport appear to be the cause of early ischaemia-induced electrocardiographic changes.[29]

Key changes are thought to be the loss of membrane control, possibly as a result of an energy shortage, and the net cellular accumulation of sodium ions, chloride ions, and water (loss of cell volume control) and a net cellular loss of potassium ions. The increase in extracellular potassium and/or the reduction of intracellular potassium is thought[39] to play a critical role in the genesis of the electrocardiographic changes. Reflecting the redistribution of potassium and other ions is the early ischaemia-induced reduction in amplitude and duration of the action potential. Although controversial[11,33,56] it is possible that these changes may be related to reduced glycolytic ATP production which may reduce calcium chelation in the subsarcolemma, thus increasing local concentrations of free calcium and reducing the slow inward calcium current. The early sodium, potassium, and calcium changes described above, together with other ionic changes such as the loss of intracellular magnesium,[79] undoubtedly reflect the progressive loss of cell membrane control and the development of cell leakiness.

Vascular changes also occur during the early minutes of ischaemia and anoxia. In an attempt to increase the delivery of oxygen to the tissue there is a marked coronary vasodilation together with an increase in the efficiency of extraction of the limited amount of oxygen which may be available. The vasodilation is thought to be mediated by a number of factors including prostaglandin release[2,30] and also the release of intracellular adenosine[3] which accumulates in and leaks from the cell as a result of the net breakdown of ATP. Also associated with net nucleotide degradation is the cellular accumulation and leakage of inosine and hypoxanthine.

Despite the conservative effects of contractile failure, the stimulation of glycolysis and anaerobic energy production and the marked vasodilation and opening of collateral vessels, the myocardium during severe ischaemia is unable to produce sufficient energy for the correct maintenance of basal metabolism and cellular integrity. Thus during the early minutes of ischaemia or anoxia a progressive fall in myocardial energy reserves is observed.[4,23,48,79] This is characterized by the very rapid breakdown of creatine phosphate, the abrupt fall of myocardial ATP followed by a transient rise then fall of ADP, AMP, and adenosine. During this period there is a progressive loss of cellular glycogen reserves.

Ultrastructural changes[37] which will be discussed in detail in Chapter 2 are not readily apparent during the early minutes of ischaemia and are limited to the loss of glycogen granules, development of intracellular oedema, and possible swelling of the mitochondria and T-tubules.

As described previously, during the first fifteen minutes or so of ischaemia there is a stimulation and subsequent slowing of glycolytic activity[75] the latter resulting from increasing cytoplasmic concentrations of protons, NADH and lactate. The decline of glycolytic activity continues and under conditions of severe ischaemia this inhibition may be total. While this inhibition may be attributed to the above factors it may in part be due to severe depletion of available glycogen.[74]

As the duration of ischaemia or hypoxia increases, more severe cellular damage can be detected together with the intensification of existing damage. Mitochondrial malfunction (which may occur despite a continuing limited supply of oxygen) will act[25] to widen the gap between energy demand and energy supply. Studies[38,40,52] with mitochondria isolated from ischaemic tissue reveal that at the early stages of the ischaemic process ADP:O values are near normal, but that there is a decreased capacity for oxygen consumption suggesting an inhibition of phosphorylation (possibly a block at site 1 on the electron transport chain) rather than an uncoupling. However, as the ischaemic damage progresses, decreases in ADP:O ratios can be observed.[52,70] Dramatic decreases in state 3 and increases in state 4 respiration, and hence, electron transport capability have been reported[38,78] as has the loss of electron transport chain components.[36] The final result is the total loss of respiratory control and the inhibition of respiration. Mitochondrial calcium loss probably[25] occurs during the relatively early stages of the ischaemic process. This mitochondrial calcium would contribute to increasing cytoplasmic ionized calcium levels, the latter arising partly as a result of dechelation of calcium from ATP (as a result of net ATP utilization) and the reduction of calcium sequestration at various subcellular sites.

During the first half hour or so of ischaemia, lipid metabolism continues to be disrupted and various long-chain fatty acyl CoA compounds and their derivatives accumulate in the cytoplasm and may exert specific toxic effects, in particular the inhibition of adenine nucleotide transferase activity,[81] a process which is vital for the translocation of ATP (arising from residual mitochondrial activity) from the mitochondria to the cytoplasm. In addition to this and other aspects of fatty acid toxicity,[66] the accumulation of fatty acid derivatives may contribute to the genesis of dysrrhythmias during the ischaemic period.[50,60,65] As well as major changes in lipid and carbohydrate metabolism, ischaemia and hypoxia cause a marked impairment in protein synthesis and protein degradation.[71]

Dependent upon the severity of the ischaemia, total contractile failure with diastolic arrest may occur. With increasing durations of anoxia[26] and possibly ischaemia,[10] myocardial contracture (a form of systolic arrest resembling rigor in skeletal muscle) may develop. This condition[26,43,45] is though to be due to the accumulation of rigor complexes as a direct result of ATP deficiency. The development of contracture or advanced ultrastructural damage[37] is evidenced by contraction bands, distorted Z bands and myofibrillar disruption. By this stage mitochondria are distorted and may appear vesiculated, glycogen granules are lost and the sarcoplasmic reticulum may be vesiculated. Extensive cellular oedema may have developed and there may be some evidence of distortion or damage to the cell membrane. Cell membrane damage may become apparent and intensify after the first hour or so of ischaemia. During this period, control of membrane activity deteriorates and there is a progressive increase in non-specific permeability. The changes in membrane characteristics (see Chapter 3) would account for the early disturbances of cellular ionic balance with the loss of a number of intracellular ions, the leakage of lactate, adenosine, and other intracellular metabolites and at later stages the leakage of macromolecules such as cytoplasmic proteins and enzymes. The appearance of these macromolecules in the interstitial space and cardiac lymph has been observed[83,85] as early as one hour after the onset of ischaemia.

During the first half to four hours of severe ischaemia or hypoxia, cardiac lysosomal changes have been reported.[18,31,72,88,89] As mentioned previously, the stimulation of lysosomal activity and lysosomal enzymes may play a key role in the disruption of cellular ultrastructure and may contribute to the development of membrane damage through the activation of lipoprotein lipases and other lysosomal hydrolytic enzymes.

As ischaemia progresses there is a loss of vascular responsiveness and although this may in part be compensated for by the development of collateral vessels,[77] there is often a secondary reduction in the remaining coronary perfusion. Two factors[6] may contribute to this secondary reduction of flow: firstly, swelling of capillary endothelial cells may reduce capillary diameter and also cause trapping of blood elements. Secondly, the development of contracture may compress the capillary vessels thereby further reducing flow. These combined effects are likely to contribute to the 'no-reflow' phenomenon[16] which is characteristic of more advanced ischaemic damage.

In later stages of ischaemia, and despite the possibility of a continuing limited supply of oxygen, cellular metabolism and ATP production virtually cease, glycogen stores may be depleted and glycolysis and mitochondrial function become totally inhibited. Anabolic and catabolic pathways for proteins, fatty acids, lipids, and nucleic acids are no longer functional, cellular autolysis occurs and large quantities of various cellular constituents leak to the extracellular space. Electrical activity is lost, extensive ultrastructural damage occurs and following cell death the process of fibrous infiltration and tissue necrosis occurs.

Examination of the continuum of damage which is characteristic of ischaemia or anoxia reveals that the damage is progressive in nature. As will be discussed in greater detail in Chapter 2 this gives rise to the concept of the reversibility of damage.[38] Early ischaemic changes are generally fully reversible, such that if adequate coronary perfusion is restored, then a rapid and complete recovery of contractile and metabolic activity is observed. However, as the severity or duration of ischaemia increases then the damage intensifies until beyond a certain point, as yet unidentified, the restoration of adequate flow no longer consistently reverses injury and the cells enter a phase of irreversible damage. Once in this state, and under some conditions, reperfusion may cause a major extension of damage.[19,24,25,34,38,79,80] At the present time it would seem likely that *significant* enzyme leakage does not occur until *after* the onset of irreversible damage. This is in fact a crucial assumption which is made in the studies of infarct sizing which are discussed later in this book. However, at this time it is by no means proven and since there have been some reports[76,93] of enzyme leakage from tissue which is thought to have suffered reversible damage only, the exact point of transition between reversible and irreversible damage and its relationship to enzyme leakage is clearly an important area requiring investigation.

An important feature of the process of ischaemic damage is that certain stages are characterized by the leakage of various cytoplasmic constituents to the extracellular space. It is this leakage that forms the basis of various methods for detecting and assessing myocardial damage.

THE LEAKAGE OF CYTOPLASMIC CONSTITUENTS

The loss of intracellular constituents to the extracellular space and ultimately to the circulating blood can be divided into three relatively distinct phases (ions, metabolites, and macromolecules) each of which may reflect some different aspect of cellular and metabolic damage (Figure 2).

In early stages of ischaemia, while cellular damage is still reversible the leakage of small ions such as potassium is observed. This leakage may represent the impairment or failure of energy-requiring ion pumps which normally maintain thermodynamically unfavourable transmembrane ionic gradients. The malfunction of these pumps may result in the rapid re-equilibration of otherwise relatively freely diffusible ions in either direction across the cell membrane. Early ischaemic damage is also characterized by a loss of intracellular inorganic phosphate.[62] The transport of this anion is normally very restricted and its leakage from the cell reflects its increasing intracellular concentration (as a consequence of the ischaemia-induced net breakdown of ATP, creatine phosphate, and other high-energy phosphates) and the progressive loss of membrane control.

At later stages of ischaemic damage the leakage of larger and sometimes less freely diffusible intermediates such as adenosine, inosine, and lactate occurs. Here, due to metabolic imbalance compounds such as lactate, which would normally be completely degraded accumulate in, and leak from, the cell.

During advanced stages of ischaemic damage, when some irreversible injury has most probably occurred, severe membrane damage is observed. This damage, which will be discussed in detail in Chapter 3, causes major

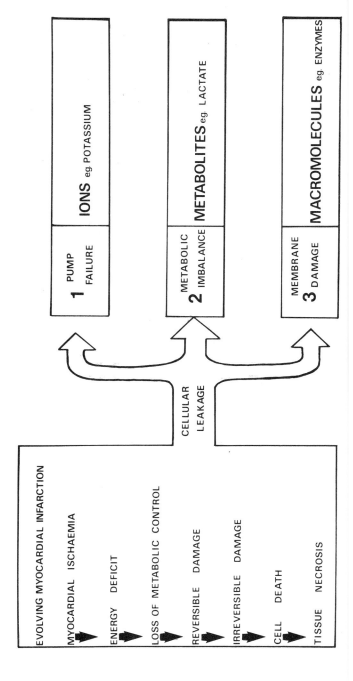

Figure 2. The leakage of cellular constituents during the progression of myocardial ischaemia. [Reproduced by permission of Elsevier North Holland from *Journal of Molecular Medicine*, **2**, 185–200 (1977).]

permeability changes such that molecules which are normally too large to cross the membrane now pass readily through it and it is possible to detect the leakage of macromolecules such as proteins and enzymes.

THE FATE OF LEAKED MATERIALS (FIGURE 3)

Current evidence would suggest that ions and metabolites are transported freely from the interstitial space to the coronary blood and thus rapidly appear in the general circulation. In contrast, and as will be discussed in Chapter 4, macromolecules are not able to pass readily across the capillary membrane and it would appear[83,85] that the cardiac lymphatics are responsible for the clearance of a substantial proportion of the leaked macromolecules. The cardiac lymphatics[46,57] drain into several efferent collecting trunks that usually follow the pathways of the coronary arteries, these

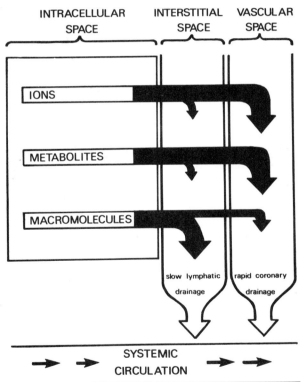

Figure 3. Possible routes for the translocation of leaking cellular constituents from the myocardial cell to the systemic circulation. [Reproduced by permission of Elsevier North Holland from *Journal of Molecular Medicine*, **2**, 185–200 (1977).]

eventually terminate in the lymph nodes positioned between the superior vena cava and the ascending aorta. It is possible to demonstrate very high levels of enzyme activity in cardiac lymph 30–60 minutes after the onset of ischaemia.[83,85]

Analysis of the circulating blood for a period of time following the onset of acute myocardial infarction reveals characteristic profiles for ions, metabolites, and macromolecules (Figure 4).

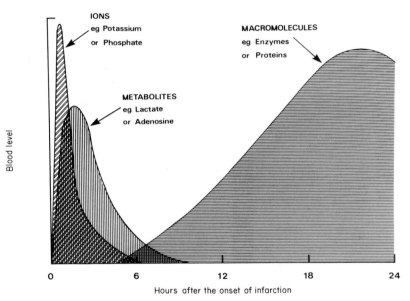

Figure 4. Approximate plasma profiles for small ions, metabolites, and macromolecules following the onset of severe myocardial ischaemia. [Reproduced by permission of Elsevier North Holland from *Journal of Molecular Medicine*, **2**, 185–200 (1977).]

Ions and metabolites, by virtue of their early loss from the cell and unimpeded transport to the rapidly flowing coronary vascular drainage may be detected in the circulation minutes after the onset of ischaemia. However, their equally rapid metabolism and clearance results in only transiently elevated blood levels. In contrast, macromolecules by virtue of their later leakage from the cell and their clearance via the slow-flowing lymphatics may not be detected in the circulating blood for several hours. However, once there (as will be described in detail in Chapter 5), clearance is relatively slow and they may remain and be detected in the circulation for some time.

CHOICE OF MARKERS OF ISCHAEMIC DAMAGE

Clearly, the appearance of ions, metabolites, or macromolecules may provide evidence of myocardial damage. However, since ions and metabolites such as potassium and lactate may arise from any tissue and are only elevated in the blood for a short time, they are of little value for the determination of injury in a specific tissue. In contrast macromolecules, which may persist in the blood for a long time and may be relatively cardiospecific, provide an effective means for the detection of specific damage.

Of all the types of macromolecule known to leak from damaged cells, enzymes are undoubtedly the best markers of damage. There are two reasons for this, the first relies upon the concept of tissue *specificity* and the second relies upon enzyme *activity*.

As will be discussed in detail in Chapter 6 of this book, the concept of tissue specificity forms the basis of all diagnostic enzymology. Each organ has its own biochemical composition which is quantitatively geared to, and reflects, the major functions of that tissue. Thus while the enzyme profile of the liver would reveal large amounts of many enzymes associated with drug metabolism, detoxication, etc. the enzyme profile of the heart would be dominated by enzymes associated with muscle contraction and energy production. In this way heart cells may be enzymatically distinguished from liver cells and if tissue damage occurred in either of these organs very different plasma enzyme profiles would result.

The second important characteristic of enzymes which makes them particularly suited to diagnostic procedures is their activity. In contrast to other macromolecules like proteins, where detection is usually limited to the measurement of mass, with enzymes, by virtue of their catalytic properties and extraordinary substrate specificity, it is possible to measure the activity of a minute amount of a single enzyme even in the presence of many others.

CONCLUDING REMARKS

This chapter has attempted to show how the highly complex process of ischaemic damage leads to the leakage of cytoplasmic enzymes to the extracellular space and ultimately to the systemic circulation. Clearly, the appearance of these enzymes in the circulation is a useful index of tissue damage, but before this can be fully exploited it is necessary to understand more about the factors governing the leakage of the enzymes across the damaged cell membrane, the routes by which these enzymes reach the general circulation, the distribution of the enzymes in various fluid compartments, and finally, the inactivation and clearance of the enzymes from the blood. These critical events are the subject of the following chapters.

REFERENCES

1. Barlow, C. H. and Chance, B. Ischemic areas in perfused rat hearts: measurement by NADH fluorescence photography. *Science*, **193,** 909–10 (1976).
2. Berger, H. J., Zaret, B. L., Speroff, L. Cohen, L. S., and Wolfson, S. Cardiac prostaglandin release during myocardial ischemia induced by atrial pacing in patients with coronary artery disease. *American Journal of Cardiology*, **39,** 481–6 (1977).
3. Berne, R. M. and Rubio, R. Adenine nucleotide metabolism in the heart. *Circulation Research Supplement 3*, **34** and **35,** 109–20 (1974).
4. Braasch, W., Gudbjarnason, S., Puri, P. S., Ravens, K. G., and Bing, R. J. Early changes in energy metabolism in the myocardium following acute coronary artery occlusion in anaesthetized dogs. *Circulation Research*, **23,** 429–38 (1968).
5. Braunwald, E. and Maroko, P. R. ST segment mapping, realistic and unrealistic expectations. *Circulation*, **54,** 529–32 (1976).
6. Brachfeld, N. Ischemic myocardial metabolism and cell necrosis. *Bulletin of the New York Academy of Medicine*, **50,** 261–93 (1974).
7. Brachfeld, N. Maintenance of cell viability. *Circulation Supplement 4*, **39** and **40,** 202–19 (1969).
8. Case, R. B. Ion alterations during myocardial ischemia. *Cardiology* **56,** 245–62 (1971/72).
9. Chance, B. In discussion of Opie, L. H: The effects of regional ischemia on metabolism of glucose and fatty acids. *Circulation Research Supplement 1*, **38,** 52–74 (1976).
10. Cooley, D. A., Reul, J., and Wukasch, D. C. Ischemic contracture of the heart: stone heart. *American Journal of Cardiology*, **29,** 575–7 (1972).
11. Coraboeuf, E., Deroubaix, E., and Hoerter, J. Control of ionic permeabilities in normal and ischemic heart. *Circulation Research Supplement 1*, **38,** 92–8 (1976).
12. Cox, J. L., McLaughlin, V. W., Flowers, N. C., and Horan, L. G. The ischemic zone surrounding acute myocardial infarction. Its morphology as detected by dehydrogenase staining. *American Heart Journal*, **76,** 650–9 (1968).
13. Crass, M. F., Shipp, J. C., and Pieper, G. M. Effects of catecholamines on myocardial endogenous substrates and contractility. *American Journal of Physiology*, **228,** 618–27 (1975).
14. Dobson, J. G. and Mayer, S. E. Mechanism of activation of cardiac glycogen phosphorylase in ischemia and anoxia. *Circulation Research*, **33,** 412–20 (1973).
15. Evans, J. R. Cellular transport of long chain fatty acids. *Canadian Journal of Biochemistry*, **42,** 955–68 (1964).
16. Fabiani, J. N. The no reflow phenomenon following early reperfusion of myocardial infarction and its prevention by various drugs. *Heart Bulletin (La Haye)*, **5,** 134–42 (1977).
17. Fozzard, H. A. and DasGupta, D. S. ST segment potentials and mapping, theory and experiments. *Circulation*, **54,** 533–7 (1976).
18. Fox, A. C., Hoffstein, S., and Weissmann, G. Lysosomal mechanisms in production of tissue damage during myocardial ischemia and the effects of treatment with steroids. *American Heart Journal*, **91,** 394–7 (1976).
19. Ganote, C. E., Seabra-Gomes, R., Nayler, W. G., and Jennings, R. B. Irreversible myocardial injury in anoxic perfused rat hearts. *American Journal of Pathology*, **80,** 419–50 (1975).
20. Gevers, W., Generation of protons by metabolic processes in heart cells. *Journal of Molecular and Cellular Cardiology*, **9,** 867–74 (1977).
21. Gloster, J. and Harris, P. Effect of anaerobiosis on the incorporation of (^{14}C) acetate into lipid in the perfused rat heart. *Journal of Molecular and Cellular Cardiology*, **4,** 213–28 (1972).
22. Harris, P. A theory concerning the course of events in angina and myocardial infarction. *European Journal of Cardiology*, **3,** 157–63 (1975).
23. Hearse, D. J. and Chain, E. B. The role of glucose in the survival and recovery of the anoxic isolated perfused rat heart. *Biochemical Journal*, **128,** 1125–33 (1972).
24. Hearse, D. J., Humphrey, S. M., and Chain, E. B. Abrupt reoxygenation of the anoxic potassium arrested perfused rat heart: a study of myocardial enzyme release. *Journal of Molecular and Cellular Cardiology*, **5,** 395–407 (1973).

25. Hearse, D. J. Reperfusion of the ischaemic myocardium. *Journal of Molecular and Cellular Cardiology*, **9**, 605–16 (1977).
26. Hearse, D. J., Garlick, P. B., and Humphrey, S. M. Ischemic contracture of the myocardium: mechanisms and prevention. *American Journal of Cardiology*, **39**, 986–93 (1977).
27. Hearse, D. J., Opie, L. H., Katzeff, I., Lubbe, W. F., Van der Werff, T. J., Peisach, M., and Boulle, G. Characterization of the 'border zone' in acute regional ischemia in the dog. *American Journal of Cardiology*, **40**, 716–26 (1977).
28. Heng, M. K., Singh, B. N., Norris, R. M., John, M. B., and Elliot, R. Relationship between epicardial ST-segment elevation and myocardial ischemic damage after experimental coronary artery occlusion in dogs. *Journal of Clinical Investigation*, **58**, 1317–26 (1976).
29. Hillis, L. D. and Braunwald, E. Myocardial ischemia. *New England Journal of Medicine*, **296**, 971–8 (1977).
30. Hintz, T. H. and Kaley, G. Prostaglandins and the control of blood flow in the canine myocardium. *Circulation Research*, **40**, 313–20 (1977).
31. Hoffstein, S., Weissmann, G., and Fox, A. C. Lysosomes in myocardial infarction: studies by means of cytochemistry and subcellular fractionation, with observations on the effects of methyl prednisolone. *Circulation Supplement 1*, **53**, 34–40 (1976).
32. Hough, F. S. and Gevers, W. Catecholamine release as a mediator of intracellular enzyme activation in ischaemic perfused rat hearts. *South African Medical Journal*, **49**, 538–43 (1975).
33. Hyde, A., Chevnal, J. P., Blondel, B., and Girardier, L. Electrophysiological correlates of energy metabolism in cultured rat heart cells. *Journal of Physiology (Paris)*, **64**, 269–92 (1972).
34. Jennings, R. B., Sommers, H. M., Smyth, G. A., Flack, H. A., and Linn, H. Myocardial necrosis induced by temporary occlusion of a coronary artery in the dog. *Archives of Pathology*, **70**, 68–78 (1960).
35. Jennings, R. B. Early phase of myocardial ischemic injury and infarction. *American Journal of Cardiology*, **24**, 753–65 (1969).
36. Jennings, R. B., Herdson, P. B., and Sommers, H. M. Structural and functional abnormalities in mitochondria isolated from ischemic dog myocardium. *Laboratory Investigation*, **20**, 548–57 (1969).
37. Jennings, R. B. and Ganote, C. E. Structural changes in myocardium during acute ischemia. *Circulation Research Supplement 111*, **34** and **35**, 156–68 (1974).
38. Jennings, R. B. and Ganote, C. E. Mitochondrial structure and function in acute myocardial ischemic injury. *Circulation Research Supplement 1*, **38**, 80–9 (1976).
39. Johnson, E. A. First electrocardiographic sign of myocardial ischemia: an electrophysiological conjecture. *Circulation Supplement 1*, **53**, 82–4 (1976).
40. Kane, J. J., Murphy, M. L., Bisset, J. K., de Soyza, N., Doherly, J. E., and Straub, K. D. Mitochondrial function, oxygen extraction, epicardial ST segment changes and tritiated digoxin distribution after reperfusion of ischemic myocardium. *American Journal of Cardiology*, **36**, 218–24 (1975).
41. Kardesch, M., Hogancamp, C. E., and Bing, R. J. The effect of complete ischemia on the intracellular electrical activity of the whole mammalian heart. *Circulation Research*, **6**, 715–20 (1958).
42. Katz, A. M. and Hecht, H. H. The early pump failure of the ischemic heart. *American Journal of Medicine*, **47**, 497–502 (1969).
43. Katz, A. M. and Tada, M. The stone heart, a challenge to the biochemist. *American Journal of Cardiology*, **29**, 578–80 (1972).
44. Katz, A. M. Effects of ischemia on the contractile processes of heart muscle. *American Journal of Cardiology*, **32**, 456–60 (1973).
45. Katz, A. M. and Tada, M. The stone heart and other challenges to the biochemist. *American Journal of Cardiology*, **39**, 1073–7 (1977).
46. Kline, I. K. Lymphatic pathways in the heart. *Archives of Pathology*, **88**, 638–44 (1969).
47. Kloner, R. A., Ganote, C. E., Reimer, K. A., and Jennings, R. B. Distribution of coronary arterial flow in acute myocardial ischemia. *Archives of Pathology*, **99**, 86–94 (1975).
48. Kubler, W. and Spieckermann, P. G. Regulation of glycolysis in the ischaemic and the anoxic myocardium. *Journal of Molecular and Cellular Cardiology*, **1**, 351–77 (1970).

49. Kubler, W. Comparative metabolism of contractile and conductive tissue. *Proceedings of the 7th European Congress of Cardiology, Amsterdam*, June 1976, Vol. 2, pp. 127.
50. Kurien, V. A. and Oliver, M. F. A metabolic cause for arrhythmias during acute myocardial hypoxia. *Lancet*, **1,** 813–5 (1970).
51. Leaf, A. Cell swelling, a factor in ischemic injury. *Circulation*, **48,** 455–8 (1973).
52. Lochner, A., Opie, L. H., Owen, P., Kotze, J. C. N., Bruyneel, K., and Gevers, W. Oxidative phosphorylation in infarcting baboon and dog myocardium. Effects of mitochondrial isolation and incubation media. *Journal of Molecular and Cellular Cardiology*, **7,** 203–17 (1975).
53. Lubbe, W. F., Peisach, M., Pretorius, R., Bruyneel, K. J. J., and Opie, L. H. Distribution of myocardial blood flow before and after coronary artery ligation in the baboon; relation to early ventricular fibrillation. *Cardiovascular Research*, **8,** 478–87 (1974).
54. Marcus, M. L., Kerber, R. E., Ehrhardt, J., and Abboud, F. M. Three dimensional geometry of acutely ischemic myocardium. *Circulation*, **52,** 254–63 (1975).
55. Mayer, S. E. Effect of catecholamines on cardiac metabolism. *Circulation Research Supplement*, **34** and **35,** 129–35 (1974).
56. McDonald, T. F., Hunter, E. G., and McLeod, D. P. Adenosine triphosphate partition in cardiac muscle with respect to transmembrane electrical activity. *Pflugers Archiv*, **322,** 95–108 (1971).
57. Miller, A. J. The lymphatics of the heart. *Archives of Internal Medicine*, **112,** 97–107 (1963).
58. Neely, J. R. and Morgan, H. E. Relationship between carbohydrate and lipid metabolism and the energy balance of the heart muscle. *Annual Reviews of Physiology*, **36,** 413–59 (1974).
59. Neely, J. R., Whitmer, J. T., and Rovetto, M. J. Effect of coronary blood flow on glycolytic flux and intracellular pH in isolated rat hearts. *Circulation Research*, **37,** 733–41 (1975).
60. Oliver, M. F. The influence of myocardial metabolism on ischemic damage. *Circulation Supplement 1*, **53,** 168–70 (1976).
61. Opie, L. H. Metabolism of the heart in health and disease. *American Heart Journal*, **76,** 658–98 (1968); **77,** 101–22 (1969); **77,** 383–410 (1969).
62. Opie, L. H., Thomas, M., Owen, P., and Shulman, G. Increased coronary venous inorganic phosphate concentrations during experimental myocardial ischemia. *American Journal of Cardiology*, **30,** 503–13 (1972).
63. Opie, L. H., Owen, P., Thomas, M., and Samson, R. Coronary sinus lactate measurements in assessment of myocardial ischemia. *American Journal of Cardiology*, **32,** 295–305 (1973).
64. Opie, L. H. Metabolism of free fatty acids, glucose and catecholamines in acute myocardial infarction. *American Journal of Cardiology*, **36,** 938–53 (1975).
65. Opie, L. H. and Lubbe, W. F. Are free fatty acids arrhythmogenic? *Journal of Molecular and Cellular Cardiology*, **7,** 155–9 (1975).
66. Opie, L. H. and Stubbs, W. A. Carbohydrate metabolism in cardiovascular disease. *Clinics in Endocrinology and Metabolism*, **5,** 703–29 (1976).
67. Opie, L. H. Effects of regional ischaemia on metabolism of glucose and fatty acids. *Circulation Research Supplement 1*, **38,** 52–74 (1976).
68. Opie, L. H. and Owen, P. The effect of glucose–insulin–potassium infusions on arteriovenous differences of glucose and of free fatty acids and on tissue metabolic changes in dogs with developing myocardial infarction. *American Heart Journal*, **38,** 310–321 (1976).
69. Pardee, H. E. B. An electrocardiographic sign of coronary artery obstruction. *Archives of Internal Medicine*, **26,** 244–57 (1920).
70. Patterson, R. A. Metabolic control of rat heart glycolysis after acute ischaemia. *Journal of Molecular and Cellular Cardiology*, **2,** 193–210 (1971).
71. Rabinowitz, M. Control of metabolism and synthesis of macromolecules in normal and ischaemic heart. *Journal of Molecular and Cellular Cardiology*, **2,** 277–92 (1971).
72. Ricciutti, M. A. Lysosomes and myocardial cellular injury. *American Journal of Cardiology*, **30,** 498–502 (1972).
73. Ross, J. Electrocardiographic ST-segment analysis in the characterization of myocardial ischaemia and infarction. *Circulation Supplement 1*, **53,** 73–81 (1976).
74. Rovetto, M. J., Whitmer, J. T., and Neely, J. R. Comparison of the effects of anoxia and

whole heart ischemia on carbohydrate utilization in isolated working rat hearts. *Circulation Research*, **32**, 699–711 (1973).
75. Rovetto, M. J., Lamberton, W. F., and Neely, J. R. Mechanisms of glycolytic inhibition in ischemic rat heart. *Circulation Research*, **37**, 742–51 (1975).
76. Sakai, K., Gebhard, M. M., Spieckermann, P. G., and Bretschneider, H. J. Enzyme release resulting from total ischaemia and reperfusion in the isolated, perfused guinea pig heart. *Journal of Molecular and Cellular Cardiology*, **7**, 827–40 (1975).
77. Schaper, W. and Pasyk, S. Influence of collateral flow on the ischemic tolerance of the heart following acute and sub-acute coronary occlusion. *Circulation Supplement 1*, **53**, 57–65 (1976).
78. Schwartz, A., Wood, J. M., Allen, J. C., Bornet, E. P., Entman, M. L., Goldstein, M. A., Sordahl, L. A., and Suzuki, M. Biochemical and morphological correlates of cardiac ischemia. *American Journal of Cardiology*, **32**, 46–61 (1973).
79. Shen, A. C. and Jennings, R. B. Myocardial calcium and magnesium in acute ischaemic injury. *American Journal of Pathology*, **67**, 417–40 (1972).
80. Shen, A. C. and Jennings, R. B. Kinetics of calcium accumulation in acute myocardial ischemic injury. *American Journal of Pathology*, **67**, 441–52 (1972).
81. Shrago, E., Shug, A. L., Sul, H., Bittar, N., and Folts, J. D. Control of energy production in myocardial ischemia. *Circulation Research Supplement 1*, **38**, 75–9 (1976).
82. Sobel, B. E. Salient biochemical features in ischemic myocardium. *Circulation Research Supplement 111*, **34** and **35**, 173–80 (1974).
83. Sobel, B. E., Roberts, R., and Larson, K. B. Considerations in the use of biochemical markers of ischemic injury. *Circulation Research Supplement 1*, **38**, 99–106 (1976).
84. Sonnenblick, E. H. and Kirk, E. S. Effects of hypoxia and ischemia on myocardial contraction. *Cardiology*, **56**, 302–13 (1971/72).
85. Spieckermann, P. G., Nordbeck, H., Knoll, D., Kohl, F. V., Sakai, K., and Bretschneider, H. J. The role of cardiac lymph in the transport of cardiac enzymes into the blood stream after myocardial infarct. *Deutsche Medizinische Wochenschrift*, **99**, 1143–4 (1974).
86. Tennant, R. and Wiggers, C. J. The effect of coronary occlusion on myocardial contraction. *American Journal of Physiology* **1123**, 351–61 (1935).
87. Tsien, R. W. Possible effects of hydrogen ions in ischemic myocardium. *Circulation Supplement 1*, **35**, 14–6 (1976).
88. Weglicki, W. B., Owens, K., Ruth, R. C., and Sonnenblick, E. H. Activity of endogenous myocardial lipases during incubation at acid pH. *Cardiovascular Research*, **8**, 237–42 (1974).
89. Welman, E. and Peters, T. J. Enhanced lysosomal fragility in the anoxic perfused guinea pig heart: effects of glucose and mannitol. *Journal of Molecular and Cellular Cardiology*, **9**, 101–20 (1977).
90. Williamson, J. R., Schaffer, S. W., Ford, C., and Safer, B. Contribution of tissue acidosis to ischaemic injury in the perfused rat heart. *Circulation Supplement 1*, **53**, 3–14 (1976).
91. Wollenberger, A. and Shahab, L. Anoxia induced release of noradrenaline from isolated perfused heart. *Nature*, **207**, 88–9 (1965).
92. Zak, R. and Rabinowitz, M. Metabolism of the ischemic heart. *Medical Clinics of North America*, **57**, 93–103 (1973).
93. Zierler, K. L. Muscle membrane as a dynamic structure and its permeability to aldolase. *Annals of the New York Academy of Sciences*, **75**, 227–34 (1958/59).

CHAPTER 2

Biology of experimental, acute myocardial ischaemia and infarction

R. B. Jennings and K. A. Reimer

INTRODUCTION	21
GENERAL BIOLOGY OF MYOCARDIAL BLOOD FLOW IN THE DOG	23
Coronary Anatomy	23
Regulation of Coronary Blood Flow	23
EFFECTS OF ACUTE CORONARY OCCLUSION ON MYOCARDIAL BLOOD FLOW	24
PHASES OF ISCHAEMIC INJURY—REVERSIBLE AND IRREVERSIBLE ISCHAEMIC INJURY	27
Reversible Phase of Ischaemic Injury	28
Metabolic Changes	28
Ultrastructural Changes	35
Effect of Reperfusion	35
Irreversible Phase of Ischaemic Injury	36
Metabolic Changes	36
Ultrastructural Changes	37
Effect of Reperfusion	40
TRANSMURAL PROGRESSION OF IRREVERSIBLE ISCHAEMIC INJURY	47
FINAL EVENTS IN MYOCARDIAL ISCHAEMIC CELL INJURY	51
ENZYME LOSS	51
TEMPORAL FEATURES OF ISCHAEMIA AND INFARCTION	56
REFERENCES	56

INTRODUCTION

Acute occlusion of a major coronary artery in man is followed by the development of a myocardial infarct which evolves through phases of ischaemic injury, coagulation necrosis, and finally of repair and replacement by scar tissue. The development of a myocardial infarct in man often is influenced by a multiplicity of factors including the pre-existing severity of coronary atherosclerotic disease, the extent of collateral anastomoses, and the haemodynamic consequences of the ischaemia, *per se.* Nevertheless, despite the potentially complex pathogenesis of human myocardial infarction, the pathological changes seen in human infarcts can be reproduced readily in experimental animals. Ischaemia produced by sudden occlusion of a major coronary artery in the dog is followed by several phases of injury including a phase of reversible ischaemic injury, a phase during which an increasing number of ischaemic cells become irreversibly injured and die, and finally a phase during which repair occurs through the removal of the dead muscle and replacement by scar tissue. Many of the general principles of myocardial ischaemic injury have been developed and are best defined in the dog model. For this reason, this chapter will concentrate on the phases of ischaemic injury and cell death as they occur in myocardial infarction induced in the dog by acute proximal occlusion of the circumflex artery.

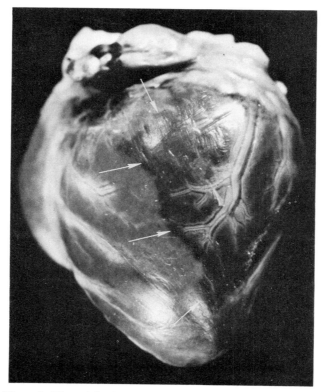

Figure 1. View of the lateral surface of the left ventricle of a dog heart to show the sharp junction of myocardium supplied by the circumflex, and anterior descending arteries, respectively. The animal was heparinized before excision of the heart. Both vessels were perfused simultaneously at 100 mmHg pressure with dyes suspended in 6% dextran 70 in normal saline. The circumflex bed, delineated by blue dye (dark areas arrowed) includes the posterolateral wall of the left ventricle, and in this dog extends almost to the apex

GENERAL BIOLOGY OF MYOCARDIAL BLOOD FLOW IN THE DOG

Coronary Anatomy

Dogs have a left dominant anatomic distribution of the coronary arteries. Thus the anterior free wall of the left ventricle is supplied by the anterior descending branch and the posterior and lateral free walls and posterior one third of the septum are supplied by the circumflex branch of the left coronary artery. The anterior two thirds of the septum is supplied by a septal artery which usually arises from either the proximal anterior descending or the left main coronary artery. The right coronary artery supplies the right ventricle but does not contribute to perfusion of the left ventricle.

The myocardium supplied by a coronary artery can be identified postmortem by injecting it with a coloured dye in a fluid such as latex or dextran. For example, in the dog heart shown in Figure 1, dye dissolved in 6% dextran in saline was injected into the circumflex artery 10 mm from its origin immediately after excision of the heart. The bed supplied by this vessel is filled with blue dye and is clearly demarcated from the non-ischaemic beds, which were simultaneously injected with a pink dye. Although not illustrated, the endocardial surface distribution of the circumflex bed also can be identified easily by this technique. The amount of muscle supplied by the circumflex artery ranges from 30–55% of the left ventricle, but the projecting basal portion of the posterior papillary muscle always is supplied by this artery. In dogs with small circumflex beds, the apex of the left ventricle may be supplied entirely by the anterior descending artery and vice versa. The frequency distribution of circumflex bed size is shown in Figure 2.

Regulation of Coronary Blood Flow

Myocardium is highly dependent on aerobic metabolism; consequently, it is a richly vascular tissue. Each myocardial cell is in close contact with two to four capillaries. The heart normally extracts about 75% of the oxygen passing through it and alterations in metabolic demand are met largely by increasing or decreasing coronary blood flow.

During systole, tissue pressure in the subendocardial myocardium approximates the blood pressure within the left ventricular cavity and causes capillary compression such that perfusion of the inner zone ceases. Tissue pressure is less in the subepicardial region and this zone receives some flow thoughout the cardiac cycle. Despite this difference in phasic coronary flow, there is no transmural flow gradient in non-ischaemic myocardium. Flow

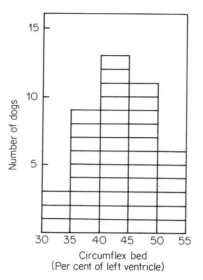

Figure 2. The distribution of the size of the circumflex bed among 42 dogs with four day old infarcts is illustrated. In each case, the artery was ligated 10–15 mm from its origin at the aorta. The circumflex artery supplied, on the average 44±10% of the left ventricle, but the distribution varied from about 30 to 55%. This distribution was established from morphometric analysis of serial transverse slices of the LV after coronary injections as illustrated in Figure 1. (Because the infarcted tissue was swollen by oedema and cellular infiltrates these estimates of circumflex bed size were 5–8% larger than estimates obtained in normal hearts)

during diastole is preferentially shunted toward the inner wall of the left ventricle by compensatory dilatation of vessels supplying this zone,[6] so that overall perfusion is similar in all regions of the left ventricle.

EFFECTS OF ACUTE CORONARY OCCLUSION ON MYOCARDIAL BLOOD FLOW

Sudden proximal occlusion of a major coronary artery in the dog results in severe reduction in arterial flow to the myocardium supplied by this vessel. The flow which persists is collateral flow coming from small epicardial arterial connections between the occluded and non-occluded beds. The size of these connections varies from 25–200 μm in diameter. In some dogs, there are numerous relatively large collaterals whereas other dogs have relatively few collateral connections. Thus the amount of collateral flow available varies among dogs.[3]

Following coronary occlusion, vasodilation of resistance vessels occurs throughout all layers of the ischaemic region. Because subendocardial flow is inhibited during systole, collateral flow is not evenly distributed to all regions of the ischaemic myocardium. Rather, there is always a transmural gradient of flow such that ischaemia is most severe in the subendocardial myocardium and least severe in the subepicardial zone. This transmural gradient of flow can be measured with radioactive microspheres. For example, $9 \pm 1 \mu m$ sized microspheres labelled with a gamma-emitting isotope can be injected into the left atrium. They lodge in capillaries throughout the dog, including the heart, in proportion to flow. Figure 3 shows the transmural gradient of collateral flow within the central portion of the ischaemic zone measured at 20 minutes after circumflex occlusion in 31 dogs. Flow to the subendocardial zone is nearly always severely reduced to 10% or less and averages 4.5% of flow to the non-ischaemic myocardium (Figure 3). Subepicardial flow is much more variable and averages 20% of non-ischaemic blood flow.

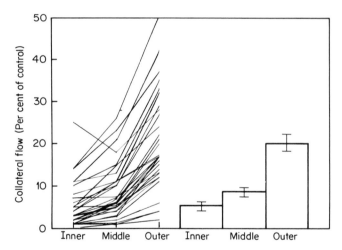

Figure 3. The transmural distribution of collateral flow found 20 minutes after circumflex occlusion in 31 dogs is illustrated. Flow was measured with 9 ± 1 μm microspheres before and after coronary occlusion. Collateral flow is expressed as a per cent of pre-occlusion flow to the same samples. The individual dogs are illustrated on the left and the group means ±SEM are shown on the right. I, M, and O = inner, middle, and outer thirds of the transmural wall in the circumflex bed. There is always a transmural gradient of flow such that flow to the outer wall is greater than flow to the inner wall. Subendocardial flow is almost always severely depressed (<15%) and averages 4.5% of control. Subepicardial flow is greater (averages 20% of control) and much more variable than subendocardial flow

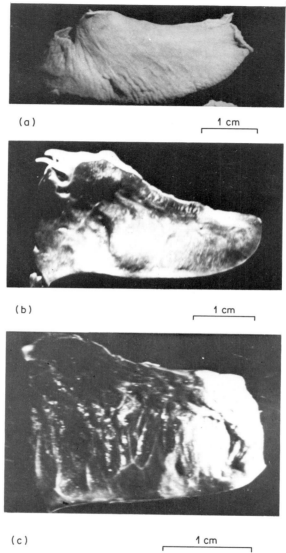

Figure 4. Myocardium viewed under ultraviolet light. The fluorescent dye thioflavin S was injected intravenously, 7–10 seconds before excision of the heart. (a) Non-ischaemic control myocardium from anterior papillary muscle. The myocardium shows homogeneous yellow–green fluorescence under ultraviolet light. (b) Transmural slice through posterior papillary muscle after circumflex occlusion. Severely ischaemic myocardium in the subendocardial region is non-fluorescent (dark) but the mid- and subepicardial myocardium are fluorescent indicating mild or moderate ischaemia. (c) Transmural slice through the posterior papillary muscle of another dog after circumflex occlusion. In this animal severely ischaemic myocardium extends across the transmural wall (left-hand portion of the slice). The apical part of this papillary muscle, supplied by the non-occluded anterior descending artery, is uniformly fluorescent. [Reprinted by permission of publishers, from *Archives of Pathology*, **99**, 86–94 (1975).]

The transmural distribution of flow from severe to more moderate ischaemia also can be visualized with the fluorescent dye, thioflavin S, a dye that binds to endothelium.[18] Injection of this dye one circulation time (7 seconds) prior to excision of the heart, results in staining of the endothelium of all capillaries and other vessels which have received significant arterial flow. Myocardium perfused with thioflavin S fluoresces when illuminated by ultraviolet light (Figure 4). However, when collateral flow is reduced to less than 20% of pre-occlusion flow, there is insufficient staining to cause detectable fluorescence.[19] Thus this technique can be used to demonstrate the distribution of the zone of severe ischaemia. Variation in the distribution of severe ischaemia is illustrated in Figures 4b and 4c. The subendocardial region was non-fluorescent (severely ischaemic) in both instances. However, the dog shown in Figure 4b exhibited a larger zone of significant mid- and subepicardial collateral flow compared to the dog in Figure 4c.

PHASES OF ISCHAEMIC INJURY—REVERSIBLE AND IRREVERSIBLE ISCHAEMIC INJURY

Coronary occlusion is followed by rapid desaturation of capillary haemoglobin which is made visible by the prompt appearance of cyanosis. The marked drop in oxygen supply is followed by rapid conversion of myocardial metabolism from aerobic respiration to anaerobic glycolysis and by rapid depletion of high-energy phosphate compounds, particularly creatine phosphate. This metabolic conversion is accompanied by cessation of contraction, and by the electrocardiographic changes of ischaemia. These changes occur within the first minute following acute coronary occlusion. Even the most severely ischaemic myocardium remains viable, however, for at least 15 to 18 min following occlusion. Termination of the ischaemic injury by reperfusion results in restoration of function. Gross and histologic evaluation of such hearts after one to four days of reperfusion shows that no cell death (necrosis) has occurred. Thus this phase of ischaemic injury is defined as the phase of 'reversible' injury.[16]

However, if the period of ischaemia is prolonged, reperfusion fails to prevent the death of some myocardial cells. Instead of resuming function, these cells undergo rapid structural disintegration and subsequently become necrotic. Thus these cells have passed the 'point of no return' and are 'irreversibly' injured. The number of irreversibly injured cells progressively increases as the duration of ischaemia is prolonged. By 40–60 minutes of ischaemia, most cells in the inner zone of severe ischaemia have died. Cells in the mid- and subepicardial zones of moderate ischaemia remain viable for longer periods of time. Thus cell death spreads as a transmural wavefront involving first the subendocardial myocardium with gradual transmural

Table 1. Severity of Ischaemia in the Centre of the Circumflex Bed of the Open Chest Dog after Proximal Arterial Occlusion

Degree of ischaemia	Qualitative description	Location	Degree of reduction of arterial flow in (%)	Speed of development of cell death
Severe	Low-flow ischaemia	Subendocardial	>85%	Prompt
Moderate	Moderate-flow ischaemia	Midmyocardial	70–85%	Slow
Mild	High-flow ischaemia	Subepicardial	<70	Probably survives

The subendocardial myocardium always is severely ischaemic. However, severe ischaemia may be transmural in which case no areas of moderate- or high-flow ischaemia will be present. When moderate- or high-flow ischaemia is present, it is found in the mid- or subepicardial myocardium.

extension.[23] By 6 hours after coronary occlusion, most cells which are going to die have died and the infarct reaches its full extent. The data in Table 1 summarize these concepts.[13]

Reversible Phase of Ischaemic Injury

Metabolic changes

Acute ischaemia occurs when arterial flow becomes inadequate to provide the oxygen required to support aerobic metabolism and myocardial function. Tissue Po_2 falls quickly following coronary occlusion.[24] Thus hypoxia (or anoxia) is an essential component of ischaemia and is directly related to the observed shift to anaerobic metabolism. Because of the reduction in arterial flow, transport of metabolites into and out of the ischaemic area also is impaired. This impairment is a second implicit part of the definition of ischaemia. In this respect, ischaemia differs from anoxia or hypoxia with continued flow. Ischaemia can be 'total', in which case no arterial flow is present or it can exist under high- or low-flow conditions[11] (Table 1).

Sudden coronary occlusion results in a dramatic series of metabolic events. During the first 5–15 seconds, there may be sufficient oxygen, associated with myoglobin or red cells trapped in the microvasculature to maintain aerobic metabolism. However, when oxygen no longer is available to accept the hydrogen removed from substrates, all parts of the terminal electron transport system including NAD^+, FAD, and the cytochromes become reduced. This reduction occurs quickly[18] and can be demonstrated visually by injecting the dog with methylene blue (prior to occlusion), and observing the conversion in the ischaemic region of deep-blue oxidized methylene blue to its colourless reduced form, leukomethylene blue (see Figure 3 in reference 8). Williamson et al.[28] also have demonstrated this transition by measuring the conversion of NAD^+ to the fluorescent NADH. This transition occurs about 10–12 seconds after coronary occlusion.

The degree of reduction in Po_2 found on the epicardial surface of the occluded arterial bed varies significantly. The central portion of the zone of epicardial cyanosis has the lowest Po_2 and the Po_2 rises as one approaches the junction with non-ischaemic myocardium.[24] Although no Po_2 measurements are available in a transmural direction, it seems likely that the transmural gradient of collateral flow demonstrated in Figure 3 is associated with a similar transmural Po_2 gradient, especially in the zones of moderate- to high-flow ischaemia. This consideration is important because it is possible that metabolism in zones of high collateral flow may be intermittently aerobic, a possibility which may explain why the injury is less severe in these zones (Table 1). However, net metabolism throughout the region of ischaemia is anaerobic in the sense that the Po_2 is low, oxidation–reduction indicators are reduced, cyanosis and contractile failure are present, and tissue lactate is high. In fact, from the point of view of the tissue, anaerobic metabolism is the *sine qua non* of ischaemia. By definition, ischaemia is present when the arterial flow is reduced to the point that O_2 availability exceeds the O_2 supply and the tissue therefore converts to a state of net anaerobic metabolism.[11]

As soon as the electron transport chain becomes saturated with electrons, the metabolism of the myocardium converts to anaerobic glycolysis. In glycolysis, glucose-6-phosphate, from circulating glucose or endogenous glycogen, is converted to lactate. Three net moles of ATP can be produced per mole of glucose from glycogen converted to lactate, while glucose from the plasma pool yields only 2 net moles of ATP per mole of glucose metabolized. The latter process is of negligible significance in areas of total or low-flow ischaemia but may be an important source of ATP in moderate- or high-flow ischaemic myocardium.

Lactate accumulates within the ischaemic tissue both because lactate cannot be metabolized further by ischaemic cells and because washout is slow. The speed with which lactate accumulates is shown by the data of Braasch et al. (Figure 5).[4] These investigators measured various metabolites extracted from instantly frozen biopsies of severely ischaemic myocardium. Myocardial lactate content did not change much during the first 15 seconds, but doubled after 30 seconds and by 60 seconds, was four times the level of non-ischaemic control tissue. Thus using lactate as a measure of the conversion to anaerobic glycolysis, ischaemia was detected by 15 to 30 seconds after the occlusion.

Depression in the level of high-energy phosphate within the affected myocardium is an early consequence of the conversion to anaerobic

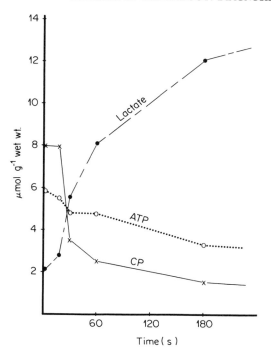

Figure 5. The effect of the first 180 seconds of ischaemia on ATP, CP, and lactate of myocardium is shown in this graph. The data is plotted from the results of Braasch et al., Circulation Research, **23**, 249 (1968). Samples were obtained by bone rongeurs cooled in liquid nitrogen from the cyanotic epicardium of the anterior surface of the heart of the dog. Note that little change occurs in CP or lactate during the first 15 seconds of ischaemia but that CP decreases and lactate increases markedly during the next 15 seconds. The increase in lactate marks the appearance of anaerobic glycolysis and shows that depression of aerobic metabolism is associated with depression in total high-energy phosphates as ATP+CP. Note that the CP decreases relatively little during the first 3 minutes of ischaemia

glycolysis. At the onset of ischaemia, high-energy phosphate continues to be used, first in maintaining contraction and, even after contraction ceases, in other cellular metabolic processes such as maintenance of ionic gradients. The transition to anaerobic glycolysis is associated with rapid catabolism of creatine phosphate (CP) (Figure 5). CP depletion begins within 15 seconds and by 3 minutes after coronary occlusion, only 10–15% of the initial CP remains. ATP on the other hand, is partially preserved during this early phase by the transfer of phosphate from CP to ADP via creatine kinase (CK).

Tissue acidity increases in ischaemic myocardium.[20] Increased H^+ ion content results from the accumulation of acidic glycolytic intermediates, of fatty acids from lipolysis, and of protons from ADP and ATP. Proton generation from the latter is related to pH and the availability of Mg^{2+}; however, it seems clear that much of the increase in H^+ in ischaemic tissue can come from adenine nucleotides.[9] The magnitude and rate of decrease in pH have not been established accurately in the dog because of limitations in the currently available techniques for measuring intracellular pH. However, several laboratories[22,28] have shown that the pH drops quickly in isolated hearts, made ischaemic *in vitro*.

The decrease in pH has two direct consequences: (i) There is good evidence that the increased H^+ ion content in the cell is the cause of the early contractile failure (Figure 6) observed in ischaemia.[28] (ii) It causes a prompt increase in the P_{CO_2} of the affected myocardium as protons exported from the cell are neutralized by bicarbonate in the extracellular fluid. (Presumably, little of the observed increase in tissue P_{CO_2} is due to metabolism, because CO_2 is not produced by anaerobic glycolysis and because the pentose phosphate shunt is a minor component of myocardial metabolism.)

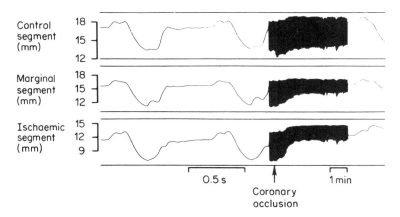

Figure 6. Regional segment function measured in the unanaesthetized dog by pairs of ultrasonic crystals positioned subendocardially in three regions of the left ventricle: in a normal control segment, a segment in the middle of a region to be rendered ischaemic by coronary occlusion (ischaemic segment), and in a region on the margin of the ischaemic zone (marginal segment). Normal shortening of the three segments is demonstrated in the left panel. After circumflex coronary artery occlusion (arrow) by inflation of a balloon cuff, there is rapid development of hypokinesis in the marginal segment, and a holosystolic bulge in the ischaemic segment. [Reprinted by permission of publishers from *Circulation*, **53**, 188–192 (1976).]

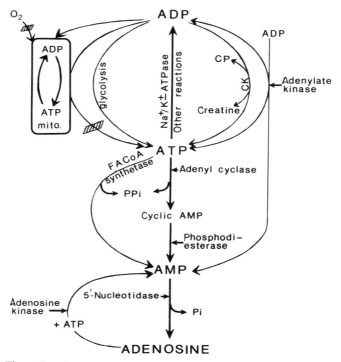

Figure 7. Diagram of nucleotide metabolism in severely ischaemic myocardium. The reactions which take place in the markedly hypoxic environment of the zone of severe ischaemia are emphasized. In this zone, mitochondrial respiration is assumed to be inoperative because of the virtual absence of oxygen. After contraction ceases, the reactions utilizing ATP to produce ADP include the Na^+/K^+-ATPase of the cell membrane and a variety of ATPases and protein kinases of the cell interior. The principal means of generating ATP from ADP, after CP supplies for the action of creatine kinase (CK) are exhausted, is anaerobic glycolysis or adenylate kinase (myokinase). The routes by which AMP is generated are emphasized because of the crucial role of 5'-nucleotidase in the loss of adenine nucleotides from ischaemic cells. This enzyme of the sarcolemma dephosphorylates adenosine, which then diffuses to the extracellular space. There is little reutilization of the adenosine because it seems unlikely that enough ATP is available, in areas of severe ischaemia, to support the action of adenosine kinase to convert adenosine to AMP and then to ATP via adenylate kinase. The principal reactions producing AMP include fatty acid acyl CoA synthetase (FA CoA synthetase), adenylate kinase, and the formation and degradation of cyclic AMP via the action of adenyl cyclase and phosphodiesterase. About 50% of the adenine nucleotides of a myocyte are in the mitochondria and the means by which these are degraded is unknown. Since the cellular ATP drops quickly in ischaemia, they presumably reach the sarcoplasm and participate in the reactions shown. [Reprinted by permission of publishers from *American Journal of Pathology*, **92**, 187–214 (1978).]

Another early consequence of ischaemia is the release of endogenous catecholamines. Wollenberger[29] has shown that a 2.5 minute period of ischaemia causes release of norepinephrine from endogenous stores within ischaemic myocardium. This stimulates formation of cyclic AMP from ATP via the sarcolemmal enzyme adenyl cyclase. The formation and subsequent breakdown of cyclic AMP occur via the reactions

$$ATP \xrightarrow[\text{adenyl cyclase}]{Mg^{2+}} \text{cyclic AMP} + \text{pyrophosphate}$$

$$\text{cyclic AMP} + H_2O \xrightarrow[\text{phosphodiesterase}]{} AMP$$

Cyclic AMP stimulates the conversion of phosphorylase b to phosphorylase a and thereby accelerates glycolysis. Also, it may stimulate lipolysis with the consequent increased production of fatty acids. Both processes utilize ATP and contribute to the depletion of ATP and the accumulation of cyclic AMP, AMP, and pyrophosphate.

Some of the reactions involving ATP which take place under ischaemic conditions are shown in Figure 7. Once contraction has ceased, the reactions utilizing ATP include the Na^+/K^+-ATPase of the cell membrane, adenyl cyclase, fatty acid CoA synthetase, and unidentified ATPases within the cell. ATP is regenerated to support this metabolism by anaerobic glycolysis and by the action of adenylate kinase.

By three minutes CP has been depleted to 10% of its original level in severely ischaemic tissue. The ATP content then begins to decrease rapidly and reaches 38% by 15 minutes and is less than 10% by 40 minutes (Figure 8).[14] The changes in the distribution of adenine nucleotides and the total adenine nucleotide pool ($\sum AD$) during low-flow ischaemia are summarized in Table 2. There is no net increase in ADP but AMP increases 631% by 40 minutes. However, this increase in AMP accounts for anly a small portion of the ATP lost.[14] Thus the total adenine nucleotide content decreases in a curve which parallels ATP depletion. Much of the depletion of the $\sum AD$ pool occurs via further degradation of AMP. AMP is degraded to adenosine by 5'-nucleotidase located in the sarcolemma. Berne's group has shown, in totally ischaemic myocardium *in vitro*, that one can account for all of the $\sum AD$ loss by a large increase in adenosine and its metabolites, i.e. inosine and hypoxanthine.[26] Degradation probably follows similar pathways during severe ischaemia *in vivo* but this has not been confirmed. Since neither *de novo* nor salvage synthesis of adenine nucleotides is possible under anaerobic conditions, destruction of adenine nucleotides is a very important feature of ischaemic injury.[14]

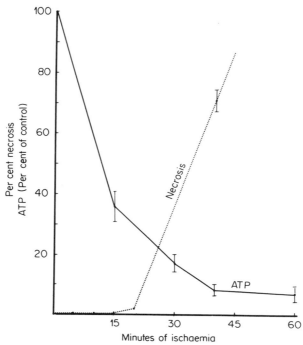

Figure 8. The relationship between the level of ATP remaining in the posterior papillary muscle and the amount of irreversible injury in this muscle after various periods of ischaemia is plotted in this graph. The ATP data is from groups of 4 to 6 dogs subjected to 15, 30, 40, or 60 minutes of ischaemia. The posterior papillary muscle was used for the analyses. The per cent necrosis data is from a series of experiments in which the amount of necrosis found after 5 to 60 minutes of temporary ischaemia and 2–4 days of arterial reperfusion was quantitated by histologic techniques (see Figure 25). Intervals of ischaemia of 15 minutes or less duration result in no irreversible injury. After 20 minutes of ischaemia, only 2%, on the average, of the portion of the posterior papillary muscle used to obtain the ATP curve is necrotic. After 40 minutes of ischaemia, an average of 72% of the cells in the projecting portion of the posterior papillary muscle become necrotic. The 20 and 40 minute points are based on 15 and 24 dogs respectively. Examination of these curves shows that the myocardial cells tolerate ATP levels of 38% of control without developing necrosis. Necrosis first appeared when the ATP reached 20% and depletion to <10% of control ATP was associated with extensive necrosis. [Reprinted by permission of publishers from *American Journal of Pathology*, **92**, 187–214 (1978).]

Table 2. Adenine Nucleotide Distribution of Normal and Ischaemic Myocardium[a]

Nucleotide[a]	Control AP (μmol g^{-1} wet)	Period of ischaemia PP		
		15 min	30 min (μmol g^{-1} wet)	40 min
ATP	6.02	2.12***	0.82***	0.37***
% Reduction	±0.21	±0.45	±0.18	±0.07
	(12)[b]	(3)	(5)	(4)
in ATP content		65	85	94
ADP	1.23	0.83***	0.96**	0.74***
% Reduction	±0.06	±0.01	±0.05	±0.09
	(11)	(3)	(5)	(4)
in ADP content		33	22	40
AMP	0.16	0.37**	0.66**	1.17***
% Increase in	±0.01	±0.11	±0.23	±0.19
	(11)	(3)	(5)	(4)
AMP content		131	313	631
Total ATP+ADP+AMP	7.34	3.32***	2.50***	2.28***
	±0.27	±0.34	±0.15	±0.22
	(11)	(3)	(5)	(4)
% Decrease in				
total nucleotides		55	66	69

[a] AP and PP refer to anterior and posterior papillary muscles respectively. The control AP sample of one 40 minute dog had sufficient extract to allow determination of ATP but not ADP and AMP. The ATP results on this PP sample are included in the 40 minute mean. The total nucleotides were obtained by summing the totals of the 11 animals on which data were available.

Mean differences were assessed between AP tissue and PP tissue after 15, 30, or 40 minutes of ischaemia. Statistical comparisons were with a two-tailed non-paired t test. In each case: **$p<0.01$; ***$p<0.001$.

[b] Number of samples assayed is included in brackets.

Ultrastructural changes

Control non-ischaemic myocardium reveals abundant mitochondria, contracted myofibrils in register, even nuclear chromatin distribution, abundant sarcoplasmic glycogen and an intact sarcolemma (Figure 9).

No significant structural changes can be identified after 5–10 minutes of ischaemia. By 15 minutes, relaxation of myofibrils and depletion of glycogen granules can be observed. Occasional swollen mitochondria and mild margination of nuclear chromatin also occur.

Effect of reperfusion

Reperfusion during the first 15 minutes of ischaemic injury permits the rapid restoration of normal cellular architecture and subsequent necrosis does not occur. This fact forms our basis for defining this period of ischaemia as the period of reversible injury.[12,16]

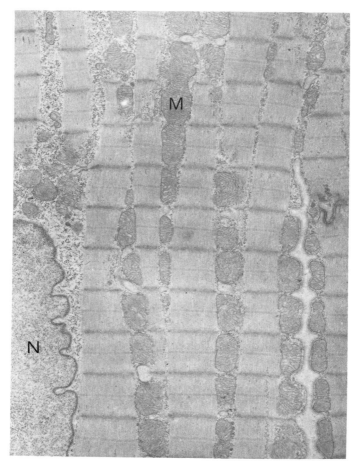

Figure 9. Control, non-ischaemic myocardium. This shows a portion of a myocardial cell with contracted myofibrils, abundant glycogen, dense mitochondria (M), and nucleus (N). ×13,600. [Reprinted by permission of publishers from *American Journal of Pathology*, **57**, 539–57 (1969).]

Irreversible Phase of Ischaemic Injury

Metabolic changes

By 20 minutes after coronary occlusion, severely ischaemic myocardium has lost 62% of its ATP (Figure 8) and has accumulated lactate levels 12 times the normal value of 2.1 $\mu mol\,g^{-1}$ wet weight. Lactate continues to accumulate during the first hour and reaches 15, 20, and 30 times control by 30, 40, and 60 minutes respectively. By 60 minutes, little or no glycogen is detectable in the severely ischaemic myocardium.[10] Less lactate accumulation occurs in areas of moderately ischaemic myocardium, presumably because washout is greater in these areas.

Although the cause of irreversibility has not been established, many of the biologic features of irreversible injury in severely ischaemic myocardium have been characterized. One of the more striking relationships is that found between tissue ATP depletion and irreversible injury (Figure 8).[14]

Ultrastructural changes

The ultrastructural features which occur early in the phase of irreversible injury are quite striking (Figures 10a and 11).[12] By 30–60 minutes

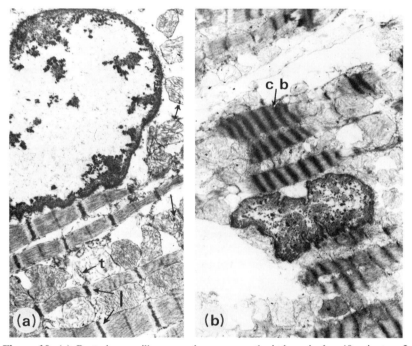

Figure 10. (a) Posterior papillary muscle permanently ischaemic for 40 minutes. I bands (I), intermyofibrillar oedema, and nuclear chromatin clumping and margination are prominent. Z bands remain in register. Many mitochondria are swollen and contain tiny amorphous dense bodies (arrow). Transverse tubules (t) do not appear to be swollen. ×7900. (b) Posterior papillary muscle temporarily ischaemic for 40 minutes with two minutes of reflow. There is marked disruption of cellular architecture. Intracellular swelling, contraction bands (cb), large vacuoles, lifting of the sarcolemmal membrane by subsarcolemmal blebs, and swollen mitochondria containing dense bodies are visible. There is marked chromatin clumping and margination. ×6000. [Reprinted by permission of publishers from *American Journal of Pathology*, **74**, 399–423 (1974).]

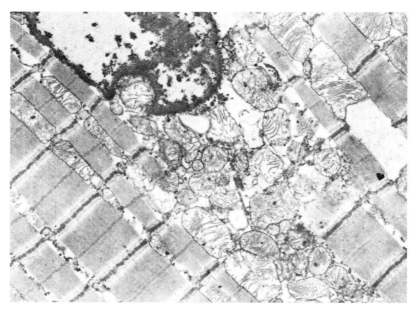

Figure 11. Sixty minutes of ischaemia. The myofibrils are very relaxed, and clearly show an N line within the I band on either side of the Z band. An M band is present within the H band. Margination of nuclear chromatin is prominent, and mitochondrial changes include loss of dense matrices, disruption of cristae, and the presence of amorphous intramitochondrial densities × 14,000. [Reprinted by permission of publishers from *Archives of Pathology* **79**, 135–43 (1965).]

mitochondria become markedly swollen with an increased matrix space. Osmiophilic amorphous densities, generally termed amorphous matrix densities, appear within the matrix space. These densities measure up to 200 Å diameter and as many as four densities may appear in a single mitochondrial profile. Nuclei show marked peripheral aggregation of chromatin (Figures 10 and 11). The sarcoplasm contains virtually no glycogen and the myofibrils become superstretched. Discontinuities in the plasma membrane of the sarcolemma also become detectable (Figure 12).

As the period of ischaemia is prolonged from 30 minutes to several hours, the mitochondrial amorphous matrix densities increase in size and number. Also, the defects in the sarcolemma increase in size and frequency. However, the myofibrils of cells in zones of severe ischaemia remain in register and, except for stretching, remain structurally intact. This latter observation explains why irreversible injury cannot be detected for several hours by routine techniques of light microscopy, despite the early devastating signs of nuclear, mitochondrial, and sarcolemmal disruptions by electron microscopy.

available energy supplies.[45] The point at which this ATP deficit seems usually to declare itself most rapidly is in the failure of cells to sustain normal rapid energy-linked processes such as contractility or secretion of macromolecules. These, like the maintenance of ion gradients by plasma membrane pumps, seem usually to draw upon a fraction of the cell's ATP that is localized in the cytosolic compartment. In the ischaemic heart one of the unsolved puzzles is the exact nature of this highly susceptible ATP compartment which sustains contractility and appears to suffer severe, but reversible, depletion almost immediately after the interruption of the blood supply.[31]

The second, but still relatively rapid, effect of ATP depletion in the cytosolic compartment is a decline in a cell's ability to pump Na^+ and Ca^{2+} out, and K^+ in, through the plasma membrane, allowing the cytosolic concentration of Na^+ and Ca^{2+} to rise and that of K^+ to fall. Although apparently a simple defect, the interactions between the different ion permeability systems of cells mean that the consequences of such a failure can be remarkably complex; it may turn out to be the onset of a sequence of events which form a self-perpetuating vicious circle. As soon as Na^+ and Ca^{2+} begin to accumulate in the cell and K^+ begins to leak out, then an obvious effect is the increased activation, by these ions, of the ATP-starved ion pumps of the plasma membrane and sarcoplasmic reticulum, so that the already depressed ATP levels tend to fall even faster. As a result of the rise in intracellular Ca^{2+} concentration, the permeability of the plasma membrane to Na^+ and to K^+ (particularly to K^+, since this ion crosses the membrane through an ion 'gate' whose opening is activated by intracellular Ca^{2+}) rises substantially to further frustrate the homeostatic efforts of the starved cells. Two further effects of the rise in intracellular Ca^{2+} concentration are (i) to inhibit Na^+ and K^+ pumping (as a direct result of the inhibitory effects on Na^+/K^+-ATPase- of Ca^{2+} present at the cytoplasmic surface of the plasma membrane),[46] and (ii) to compete with ATP synthesis for the limited mitochondrial energy supplies, so that mitochondria begin to accumulate unusual quantities of Ca^{2+} (though not enough to form defined precipitates, at least during the early phases of energy starvation).

In addition to these direct effects on the intracellular ion environment of the energy-deficient cell, the loss of the Na^+ gradient has another most important effect. In healthy cells the main osmotic influence which counteracts the natural tendency to swelling, (because of the high intracellular colloid osmotic pressure) is the opposite osmotic influence exerted by the excess of Na^+ outside the cell. With the reduction of the normal Na^+ gradient this counteracting force is also diminished and the cells tend to accumulate water and therefore swell.[51] This can have two serious consequences. The simpler effect is the occlusion of blood vessels, thus further exacerbating the cell's energy supply problem. Indeed, it is relatively soon

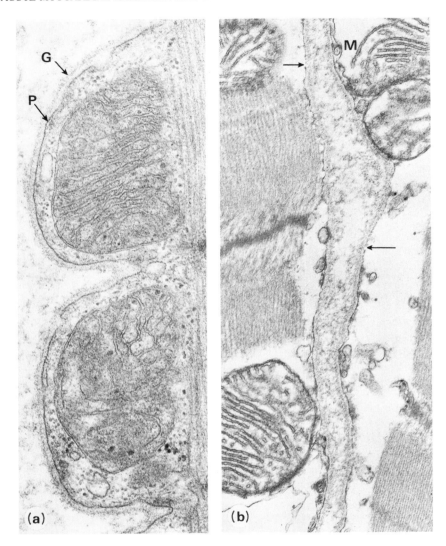

Figure 12. (a) Non-ischaemic anterior papillary muscle. The sarcolemma is scalloped over two contracted myofibrils. The plasma membrane (P) and glycocalyx (G) are typical of control myocardium. ×50,000. (b) Posterior papillary muscle after 40 minutes of ischaemia. Several small breaks (arrows) are visible in the plasma membrane although the glycocalyx is continuous. Swollen mitochondria are present (M). The myofibrils are relaxed and show I bands. ×38,000. [Reprinted by permission of publishers from *American Journal of Pathology*, **92,** 187–214 (1978).]

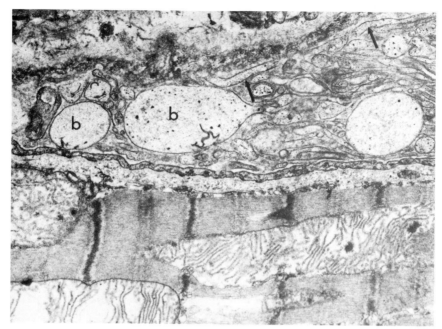

Figure 13. Posterior papillary muscle after 90 minutes of ischaemia and 10–12 seconds of coronary blood reperfusion (from an area of 'no reflow'). Endothelial pinocytotic vesicles are sparse. The capillary lumen is full of endothelial protrusions (arrows) and membrane-bound bodies (b), some of which might represent degranulated platelets. Mitochondria are swollen with amorphous matrix dense bodies. I bands and intermyofibrillar oedema are present. ×22,640. [Reprinted by permission of publishers from *Journal of Clinical Investigation*, **54**, 1496–508 (1974).]

The effects of ischaemia on cell types within the heart other than cardiac muscle have not been studied in as much detail. However, capillary endothelial cells lose their pinocytotic vacuoles and develop large cytoplasmic blebs (Figure 13). This occurs somewhat later than the changes in muscle cells but becomes prominent after 60 or more minutes of severe ischaemia.

Effect of reperfusion

After only 2 minutes of reperfusion with arterial blood, myocytes develop prominent contraction bands and subsarcolemmal blebs. The contraction bands are dense bands composed of several sarcomeres, with complete loss of identifiable A, I, or Z band structure (Figure 14), which develop by excessive shortening of sarcomeres. In Figure 10b, early contraction bands with identifiable Z bands still present are shown while fully developed contraction bands with disrupted myofibrils are shown in Figure 14. Although ischaemic myocardium shows no early changes by light microscopy,

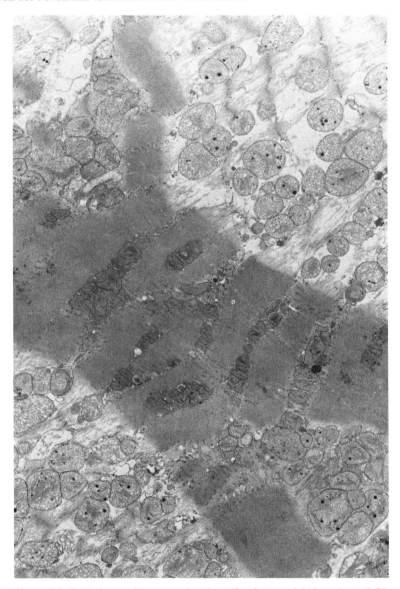

Figure 14. Posterior papillary muscle after 40 minutes of ischaemia and 20 minutes of reperfusion *in vivo*. There is a dense contraction band in which some mitochondria have become 'trapped'. Adjacent areas show myofibrillar disruption. Mitochondria containing both amorphous and calcium dense bodies. ×5500. [Reprinted by permission of publishers from *American Journal of Pathology*, **46,** 367–86 (1965).]

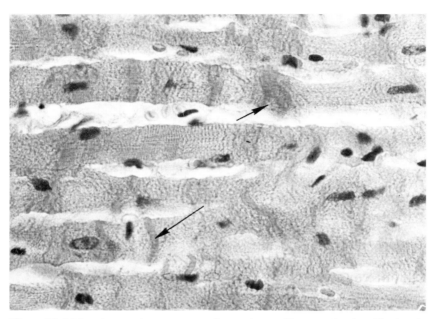

Figure 15. Light microscopic appearance of posterior papillary muscle after 40 minutes of ischaemia and 20 minutes of reperfusion. The disruption of myofibrillar architecture with formation of contraction bands, is shown at the arrows. This change is characteristic of irreversibly injured cells by ischaemia. There is separation of myofibres, presumably due to the presence of subsarcolemmal blebs as illustrated by electron micrograph in Figure 12b. ×800. [Reprinted by permission of publishers from *Laboratory Investigation*, **13**, 1491–503 (1964).]

these large contraction bands are readily apparent by light microscopy after as little as two minutes of reperfusion (Figure 15). The subsarcolemmal blebs, produced by cell swelling, cause apparent separation of cells by light microscopy, a finding which is often misinterpreted as interstitial oedema.

The sarcolemma exhibits striking changes in the region of the subsarcolemmal blebs. Here there are large defects in the plasma membrane leaving bare glycocalyx. Remnants of plasma membrane remain as circular profiles attached to the glycocalyx (Figure 16). Some plasma membrane remnants also can be identified attached to the myofibrils especially near the Z lines where the T tubules originate.

Reperfusion also results in the formation of a second type of mitochondrial granule (Figures 14 and 17). The latter is more granular (therefore referred to as granular densities) than the amorphous densities and are composed of calcium phosphate. These granular densities are small after 2 minutes of reflow but become progressively larger after 5–20 minutes of reperfusion.

Striking changes in tissue water and electrolyte content accompany the ultrastructural events associated with reperfusion.[27] Only 2 minutes of reflow following a 40 minute period of ischaemia (sufficient to kill 72% of the cells in the posterior papillary muscle) results in a 21% increase in cell water in the posterior papillary muscle (Figure 18). Much of this increased water must be located in the subsarcolemmal blebs described above. Na^+ and Ca^{2+} increase rapidly and Mg^{2+} and K^+ are washed out. By 20 minutes of reperfusion the tissue Ca^{2+} increases tenfold (Figure 19). Studies with $^{45}Ca^{2+}$ have shown that the calcium which accumulates is derived from the plasma perfusing the previously ischaemic area.[25] The massive quantity of calcium which enters irreversibly injured cells is the most likely cause of the myofibrillar contraction bands and is largely deposited in the mitochondrial granules seen by electron microscopy.

The reperfusion phenomenon can be studied, *in vitro*, with tissue slices or isolated hearts in a Langendorff apparatus. The tissue slice technique, which we have used extensively,[7] consists of thin free-hand slices cut from the

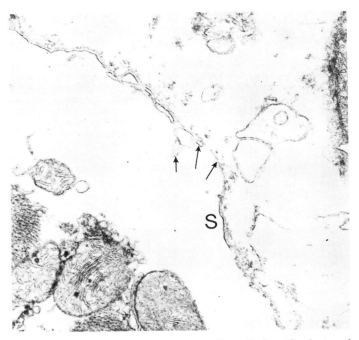

Figure 16. Sarcolemmal bleb in a myocardial cell after 40 minutes of ischaemia and 5 minutes of arterial reperfusion. The usual architecture of the sarcolemma is shown at (S). In adjacent segments (arrows) the sarcolemma is incomplete, having been converted to circular profiles with intervening gaps. ×21,000. [Reprinted by permission of publishers from *American Journal of Pathology*, **81**, 179–98 (1975).]

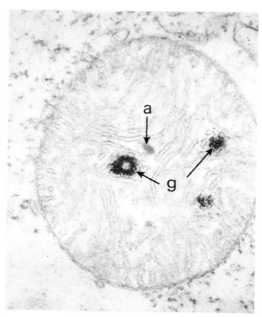

Figure 17. Mitochondrion of a myocardial cell after 40 minutes of ischaemia and 20 minutes of reperfusion. With reflow of blood, cells accumulate calcium much of which is deposited in doughnut-shaped granular dense bodies which develop in the mitochondrial matrix (g). These densities have intrinsic electron opacity and are different from the amorphous densities (a) which also are present. ×67,500. [Reprinted by permission of publishers from *American Journal of Pathology*, **81**, 179–98 (1975).]

papillary muscles parallel to the long axis of the myocytes. Slices of non-ischaemic myocardium, when incubated at 37 °C in oxygenated Krebs–Ringer phosphate media at pH 7.4, maintain ion gradients, cell volume, and ultrastructure for 3 or more hours (Figure 20). Cell volume is maintained largely by the action of the Na^+/K^+-ATPase. Inhibition of this enzyme by 10^{-3} M ouabain or inhibition of metabolism by 0–1 °C temperature causes cell swelling.

Placing slices of posterior papillary muscle, injured by ischaemia *in vivo*, into Krebs–Ringer phosphate *in vitro* is roughly equivalent to reperfusion of this same tissue *in vivo*. Fresh slices of muscle irreversibly injured by 40 or 60 minutes of ischaemia show no abnormalities in tissue electrolytes or water. However, incubation of posterior papillary muscle slices results in marked cell swelling associated with increased Na^+ and decreased K^+ and

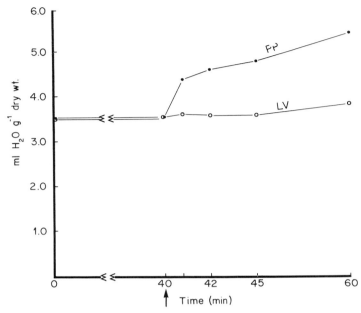

Figure 18. Water content (ml H_2O g^{-1} dry weight) of the injured posterior papillary (PP) muscle and non-ischaemic anterior wall of the left ventricle (LV) during the first 20 minutes of reperfusion following 40 minutes of ischaemia. Each point is a mean from a group of 4 dogs. Note the abrupt increase in H_2O after 2 minutes of arterial reflow and the gradual increase in H_2O thereafter as the period of reflow is extended to 4, 10, and 20 minutes. [Reprinted by permission of publishers from *American Journal of Pathology*, **74**, 381–97 (1974).]

Mg^{2+}. These incubated slices also develop subsarcolemmal blebs, membrane defects, contraction bands, and in some instances, granular mitochondrial densities of calcium phosphate (Figure 21). Thus, the effects of reperfusing irreversibly injured myocardium *in vivo* are closely mimicked by incubating slices *in vitro*.

It is of interest that incubating slices of injured myocardium at 0–1 °C allows dissociation of the cell swelling from the development of contraction bands. Injured cells incubated in the cold develop subsarcolemmal blebs and defects in the plasma membrane but do not develop contraction bands (Figure 22). Neither change occurs in cold slices of control myocardium.

Perfusion of rat hearts with anoxic substrate-free media in a Langendorff apparatus for 50 minutes followed by perfusion with oxygenated media containing glucose results in changes similar to those seen *in vivo*, following reperfusion of irreversibly injured ischaemic myocardium.[8] During the period of anoxic perfusion no change in membrane ultrastructure has been detected but reoxygenation is followed by striking changes in membrane

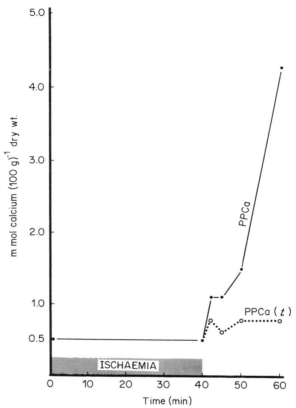

Figure 19. Plot of the content of calcium in the posterior papillary muscle (PPCa) as a function of time. The PPCa does not change during the 40 minute ischaemic period. After 2, 5, and 10 minutes of arterial reflow, it is increased markedly, and at 20 minutes, the content is about eight times the content of non-ischaemic control myocardium. The value PPCa(*t*) is the increase in Ca^{2+} which results from the increase in tissue H_2O (Figure 18) and illustrates that little of the increase in tissue Ca^{2+} is due to oedema. Other processes including the massive Ca^{2+} loading of mitochondria explain the marked increase in PPCa. [Reprinted by permission of publishers from *American Journal of Pathology*, **74**, 381–97 (1974).]

ultrastructure. As in the ischaemic dog heart, the glycocalyx appears intact but there are breaks in the plasma membrane.

The results of reperfusion *in vivo* and as well reoxygenation or slice incubation *in vitro*, all suggest that alterations in the structure of the sarcolemma may be a primary event in the development of irreversibility.

Periods of ischaemia of one hour or more result in areas of vascular

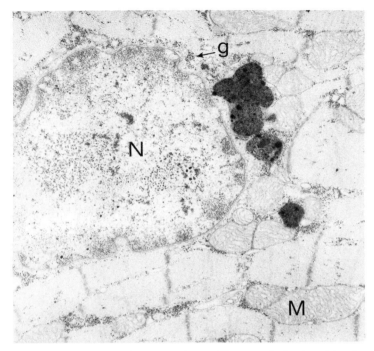

Figure 20. Free-hand slice of anterior papillary muscle incubated *in vitro* for 60 minutes at 37 °C. Part of a representative cell from the centre of the slice is illustrated. The chromatin of the nucleus (N) is dispersed evenly as it is in control non-incubated tissue. Mitochondria (M) show minimal or absent swelling. Glycogen (g) is abundant. Lysosomes are intact. The Z bands are not thickened. Tissue incubated for 60 minutes at 0–1 °C is indistinguishable from the figure illustrated here. ×16,400. [Reprinted by permission of publishers from *Journal of Molecular and Cellular Cardiology*, **8**, 173–87 (1976).]

injury, as noted above, and studies with thioflavin S, india ink, or radioactive microspheres, identify areas of poor or absent reperfusion in areas of severe ischaemia (the 'no-reflow phenomenon'). Capillaries and small arterioles in such areas are obstructed by endothelial blebs and/or red cell sludging (Figure 13).[17] More severe vascular injury results in intramyocardial haemorrhage. In general, vascular injury occurs on a slower time scale than irreversible muscle cell injury and the role of the no-reflow phenomenon in the pathogenesis of ischaemic cell death is uncertain.

TRANSMURAL PROGRESSION OF IRREVERSIBLE ISCHAEMIC INJURY

The development of irreversible injury does not occur at the same time in all ischaemic cells but rather involves progressively more cells over a period

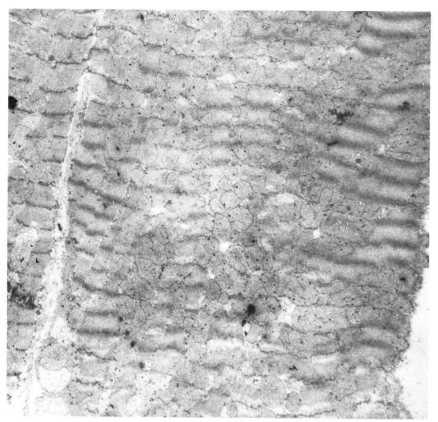

Figure 21. Slice of posterior papillary muscle injured by 60 minutes of ischaemia *in vivo* and then incubated for 60 minutes at 37 °C. Note the irregular contraction of myofibrils. I bands have disappeared; only thickened Z bands are present. ×6000. [Reprinted by permission of publishers from *Acta Medica, Scandinavica Supplement*, **587**, 83–92 (1975).]

of 3 to 6 hours.[23] No cells die during the first 15 to 18 minutes of ischaemic injury. By 20 minutes, necrosis occurs in only about half the hearts. Even in those hearts where necrosis occurs, it is confined to small areas within the severely ischaemic subendocardial zone. By 40 to 60 minutes. however, there is confluent necrosis within the subendocardial myocardium and by 60 minutes most cells in the posterior papillary muscle are dead. This subendocardial necrosis occurs throughout the entire circumflex bed at risk and extends to within a few cells of the lateral edges of the ischaemic bed following as little as 40 minutes of ischaemic injury. Thus the lateral border zone is sharp. However, the subepicardial half of the myocardium usually still is viable at 40 minutes and the transmural progression of ischaemic injury occurs more slowly. Thus overall infarct size at any time of study is

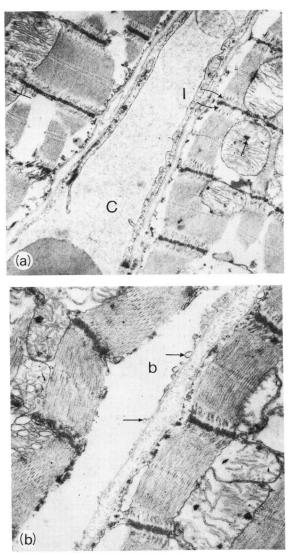

Figure 22. (a) *In vivo* myocardium after 60 minutes of permanent ischaemia. Glycogen is absent. The mitochondrial matrix space is enlarged and contains amorphous densities (arrows). I bands (I) are prominent. The sarcoplasmal membrane appears intact. A capillary (C) containing flocculent precipitates occupies the centre of the picture. ×12,700. (b) Slice of ischaemic myocardium incubated in the cold. Mitochondria are greatly swollen and contain amorphous densities. Glycogen is absent. In the cell on the right, the myofibrils have separated from the surrounding cell basement membrane forming an empty space (b). The plasma membrane of the sarcolemma has fragmented to form small vesicles lying beneath the basement membrane (arrows). ×25,200. [Reprinted by permission of publishers from *Journal of Molecular and Cellular Cardiology*, **8**, 189–204 (1976).]

determined by two parameters, i.e. the size of the circumflex bed at risk[21] and the transmural extent of necrosis within the bed. This transmural progression of cell death has been evaluated by quantitating necrosis histologically in transmural sections through the posterior papillary muscle obtained 2 to 4 days after reperfusion of myocardium made ischaemic for different periods of time (Figures 23 and 24). Necrosis within these central sections has proven to be a reliable index of overall infarct size.[23] At 40 minutes necrosis involves about 30–35% of the transmural wall through the posterior papillary muscle and 70–80% of the projecting portion of this muscle. By 3 hours, 50–60% of the transmural wall is necrotic and by 6 hours the infarct has nearly reached its full transmural extent, often involving 75% or more of the wall.[23] Permanent occlusions result in necrosis involving on the average, 75% of the circumflex bed and 80–85% of the transmural wall through the posterior papillary muscle in the centre of this bed. Thus some of the mildly or moderately ischaemic myocardium in the subepicardial zone, particularly toward the lateral edges of the circumflex bed, never becomes necrotic.

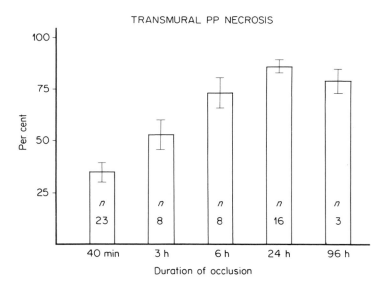

Figure 23. Transmural necrosis, calculated from the histologic sections through the posterior papillary muscle, is plotted with respect to duration of occlusion. Brackets indicate the SEM. Infarct size increased progressively with time. Reperfusion at 40 minutes and 3 hours resulted in significantly less necrosis compared with 24 hour infarcts. Six hour infarcts were intermediate between but not significantly different from either 3 or 24 hour infarcts. [Reprinted by permission of the American Heart Assn. from *Circulation*, **56**, 786–94 (1977).]

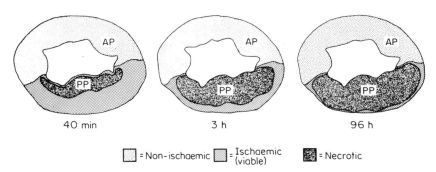

Figure 24. Diagrammatic summary of cross-sections through the left ventricle of the dog heart to show the progression of ischaemic cell death with respect to duration of coronary occlusion. Necrosis occurs first in the subendocardial myocardium and with longer durations of coronary occlusion involves progressively more of the transmural thickness of the ischaemic zone. The lateral border of the infarct is sharp but there is typically a significant amount of subepicardial myocardium which is viable and can be salvaged by reperfusion, as late as 3 hours after coronary occlusion.

FINAL EVENTS IN MYOCARDIAL ISCHAEMIC CELL INJURY

Once the injury to the myocardial cells become irreversible, the cells pass through a series of events which can be grouped together as the 'final events' or, in a funeral sense, the 'undertaking events'. During the early part of this phase, soluble macromolecular components of the cell slowly diffuse into the extracellular space and then are carried into the systemic circulation via the lymphatics or blood. The loss of intracellular enzyme occurs at or about the time the defects in plasma membrane integrity are noted, but definite proof of the relationship between enzyme loss and membrane defects is not yet available. Solubilization of the necrotic myocardium by lysosomal enzyme activity does not appear to occur to any significant extent. Rather, the cellular debris remaining in the infarct eventually is phagocytosed by polymorphonuclear neutrophils and monocytes. These first appear on the periphery of the infarct and digest the muscle from the periphery to the centre.

The appearance of necrosis 4 days after reperfusion of a temporary occlusion is shown in low-power view (Figure 25). In this case, the process of repair is well advanced throughout the area of injury because the vasculature has largely been spared. Finally, the dead myocardium is replaced by fibrous tissue.

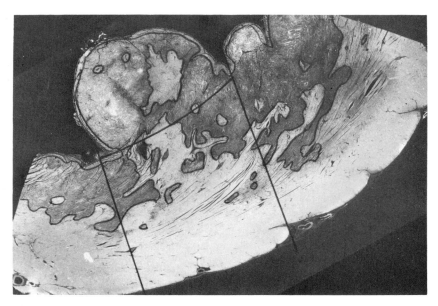

Figure 25. Negative print of a histologic section stained with Heidenhain's variant of Mallory's connective tissue stain to demonstrate the technique of quantitating necrosis. This section was prepared from a transmural slice of heart cut transversely through the circumflex bed. The posterior papillary muscle is in the centre of this section. The negative print was made by projecting the histologic section from a photographic enlarger onto black and white photographic paper. Necrotic areas identified by microscopic examination are easily distinguished (dark areas) from viable areas on the low-power magnification shown here. The correlation presented in Figure 13 was based on necrosis and ATP content within the projecting portion of the posterior papillary muscle, i.e. the myocardium above the curved line. The transmural progression of necrosis plotted in Figure 23 was based on transmural sections through the posterior papillary muscle.

ENZYME LOSS

Currently available data indicates that all enzymes of the myocardial cell remain in their intracellular position during the reversible phase of ischaemic injury. This includes soluble enzymes of the sarcoplasm such as creatine kinase (CK), aspartate aminotransferase (AST), lactate dehydrogenase (LD), aldolase, myokinase, and alanine aminotransferase (ALT) as well as bound or insoluble enzymes such as the pyruvate or α-ketoglutarate dehydrogenase enzyme complexes, succinate dehydrogenase (SD) and β-hydroxybutyrate dehydrogenase. However, the soluble enzymes will disappear from the tissue and appear in the circulation following the development of irreversible ischaemic injury, and circulating CK, LD, AST, and ALT have been used extensively to diagnose the presence of irreversible injury in the tissue. Other macromolecular components such as myoglobin follow a similar pattern of release. In contrast, insoluble enzymes such as SD, have not been detected in an active form in the plasma.

Much is known about the nature of the process of enzyme release and the factors which affect it. It seems very likely that the rate of appearance of an enzyme in the circulation is flow dependent. Thus enzymes released from cells dying in areas of low-flow ischaemia will reach the general circulation more slowly than enzymes released from cells dying in areas of moderate- or high-flow ischaemia. Some of the evidence on which these general concepts are based will be presented in the next few paragraphs.

The phenomenon of enzyme loss has been examined by three principal means, by comparing enzyme activity in injured and control tissue to ascertain if enzyme has been lost, by following the temporal course of the appearance of enzyme in the circulation, and by enzyme histochemistry. The data in Figure 26 show the course of disappearance of various enzyme activities from the subendocardial zone of severe ischaemia in dogs with occluded circumflex arteries. This tissue contains a high proportion of irreversibly injured cells.[15] Two of the enzymes studied, aspartate aminotransferase (AST) and lactate dehydrogenase (LD) are sarcoplasmic and soluble. The activity of AST begins to decrease after about 60 minutes of ischaemia or at about the same time most of the severely ischaemic cells have passed the point of no return. The activity of LD begins to decrease after 120 minutes of ischaemia. Enzyme depletion occurs gradually and by 8–10 hours AST and LD respectively reach values 33 to 50% of control. The prompt decrease in AST activity and slightly slower loss of LD activity may be related to differences in molecular weight; AST weighs about 60,000 *vs* 135,000 for LD. Both enzymes appear in an active form in the circulation of an open chest anaesthetized dog. AST and CK curves from a representative dog with a circumflex occlusion are compared in Figure 27. The peak activity of both is noted 18–20 hours after occlusion. SD, on the other hand, is a structural part of the mitochondrial inner membrane. Its activity in the tissue is not changed from control during the first 5 hours of ischaemia. SD enzyme activity has not been detected in the plasma space.

The rate at which enzyme is released and enters the blood and the eventual serum activity depend on a number of factors. The transmural progression of cell death must affect enzyme release. As increasing numbers of cells die over a 3 to 6 hour period following coronary occlusion, the total amount of enzyme potentially available to the circulation progressively increases. Although cell death occurs early in the subendocardial zone of severe ischaemia, enzyme release to the blood stream should be delayed in this zone because of the low collateral flow and consequent longer diffusion distances. Although cell death occurs later in areas of moderate to mild ischaemia, enzyme released from such cells may reach the circulation relatively quickly. Also efflux of enzyme from dying cells to the extracellular

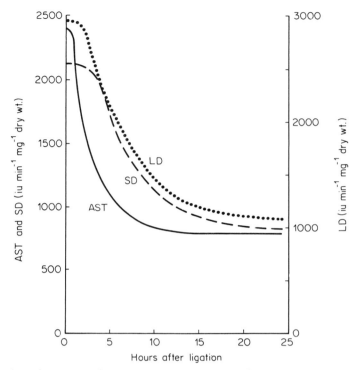

Figure 26. Progressive loss of aspartate aminotransferase, lactate dehydrogenase (LD), and succinate dehydrogenase (SD) respectively, from the infarct with increasing duration of circumflex occlusion. All three enzyme contents are expressed as units per milligram of fat-free dry tissue. Note that significant enzyme activity remains in the zone of the severe ischaemia (posterior papillary muscle) even though histologic data indicate that 100% of the myocytes are necrotic. In the case of AST and LD, which are soluble enzymes, the reason for the persistence of significant activity in a zone of 100% cell death remains unknown. Perhaps it is related to further depression of collateral flow in the zone of severe ischaemia. [Reprinted by permission of publishers from *Archives of Pathology*, **64**, 10–6 (1957).]

space appears to depend on the development of membrane defects large enough to allow passive loss of the enzyme.

Reperfusion of irreversibly injured myocardium results in immediate and massive release of enzyme into the general circulation. This rapid appearance of enzyme may simply be due to washout of extracellular enzyme from the tissue by the restoration of vascular flow. It is also possible that the rapid cell swelling with development of subsarcolemmal blebs and obvious defects in the plasmalemma which accompany reperfusion of irreversibly injured myocardium, may faciltiate loss of intracellular enzyme.[8] Reoxygenation of

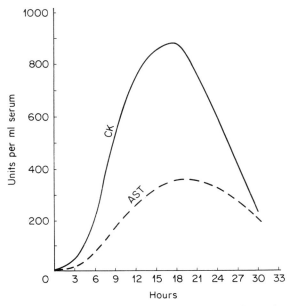

Figure 27. This graph shows the appearance of creatine kinase (CK) and aspartate aminotransferase (AST) in the serum of a dog with an acute infarct induced by left circumflex coronary artery ligation. About 40% of the left ventricle of this dog became necrotic. Note that the peak serum activity of these enzymes occurred between 18 and 24 hours after ligation. In the case of AST, by the time the serum activity of these enzymes has reached its maximum, the tissue activity within the infarct is about one third of control. [Data of Rasmussen, M. M., Reimer, K. A., and Jennings, R. B.; published with permission of McMillan & Co. and Mason, H. and Buist, N. from *Functional Biochemistry*.]

anoxic perfused hearts, *in vitro*, also results in rapid cell swelling with an associated spike of CK release. This observation supports the likelihood that rapid enzyme release at the time of reperfusion *in vivo* is due to acute membrane damage as well as to washout of extracellular enzyme.

CK has been extensively studied because of its clinical application to the problems of diagnosing and sizing myocardial infarcts.[2] CK has a molecular weight of 85,000 and temporary coronary occlusion studies in dogs have demonstrated a close concordance between the development of irreversible ischaemic injury and the appearance of this enzyme in the general circulation.[1] It is a relatively specific marker of myocardial injury since it is found only in muscle and not in the inflammatory components of an infarct (as is LD) or in other organs such as the lungs (ALT) and liver (AST) which may release enzyme in settings where clinical interpretation of the source may be difficult.[2]

Further specificity has been achieved by isoenzyme analysis. The MB isoenzyme of CK, although it accounts for only 15% of the total human myocardial CK, is not detectable in other tissues which contain CK such as muscle (MM isoenzyme) or brain (BB).[2] In clinical studies, CK-MB has proven to be a sensitive as well as specific measure of myocardial infarction even in patients with skeletal muscle damage, e.g. produced by i.m. injections, coronary catheterization procedures, or electrical defibrillation. It has also become useful as an index of perioperative myocardial infarction in patients undergoing cardiac surgery.[5]

TEMPORAL FEATURES OF ISCHAEMIA AND INFARCTION

The data presented in this chapter show clearly that myocardial ischaemic injury and infarction is a dynamic process in which the proportion of injured, dying, and dead cells varies greatly as a function of time. Acutely, the ischaemic tissue exhibits contractile failure, a variety of metabolic changes including glycogen and adenine nucleotide depletion, and lactate and proton accumulation. Later, some cells die and the final events of cell death then begin. These include enzyme loss, phagocytosis, and repair. Finally, the dead cells are replaced by scar tissue. The surviving myocardium, i.e. the myocardium which was the seat of mild ischaemia initially, is presumed to resume function 3–5 days after the onset when collateral arterial flow increases. Only when healing is complete does the heart assume a new, steady, stable state. Appropriate use of this model of ischaemic injury requires that the temporal features of this sequence of events be well understood.

REFERENCES

1. Ahmed, S. A., Williamson, J. R., Roberts, R., Clark, R. E., and Sobel, B. E. The association of increased plasma MB CPK activity and irreversible ischemic myocardial injury in the dog. *Circulation*, **54,** 187–93 (1976).
2. Ahumada, G., Roberts, R., and Sobel, B. E. Evaluation of myocardial infarction with enzymatic indices. *Progress in Cardiovascular Diseases*, **18,** 405–20 (1976).
3. Bloor, C. M. Functional significance of the coronary collateral circulation. *American Journal of Pathology*, **76,** 562–88 (1974).
4. Braasch, W., Gudbjarnason, S., Puri, P. S., Ravens, K. G., and Bing, R. J. Early changes in energy metabolism in the myocardium following acute coronary artery occlusion in anesthetized dogs. *Circulation Research*, **23,** 429–38 (1968).
5. Dixon, S. H., Limbird, L. E., Roe, C. R., Wagner, G. S., Oldham, H. N., Jr., and Sabiston, D. C., Jr. Recognition of postoperative acute myocardial infarction. Application of isoenzyme techniques. *Circulation*, **47** and **48,** III-137–III-140 (1973).
6. Downey, J. M., Downey, H. F., and Kirk, E. S. Effects of myocardial strains on coronary blood flow. *Circulation Research*, **34,** 286–92 (1974).
7. Ganote, C. E., Jennings, R. B., Hill, M. L., and Grochowski, E. C. Experimental myocardial ischemic injury. II. Effect of *in vivo* ischemia on dog heart slice function *in vitro*. *Journal of Molecular and Cellular Cardiology*, **8,** 189–204 (1976).

8. Ganote, C. E., Seabra-Gomes, R., Nayler, W. G., and Jennings, R. B. Irreversible myocardial injury in anoxic perfused rat hearts. *American Journal of Pathology*, **80**, 419–50 (1975).
9. Gevers, W. Generation of protons by metabolic processes in heart cells (Editorial). *Journal of Molecular and Cellular Cardiology*, **9**, 867–73 (1977).
10. Herdson, P. B., Kaltenbach, J. P., and Jennings, R. B. Fine structural and biochemical changes in dog myocardium during autolysis. *American Journal of Pathology*, **57**, 539–57 (1969).
11. Jennings, R. B. Myocardial ischemia—observations, definitions and speculations (Editorial). *Journal of Molecular and Cellular Cardiology*, **1**, 345–9 (1970).
12. Jennings, R. B. and Ganote, C. E. Structural changes in myocardium during acute ischemia. *Circulation Research Supplement*, **34** and **35**, III-156-III-172 (1974).
13. Jennings, R. B., Ganote, C. E., and Reimer, K. A. Ischemic tissue injury. *American Journal of Pathology*, **81**, 179–98 (1975).
14. Jennings, R. B., Hawkins, H. K., Lowe, J. E., Hill, M. L., Klotman, S., and Reimer, K. A. Relation between high energy phosphate and lethal injury in myocardial ischemia in the dog. *American Journal of Pathology*, **92**, 187–214 (1978).
15. Jennings, R. B., Kaltenbach, J. P., and Smetters, G. W. Enzymatic changes in acute myocardial ischemic injury. *Archives of Pathology*, **64**, 10–6 (1957).
16. Jennings, R. B., Kaltenbach, J. P., Sommers, H. M., Bahr, G. F., and Wartman, W. B. Studies of the dying myocardial cell. In *Henry Ford Hospital Symposium on the Etiology of Myocardial Infarction* (Eds. James, T. N. and Keyes, J. W.), Boston, Little, Brown, and Co., 1963, pp. 189–205.
17. Kloner, R. A., Ganote, C. E., and Jennings, R. B. The 'no-reflow' phenomenon after temporary coronary occlusion in the dog. *Journal of Clinical Investigation*, **54**, 1496–508 (1974).
18. Kloner, R. A., Ganote, C. E., Reimer, K. A., and Jennings, R. B. Distribution of coronary arterial flow in acute myocardial ischemia. *Archives of Pathology*, **99**, 86–94 (1975).
19. Kloner, R. A., Reimer, K. A., and Jennings, R. B. Distribution of collateral flow in acute myocardial ischemic injury—effect of propranolol therapy. *Cardiovascular Research*, **10**, 81–90 (1976).
20. Krug, A. The extent of ischemic damage in the myocardium of the cat after permanent and temporary coronary occlusion. *Journal of Thoracic and Cardiovascular Surgery*, **60**, 212–47 (1970).
21. Lowe, J. E., Reimer, K. A., and Jennings, R. B. Infarct size as a function of the amount of myocardium at risk. *American Journal of Pathology*, **90**, 363–78 (1978).
22. Neely, J. R., Rovetto, M. J., and Whitmer, J. T. Rate-limiting steps of carbohydrate and fatty acid metabolism in ischemic hearts. *Acta Medica Scandinavica*, **587**, 9–13 (1976).
23. Reimer, K. A., Lowe, J. E., Rasmussen, M. M., and Jennings, R. B. The wavefront phenomenon of ischemic cell death. I. Myocardial infarct size *vs.* duration of coronary occlusion in dogs. *Circulation*, **56**, 786–94 (1977).
24. Sayen, J. J., Sheldon, W. F., Pierce, G., and Kuo, P. T. Polarographic oxygen; the epicardial electrocardiogram and muscle contraction in experimental acute regional ischemia of the left ventricle. *Circulation Research*, **6**, 779 (1958).
25. Shen, A. C. and Jennings, R. B. Kinetics of calcium accumulation in acute myocardial ischemic injury. *American Journal of Pathology*, **67**, 441–52 (1972).
26. Thomas, R. A., Rubio, R., and Berne, R. M. Comparison of the adenine nucleotide metabolism of dog arterial and ventricular myocardium. *Journal of Molecular and Cellular Cardiology*, **7**, 115–23 (1975).
27. Whalen, D. A., Hamilton, D. G., Ganote, C. E., and Jennings, R. B. Effect of a transient period of ischemia on myocardial cells. I. Effects on cell volume regulation. *American Journal of Pathology*, **74**, 381–97 (1974).
28. Williamson, J. R., Schaffer, S. W., Ford, C., and Safer, B. Contribution of tissue acidosis to ischemic injury in the perfused rat heart. *Circulation*, **53**, I-3-I-14 (1976).
29. Wollenberger, A. and Shahab, L. Anoxia-induced release of noradrenaline from the isolated perfused heart. *Nature*, **207**, 88–9 (1965).

CHAPTER 3

Structure and permeability of normal and damaged membranes

R. H. Michell and R. Coleman

INTRODUCTION	59
MEMBRANE STRUCTURE AND TURNOVER	60
MEMBRANES AS SELECTIVE PERMEABILITY BARRIERS AND ION PUMPS	63
NORMAL MECHANISMS FOR PROTEIN MOVEMENTS ACROSS MEMBRANE BARRIERS	64
ALTERED MEMBRANE PERMEABILITIES	68
THE SITUATION IN THE ISCHAEMIC MYOCARDIUM	72
REFERENCES	76

INTRODUCTION

Many of the cellular changes which occur during clinical or experimental myocardial infarction are related to the disturbance of various processes which are carried out or controlled by membranes. These membranes include the myocardial plasma membrane and the membranes of the mitochondria, the sarcoplasmic reticulum and the lysosomes. For general reviews of the structure and function of these and other membranes in animal cells the reader is referred to references 20 and 81.

In this chapter we shall consider, in a fairly general way, the nature and permeability characteristics of biological membranes and the factors which might lead to impairment of their normal function. Our hope is that such a treatment may complement the discussions of the biochemical and morphological changes which occur in damaged myocardium and which are covered in other chapters of this book. The early changes which involve membranes in the damaged myocardium appear to be related mainly to reversible effects upon ion and small organic solute permeabilities, but there may be some changes involving a slight loss of cytosolic enzymes. The later changes, which appear to be irreversible, involve much greater leakage of enzymes to the exterior both from the cytosol and from other intracellular compartments and also the irreversible failure of mitochondrial function. Our emphasis will be more on the early events, since it is amongst these that one would hope to find the progenitors of the later irreversible decline in cell function. Since many possible mechanisms of ischaemic membrane damage are not yet amenable to direct study in the myocardium, this article will draw widely on studies of a variety of membranes, in the hope that such considerations might shed some light on the aetiology of myocardial ischaemic damage.

MEMBRANE STRUCTURE AND TURNOVER

It has long been known that the major building blocks of cellular membranes are proteins and lipids (mainly phospholipids and, in plasma membranes, Golgi membranes and lysosomes, considerable quantities of cholesterol and glycosphingolipids); the proteins usually constitute from about half to three-quarters of the dry mass. Until about 1970 there was a fierce debate as to whether all membranes were built on a relatively standard pattern and, if they were, whether this pattern was a lipid bilayer modified in its properties by the presence of particular proteins, or a structure consisting of some type of two-dimensional array of interlocking lipoprotein subunits.[19] There were suggestions that a substantial proportion of membrane proteins might have a largely structural, rather than catalytic role. The outcome of this debate has been the agreement that a lipid bilayer does indeed constitute the main organizing structural framework and permeability barrier in all membranes, and that membrane proteins generally display both catalytic and structural roles.

A typical modern view of membrane organization would be of a 'fluid mosaic' structure akin to that described by Singer and Nicholson (see, for example, references 8, 20, and 77 and Figure 1). The basis of this is a lipid bilayer in which the hydrocarbon chains at the centre of the membrane are in a fluid state and the majority of the neighbouring lipid molecules are free to exchange places and move laterally quite rapidly. At least in some membranes, the two leaflets of the bilayer have been shown to contain different types of lipids.[63,89]

Into this bilayer are inserted the intrinsic (or integral) proteins (Figure 1). These are proteins which are tightly associated with the membranes and are only released from the structure by disruption of the lipid organization. Once they have been released from the membranes by detergents, they usually retain their functional conformation in solution only if detergent molecules shield them from aggregation. Intrinsic proteins appear to have regions of their surfaces in which the side chains of hydrophobic amino acids predominate: these regions interact with the hydrocarbon region in the centre of the membranes. The proteins are therefore bound tightly into the membrane structure, with their hydrophilic regions projecting into the adjacent aqueous phase.[11,24,26,30,37] Some of these intrinsic proteins are present only at one surface of a membrane whilst others, particularly

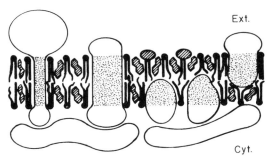

Figure 1. A diagrammatic impression of the organization of the mammalian erythrocyte membrane (hydrophilic reas of proteins are white, whilst hydrophobic areas are stippled). The membrane consists essentially of an asymmetric lipid bilayer with its associated proteins. Some intrinsic proteins span the lipid bilayer (e.g. glycoproteins in which large hydrophilic glycopeptide sequences are exposed at the extracellular surface (Ext.) of the membrane, and transport proteins which are to a substantial extent within the hydrocarbon region of the bilayer). Other intrinsic proteins have hydrophilic regions exposed only at one surface of the membrane and are tethered to the lipid bilayer by areas of hydrophobic surface. Extrinsic proteins are abundant at the cytoplasmic (Cyt.) surface of the membrane, where they are attached through polar interactions such as charge interactions and hydrogen bonds. Reproduced by permission of Blackwell Scientific Publications, Oxford from *Membranes and Their Cellular Function*, 2nd edn., 1978, p. 37.]

proteins responsible for transport processes, appear to span membranes and have different hydrophilic regions exposed at each membrane surface.[14,20,24,63]

The other proteins of the membrane are designated extrinsic (or peripheral) proteins. These are usually removed quite easily by exposure of membranes to high ionic strength solutions, to chelators of divalent cations, or to reagents which break hydrogen bonds, e.g. urea. It seems likely that they are attached to the membranes mainly through polar interactions with the hydrophilic regions of intrinsic proteins and with the polar headgroups of membrane lipids (see references 24, 57, 76, 77 and Figure 1). In addition to the extrinsic proteins which are likely associated with membranes when they are isolated, there are also indications that membranes, particularly plasma membranes, form functional links with cytoskeletal proteins.[57] In many cells these elements, which include microtubules and actin filaments, do not form very regular structures, but in the myocardial cell the contractile elements are of course highly ordered, and they probably interact most with membranes at the intercalated discs.

Although general impressions of membranes often arise from static morphological images, metabolic studies belie this static impression, and studies of the turnover rates of radioactively labelled membranes show that individual membrane components are renewed every few hours or days (see, for example, references 20, 73 and also Siekevitz in reference 81). It is assumed that this turnover has two main purposes; (i) to allow relatively rapid and specific alterations to be made to the chemical or enzymic composition of membranes whose overall nature remains unchanged, and (ii) to allow rapid replacement of components whenever damage occurs. The result is that membranes are long-lived structures whose overall consistency of structure and composition is sustained by continuous deletion and addition of individual components. The biosynthesis of most of these components occurs mainly at the endoplasmic reticulum,[20,44] even though chemical modifications may occur elsewhere (e.g. glycosylation of proteins in the Golgi complex), and the final destinations of the molecules may be other membranes (for further detail, see reference 20). The mechanisms for the necessary intermembrane transfers are not fully understood, but include exchange of lipid molecules across the cytosol mediated by specific transfer proteins and transfer of membrane material in vesicles from membrane to membrane.[56,85]

The degradation of membrane components is even less well understood than their biosynthesis and integration into membranes, particularly because many of the necessary enzymes are sequestered within lysosomes. Unfortunately, there is little good information on the factors which control these activities, and we do not know how they are able selectively to degrade different components of the same membranes at very different rates. In addition, proteolytic and lipolytic enzymes potentially capable of inflicting damage on membrane components exist outside lysosomes.[3,13] It may be important to the onset of membrane damage processes that the main type of reaction involved in phospholipid degradation is the removal of fatty acids from diacylglycerophospholipids by phospholipase A. This gives rise to lysophospholipids and unesterified fatty acids, either of which could cause membrane damage by disorganization of the lipid bilayer if allowed to accumulate (see, for example, reference 47).

It is obvious that the biosynthetic reactions necessary for the maintenance of membranes can only continue for as long as there is a sustained input of metabolic energy in the form of nucleoside triphosphates and reduced coenzymes. However, studies of the rates of degradation of cellular components indicate that there is also an energy requirement for these processes: inhibition of energy metabolism leads to inhibition of cellular protein degradation rates. The nature of this latter energy dependence is uncertain, but suggestions include the possibility that there may be an ATP-driven proton pump responsible for maintaining the low internal pH of lysosomes and that energy depletion reduces the rates of biosynthesis of proteolytic enzymes.[3,29] It is not yet clear whether the reduced rates of degradation during energy deprivation apply equally to all protein species or whether the overall fall in degradation rate masks sharp changes in the absolute and relative degradation rates of different individual proteins. On balance, it seems most likely that the usual net effect of energy deprivation will be a loss of cellular protein and lipid components arising from simultaneous, but unequal, decreases in their biosynthesis and degradation rates.

MEMBRANES AS SELECTIVE PERMEABILITY BARRIERS AND ION PUMPS

Since the lipid bilayer is the major structural framework around which membranes are constructed, many of the permeability properties of biological membranes are very similar to those of artifically prepared lipid bilayers such as liposomes or black lipid membranes.[20,30] This applies in particular to large molecules such as proteins.[12] These normally do not pass through membranes at any appreciable rate (except as a result of special mechanisms such as endocytosis, exocytosis, or during protein biosynthesis by membrane-bound ribosomes, see later). As a result, the extracellular fluid, the cytosol, and the internal compartments of organelles (such as lysosomes, endoplasmic reticulum, or mitochondria), which are only separated from each other by membrane barriers, can maintain distinct protein complements with virtually no mixing. The leakage of cytosolic proteins from cells therefore becomes a sensitive, and diagnostically specific, indicator of membrane damage.

With smaller molecules the situation is more complex. Molecules with a moderate degree of hydrophobicity usually pass quite freely through lipid bilayers, but hydrophilic solutes are usually too polar to permeate rapidly. The permeability characteristics of biological membranes, although broadly similar, deviate from this pattern in several important ways: (i) their passive permeabilities to water and small ions are much higher than those of lipid bilayers; (ii) they exhibit highly selective permeabilities to a relatively small number of hydrophilic organic molecules, e.g. sugars, amino acids, and (iii) they generate concentration gradients of some permeants, especially ions such as H^+, Na^+, K^+, and Ca^{2+}, and the maintenance of these gradients fails in the absence of a source of metabolic energy.[10,15,20,83] The distribution of such selective permeation processes, each of which appears to be mediated by a particular membrane protein, differs amongst different cellular membranes, so that, for example, Na^+ is pumped out and K^+ in at the plasma membrane, mitochondria pump out protons, and the Ca^{2+} concentration of the cytosolic compartment is kept low by the action of pumps in several membranes such as the plasma membrane, mitochondrial membranes, and sarcoplasmic reticulum.[10,20,43,45,65]

The appreciable, but relatively low, passive permeabilities of membranes to cations are normally counteracted by the rapid uphill pumping of most of these ions, with the well-known result that intracellular cation concentrations are normally higher (K^+), lower (Na^+), or very much lower (Ca^{2+}) than those outside cells: Mg^{2+} does not appear to be pumped rapidly between different cell compartments, and its concentration is probably much the same both inside and outside the cell. However, there are also some additional and relatively specific contributors to passive ion permeabilities which can exist either in ion-permeable (open) and ion-impermeable (closed) states. These routes, through which the usual permeabilities of membranes to particular ionic species can be temporarily increased in a controlled way, are often referred to as ion 'gates': examples include Na^+ channels in nerve and muscle which are potential-sensitive and are inhibited by tetrodotoxin,[20] Ca^{2+} 'gates' controlled by activation of some cell surface receptors for hormones and neurotransmitters,[20,55] and plasma membrane K^+ 'gates' which are activated when the intracellular Ca^{2+} concentration rises.[5,42,55,61]

Although only a few ion transport systems have been studied in molecular detail (particularly the ATP-driven Na^+/K^+ pump of the plasma membrane, the ATP-driven Ca^{2+} pump of skeletal muscle sarcoplasmic reticulum, and the reversible proton gradient-driven ATP synthase of energy-coupling membranes), the available evidence indicates that most important ion permeation processes in natural membranes are functions catalysed by intrinsic proteins which span the relevant membranes. This conclusion comes both from direct biochemical evidence and from evidence that points to different functional interactions between transport systems and substrates or inhibitors at the two surfaces of membranes.[10,20,52,68]

NORMAL MECHANISMS FOR PROTEIN MOVEMENTS ACROSS MEMBRANE BARRIERS

Although it has been emphasized above that membranes are not appreciably permeable to proteins, it is clear that different intracellular compartments all contain proteins, that secreted proteins can leave cells, and that external proteins can be taken up by cells. However, neither the exit of secreted proteins nor the entry of foreign proteins appear to require mechanisms which involve the penetration of individual protein molecules through the lipoprotein barrier offered by the plasma membrane. Instead, they occur by the processes of exocytosis and endocytosis, in which either an intracellular protein-filled vesicle fuses with the plasma membrane to release its contents or an extracellular protein is taken into a pinocytotic vesicle consisting of internalized plasma membrane (see Figure 2 and references 20, 60, 75).

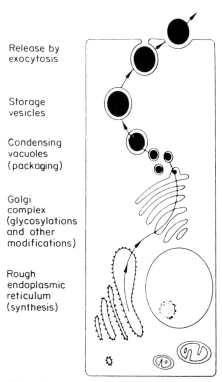

Figure 2. A schematic representation of the normal cellular progress of secretory proteins, from synthesis at the endoplasmic reticulum to release by exocytosis. Other proteins destined for sites outside the cytosolic compartment of the cells, such as those inside lysosomes or on the external surface of the plasma membrane, probably traverse essentially similar routes, i.e. synthesis at the endoplasmic reticulum, modification at the Golgi complex, and transport in vesicular form to their final destinations. For further information, see references 20, 58, 60. [Reproduced by permission of Blackwell Scientific Publications, Oxford from *Membranes and Their Cellular Function*, 2nd edn., 1978, p. 103.]

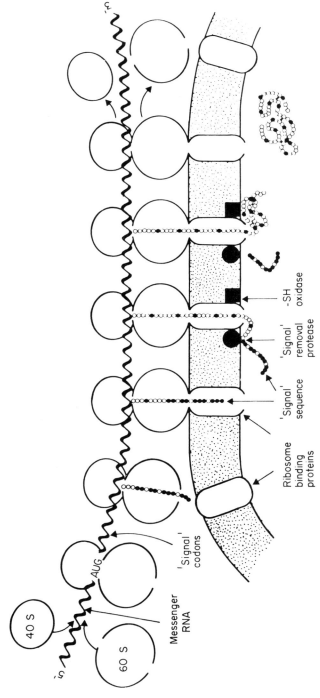

Figure 3. The probable sequence of events in the synthesis of most secretory proteins by membrane-bound ribosomes at the rough endoplasmic reticulum. After the appropriate messenger RNA has bound to the ribosomes, the hydrophobic 'signal' peptide sequence is translated and this binds the ribosomes to the endoplasmic reticulum membrane. The growing polypeptide then passes through the membrane, its 'signal' peptide is removed, appropriate sulphydryl bridges are inserted and it folds to its native configuration. It may, of course, be further modified (e.g. by glycosylation, selective proteolysis, etc.) during its progression through the cell, prior to its release as a secretory material (see reference 20 for further information). [Reproduced by permission of Blackwell Scientific Publications, Oxford from *Membranes and Their Cellular Function*, 2nd edn., 1978, p. 114]

The only normal situation in which individual protein molecules traverse membranes in quantity appears to be in getting newly synthesized proteins from the cytosolic compartment of the cell, where essentially all protein synthesis takes place, into other intracellular spaces that are separated from the cytosol by a membrane barrier, e.g. to the lumen of the endoplasmic reticulum and thence to the interiors of the Golgi system or lysosomes. It has been thought for many years that this is achieved by membrane-bound ribosomes at the endoplasmic reticulum, but with little understanding of what might dictate which proteins are to be made on membrane-bound ribosomes and to traverse the membrane, and which are to be made on free ribosomes and retained within the cytosolic compartment. It now appears likely that this discrimination is often achieved by 'signal peptides', which are the initial sequences of amino acids to be assembled during the translation of proteins made by membrane-bound ribosomes[20,70] (Figure 3). The signal sequences for different proteins are about 15–25 amino acid residues long and seem always to have a marked predominance of hydrophobic amino acids, but these are not identical. They often appear to be removed even before completion of the synthesis of the protein molecules, suggesting that their role is complete by that time. It is thought that the initiation of synthesis for all proteins occurs on free polyribosomes, but that when the signal sequences have been translated they interact specifically with proteins of the endoplasmic reticulum. As a result of this interaction, the ribosomes bind to the membranes and a channel of some type is formed, through which the protein passes into the lumen of the endoplasmic reticulum as it is translated.[20,70]

Once a newly synthesized protein has crossed into the lumen of the endoplasmic reticulum, it is usually able to move on to its final destination (whether this is the interior of a lysosome, the extracellular fluid, etc.) without directly traversing another membrane barrier. Instead, the necessary movements from compartment to compartment can be achieved by membrane vesiculation and fusion processes, e.g. transfer from endoplasmic reticulum to Golgi complex, exocytosis, mixing of lysosomal enzymes with material ingested by pinocytosis or phagocytosis.[20,56,58]

In some situations a combination of endocytosis and exocytosis acts as a shuttle system by which macromolecules can pass through an otherwise impervious cellular sheet without becoming mixed with the contents of the cells through which they pass. A combination of morphological and biochemical studies have indicated an important role for pinocytotic vesicles as a protein-carrying transcellular shuttle system in several tissues, amongst which are vascular endothelium and neonatal intestine.[75]

Although proteins which leave cells are usually secretory products which follow the endoplasmic reticulum → Golgi → exocytosis route, there are a small number of situations in which macromolecules *from the cytosolic compartment* are lost from cells by a process of outward budding of the plasma membrane, e.g. secretion of milk fat and the release of some enveloped viruses.[20] An experimentally induced and pathological variant of this process has been recognized in mammalian erythrocytes, which release small membrane vesicles filled with cytosol to the exterior when stored for long periods *in vitro*,[64,79] when incubated at 37 °C in substrate-free medium for several hours,[49] when treated briefly with an ionophore (A23187) and Ca^{2+} so as to raise the intracellular Ca^{2+} concentration,[1] or when insulted mildly with the bile salt glycocholate.[6] In these situations, with the exception of glycocholate attack, the release of vesicles does not occur until the cells become substantially energy starved. Such cytosol-filled vesicles are relatively rapidly sedimented by ultracentrifugation, thus distinguishing this process from the leakage of individual soluble enzyme molecules through abnormally permeable membranes.

ALTERED MEMBRANE PERMEABILITIES

The appearance in the extracellular space, in abnormal quantities, of materials which are usually found inside cells must indicate some overall change in membrane permeability. This can be a result of a failure in an homeostatic mechanism (such as maintenance of cellular ATP levels), so that the rates of normal membrane permeation processes are altered. Alternatively, it can be a direct result of membrane leakiness or vesiculation: such lesions may themselves arise either as a result of homeostatic failures or of some direct physical insult. Although it seems certain that the permeability changes which occur during cardiac ischaemia are initiated by homeostatic problems which arise either directly or indirectly from energy deprivation and only later progress to the stage of overt physical lesions, the complexity of the system is such that it is difficult to untangle the exact causes of the various stages in the progression towards irreversible cellular injury. For this reason, much of the next section of our discussion will concentrate on the factors that are essential to the maintenance of normal structure and permeability of cells in general, in the hope that consideration of these general ideas may provoke experiments which can illuminate the aetiology of cell damage during cardiac ischaemia. The most common of these models are energy-starved, but perfused, cells of various types; these tend not to show those effects which in the heart are substantially attributable to the accumulation of metabolic wastes.

Ischaemic oxygen deprivation initially makes itself felt in terms of reduced mitochondrial function. The first evidence of such mitochondrial dysfunction is likely to be a fall in the rate of ATP synthesis. This reduced rate is incapable of meeting the continuing energy demand, since active uptake of Ca^{2+} (the other important, normal and energy-coupled function of mitochondria) takes precedence over ATP synthesis in the utilization of after the onset of this swelling process that cellular damage, at least in the myocardium, begins to assume an irreversible nature.[25,35] The dramatic nature of the circulatory obstruction that can be caused by cell swelling was well illustrated in studies of kidney in which latex casts were obtained of the unoccluded parts of the circulation in normal and ischaemic kidney: substantial regions of the ischaemic tissue received essentially no perfusion.[23] These studies also provided direct evidence of the importance of this swelling process to the osmotic imbalance caused by Na^+ loss, in that perfusion of ischaemic tissues with buffers made hyperosmotic by the addition of mannitol could correct both the swelling of the cells and the circulatory obstruction.[23]

The second, and less well-understood, effect of cell swelling may be to stress the plasma membrane, always provided that the membrane is not so convoluted in the unswollen cell that it can readily accommodate the increased cell volume without stress. If such stresses do occur, however, they are potentially capable of causing a selective increase in the permeability, particularly towards small molecules. There is little difficulty in envisaging that such stretching could sufficiently expand the packing of membrane constituents to enhance permeability to small molecules, but suggestions of stretching to a degree that would afford appreciable permeability to macromolecules are more surprising. In one controlled study of this type solutions of steadily declining osmotic pressure were flowed past erythrocytes and the sequence of leakage of the intracellular components was followed: the hypo-osmotic conditions chosen were not sufficiently severe to bring about substantial haemolysis until after the initial observations of leakage of intracellular components. In these experiments it was found that K^+ loss occurred first, followed by protein loss, and that small proteins were lost before large ones. The interpretation put upon these results was therefore that the stressed membrane acted as some form of molecular sieve.[50] Similar results have been obtained with thymocytes.[4] Studies of the losses of different enzymes from ischaemic heart tissue have also pointed towards an enhanced loss of small proteins relative to larger ones, and this has again been interpreted as an indication that during the reversible phase of ischaemic damage small quantities of proteins might be lost by a molecular sieving action rather than through substantial holes in the plasma membrane.[78]

This may, however, be too simple a viewpoint. Analysis of the events involved in acute haemolysis of erythrocytes has emphasized both the occurrence of substantial holes in the erythrocyte membrane during lysis, and the remarkable resealing abilities of membranes. During gentle hypotonic haemolysis small 'slits' about 20–1000 nm long appear in the erythrocyte membrane but reseal after only a few seconds.[67] Provided that the cell 'ghosts' are kept in a medium of relatively physiological ionic strength, this morphological sealing is reflected in a rapid return of the membrane to a state impermeable to macromolecules: provided that haemolysis and resealing are rapid, the 'ghosts' can also retain a substantial proportion of their original intracellular complement of macromolecules.[5,66] Resealing of the lysed membrane to small molecules, and particularly to ions such as K^+, is not so rapid, but a substantial degree of impermeability can be restored under appropriate conditions *in vitro*, even in 'ghosts' which have lost much of their contents.[5,66] Given these observations, one may then ask whether the 'molecular sieving' behaviour during hypotonic treatment of erythrocytes (or during myocardial anoxia or ischaemia) arises from a structural reorganization of the membrane so as to give it appropriate molecular sieving properties or, alternatively, whether the greater diffusion coefficients of small rather than of large proteins might mean that small proteins escape more rapidly through transient holes in membranes than do large ones. An indication that the latter is a real possibility comes from the observation that during a single rapid lysis and resealing of erythrocytes in the presence of polydisperse extracellular dextran the small dextran molecules entered the cells (and were retained when the membranes resealed) to a greater extent than the larger dextran molecules.[54]

Thus it seems possible that the limited release of enzymes, with selectivity for smaller molecules, which can occur during early phases of cell damage is a consequence of the appearance, albeit transiently, of substantial lesions in the membrane rather than to some more subtle molecular sieving mechanism. If this is correct, then a key element in irreversible myocardial damage might be the transition of the sarcolemma from a healthy, self-sealing state to a damaged state in which the membrane's self-sealing capabilities are impaired.

Ion imbalance and cell swelling may be major factors that lead to added stresses on ischaemic tissue, and might even lead to the transient permeability of cells to macromolecules, but they are probably not sufficient alone to lead to the irreversible phase of cellular damage and enzyme leakage. A factor which favours the move towards this irreversible phase may, however, be the development of cellular acidosis due largely to the hydrolysis of nucleoside di- and triphosphates.[27] In a recent study in which erythrocyte ghosts were prepared under conditions approximately mimicking intracellular ionic conditions, it was found that the resealing of the membrane after lysis was substantially less efficient if the pH of the lysing medium was reduced from 7.0 to 6.0.[5] In the same study it was also shown that elevation of the cytosolic Ca^{2+} ion concentration above approximately 10 μM (which should only occur when the cells are in a substantially energy-deprived or leaky condition) also reduces the resealing capability of the lysed membrane.[5] A decrease in the availability of cytoplasmic ATP may also contribute directly to increased membrane permeability.[84]

Although the previous discussion might have identified some of the factors involved particularly in the reversible phases of membrane damage and altered permeability, it offers little in the way of suggestions as to how these changes become irreversible. Here it seems that the occurrence of covalent changes in membrane components, and their persistence due to the lack of energy supplies for normal biosynthetic and repair processes, probably play a part. The balance between synthesis and breakdown of cellular proteins will probably be disturbed in such situations, since both of these processes are influenced by the cellular energy supply. The reduced pH of the cell interior is likely both to labilize lysosome membranes directly and also to greatly enhance the activities of any hydrolases liberated from lysosomes. Of these, the most potentially damaging to membranes, in the short term, will be proteases and phospholipase A. In particular, any activation of phospholipase A will generate lysophospholipids and unesterified (often polyunsaturated) fatty acids; both of these are membrane-disorganizing agents which perturb the lipid bilayer. High levels of these lipids, whether exogenous or generated within the membrane by phospholipases, can enhance membrane fusion, fragility, and permeability, and can lead to cell lysis.[37,47,48] The concentrations of these lipids that are generated within membranes by endogenous phospholipases have been little studied, but it seems possible that high local concentrations might sometimes be generated. There is also evidence of enhanced fatty acid levels in plasma and of the accumulation of acyl CoA derivatives within the cells in the ischaemic situation.[59] Although a defined mode of action of acyl CoA's is by inhibition of mitochondrial adenine nucleotide translocation[71,72] it may be just as important that these fatty acids and acyl CoA's are agents which can affect the organization of the lipid bilayer in membranes.

THE SITUATION IN THE ISCHAEMIC MYOCARDIUM

To what extent do the discussions in the previous section emphasize processes that are important in the ischaemic myocardium? One of the most important areas of uncertainty is in the details of the disturbed Ca^{2+} homeostasis of ischaemic tissue, particularly any changes in the ionized calcium activity in the cytosolic compartment. The heart stops beating, a change which might indicate a fall in cytosolic Ca^{2+} concentration, but which might equally be caused by some other change in cellular conditions (e.g. loss of membrane excitability or a deficiency of ATP in the immediate vicinity of the contractile apparatus). By contrast, the bulk of the remaining observations appear to suggest that a major abnormality of the ischaemic cell is an elevation of cytosolic Ca^{2+} concentration. A part of this increase in Ca^{2+} might arise as a result of the ionization of calcium that had previously been chelated to molecules whose concentrations decline during ischaemia (e.g. ATP, citrate). However, the bulk of these molecules probably are chelated with Mg^{2+}, since this ion is present in the cytosol at a much higher total concentration than Ca^{2+}. Another obvious route is through increased cell surface permeability to Ca^{2+}, possibly involving the normal Ca^{2+} channels which allow controlled access of Ca^{2+} to the cell interior in response to stimuli such as action potentials or activation of receptors for hormones or neurotransmitters. In favour of this view is the observation that drugs which inhibit Ca^{2+} movements through these Ca^{2+} channels, e.g. verapamil, cinnarizine, nifedipine, protect the myocardium from the effects of ischaemia[21,69] and of other myopathic situations, e.g. the hereditary cardiomyopathy of strain B10 8262 syrian hamsters or massive β-adrenergic stimulation.[21,22,28,40,74] Ideally, changes in cytosolic Ca^{2+} in the perturbed myocardium would be assessed by some method capable of continuously monitoring the ionized calcium concentration in the cytosolic compartment of the myocardial cells (e.g. by methods such as those discussed in reference 32). So far, however, this has not been achieved, and in the myocardium the experimental observation that most often indicates excessive cell permeability to Ca^{2+} is the detection of an abnormally high rate of mitochondrial Ca^{2+} accumulation: this produces Ca^{2+}-loaded mitochondria which can subsequently be isolated from the ischaemic tissue.[16,38,41,53,69,86,87] This increased availability of Ca^{2+} to mitochondria, and the resulting decline in the proportion of the mitochondrial energy supply that is available for ATP synthesis, are also seen in the isoprenaline-induced lesions and in hereditary cardiomyopathy.

Although these mitochondrial changes might be major factors in the aetiology of ischaemic necrosis,[7,87] it seems more likely that they are not primary defects, but simply accessible ways of demonstrating the occurrence of a relatively substantial and sustained elevation of cell permeability to Ca^{2+} and of the Ca^{2+} concentration in the cytosol.

The primary cause of these mitochondrial lesions is presumably an increased net Ca^{2+} influx into the ischaemic cell through the plasma membrane, but this excess of intracellular Ca^{2+} may be accentuated by a decrease in the cell's ability to pump Ca^{2+} out of the cytosol caused by lack of ATP. Since the endoplasmic reticulum and mitochondria are both closed membrane systems of limited volume and Ca^{2+}-accumulating capacity, they can only serve as temporary intracellular Ca^{2+}-buffering systems: net movement out of the cell of any excess Ca^{2+} must be through the plasma membrane, catalysed by its Ca^{2+} pump(s) (Figure 4). In view of this central position of the sarcolemma in Ca^{2+} homeostasis, there appears to be surprisingly little information on the properties of its Ca^{2+} pump. The sarcolemma possesses a Ca^{2+}-ATPase, but in the published[17,33] experiments this appears only to be activated by Ca^{2+} in the millimolar range, i.e. at concentrations too high for this enzyme to effectively sustain the physiological cytosolic Ca^{2+} concentrations which are in the micromolar range. In the erythrocyte, where plasma

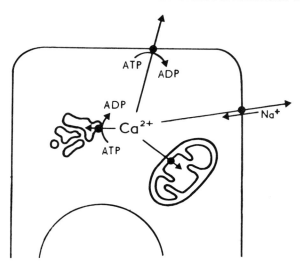

Figure 4. Transport mechanisms which pump Ca^{2+} out of the cytosol and thus maintain its low Ca^{2+} ion concentration. [Reproduced by permission of Blackwell Scientific Publications, Oxford from *Membranes and Their Cellular Function*, 2nd edn., 1978, p. 115.]

membrane Ca^{2+}-ATPase activities have been studied in greatest detail, two distinct phases of Ca^{2+} activation are observed, with the more sensitive activation in the micromolar range being regarded as the phase with the greater significance in terms of cellular homeostasis.[65] A careful search for an ATPase of similar Ca^{2+} sensitivity in the sarcolemma might be profitable.

Possibly the most direct evidence for the myopathic effects of rapidly elevating the cytosolic Ca^{2+} concentration to an abnormal degree comes from the experiments which give rise to the so-called 'calcium paradox'[34] (see Chapters 17, 18). Physiological studies in other fields have shown that a frequent consequence of exposing cells to Ca^{2+}-free extracellular media is a subsequent rise in plasma membrane permeability to this ion,[18] and it seems probable that in the Ca^{2+}-reperfused heart the major, if not the only, primary trigger for myopathic events is a precipitous rise in the cytosolic Ca^{2+} concentration which overwhelms the cells' normal mechanisms for pumping Ca^{2+} out of the cytosolic compartment;[88] as a result the cellular energy stores (ATP, creatine phosphate) are rapidly depleted[7,34] and mitochondrial function is directed away from ATP synthesis and into Ca^{2+} accumulation. If Ca^{2+}-free perfusion and Ca^{2+} reperfusion are performed at 4 °C rather than at 37 °C then the reperfusion does not damage the heart: under these circumstances only a small quantity of Ca^{2+} enters the cells.[39] Perfusion of hearts with a physiological Ca^{2+}-containing medium including a small quantity of a divalent cation ionophore such as A23187 might provide a direct method by which to test whether a very rapid rise in cytosolic Ca^{2+} concentration is really sufficient on its own to provoke myocardial lesions similar to those in ischaemia, isoprenaline toxicity, Ca^{2+} reperfusion, and hereditary cardiomyopathy. Situations in which this compound has been invaluable in demonstrating the ability of Ca^{2+} alone to provoke particular cellular responses include control of secretion, glycogenolysis, and K^+ efflux (e.g. references 1, 42).

A second major question relates to the nature of the enzyme leakage process or processes. After 30 minutes of ischaemia, when irreversible damage is likely to have begun, lesions are apparent in the cell surface, and enzymes presumably leak out through these. Such lesions are not, however, apparent in ischaemic heart at earlier times, even though some enzyme leakage may occur. In contrast, there is direct evidence in the isoprenaline-stressed heart that the interior of the cells can become accessible to extracellular proteins within 10 minutes,[62] suggesting the presence at least of transient membrane lesions by that time. There are some studies which show an early and minor phase of ischaemic enzyme loss after 10–20 minutes, when ion imbalances and cell swelling are well advanced, but when cell damage is still fully reversible.[36] No lesions are apparent at the cell surface at this time and there appears to be no evidence to indicate whether this enzyme loss occurs through short-lived lesions of the plasma membrane which rapidly reseal or through vesiculation. The latter possibility seems worth testing, since a rise in intracellular Ca^{2+} concentration and fall in ATP levels are the conditions which bring about such vesiculation in the erythrocyte:[1] it could easily be tested by ultracentrifugation of the enzyme-containing perfusates to determine whether the enzymes were released in molecular dispersion or in relatively readily sedimented structures. If enzymes were found in vesicles it might also transpire that their amounts had been underestimated due to structure latency (i.e. the limitation of free diffusion of substrate to the enzyme across the intervening membrane barrier). If vesiculation was observed during this early phase, it could provide an explanation for the simultaneous loss of creatine kinase, of cytosolic enzyme, and of β-naphthylamidase:[36] the latter is an enzyme that occurs in the plasma membrane in some tissues.

There appears to be very little direct information on the origin of persistent membrane lesions and the mechanism of irreversible damage. There is evidence in the anoxic heart of a substantial enhancement of lysosomal fragility, suggesting quite rapid lysosomal activation.[82] Another source of degradative enzymes might even be the plasma membrane itself, since membrane fractions enriched in sarcolemma show a Ca^{2+}-activated phospholipase A activity,[80] and exposure of erythrocyte ghosts to Ca^{2+} activates both lipolytic[1] and proteolytic enzymes.[2] The chief problem lies in the identification of the molecular lesions produced by such activities. Perhaps a fruitful approach, though a technically difficult one, would be to try to isolate purified sarcolemma preparations from normal and ischaemic heart and make detailed comparisons of their polypeptide and lipid profiles: detected differences might point to whether the key events leading to sarcolemmal damage and increased permeability are lipolytic or proteolytic. Another sensitive indicator, if it proved technically feasible, might be to take pulse-labelled radioactive heart tissue and determine the effects of ischaemic and other myopathic insults upon the radiochemical half-lives (i.e. degradation rates) of the individual components of the sarcolemma.

REFERENCES

1. Allan, D. and Michell, R. H. Calcium ion-dependent diacylglycerol accumulation in erythrocytes is associated with microvesiculation but not with efflux of potassium ions. Biochemical Journal, **166,** 495–9 (1977).
2. Anderson, D. R., Davis, J. L., and Carraway, K. L. Calcium-promoted changes of the human erythrocyte membrane. Journal of Biological Chemistry, **252,** 6617–23 (1977).
3. Ballard, F. J. Intracellular protein degradation. Essays in Biochemistry, **13,** 1–37 (1977).
4. Bauer, H. G., Speth, V., and Brunner, G. Osmotic effects on the plasma membrane of thymocytes. Biochimica et Biophysica Acta, **507,** 408–18 (1978).
5. Billah, M. M., Finean, J. B., Coleman, R., and Michell, R. H. Permeability characteristics of erythrocyte ghosts prepared under isoionic conditions by a glycol-induced osmotic lysis. Biochimica et Biophysica Acta, **465,** 515–26 (1977).
6. Billington, D. and Coleman, R. Effects of bile salts on human erythrocytes: plasma membrane vesiculation, phospholipid solubilization and their possible relationships to the bile secretion. Biochimica et Biophysica Acta, **509,** 33–47 (1978).
7. Boink, A. B. T., Ruigrok, T. J. C., Maas, A. H. J., and Zimmerman, A. N. E. Changes in high-energy phosphate compounds of isolated rat hearts during Ca^{2+}-free perfusion and reperfusion with Ca^{2+}. Journal of Molecular and Cellular Cardiology, **8,** 973–9 (1976).
8. Bretscher, M. S. and Raff, M. C. Mammalian plasma membranes. Nature, **258,** 43–6 (1975).
9. Carafoli, E. Mitochondria, Ca^{2+} transport and the regulation of heart contraction and metabolism. Journal of Molecular and Cellular Cardiology, **7,** 83–9 (1976).
10. Christensen, H. N. Biological Transport, 2nd edn., W. H. Benjamin Inc., New York 1975.
11. Coleman, R. Solubilization of membrane components. Biochemical Society Transactions, **2,** 813–6 (1974).
12. Colley, C. M. and Ryman, B. E. The liposome: from membrane model to therapeutic agent. Trends in Biochemical Sciences, **1,** 203–5 (1976).
13. Dean, R. T. and Barrett, A. J. Lysosomes. Essays in Biochemistry, **12,** 1–40 (1976).
14. DePierre, J. W. and Ernster, L. Enzyme topology of intracellular membranes. Annual Review of Biochemistry, **46,** 201–62 (1977).
15. Deuticke, B. Properties and structural basis of simple diffusion pathways in the erythrocyte membrane. Reviews of Physiology, Biochemistry and Pharmacology, **78,** 1–97 (1977).
16. Dhalla, N. S. Involvement of membrane systems in heart failure due to calcium overload and deficiency. Journal of Molecular and Cellular Cardiology, **8,** 661–7 (1976).
17. Dietze, G. and Hepp, K. D. Calcium stimulated ATPase of cardiac sarcolemma. Biochemical and Biophysical Research Communications, **44,** 1041–9 (1971).
18. Douglas, W. W. Stimulus-secretion coupling: the concept and clues from chromaffin and other cells. British Journal of Pharmacology, **34,** 451–74 (1968).
19. Finean, J. B. The development of ideas on membrane structure. Subcellular Biochemistry, **1,** 363–73 (1972).

20. Finean, J. B., Coleman, R., and Michell, R. H. *Membranes and their Cellular Functions*, 2nd edn., Blackwell Scientific, Oxford, 1978.
21. Fleckenstein, A. and Rona, G. (Eds.). *Pathophysiology and Morphology of Myocardial Cell Alterations*, University Park Press, Baltimore, 1975.
22. Fleckenstein, A., Janke, J., Doring, H. J., and Leder, O. Key role of Ca in the production of noncoronargenic myocardial necroses. In ref. 21, pp. 21–32 (1975).
23. Flores, J., Dibona, D. R., Frega, N., and Leaf, A. Volume regulation and ischaemic tissue damage. *Journal of Membrane Biology*, **10**, 331–43 (1972).
24. Furthmayr, H. Erythrocyte proteins. In *Receptors and Recognition* (Eds. Cuatrecasas, P. and Greaves, M. F.). Vol. 3A, Chapman and Hall, London, 1977, pp. 101–32.
25. Ganote, C. E., Jennings, R. B., Hill, M. L., and Grochowski, E. C. Experimental myocardial ischaemic injury: II. Effect of *in vivo* ischaemia on dog heart slice function *in vitro*. *Journal of Molecular and Cellular Cardiology*, **8**, 189–204 (1976).
26. Gennis, R. and Jonas, A. Lipid–protein interactions. *Annual Reviews of Biophysics and Bioengineering*, **6**, 195–238 (1977).
27. Gevers, W. Generation of protons by metabolic processes in heart cells. *Journal of Molecular and Cellular Cardiology*, **9**, 867–74 (1977).
28. Godfraind, T. and Sturbos, X. Inhibition by cinnarizine of ionic changes induced by isoprenaline. In ref. 21, pp. 127–34 (1975).
29. Goldberg, A. L. and St. John, A. C. Intracellular protein degradation in mammalian and bacterial cells. Part 2. *Annual Review of Biochemistry*, **43**, 835–69 (1974).
30. Gomperts, B. D. *The Plasma Membrane: Models for Structure and Function*, Academic Press, London, 1977.
31. Gudbjarnason, S., Mathes, P., and Ravens, K. G. Functional compartmentation of ATP and creatine phosphate in heart tissue. *Journal of Molecular and Cellular Cardiology*, **1**, 325–39 (1970).
32. Hales, C. N., Campbell, A. K., Luzio, J. P., and Siddle, K. Calcium as a mediator of hormone action. *Biochemical Society Transactions*, **5**, 866–72 (1977).
33. Harris, P. A theory concerning the course of events in angina and myocardial infarction. *European Journal of Cardiology*, **3**, 157–63 (1975).
34. Hearse, D. J. Reperfusion of ischaemic myocardium. *Journal of Molecular and Cellular Cardiology*, **9**, 605–16 (1977).
35. Hearse, D. J. This book—Chapter 1.
36. Hearse, D. J., Humphrey, S. M., and Garlick, P. B. Species variation in myocardial anoxic enzyme release, glucose protection and reoxygenation damage. *Journal of Molecular and Cellular Cardiology*, **8**, 329–39 (1976).
37. Helenius, A. and Simons, K. Solubilization of membranes by detergents. *Biochimica et Biophysica Acta*, **415**, 29–79 (1975).
38. Henry, P. D., Schuchlieb, R., Davis, J., Weiss, E. S., and Sobel, B. E. Myocardial contracture and accumulation of mitochondrial calcium in ischaemic rabbit heart. *American Journal of Physiology*, **233**, H677–H684 (1977).
39. Holland, C. E. and Olson, R. E. Prevention by hypothermia of paradoxical calcium necrosis in cardiac muscle. *Journal of Molecular and Cellular Cardiology*, **7**, 917–28 (1975).
40. Jasmin, G. and Bajusz, E. Prevention of myocardial degeneration in hamsters with hereditary cardiomyopathy. In ref. 21, pp. 219–29 (1975).
41. Jennings, R. B. and Ganote, C. E. Mitochondrial structure and function in acute myocardial ischaemic injury. *Circulation Research Supplement 1*, 1–80–1–91 (1976).
42. Jones, L. M. and Michell, R. H. Stimulus-secretion coupling at α-adrenergic receptors. *Biochemical Society Transactions*, **6**, 673–88 (1978).
43. Kozlov, I. A. and Skulachev, V. P. H^+-Adenosine triphosphatase and membrane energy coupling. *Biochimica et Biophysica Acta*, **463**, 28–89 (1977).
44. Lands, W. E. M. and Crawford, C. G. Enzymes of membrane phospholipid metabolism in animals. In *The Enzymes of Biological Membranes*, (Ed. Martonosi, A.), Vol. 2, John Wiley and Sons, London, 1976, pp. 3–85.
45. Lehninger, A. L. Mitochondria and calcium ion transport. *Biochemical Journal*, **119**, 129–38 (1970).
46. Lindenmayer, G. and Schwartz, A. A kinetic characterisation of calcium on (Na^+ and

K$^+$)-ATPase and its potential role as a link between extracellular and intracellular events: hypothesis for digitalis-induced inotropism. *Journal of Molecular and Cellular Cardiology*, **7**, 591–612 (1975).
47. Lucy, J. A. The fusion of cell membranes. In ref. 81, pp. 75–86 (1975).
48. Lucy, J. A. Lipids and membranes. *FEBS Letters*, **40**, 1055–115 (1974).
49. Lutz, H. U., Barber, R., and McGuire, R. F. Glycoprotein enriched vesicles from sheep erythrocyte ghosts obtained by spontaneous vesiculation. *Journal of Biological Chemistry*, **251**, 3500–10 (1976).
50. MacGregor, R. and Tobias, C. A. Molecular sieving of red cell membranes during gradual osmotic lysis. *Journal of Membrane Biology*, **10**, 345–56 (1972).
51. MacKnight, A. D. C. and Leaf, A. Regulation of cellular volume. *Physiological Reviews*, **57**, 510–73 (1977).
52. MacLennan, D. H. and Holland, P. C. The calcium transport ATPase of sarcoplasmic reticulum. In *The Enzymes of Biological Membranes* (Ed. Martonosi, A.), Vol. 3, J. Wiley, London, 1976, pp. 221–59.
53. Mansford, K. R. L. and Hearse, D. J. Metabolic approaches to myocardial infarction. In *Biological Substances: Exploration and Exploitation* (Ed. Hems, D. A.,) J. Wiley, London, 1977, pp. 239–64.
54. Marsden, N. V. B. and Ostling, S. G. Accumulation of dextran in human red cells after haemolysis. *Nature*. **184**, 723–4 (1959).
55. Michell, R. H. Cell surface receptors for hormones and neurotransmitters. In *Companion to Biochemistry* (Eds. Bull, A. T., *et al.*), Vol. 2, Longmans, London, 1978, pp. 205–228.
56. Morré, D. J. Membrane Flow and its contribution to surface formation. In *Membrane Alterations as Basis of Liver Injury*, (Eds. Popper, H., Bianchi, L., and Reutter, W.) MTP Press Ltd., Lancaster, U.K., 1977, pp. 15–28.
57. Nicolson, G. L. Transmembrane control of the receptors on normal and tumour cells: I. Cytoplasmic influence over cell surface components. *Biochimica et Biophysica Acta*, **457**, 57–103 (1976).
58. Novikoff, A. B. and Holtzmann, E. *Cells and Organelles*, 2nd edn., Holt, Rinehart, and Winston, Inc., New York, 1976.
59. Opie, L. H. and De Leiris, J. This Book, Chapter 21.
60. Palade, G. Intracellular aspects of the process of protein secretion: *Science*, **189**, 347–58 (1975).
61. Romero, P. J. Is the Ca^{2+}-sensitive K$^+$ channel under metabolic control in human red cells? *Biochimica et Biophysica Acta*, **507**, 178–81 (1978).
62. Rona, G., Boutet, M., and Huttner, Membrane permeability alterations as manifestation of early cardiac muscle cell injury. In ref. 21, pp. 439–52 (1975).
63. Rothman, J. E. and Lenard, J. Membrane asymmetry. *Science*, **195**, 743–53 (1977).
64. Rumsby, M. G., Trotter, L., Allan, D., and Michell, R. H. Recovery of membrane micro-vesicles from human erythrocytes stored for transfusion: a mechanism for the erythrocyte discocyte-to-spherocyte transformation. *Biochemical Society Transactions*, **5**, 125–8 (1977).
65. Schatzmann, H. J. Active calcium transport and Ca^{2+}-activated ATPase in human red cells. *Current Topics in Membranes and Transport*, **6**, 125–68 (1975).
66. Schwoch, G. and Passow, H. Preparation and properties of human erythrocyte ghosts. *Molecular and Cellular Biochemistry*, **2**, 197–218 (1973).
67. Seeman, P. Ultrastructure of membrane lesions in immune lysis, osmotic lysis and drug-induced lysis. *Federation Proceedings*, **33**, 2116–24 (1974).
68. Semenza, E., Semenza, G., and Carafoli, E. (Eds). *Biochemistry of Membrane Transport*. *FEBS Symposium No. 42*, Springer-Verlag, Berlin, 1977.
69. Shen, A. C. and Jennings, R. B. Kinetics of calcium accumulation in acute myocardial ischaemic injury. *American Journal of Pathology*, **67**, 417–40 (1972).
70. Shore, G. and Tata, J. R. Functions for polyribosome–membrane interactions in protein synthesis. *Biochimica et Biophysica Acta*, **472**, 197–236 (1977).
71. Shrago, E. Mitochondrial adenine nucleotide translocase. *Journal of Molecular and Cellular Cardiology*, **8**, 497–500 (1976).
72. Shrago, E., Shug, A. L., Sul, H., Bittar, N., and Folts, J. D. Control of energy production in myocardial ischaemia. *Circulation Research Supplement 1*, **38**, 75–9 (1976).

73. Siekevitz, P. Biological membranes: the dynamics of their organisation. *Annual Reviews of Physiology*, **34**, 117–40 (1972).
74. Sigel, H., Janke, J., and Fleckenstein, A. Restriction of isoproterenol-induced myocardial Ca uptake and necrotization in rats by a new Ca-antagonistic compound. In ref. 21, pp. 121–216 (1975).
75. Silverstein, S. C., Steinman, R. M., and Cohn, Z. A. Endocytosis. *Annual Review of Biochemistry*, **46**, 669–722 (1977).
76. Singer, S. J. The molecular organisation of membranes. *Annual Review of Biochemistry*, **43**, 805–33 (1974).
77. Singer, S. J. and Nicolson, G. L. The fluid mosaic model of the structure of cell membranes. *Science*, **175**, 720–31 (1972).
78. Spieckermann, P. G., Gebhard, N. M., and Nordbeck, H. This Book, Chapter 4.
79. Weed, R. I. and Reed, C. F. Membrane alterations leading to red cell destruction. *American Journal of Medicine*, **41**, 681–98 (1966).
80. Weglicki, W. B., Waite, B. M., and Stam, A. C. Association of phospholipase A with a myocardial membrane preparation containing the $(Na^+ + K^+)-Mg^{2+}$-ATPase. *Journal of Molecular and Cellular Cardiology*, **4**, 195–201 (1972).
81. Weissman, G. and Claiborne, R. (Eds.). *Cell Membranes*. H. P. Publishing Co., New York, 1975.
82. Welman, E. and Peters, T. J. Enhanced lysosome fragility in the anoxic perfused guinea pig heart: effects of glucose and mannitol. *Journal of Molecular and Cellular Cardiology*, **9**, 101–20 (1977).
83. Whittam, R. and Wheeler, K. P. Transport across cell membranes. *Annual Reviews of Physiology*, **32**, 21–60 (1970).
84. Wilkinson, J. H. and Robinson, J. M. Effect of ATP on release of intracellular enzymes from damaged cells. *Nature*, **249**, 662–3 (1974).
85. Wirtz, K. W. A. Transfer of phospholipids between membranes. *Biochimica et Biophysica Acta*, **344**, 95–117 (1974).
86. Wrogemann, K. and Nylene, E. G. Mitochondrial calcium overloading in cardiomyopathic hamsters. *Journal of Molecular and Cellular Cardiology*, **10**, 185–93 (1978).
87. Wrogemann, K. and Pena, S. D. J. Mitochondrial Ca^{2+} overload: a general mechanism for cell necrosis in muscle diseases. *Lancet*, **i**, 672–4 (1976).
88. Yates, J. C. and Dhalla, N. S. Structural and functional changes associated with failure and recovery of hearts after perfusion with Ca^{2+}-free medium. *Journal of Molecular and Cellular Cardiology*, **7**, 91–103 (1975).
89. Zwaal, R. F. A. The use of pure phospholipases in the study of membrane structure and function. *Biochemical Society Transactions*, **2**, 821–5 (1974).

CHAPTER 4

From heart to plasma

P. G. Spieckermann, H. Nordbeck, and C. J. Preusse

INTRODUCTION	81
THE FUNCTIONAL ANATOMY OF MYOCARDIAL CAPILLARIES	82
THE FUNCTIONAL ANATOMY OF THE INTERSTITIUM	83
THE ANATOMY OF THE CARDIAC LYMPHATICS AND DIFFERENCES BETWEEN LYMPH AND BLOOD CAPILLARIES	84
EXPERIMENTAL STUDIES OF CARDIAC ENZYME CLEARANCE	86
ENZYMES IN CARDIAC LYMPH UNDER CONTROL CONDITIONS	86
ENZYMES IN CARDIAC LYMPH DURING INFARCTION	89
RELATIVE ENZYME TRANSPORT RATES	91
THE PERMEABILITY OF THE ENDOTHELIUM TO ENZYMES AND OTHER MACROMOLECULES	91
THE TRANSIT TIME OF ENZYMES TRANSPORTED VIA THE LYMPHATICS FROM THE HEART TO THE PLASMA	92
CONCLUDING COMMENTS	92
REFERENCES	93

INTRODUCTION

In their passage from the myocardial cell to the circulating blood, enzyme molecules encounter a number of barriers. The preceding chapter discussed the translocation of the enzymes from the cytoplasm to the interstitial space and the importance of intracellular and plasma membranes in the process of enzyme permeation. The aim of this chapter is to describe the further passage of these macromolecules from the interstitial space to the systemic circulation.

Two potential pathways exist for the transport of molecules from the heart to the systemic circulation: the coronary system and the myocardial lymphatics. Since the rate and route of clearance of enzymes from the myocardial cell is critical to the implementation of various equations for calculating infarct size (see Chapters 5, 11, and 12) it is important to ascertain whether one or both of these pathways is responsible for enzyme clearance. If both pathways are involved we should know the proportion of the enzyme cleared by each pathway and the characteristics of that clearance. Since enzyme molecules are very large, a number of investigators[27,50] believe the lymphatics to be the predominant route of clearance. However, clarification of this issue requires a careful consideration of the two pathways and the structures and membrane barriers associated with them.

THE FUNCTIONAL ANATOMY OF MYOCARDIAL CAPILLARIES

The myocardial blood capillary (see Figure 1) is a tubular vessel approximately 8 μm in diameter which is lined by a single layered 'continuous' endothelium. Surrounding this is a continuous basement membrane.[18,19,26] The endothelial cells have a thickness ranging from 0.1 μm in the flat part to approximately 3 μm in the nuclear region. Their cytoplasm is filled with numerous micropinocytotic vesicles which account for approximately 50 per cent of the cell volume. The intercellular clefts between adjacents cells are not readily apparent but are known to represent special structures. Recent electron microscopic findings show that these regions are maculae occludentes with a small gap of 60 Å[5] through which smaller molecules are thought to be able to pass, thus crossing the capillary wall. These endothelial cell junctions are thus thought to represent the 'small-pore transport system'. In contrast to this, the pinocytotic vesicles may be the morphological equivalent of the so called 'large-pore system'. This latter system, through Brownian motion, is thought to be capable of transporting molecules with diameters of greater than 40 Å in both directions.[18] The slight molecular sieving that occurs with molecules of increasing molecular size (in the range

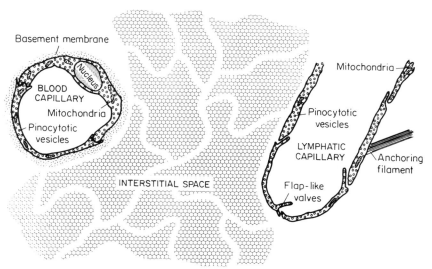

Figure 1. Diagrammatic representation of a blood capillary, a lymph capillary, and the interstitial space.

100,000 to 400,000 daltons) would also be suggestive of the existence of a vesicular transport mechanism for large molecules.[13,31]

The basement membrane (basal lamina) is a continuous layer surrounding the capillary, it consists of fine fibrillar material embedded in an amorphous matrix. This structure may act as a filtration barrier for transendothelial exchange but since it is permeable to relatively large molecules, it would appear to represent only a relatively coarse form of filter or barrier.

In considering the permeability of the myocardial capillaries it should be appreciated that different regions of the capillary system exhibit differential permeabilities to various solutes.[49] Thus water-soluble dyes pass more rapidly from venous capillaries and venules[41] than they do from arterial capillaries, an observation that may be explained by the microscopic finding that 'large pores' are more abundant in the walls of the venous capillaries.[21] Furthermore, the surface area of the venous capillaries is between 4 and 6 times greater than that of the arterial capillaries. This greater permeability may be further supported by extremely thin regions of the venous endothelium which are known to occur and which have abundant pinocytotic vesicles.[2] Furthermore, intramyocardial lymph capillaries frequently lie in close proximity to these venules and the distance between the adjacent plasma membranes is often very small and of the order of 0.12–0.25 μm.

THE FUNCTIONAL 'ANATOMY' OF THE INTERSTITIUM

All capillaries together with blood and lymph vessels and the muscle cells themselves exist in a very complex environment: the extravascular connective tissue space. The interstitial fluid within this space does not appear to exist as a free fluid pool but is thought to be, at least in part, immobilized in a relatively dense fibrous structure interspersed with a gel-like pseudoamorphous structure called the ground substance. This substance is composed mainly of mucopolysaccharides, e.g. hyaluronic acid and chondroitin sulphates, which are covalently bound to proteins to form enormous complexes with molecular weights of several million. These macromolecular mucopolysaccharide complexes form an entangled meshwork with gel-like properties such as swelling and syneresis. Colloid-chemical and electron microscopic investigations of this complex would suggest that the interstitium exists as a two-phase system with fluid bound in the gel-like colloid phase and a free fluid phase in which proteins are mainly dispersed (see reference 6). These two phases are thought to be in a state of dynamic equilibrium. The free fluid phase may well correspond to pre-lymphatic channels of the interstitium. These channels with diameters of between 600 and 1000 Å would penetrate the ground substance and the resistance to flow in this free fluid phase would be small compared with that in the colloid phase.

The transport of molecules, especially macromolecules, through the interstitium from the cell to the capillaries therefore depends mainly upon the physicochemical characteristics of this mucopolysaccharide network.[22] Two important and competing factors contribute to this transport process; the first is called the sieve effect and the second the exclusion phenomenon. The network itself acts as a sieve thus retarding the movement of large molecules to a greater extent than small molecules. On the other hand, when macromolecules are transported in the free fluid phase between polysaccharide compartments there is a tendency to diffuse into an adjacent compartment. However, the larger the molecule the smaller is this diffusion tendency and the more likely is the molecule to stay in the fluid phase. The net effect of these two phenomena is the surprising situation that under steady-state conditions large molecules can theoretically move faster through the interstitium than can small molecules.

THE ANATOMY OF THE CARDIAC LYMPHATICS AND DIFFERENCES BETWEEN LYMPH AND BLOOD CAPILLARIES

The lymphatics originate as blind terminated sacs in the interstitium, they are arranged in the heart as three plexuses: the subendocardial, the myocardial, and the subepicardial. The fluid from all of these plexuses ultimately reaches the subepicardial lymph network where it flows together forming higher-order lymphatics, the drainage trunks of which accompany the coronary vessels. These greater lymphatics unite to form one or two vessels which enter the cardiac lymph node between the superior vena cava and the brachiocephalic vein (very often, a second (pre-tracheal) node being intercalated). The cardiac lymph node then drains into the right lymphatic duct which finally empties into the external jugular vein.[8,16,17,32,38] A general overview of the gross anatomy of the myocardial lymphatics is shown in Figure 2.

There is disagreement about the abundance of cardiac lymphatic capillaries. Some investigators[1,17,20,32,39] have suggested a dense network whereas others[3,15,25,37] have reported only a scattered distribution of lymph capillaries between the myocardial fibres. In a recent and very detailed histological and electron microscope study Böger and Hort[2] have demonstrated that mouse heart lymphatics are very sparsely distributed and that the ratio of muscle fibre to lymphatic capillary is of the order of 1000:1 or 2000:1 whereas the ratio muscle fibre to blood capillary is approximately 1:1.[14,23,24,48] It remains to be shown whether this situation prevails in other and larger hearts.

Lymph capillaries have distinct morphological differences (see Figure 1) from blood capillaries.[4,5,50] The endothelial cells (which also contain

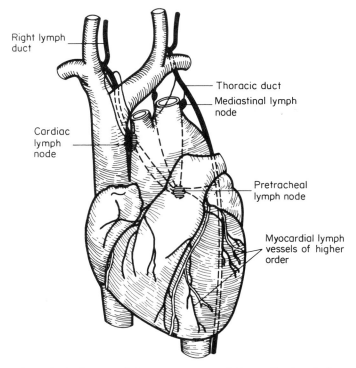

Figure 2. A diagram of the gross anatomy of the cardiac lymphatics.

pinocytotic vesicles) are attached to the surrounding connective tissue by anchoring filaments which serve to hold the walls apart. Between the endothelial cells in the cardiac lymphatics large clefts may exist which may be as large as 5 μm. The intercellular clefts are overlapped with 'flap'-like valves. Under certain conditions these valves may close thereby preventing fluid from leaving the lymph capillary and channelling it to a region of lower pressure. Lymphatic flow is additionally controlled by the true lymph valves which occur in the larger lymph vessels. The basement membrane which is characteristic of blood capillaries is absent from lymph capillaries and is discontinuous in the greater lymph vessels. In this way the interior of the lymphatic capillary has relatively unrestricted communication with the interstitial space and this facilitates the clearing of fluid, small and large molecules, and cellular elements from the tissue.

As mentioned previously, knowledge of enzyme clearance via the coronaries and the lymphatics, and in particular the relative importance of each of these pathways, is critical to a number of clinical and experimental uses of myocardial enzyme leakage. Although we now have a reasonable working knowledge of the functional anatomy of the coronaries and the lymphatics it is not possible to predict with any great degree of confidence, which is the dominant pathway for enzyme clearance. We have therefore recently undertaken a series of experimental studies aimed at clarifying this important point.

EXPERIMENTAL STUDIES OF CARDIAC ENZYME CLEARANCE

In these open chest dog experiments[36] cardiac lymphatic channels were visualized by an intramuscular injection of dye (0.2 ml of 2% Evans blue) into the left ventricular wall. A cardiac lymphatic, reaching the cardiac lymph node between the superior vena cava and the truncus brachiocephalicus, was then cannulated using a thin heparinized catheter. The other lymphatics passing cephalad from the cardiac node were also ligated in order that all cardiac lymph could be collected and the lymph flow rate be continuously monitored. In addition we also measured arterial and central venous pressure, dP/dt, and the median and pulsatile coronary blood flow through the ramus circumflexus of the left coronary artery. Coronary sinus blood was collected via a suitable catheter. After a control period, myocardial infarction was produced by ligation of the ramus interventricularis anterior of the left coronary artery. After 6 hours the ligation was released and the ischaemic area was reperfused for a 90 minute period. In some groups of dogs the heart was stimulated by the infusion of isoprenaline and noradrenaline (3 μg min^{-1} kg^{-1} body wt of both compounds together with 0.5 mg atropine).

ENZYMES IN CARDIAC LYMPH UNDER CONTROL CONDITIONS

At the beginning of the experiment, prior to any intervention, the activity of a number of enzymes (see Figure 3a) was significantly higher ($p < 0.001$) in the lymph than in serum, with some activities being as much as ten times greater.[30] The high enzyme activity in cardiac lymph could have originated either directly from the myocardial cells or from the coronary blood passing through the heart. In order to clarify this possibility, lactate dehydrogenase isoenzyme profiles were prepared for the control blood and lymph. The results (Figure 3b) reveal different patterns with the 'cardiospecific' isoenzymes 1 and 2 predominating in the cardiac lymph. This isoenzyme distribution pattern together with the high enzyme activities of cardiac lymph and its relatively low protein content when compared to plasma would indicate firstly, that a substantial proportion of the lymphatic enzymes originate from the heart and secondly, that this leakage occurs even under normal or

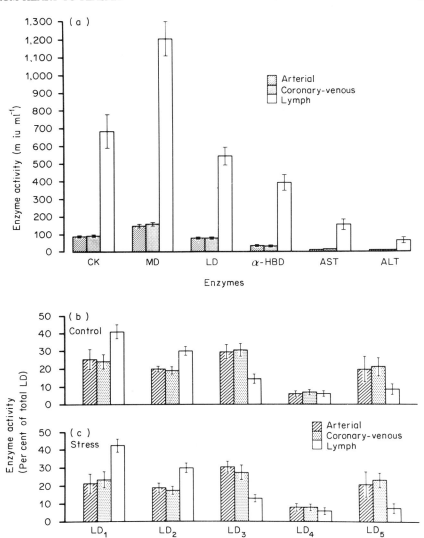

Figure 3. Enzymes in cardiac lymph and coronary venous and arterial plasma. (a) Under control conditions showing 6 different enzymes (mean±SEM, n = between 30 and 70). (b) Under control conditions showing LD isoenzyme profiles. (c) Under stress conditions (see text) showing LD isoenzyme profiles. Each isoenzyme result is the mean of 7 determinations and the bars represent the standard error of the mean.

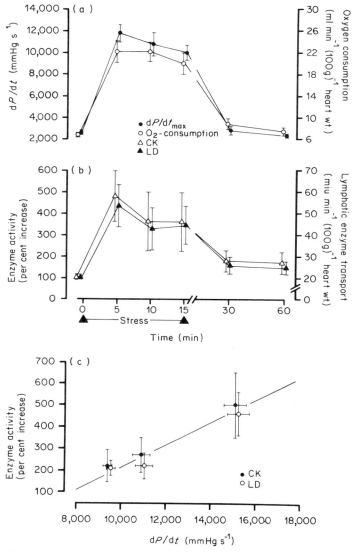

Figure 4. Lymphatic enzyme release rates before, during and after a catecholamine-induced increase in heart work. (a) dP/dt and myocardial oxygen consumption. (b) corresponding patterns of enzyme release. (c) The correlation between lymphatic enzyme transport and dP/dt (mean ± SEM, $n = 13$).

control conditions. Similar conclusions can be drawn from a series of guinea-pig experiments.[10,11,43] Studies[7] of enzyme loss have yielded Q_{10} values of 1.2 to 1.5 and may indicate that the 'physiological' enzyme loss which occurs under normal conditions may be mediated by physicochemical factors. Since the heart has a fixed complement of cells[23,24] it is certainly not possible to explain the basal level of leakage by cell breakdown and replacement such as occurs in the liver. Certainly, a low level of myocardial enzyme leakage would appear to represent a normal cellular phenomenon and as such cannot be restricted to pathological conditions. This point is further reinforced by the studies illustrated in Figure 4, which reveal that during mechanical stress this basal level of enzyme release increases to an extent which is dependent upon the degree of work. Thus during catecholamine-induced cardiac stress, with dP/dt values of about 10,000 mmHg s^{-1}, the amount of enzyme activity in the lymph increases by 3- to 5-fold. If the stress is terminated after 15 minutes, then the enzyme activities return to their basal levels within a further 15 minutes, presumably indicating that no irreversible damage has been sustained. These studies would indicate that enzyme activity in cardiac lymph begins to increase when the dP/dt value is about 8000 mmHg s^{-1}, thereafter the increase would appear to be directly proportional to the energy demand of the heart muscle as also indicated by other experiments.[7,10,11,42,44,46] Arterial–coronary venous differences of enzyme activities could be detected but as a consequence of the increased coronary blood flow during stress these changes were small and not statistically significant.[34,36] Thus only in the lymph, with its slow flow and its ability to 'concentrate' macromolecules are these small effects detectable. In this way the myocardial lymphatic system can be considered to act as a biological enzyme multiplier which allows us to detect small changes in myocardial enzyme leakage. These small changes are unlikely to represent irreversible damage and could not easily be detected by conventional measurement with coronary sinus blood.

ENZYMES IN CARDIAC LYMPH DURING INFARCTION

During the infarction experiments (infarct size approximately 30 g) we found that 15 to 20 minutes after ligation there was a statistically significant ($p<0.01$) increase of enzyme activity in the lymph (see Figure 5a). Between 90 and 120 minutes after ligation the enzyme activity in the lymph plateaued at a mean maximum value of 16,000 miu ml^{-1} for the CK and 8000 miu ml^{-1} for the LD.[33,35,36,44,45] The highest lymphatic CK activity measured during the ligation period was approximately 80,000 miu ml^{-1}, in other words, the enzymes in cardiac lymph can be up to 100 times greater than those in the corresponding arterial or coronary venous blood. On average, however, the activity was about 30 times greater. Comparable

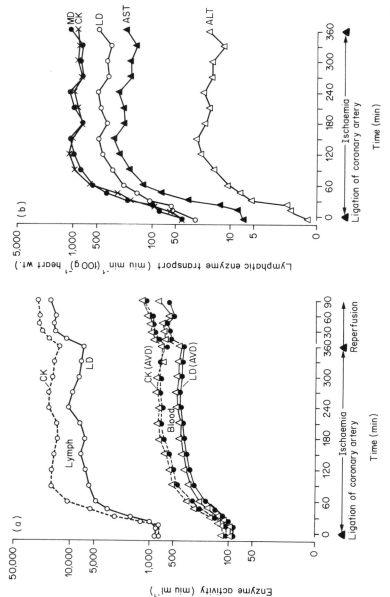

Figure 5. (a) Enzyme activity in coronary venous (△) and arterial (●) plasma and in cardiac lymph (○) during myocardial ischaemia and reperfusion. (b) Lymphatic transport rate for a number of enzymes during ischaemia. Each result is the mean of 9 determinations.

findings have also been reported by Malek et al.,[27] by Malmberg,[28,29,30] by Gervin et al.,[12] and by Feola and Glick.[9] Figure 5a also shows the effect of reperfusion upon the level of enzyme activity in lymph and blood.

Despite the collection of cardiac lymph (thereby preventing the leaked enzyme from reaching the general circulation) the ligation of the coronary artery (Figure 5a) also leads to a 5- to 8-fold increase in the enzyme activity of the coronary blood (note in Figure 5a that enzyme activity is shown on a logarithmic scale) and an arteriovenous difference is detectable. Figure 5b shows the rate of release of 5 different enzymes in cardiac lymph following coronary artery ligation. The rates, which are on a logarithmic scale are expressed as $\text{miu min}^{-1} (100 \text{ g})^{-1}$ heart weight thus compensating for the increase (30%) in lymph flow rate which occurs after coronary artery ligation. The differing profiles for each enzyme probably reflect differences in molecular weight and differences in intracellular content.

RELATIVE ENZYME TRANSPORT RATES

Despite the very high level of enzyme activity in cardiac lymph, the flow rate[30] of the lymph is very low $(10-15 \, \mu\text{l min}^{-1} (100 \text{ g})^{-1}$ heart weight). Thus the overall enzyme transport rate in the lymph is relatively low when compared with that in the coronary blood. From our studies it would appear that no more than 30% of the circulating enzymes reach the blood via the lymph and our mean calculated value was $15 \pm 2\%$.[33,35,36,44,45] We therefore conclude that the major portion of the enzymes reaching the blood stream do so by crossing the coronary vascular membrane barrier. In view of this, it is perhaps appropriate to consider briefly some aspects of the permeability of this barrier.

THE PERMEABILITY OF THE ENDOTHELIUM TO ENZYMES AND OTHER MACROMOLECULES

The findings presented thus far would indicate that the capillary membrane is relatively permeable, in both directions, to macromolecules.

Enzyme proteins which escape from the myocardial cells pass directly into the interstitium where they come in close contact with microvessels (this is favoured by the 1:1 ratio of muscle cell to blood capillary). The extent and rate of transport of the enzyme into the vessel will be dependent upon a number of factors including the enzyme gradient across the membrane. Preferential transport to the blood capillaries will be further supported by the generally greater distances between muscle cells and lymph capillaries. This conclusion based on our studies would be supported by the recent findings of Szabo[47] who administered radiolabelled albumin and ^{198}Aucolloid to the myocardium and concluded that the blood capillaries were responsible for clearing the bulk of the myocardial macromolecules transporting about three times as much as the lymph.

The transport of proteins and macromolecules is not unidirectional and it is possible to demonstrate the passage of macromolecules from the blood, via the interstitial fluid to the lymph. This can be done by determining the ratio of plasma and lymph concentrations for a series of FITC-dextrans (fluoresceine-isothiocyanate conjugated dextrans) of differing molecular weight 1 hour after a single injection. Lymph:plasma concentration ratios of 0.35, 0.13, and 0.05 have been observed for dextrans with molecular weights of 20,000, 70,000, and 150,000. These ratios, in addition to indicating multidirectional transport also underline the important relationship between transport rates and molecular size. Further evidence for the transport of macromolecules from the blood to the lymph derives from the appearance of fibrinogen (molecular weight 340,000) in the lymph and its ability to cause lymph coagulation. It is thought that these macromolecules are transported predominantly by the pinocytotic vesicles of the capillary endothelium.

THE TRANSIT TIME OF ENZYMES TRANSPORTED VIA THE LYMPHATICS FROM THE HEART TO THE PLASMA

The transit time for enzymes in the cardiac lymph is relatively short. From our studies where the lymph flow rate was approximately $50\ \mu l\ min^{-1}$ and the volume of the greater lymphatic vessels was estimated as approximately 1 ml it is possible to calculate a transit time of about 20 minutes. This would agree well with our observation of increased enzyme activity in the lymph 20 or so minutes after coronary artery ligation. These calculations and observations would also support our earlier arguments that the passage of enzyme molecules through the fluid phase of the interstitial space is relatively fast and unrestricted.

One important consequence of a short transit time relates to the problem of enzyme inactivation in the cardiac lymph. It is thought that various proteolytic enzymes may be released into the lymph together with other cytoplasmic enzymes and we and others[40] have demonstrated a selective inactivation of CK in lymph. However, in these *in vitro* studies 1 to 2 hours incubation at 35 °C was required to achieve a 60 to 70% inactivation of CK. Thus in view of the short transit time for enzymes in the lymph and the low transport rates it could be argued that enzyme inactivation may be quantitively unimportant.

CONCLUDING COMMENTS

Two pathways exist for the transport of enzymes from the cell and interstitial space to the systemic circulation: the coronary system and the lymphatics. Under normal non-pathological conditions there appears to be a basal level of enzyme leakage from the cell to both the blood and the lymph. This basal leakage would not appear to be associated with an irreversible damage but can be increased by a number of stresses, e.g. increased heart work.

Major enzyme release occurs very soon after the onset of coronary artery ligation. This results in an increase in the level of enzyme activity in both the coronary blood and lymphatic fluid. Very high activities can be detected in the lymph but due to its very slow flow rate it is only responsible for clearing approximately 15% of the enzyme from the heart to the general circulation. The bulk of the enzyme activity is thus transported by the high-flow coronary system. The transport of enzymes and macromolecules is not unidirectional and there is evidence for the movement of plasma proteins via the interstitial space to the lymph. The transport of enzymes from the heart is limited by a number of barriers including the coronary and lymphatic endothelium and the complex interstitial space and these, together with a number of physicochemical and structural factors may determine the rate and direction of transport.

REFERENCES

1. Bock, H. Die Lymphgefäße des Herzens. *Anatomischer Anzeiger*, **27**, 33–41 (1905).
2. Böger, A. and Hort, W. Qualitative und quantitative Untersuchung am Lymphgefäßsystem des Mäuseherzens, *Basic Research in Cardiology*, **72**, 510–29 (1977).
3. Bullon, A. and Huth, F. Fine structure of lymphatics in the myocardium, *Lymphology* **5**, 42–8 (1972).
4. Casley-Smith, J. R. The lymphatic system in inflammation. In *The Inflammatory Process* (Eds. Zweifach, B. W., Grant, L., and McClusky, R. C.), 2nd. edn., Vol. 2, Academic Press, New York, San Francisco, London, 1973, pp. 161–204.
5. Casley-Smith, J. R. The fine structure and functions of blood capillaries, the interstitial tissue and the lymphatics. In *Ergebnisse der Angiologie*, Vol. 12, *Basic Lymphology* (Ed. Földi, M.), F. K. Schattauer-Verlag, Stuttgart–New York, 1976, pp. 1–29.
6. Chvapil, M. *Physiology of Connective Tissue*, Butterworths, London, 1967.
7. Denkhaus, H. Temperaturabhängigkeit der myokardialen Enzymfreisetzung während Anoxie. *Thesis*, Göttingen, 1976.
8. Eliskova, M. and Eliska, O. Lymph drainage of the dog heart, *Folia morphologica*, **22**, 320 (1974).
9. Feola, M. and Glick, G. Cardiac lymph flow and composition in acute myocardial ischemia in dogs. *American Journal of Physiology*, **229**, 44–8 (1975).
10. Gebhard, M. M. Das Langendorff-Präparat des Meerschweinchens als Modell zur Untersuchung der myokardialen Enzymfreisetzung bei Ischämie, Hypoxie und Anoxie. *Thesis*, Göttingen, 1976.
11. Gebhard, M. M., Denkhaus, H., Sakai, K., and Spieckermann, P. G. Relations between energy metabolism and enzyme release. *Journal of Molecular Medicine*, **2**, 271–83 (1977).
12. Gervin, C. A., Hackel, D. B. and Roe, C. R. Enzyme content of canine cardiac lymph during acute myocardial infarction. *Surgical Forum*, **XXV**, 175–6 (1974).
13. Grotte, G. Passage of dextran molecules across the blood–lymph barrier. *Acta Chirurgica Scandinavica, Suppl.*, **211**, 1–83 (1956).
14. Hort, W. Quantitative Untersuchungen über die Kapillarisierung des Herzmuskels im Erwachsenen- und Greisenalter bei Hypertrophie und Hyperplasie. *Virchow's Archiv für pathologische Anatomie*, **327**, 560–76 (1955).

15. Jacobs, G., Kleinschmidt, F., Benesch, L., Lenz, W., Uhlig, G., and Huth, F. Tierexperimentelle Untersuchungen des kardialen Lymphgefäßytems. *Thoraxchirurgie*, **24**, 453–67 (1976).
16. Johnson, R. A. The lymphatic system of the heart. *Lymphology*, **2**, 95–108 (1969).
17. Johnson, R. A. and Blake, T. M. Lymphatics of the heart. *Circulation*, **33**, 137–42 (1966).
18. Karnovsky, M. J. The ultrastructural basis of transcapillary exchanges. *Journal of General Physiology*, **52**, 645–955 (1968).
19. Karnovsky, M. J. Morphology of capillaries with special reference to muscle capillaries. In *Capillary Permeability* (Eds. Crone, C. and Lassen, N. A.), Munksgaard, Copenhagen, 1970, pp. 341–50.
20. Kline, I. K. Lymphatic pathways in the heart. *Archives of Pathology*, **88**, 638–44 (1969).
21. Landis, E. M. Heteroporosity of the capillary wall as indicated by cinematographic analysis of the passage of dyes. *Annals of the New York Academy of Sciences*, **116**, 765–73 (1964).
22. Laurent, T. C. The structure and function of the intercellular polysaccharides in connective tissue. In *Capillary Permeability* (Eds. Crone, C. and Lassen, N. A.), Munksgaard, Copenhagen, 1970, pp. 261–77.
23. Linzbach, A. J. Mikrometrische und histologische Analyse hypertropher menschlicher Herzen. *Virchow's Archiv für pathologische Anatomie*, **314**, 534–94 (1947).
24. Linzbach, A. J. Die Muskelfaserkonstante und das Wachstumsgesetz der menschlichen Herzkammern. *Virchow's Archiv für pathologische Anatomie*, **318**, 575–618 (1950).
25. Ljungquist, A., Mandache, E., and Unge, G. Ultrastructural aspects of cardiac lymphatic capillaries in experimental cardiac hypertrophy. *Microvascular Research*, **10**, 1–7 (1975).
26. Majno, G. Ultrastructure of the vascular membrane. In *Handbook of Physiology*, Section 2, *Circulation*, Vol. III, American Physiological Society, Washington D.C., 1965, pp. 2293–375.
27. Malek, P., Knoll, J., and Skodova, Z. Importance of the lymphatic system in enzyme transport from the ischaemized heart muscle. *Review of Czechoslovakic Medicine*, **17**, 16–8 (1971).
28. Malmberg, P. Aspartate aminotransferase activity in dog heart lymph after myocardial infarction. *Scandinavian Journal of Clinical and Laboratory Investigations* **30**, 153–158 (1972).
29. Malmberg, P. Time course of enzyme escape via heart lymph following myocardial infarction in the dog. *Scandinavian Journal of Clinical and Laboratory Investigations*, **30**, 405–9 (1972).
30. Malmberg, P. Enzyme composition of dog heart lymph after myocardial infarction. *Upsala Journal of Medical Science*, **78**, 73–7 (1973).
31. Mayerson, H. S., Wolfram, C. G., Shurley Jr., H. H., and Wassermann, K. Regional differences in capillary permeability. *American Journal of Physiology*, **198**, 155–60 (1960).
32. Miller, A. J. The lymphatics of the heart. *Archives of Internal Medicine*, **112**, 501–11 (1963).
33. Nordbeck, H., Bretschneider, H. J., Knoll, D., Kohl, F. V., and Spieckermann, P. G. Freisetzung und Transport von Enzymen aus der Herzmuskelzelle nach experimentell erzeugtem Infarkt beim Hund. *Verhandlungen der Deutschen Gesellschaft für Kreislaufforschung* **40**, 292–295 (1974).
34. Nordbeck, H., Baie, W., Kahles, H., Preusse, C. J., Spieckermann, P. G., and Bretschneider, H. J. Enzymverluste des Myokards durch pharmakodynamische Belastung. *Verhandlungen der Deutschen Gesellschaft für Kreislaufforschung*, **41**, 253–7 (1975).
35. Nordbeck, H., Kahles, H., Preusse, C. J., and Spieckermann, P. G. Enzymes in cardiac lymph and coronary blood under normal and pathophysiological conditions. *Journal of Molecular Medicine*, **2**, 255–63 (1977).
36. Nordbeck, H. *Lymphphysiologie des Herzens. Experimentelle Untersuchungen zu Flußraten, Zusammensetzung und Enzymgehalt*, Thieme-Verlag, Stuttgart 1978.
37. Parker, E. F., Bradham, R. R., Henninger, G. R., and Greene, W. B. Effects of obstruction of cardiac lymphatics. *Journal of Thoracic and Cardiovascular Surgery*, **69**, 390–6 (1975).
38. Patek, P. The morphology of the lymphatics of the mammalian heart. *American Journal of Anatomy*, **64**, 203–50 (1939).
39. Ranvier, L. *Traite Technique d'Histologie*, 2nd. edn., F. Savy, Paris, 1889.

40. Robinson, A. K., Gnepp, D. R., and Sobel, B. E. Inactivation of CPK in lymph. *Circulation Supplement II*, **52,** II–5 (1975).
41. Rous, P., Gilding, H. P., and Smith, F. The gradient of vascular permeability. *Journal of Experimental Medicine*, **51,** 807–30 (1930).
42. Sakai, K. and Spieckermann, P. G. Effects of reserpine and propranolol on anoxia-induced enzyme release from the isolated perfused guinea-pig-heart. *Naunyn-Schmiedeberg's Archiv Pharmacology*, **291,** 123–30 (1975).
43. Sakai, K., Gebhard, M. M., Spieckermann, P. G., and Bretschneider, H. J. Enzyme release resulting from total ischemia and reperfusion in the isolated, perfused guinea pig heart. *Journal of Molecular and Cellular Cardiology*, **7,** 827–40 (1975).
44. Spieckermann, P. G., Gebhard, M. M., Kalbow, K., Knoll, D., Kohl, F. V., Nordbeck, H., Sakai, K., and Bretschneider, H. J. Freisetzung von Enzymen aus der Herzmuskelzelle während Sauerstoffmangel. Verhandlungen der Deutschen Gesellschaft für Kreislaufforschung, *Darmstadt*, **39,** 193–8 (1973).
45. Spieckermann, P. G., Nordbeck, H., Knoll, D., Kohl, F. V., Sakai, K., and Bretschneider, H. J. Bedeutung der Herzlymphe für den Enzymtransport ins Blut beim Myokardinfarkt. *Deutsche Medizinische Wochenschrift*, **99,** 982–3 (1974).
46. Spieckermann, P. G., Gebhard, M. M., and Nordbeck, H. Role of energy metabolism in enzyme retention. A study on isolated perfused canine hearts. *Experientia*, **31,** 1046–7 (1975).
47. Szabo, G. Movement of proteins into the blood capillaries. In *Ergebnisse der Angiologie*, Vol. 12, *Basic Lymphylogy* (Ed. Földi, M.), F. K. Schattauer-Verlag, Stuttgart–New York, 1976, pp. 31–50.
48. Wearn, J. T. Morphological and functional alterations of the coronary circulation. *Harvey Lectures*, **35,** 243–70 (1939/40).
49. Wiederhielm, C. A. Dynamics of transcapillary fluid exchange. *Journal of General Physiology*, **52,** 29s–63s (1968).
50. Yoffey, J. M. and Courtice, F. C. *Lymphatics, Lymph and the Lymphomyeloid Complex*, Academic Press, London, New York, 1970.

CHAPTER 5

The distribution, inactivation and clearance of enzymes

R. Roberts and B. E. Sobel

INTRODUCTION	97
REMOVAL OF ENZYME FROM THE SYSTEMIC BLOOD CIRCULATION	98
FALSE ASSUMPTIONS	102
VARIABILITY OF ENZYME ELIMINATION RATES	108
FACTORS INFLUENCING PLASMA ENZYME DISAPPEARANCE RATES	109
MECHANISMS OF REMOVAL OF ENZYMES FROM THE CIRCULATION	110
FUTURE DIRECTIONS	112
CONCLUDING COMMENTS	112
REFERENCES	113

INTRODUCTION

Injury or necrosis affecting specific organs or tissues results in release of enzymes into the circulation in a characteristic fashion. Nevertheless, time–activity profiles of circulating plasma enzymes vary. For example, the interval after injury to myocardium preceding the occurrence of peak plasma creatine kinase (CK) activity varies over a wide range, as does the absolute magnitude of the peak and the total duration during which CK remains elevated above the upper limit of the normal range. Although a primary determinant of the nature of plasma enzyme time–activity profiles is the subcellular distribution of enzyme and isoenzyme activity in the organ sustaining injury, several other factors influence the profiles including the distribution of injury within the organ, the mechanisms of transport of enzyme liberated from damaged cells to the systemic circulation, inactivation of liberated enzyme in lymph and interstitial fluid, and processes influencing denaturation and removal of enzymes circulating in the blood.[11,23,37]

The importance of several of these factors is underscored by recent information obtained from experimental and clinical studies. For illustrative purposes in this chapter, findings with CK will be used as an example of the general phenomena.

Only a relatively small proportion of the CK activity lost from the heart in animals subjected to coronary occlusion can be accounted for by enzyme appearing in the circulating blood.[27,34] Even when enzyme liberation is studied in a more controlled situation, with tissue incubated *in vitro*, recovery of enzymes, such as transaminase, in the medium represents only a modest proportion of enzyme lost from the tissue incubated.[4] Mechanisms responsible for the disparity have not been completely elucidated, but local degradation of enzyme protein or denaturation resulting in release of enzymatically inactive molecules into the circulation appears to contribute substantially.

The extent to which enzyme lost from tissue appears in the circulation may depend considerably on the subcellular locus of the enzyme. For example, mitochondrial isoenzymes of creatine kinase are readily detectable in extracts of myocardium from several species. However, with conventional electrophoretic techniques, mitochondrial CK is not detectable in the circulation after even profound myocardial injury in experimental animals or patients. This may be due to the lability of mitochondrial CK or to its susceptibility to inactivation when the enzyme is dissociated from mitochondrial membranes during the evolution of ischaemic injury.

One factor contributing to loss of enzyme activity appears to be inactivation in lymph.[7] After myocardial injury, several enzymes appear in the cardiac lymphatic efflux—sometimes with activities markedly in excess of those represented in the circulating blood.[20] These observations suggest that the fate of enzyme transported in lymph may influence the consequent plasma time–activity profiles. When cardiac lymphatic efflux is inhibited, the amount of enzyme appearing in the blood after experimentally induced myocardial infarction represents a smaller proportion of enzyme lost from the heart than is the case in the absence of inhibition of transport via lymph.[7] Since lymph inactivates some enzyme *in vitro*, these observations suggest that inactivation occurs *in vivo* as well, resulting in diminished appearance of enzyme lost from the heart. Naturally, the extent to which transport of enzyme in lymph will influence time–activity curves of specific enzymes in plasma will be dependent upon the lability of the specific enzyme in lymph, the overall lymphatic efflux from the organ of interest, the duration of exposure of the enzyme to lymph, and the availability of alternative pathways for transport of enzyme from the injured organ into the systemic blood circulation.

REMOVAL OF ENZYME FROM THE SYSTEMIC BLOOD CIRCULATION

Removal of substances from the blood may result from uptake by cells such as those lining the sinusoids of the reticuloendothelial system. This process is believed to play a major role in removal of enzymes and will be discussed in detail. In the case of the enzyme molecules, inactivation by inhibitors in blood may occur also, in addition to removal. Inactivation might therefore influence results of enzymatic assays independent of changes in concentration of the enzyme protein. Transport of biochemical constituents is not necessarily confined to blood. It may involve transport in lymph or interstitial fluid as well, in which case the fate of the marker may be influenced by metabolic events in one of these compartments. Loss into biological fluids such as saliva, sweat, urine, and faeces does not appear to play a significant role in removal of enzymes such as LD, AST, or CK.

Other factors that may influence removal of proteins from blood include cardiac output, haemodynamics, body temperature, and metabolic rate. The primary emphasis in this discussion will concern compartmental interchange, since considerable confusion exists regarding this particular subject.

In the material to follow, we will consider factors influencing the decline of enzyme activity from blood after intravenous injection of purified enzyme in experimental animals or after release of enzyme into the circulation resulting from injury to the myocardium.[36] When referring to the absolute number of enzyme molecules in a defined volume, we will use the term 'concentration' (C)—recognizing that some of the molecules may be enzymatically inactive. The overall amount of enzyme will be symbolized by E. The terms 'pools', 'spaces', and 'compartments' will be used interchangeably, and distinctions will be made between fractional and absolute rates of transfer or disappearance.

On the basis of studies of several proteins, including clotting factors and enzymes, it appears likely that enzyme in the vascular space is in a dynamic equilibrium reflecting removal (and/or denaturation) of enzyme from plasma and resupply of the vascular space from extravascular compartments.[14] Although some have suggested that all enzymes are initially released into the circulation at approximately the same rate after injury to an organ *in vivo*, it appears likely that the rate of release may depend on specific physical properties of the individual enzymes, such as molecular weight. On the other hand, differences in the rate of release appear to be relatively minor determinants of enzyme time–activity curves in plasma when compared to the marked differences in rates of removal or inactivation. For example, even though isocitrate dehydrogenase (ICD) is released after experimental myocardial infarction, only transitory elevations of ICD in plasma are observed because of the rapid disappearance of enzyme activity from blood.[39] On the other hand, enzymes such as lactate dehydrogenase (LD) exhibit relatively slow rates of removal from the circulation and hence corresponding amounts of LD released from the heart result in sustained elevations of plasma LD activity.[9] In normal organisms, the activity of each enzyme in plasma remains remarkably constant from day to day because of the fine balance between the rate of release of enzyme into the circulation and the rate of removal. Activity at any given time is dependent on several parameters including:

(1) the rate of release of enzyme into the vascular and/or extravascular spaces or compartments;
(2) the volumes of distribution of each of these compartments;
(3) the rates of exchange between the compartments; and
(4) the rates of elimination of enzyme activity from vascular and extravascular pools.

Elimination of enzyme activity from the vascular space appears to be first order—that is to say that the fraction of the amount of material that is removed per unit time is constant. If one assumes that each molecule is at the same risk with respect to removal and that there is no interaction between molecules, a constant fractional rate of disappearance would be anticipated. Such a process results in a plasma time–activity curve conforming to a decaying exponential function. Mathematical treatment of this process can perhaps be best understood by considering the special case in which enzyme is distributed in a single compartment with constant volume V. The concentration of enzyme activity at any time t is $C(t)$, and $g(t)$ is the input function describing the rate of release of enzyme into the compartment. If removal of enzyme activity from the compartment is assumed to be first order with a fractional disappearance rate of a, then the rate of change of the enzyme concentration is described by the following differential equation:

$$\frac{dC(t)}{dt} = \frac{g(t)}{V} - aC(t) \tag{1}$$

The solution to this equation then depends on the form of the input function $g(t)$. If $C(t) = C$ is constant, then $g(t)$ is also constant since

$$\frac{g}{V} = aC \tag{2}$$

If however g is constant, but $C(t)$ is not, the solution to equation (1) is

$$C(t) = C(0)e^{-at} + \frac{g}{Va}(1 - e^{-at}) \tag{3}$$

where $C(0)$ is the concentration at time zero.

This formulation describing the behaviour of many circulating proteins assumes that the rate of release of enzyme into the circulation (g) and the fractional disappearance rate (a) are independent of one another and of the concentration of enzyme (C). The formulation also assumes that enzyme is distributed in only one compartment and that reentry of enzyme into the compartment does not occur after its initial removal.

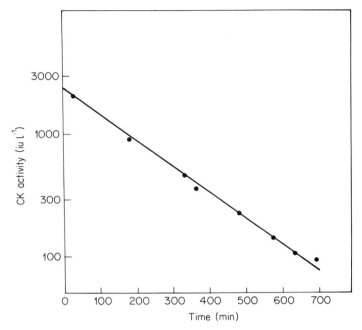

Figure 1. The time–activity curve of plasma CK after intravenous injection of purified canine myocardial enzyme in a conscious dog. The log of CK activity is plotted on the ordinate and the interval after injection on the abscissa

Based on this simple formulation, analysis of the disappearance of enzyme from blood after intravenous injection of purified enzyme is straightforward.[34] The fractional disappearance rate can be estimated from the slope of the plot of the log of enzyme activity as a function of time. Results of analysis of data from an experiment in a conscious dog are shown in Figure 1. Here, the log of plasma CK activity is plotted on the ordinate and time after intravenous injection of partially purified canine myocardial CK is plotted on the abscissa. The slope of the best-fit straight line to the data provides an estimate of the fractional disappearance rate. Assuming the distribution volume is constant and release of enzyme is zero, the concentration of enzyme is described by the exponential function,

$$C(t) = C(0)e^{-at} \qquad (4)$$

The removal rate may also be expressed in terms of the biological half-life $t_{\frac{1}{2}}$ where

$$t_{\frac{1}{2}} = \frac{\ln 2}{a} = \frac{0.693}{a} \qquad (5)$$

FALSE ASSUMPTIONS

Despite the appeal of this simplistic formulation, and despite the fact that correlation coefficients, exceeding 0.90 are obtained frequently when one analyses plots of the log of enzyme activity in plasma as a function of time after intravenous injection or release of enzyme from an organ undergoing injury,[27] it appears that most proteins entering the blood pool do not remain confined to the vascular space but rather that they are distributed in multiple compartments with exchange between the vascular and the extravascular pools in both directions.[14] The rates of exchange between compartments may be characteristic of each protein. In the case of enzymes, the amount of enzyme exchanged during any given interval may depend on the prevailing concentration gradients. These physiological factors add complexity to estimates of disappearance rates of enzyme from the blood after a single intravenous injection and to estimates of disappearance from time–activity curves of plasma enzyme activity *in vivo* associated with myocardial infarction.

One factor contributing to the complexity is readily apparent. Enzyme is not only being removed from the vascular compartment according to first-order kinetics described by the formula

$$E(t) = E(0)e^{-at} \tag{6}$$

but also disappearing because of transport into one or more extravascular compartments, presumably with characteristic fractional transfer rates. As enzyme begins to accumulate in the extravascular pools, it also begins to resupply the vascular pool—again with a characteristic transfer rate. Thus at all times transfer occurs simultaneously in both directions (between vascular and extravascular pools). At first, after intravenous injection of enzyme or release of enzyme directly into the vascular pool, net transfer to the extravascular compartment will occur. Subsequently, at some point in time, the extravascular pool will reach steady state and net transfer will be towards the vascular pool as enzyme disappears from this compartment. In addition, removal of enzyme from one or more extravascular pools may occur since catabolic processes may differ from compartment to compartment.

Considerable evidence implicates the functional importance of one or more extravascular compartments. For convenience, a single extravascular compartment has been envisaged in some instances, but in reality, the likelihood of multiple compartments is high. Both extracellular fluid and lymph appear to be biological pools represented by extravascular compartments in mathematical models pertinent to characterizing enzyme disappearance.

Characterization of rate constants describing the transport of enzyme between the vascular and extravascular compartments has been difficult. Perhaps of more importance, interpretation of plasma time–activity curves after myocardial infarction has been clouded because of the lack of consideration of the potential importance of multiple compartments. For example, estimates of disappearance rates of enzyme from the slope of the declining portion of plasma time–activity curves after myocardial infarction in conscious animals differ markedly from estimates based on multicompartmental analysis of the same data.[21,36] The main reason for the disparity is the lack of consideration of resupply of the vascular compartment, which results in a much slower net loss of enzyme activity from blood than that accounted for by the true removal rate.

Improved interpretations require analysis of disappearance rates of enzyme under conditions in which release of enzyme is virtually negligible. One approach entails experiments in which purified enzyme is injected intravenously in conscious animals, with the amount of enzyme injected markedly in excess of the amount of enzyme present in extracellular fluid, lymph, and blood. Analysis of data is facilitated when samples are obtained frequently during intervals in which plasma enzyme activity is changing rapidly. This frequently requires sampling as often as every minute during the first hours after intravenous injection of enzyme.

Figure 2 depicts the same disappearance curve as that shown in Figure 1, but in addition, it includes an analysis based on more frequent sampling of plasma after the intravenous injection of enzyme. Although the curve exhibits a biphasic form, it is not appropriate to assume that the rate of disappearance during the first portion of the curve reflects the true removal rate of enzyme, since enzyme is being transferred also to the extravascular compartment.

Estimates of distribution volume have been obtained by extrapolating enzyme disappearance curves to zero time and assuming that the concentration of enzyme activity denoted by the intercept reflects an initial concentration. Since one knows the amount of enzyme injected and one assumes the initial concentration to be defined, one can calculate the initial distribution volume by the dilution principle. However, extrapolation of the first portion of a biphasic curve such as that shown in Figure 2 to zero time (intercept of the curve) does not necessarily provide an accurate estimate of the distribution volume—even though this procedure has given rise to estimates close to plasma volume in several instances. Similarly, extrapolation of the second portion of the curve to zero time does not necessarily provide an accurate estimate of the distribution volume, although it does provide an estimate well in excess of plasma volume. Estimates of distribution volume based on extrapolation of the initial or terminal portions of such curves to zero time are spurious for the same reason that estimates of

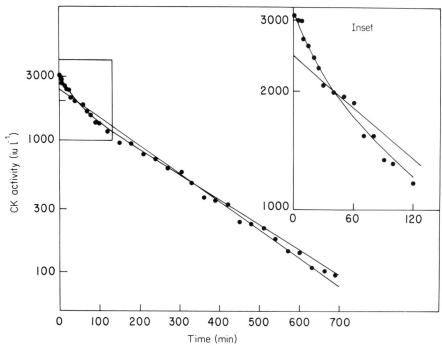

Figure 2. The time–activity curve of plasma CK after intravenous injection of purified canine myocardial CK in a conscious dog. The regression line of a single-exponential fit and the best-fit urve derived from the two-compartment model are shown. As can be seen in the inset, the curve based on a two-compartment model fits the early data much more closely than a regression line based on a single-exponential fit. [Reproduced by permission of the American Heart Association, Inc. from *Circulation Research*, **41**, 836–44 (1977).]

fractional disappearance rates based on only one or the other portion of the curve are spurious, if indeed the kinetics of the system reflect continuing exchange of enzyme between vascular and extravascular compartments. In fact, procedures using one phase of a multiexponential curve are appropriate in general only when the transfer rates between compartments are at least tenfold faster or slower than the removal (or metabolic clearance) rate. Calculation of the enzyme distribution volume from the second portion of a biphasic curve has been based on a model originally formulated by Sterling (for the special case in which removal is much slower than exchange between compartments).[38] This model assumes that the slow phase represents the metabolic clearance of enzyme from the vascular compartment. In this model $\hat{C}(0)$ is the intercept value obtained by extrapolating the slow phase back to zero and is therefore an estimate of the concentration that would have been present in the vascular space (V_v) if the exchange between V_v and the extravascular compartment (V_e) were extremely rapid. The ratio

$V_v/(V_e + V_v)$ can be calculated as follows (with V_e therefore apparent when V_v is known):

$$\frac{V_v}{V_v + V_e} = \frac{\hat{C}(0)}{C(0)} \qquad (7)$$

When the removal rate is slow with respect to the transfer rates, this model gives results that are close (within 5%) to the true values. However, when this approach is applied to inappropriate situations, it may lead to grossly inaccurate results. Thus one needs to have knowledge in advance (perhaps on the basis of biological considerations) of the relative rates of transfer and removal—a constraint that has unfortunately been disregarded too often with spurious results that appear to be 'reasonable'.

As discussed previously, both portions of the biphasic plasma enzyme time–activity curve after intravenous injection of enzyme represent removal of enzyme from the vascular compartment and resupply due to exchange between vascular and extravascular pools. The earlier portion of the curve reflects mainly removal (both irreversible and net transfer to the extravascular compartment); the latter portion represents both removal and net transfer from the extravascular pool (resupply of the vascular compartment). Accordingly, this portion of the curve reflects also effects due to resupply of the vascular space from the extravascular pools.

In Figure 3, $g(t)$ is the input function. $g(t)$ is of course zero when plasma enzyme activity is declining, in studies in which enzyme has been injected intravenously as a bolus. Characterization of enzyme disappearance requires estimation of λ_v, the catabolic rate of clearance from the vascular compartment; λ_{ev}, the transfer rate from the vascular to the extravascular compartment; λ_{ve}, the transfer rate from extravascular to vascular compartments; and the volumes, V_v and V_e, of the vascular and extravascular pools. If one assumes that catabolic clearance occurs only from the vascular space, then λ_e (not shown in Figure 3) can be neglected. This assumption seems reasonable for enzymes such as CK with relatively rapid rates of disappearance. In addition, we can also assume that λ_{ve} and λ_{ev} are constant fractional exchange rates based on the behaviour of many proteins. Accordingly, the amount of enzyme present in the vascular compartment, $E_v = C_v V_v$, and in the extravascular compartment, $E_e = C_e V_e$, as a function of time can be calculated from the following system of equations:

$$\frac{dE_v(t)}{dt} = \lambda_{ve} E_e(t) - (\lambda_{ev} + \lambda_v) E_v(t) + g(t) \qquad (8)$$

$$\frac{dE_e(t)}{dt} = \lambda_{ev} E_v(t) - \lambda_{ve} E_e(t) \qquad (9)$$

The solution to this system for the vascular pool under conditions in which CK is administered by bolus intravenous injection and $g(t)$ is the function

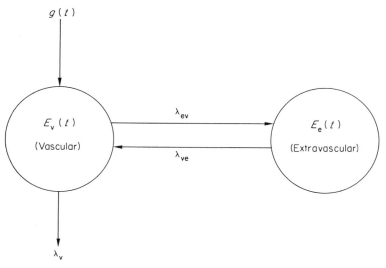

Figure 3. Representation of a two-compartment model pertaining to inactivation and clearance of enzyme. As noted in the text, $g(t)$ represents the input function, λ_v the catabolic rate of clearance from the vascular compartment, λ_{ev} and λ_{ve} the transfer rates between vascular and extravascular compartments, and $E_v(t)$ and $E_e(t)$ the concentration of enzyme in the vascular and extravascular compartments as a function of time. [Reproduced by permission of the American Heart Association, Inc. from *Circulation Research*, **41**, 836–44 (1977).]

accounting for maintenance of normal plasma CK activity is of the form:

$$E_v(t) = Ae^{-\alpha t} + Be^{-\beta t} + C \tag{10}$$

where C is the normal plasma CK level and the variables A, B, α, and β are functions of the rate constants and the initial conditions, i.e. the amount of CK present in the compartments at time zero. The four parameters of the solution can be obtained experimentally from double-exponential fits to the plasma CK time–activity curves following bolus injections. From these fits, the rate constants can then be calculated from the following formulae:

$$\lambda_v = \frac{\alpha\beta(A+B)}{A\beta + B\alpha} \tag{11}$$

$$\lambda_{ve} = \frac{A\beta + B\alpha}{A+B} \tag{12}$$

$$\lambda_{ev} = \frac{A\alpha + B\beta}{A+B} - \lambda_v \tag{13}$$

In special cases where the removal rate is either much slower or much faster than the exchange rates between compartments, there are approximations which can be used to simplify the expressions for the rate constants. In general however, one must use the above equation to compute the rate constants.[36]

Several assumptions are implicit in models underlying these calculations including: (i) λ_v is independent of $g(t)$ and of $E(t)$; and (ii) V_v and V_{ev} remain constant. Except in certain pathological states, marked changes in distribution volume probably do not occur. Whether λ_v is independent of $g(t)$ and $E(t)$ remains to be validated. However, since calculated k_d for CK based on a two-compartment model remained constant after frequently repeated intravenous bolus injections of CK with plasma activity increasing from 10- to 120-fold, the volume of distribution of the enzyme appeared to be independent of the input function.[36]

Studies with LD support the view that true rates of disappearance of plasma enzymes can be estimated more accurately by analysis based on multicompartmental models.[38] Analysis of data with the two-compartment model described above provides estimates of disappearance rates of enzyme from the circulation which are substantially greater than the observed rate of decline of activity in serial samples.[36] This is because the model takes into account continuing supply of the vascular pool with enzyme from the extravascular compartment. Comparisons of estimates of the CK disappearance rate obtained from single-compartment versus two-compartment models following injection of purified CK in conscious dogs are shown in Table 1. Since enzymatic estimates of infarct size depend on the true disappearance rate as one parameter, the two-compartment model would account for

Table 1. Comparison of CK Disappearance Estimated from One- and Two-compartment Models

Dog number	One-compartment model k_d (min^{-1})	Two-compartment model k_d (min^{-1})
1	0.005	0.006
2	0.006	0.009
3	0.005	0.006
4	0.007	0.009
5	0.005	0.012
Mean	0.005	0.008
Standard deviation	0.001	0.002

Estimates of the true catabolic disappearance rate of CK based on analysis of plasma CK curves in each of 5 conscious dogs given bolus injections of purified, canine myocardial CK intravenously. Results expressed are estimates based on analysis of the data with a one-compartment model (the slope of the regression line of log CK versus time) and those based on a two-compartment model with data analysed as outlined in the text. As can be seen, the standard deviations of the estimates are comparable, but the estimate of the true disappearance rate is substantially higher when results are analysed with a two-compartment model more CK released from the heart based on analysis of any given plasma curve than the one-compartment model.

VARIABILITY OF ENZYME ELIMINATION RATES

The disappearance rate (k_d) of a plasma enzyme is characteristic of each enzyme in the same organism. Unfortunately, comparison of enzyme disappearance rates estimated by different investigators is often difficult because the methods used to calculate k_d are seldom the same. Although most investigators recognize the biphasic nature of enzyme disappearance curves, they select the slope of either the early or late portion of the curve as an estimate of k_d rather than an estimate obtained from multicompartment analysis. In addition, the variation in elimination rate of the same enzyme from individual to individual adds complexity. For example, values obtained by Wakim et al.[11] for AST and LD elimination rates in dogs exhibit considerable variation. With purified canine myocardial CK-MM, we have found a range of variation of the observed elimination rate (k_e) in excess of 100% in conscious dogs.[27,34] Results from 10 different animals, shown in Table 2, illustrate the variation. However, k_e determined repetitively on consecutive days in the same animal varied by <10%. Although studies comparing the k_e after bolus injections of purified enzymes in man are not available, single-compartment analysis from the terminal portion of plasma CK time–activity curves after myocardial infarction indicate marked variation in CK elimination rates among patients. However, as in individual

Table 2. CK Disappearance (k_e) in Ten Conscious Dogs

Dog number	Half-life ($t_{\frac{1}{2}}$) (min)	k_e (min^{-1})
1	154	0.0045
2	187	0.0037
3	150	0.0046
4	178	0.0039
5	112	0.0062
6	161	0.0043
7	88	0.0078
8	105	0.0066
9	133	0.0052
10	110	0.0063

Estimates of the elimination rate of CK based on a one-compartment model in which the rate of disappearance was considered to be the same as the slope of the best-fit regression line of the log of CK activity versus time after intravenous injection of myocardial CK in 10 conscious dogs. As can be seen, there is a wide range of apparent elimination rate from animal to animal animals, the elimination rate in the same patient appears to be quite constant when evaluated on several occasions.[5,22]

Systematic studies in which the k_e of specific enzymes has been determined in large numbers of animals from several different species are not available. However, results in dogs and baboons underscore the wide range of variation among different species.[27,40]

Differences of k_e of isoenzyme forms of the same enzyme have been well documented for AST, LD, and CK.[12] Thus after bolus intravenous injections of purified LD isoenzymes 1, 2, 3, 4, and 5 in sheep,[5] average plasma half-lives of 48, 31, 20, 13, and 7 hours were observed. Marked variation has also been shown for k_e for cytoplasmic and mitochondrial transaminase[12] and for CK isoenzymes[26,27] with a mean elimination rate for CK-MM, CK-MB, and CK-BB in dogs of 0.0048, 0.0078, and 0.0100 min^{-1} in dogs[26] and analogous differences in baboons[40] and man.[27]

The plasma half-life of LD in man after myocardial infarction appears to be approximately 36 hours in contrast to 14 hours after liver damage.[15] This striking difference appears to be due to the different values for k_e associated with elevation of primarily LD_1 (slow elimination rate) and liver with primarily LD_5 which has a fast elimination rate. A similar explanation may apply to the generally lower plasma LD activity after skeletal muscle injury (with elevation of LD_5 and a rapid elimination rate) than that seen after myocardial necrosis (with increased LD, and a slow k_e). Lack of significant elevation of plasma CK-BB after cerebral injury may be due in part to its rapid k_e; and low plasma isocitrate dehydrogenase after myocardial infarction, despite substantial release from the heart, may result from the enzyme's extremely rapid rate of elimination.[39]

Variation in k_e among patients may account in part for different plasma enzyme time–activity curves observed after tissue necrosis and some variation in baseline plasma enzyme activity levels. Characteristic time–activity curves may be associated with the particular organ undergoing injury, reflecting the contributions of predominant isoenzymes of the organ with their characteristic elimination rates.[9]

FACTORS INFLUENCING PLASMA ENZYME DISAPPEARANCE RATES

Proper interpretation of plasma enzyme levels for diagnostic purposes and particularly for enzymatic estimation of infarct size requires consideration of factors influencing elimination rates. Haemodynamic alterations associated with disorders such as myocardial infarction could conceivably result in sustained elevations of plasma enzyme levels because of decreased clearance. However, in conscious dogs given purified myocardial CK intravenously, marked increases in heart rate (from 100 to 180 beats min^{-1}) induced by pacing or administration of isoproterenol did not influence CK clearance substantially. Furthermore, reduction of cardiac output by 67% by constriction of the inferior vena cava, reduction of coeliac flow by 70% with an occluder, or complete interruption of renal blood flow bilaterally did not affect CK clearance rate by more than 10%.[27]

Hepatectomy, bilateral nephrectomy, and splenectomy do not appear to affect the clearance of LD, AST, or ICD.[12,39] In conscious dogs given ^{14}C-labelled CK intravenously in whom k_e was determined from loss of ^{14}C-radioactivity from the blood, the CK clearance was virtually identical in the same conscious animal before, during, and after experimentally induced myocardial infarction.[31] Thus clearance of plasma enzymes does not appear generally to be sensitive to haemodynamic perturbations.

Another potentially important factor is the influence of drugs. In conscious dogs, conventional therapeutic doses of several drugs did not alter k_e after CK was injected intravenously. However, k_e did decrease in animals given large doses of 'Valium', morphine, and barbiturates.[31]

Corticosteriods do not appear to affect LD or CK clearance.[12,31] However, high doses of propranolol may decrease k_e in man.[6] Since results in these studies were based on single-compartmental analysis, it is not clear whether the drugs altered the true catabolic enzyme disappearance rate (k_d) or influenced k_e by altering intercompartmental transfer rates.

Hypermetabolic states such as hyperthyroidism and pregnancy are associated with decreased plasma CK activity,[35] and hypometabolic states such as hypothermia[17] and hypothyroidism[13] with high levels of plasma CK activity. Analogous effects on LD and AST activities are well recognized. Although it has been thought by some that the low CK levels associated with hyperthyroidism reflect inhibition of CK assays by thyroxine,[2] even high concentrations of thyroxine do not inhibit spectrophotometric assays[16] and experimentally induced hyper- and hypothyroidism are associated with low and high baseline plasma CK activity respectively. Furthermore, hyperthyroidism is associated with increased CK elimination rates and hypothyroidism with low rates. Thus altered baseline plasma CK levels appear to result from changes in the elimination rates *in vivo*.

MECHANISMS OF REMOVAL OF ENZYMES FROM THE CIRCULATION

The site and mechanism of removal of enzymes from the circulation remain enigmatic. Most enzymes appear to be catabolized, rather than excreted unchanged in significant quantities in the urine, saliva, bile, or gastrointestinal tract.[23] Amylase, which has a low molecular weight of 45,000 daltons, is an exception in that it is excreted via the kidney. However, other mechanisms may be predominant. The reticuloendothelial system appears to be involved in degradation of at least some plasma enzymes. Riley virus, which selectively infects the reticuloendothelial system of mice, causes marked elevations of plasma LD, ICD, MD, AST, CK, and amylase.[18] Reticuloendothelial system blockade induced with 'Thorotrast', cholesterol oleate, carbon, or zymosan is associated with decreased clearance of both purified and endogenous plasma enzymes.[11,12,23,47] In contrast, stimulation of the reticuloendothelial system with stilboestrol is associated with accelerated elimination.[19] However, clearance of the predominant cardiac LD isoenzyme (LD_1) and of alkaline phosphatase remain unchanged after reticuloendothelial blockade.[23] Interestingly, the enzymes whose clearance is most affected by reticuloendothelial blockade do not exhibit comparable changes in k_e after hepatectomy or decreased liver perfusion, suggesting that reticuloendothelial system reserve is such that the liver alone is not a critical determinant of k_e. It is well known that cells of the reticuloendothelial system and also parenchymal cells of most organs, particularly those with rapid protein turnover, can ingest large proteins.[33]

The disappearance rate of radioactively labelled and unlabelled rabbit LD_5 incubated with leukocytes *in vitro* is similar to k_e *in vivo*.[24] This suggests that LD may be degraded in the vascular pool *in vivo* by white blood cells, potentially explaining why removal of the liver or spleen does not affect the LD elimination rate markedly. Observed effects of zymosan may also be explained by its inhibitory effect on the binding or uptake of LD by white cells.

Enzymes released *in vivo* appear to often encounter interstitial fluid and lymph prior to entering the blood circulation.[20] Although the influence of lymph and interstitial fluid on the removal of plasma enzymes has been explored only recently, it may be substantial.[7,10,20,37] For example, selective occlusion of cardiac lymphatics in conscious dogs subjected to myocardial infarction is associated with a marked reduction of the amount of CK appearing in blood compared to that lost from myocardium. Since lymph inactivates CK rapidly *in vitro*,[7] these results suggest that inactivation *in vivo* may contribute to the relatively low proportion of enzyme lost from the heart accounted for by enzyme appearing in blood.[8]

It is not yet known whether enzyme disappearance reflects primarily inactivation, denaturation, or actual removal of enzyme molecules, in part because of the lack of assay systems capable of measuring the concentration of enzyme protein as opposed to enzymatic activity.[32] Results of some recent studies performed with radioactively labelled enzymes suggest that loss of enzymatic activity reflects primarily enzyme removal[37] although contrary results have been reported as well.[25] Unfortunately, radioactive labels are prone to exchange with other protein carriers, to separate from the labelled enzyme, or to recirculate attached to degradative products of the enzyme. Furthermore, radioactively labelled enzymes may not exhibit biological behaviour comparable to that of native molecules. To clarify the problem, we developed a radioimmunoassay for CK isoenzymes capable of measuring the concentration of plasma enzyme protein *per se*[29,32] (see Chapter 10). Results in preliminary studies indicate that the site of antibody binding to the CK isoenzyme is independent of the enzymatically active site. Antibody does not bind to dissociated subunits, and thus the intact molecule appears to be required for recognition.[29] The parallelism of CK enzymatic and radioimmunoassayable CK protein plasma time–activity curves after myocardial infarction in experimental animals and patients suggests that removal of enzyme protein is a major mechanism underlying observed elimination rates.[28,30]

The markedly different yet characteristic elimination rates for different molecular forms of the same enzyme suggest that mechanisms involved in removal are quite selective. They appear to involve receptors capable of detecting subtle differences in complex proteins.

FUTURE DIRECTIONS

We have recently shown that canine CK purified from myocardial tissue disappears two- to threefold faster *in vivo* than endogenously released cardiac CK in the same animal,[36] even though gross chemical differences between purified and endogenous enzyme have not been detectable. Clearance of some circulating proteins such as glycoproteins has been shown to be remarkably dependent on binding of a component masked by sialic acid to specific receptors in the liver.[1] The ultimate elucidation of regulation of clearance of enzymes in general may therefore depend in part on clarification of the subcellular locus involved in recognition of circulating material and identification of sites and processes involved in degradation. Elucidation of such mechanisms may be pertinent not only to improved diagnosis and estimation of the extent of tissue damage but also to potential treatment modalities for enzyme deficiency disorders, with delivery of selected enzymes to intracellular sites with the use of liposomes,[3] and with the need for prolongation of the biological life of the substituted enzyme *in vivo*.

CONCLUDING COMMENTS

Some mathematical approaches to determination of the distribution volume and disappearance of enzymes from the circulation along with consideration of some deficiencies of methods of analysis based on single compared to multicompartmental are presented. Disappearance of proteins such as enzymes depends in part on transfer rates between vascular and extravascular compartments reflecting diffusion characteristics and concentration gradients. Elimination rates for each enzyme vary not only among species but also among individuals. Mechanisms of removal of enzymes from the circulation have been explored with data documenting the relative lack of effect of marked haemodynamic perturbations, hepatectomy, or nephrectomy on the disappearance rates of several enzymes including CK, LD, and AST. Since factors influencing activity of the reticuloendothelium system such as viral infections and zymosan also alter elimination rates, it appears likely that this system plays a major role in processes of clearance. Whether disappearance of enzyme activity from the circulation reflects inactivation, denaturation, or removal of molecules is not yet clear, but the recent development of a CK isoenzyme radioimmunoassay should help to elucidate which of these processes predominates. Results of preliminary studies suggest that endogenous CK behaves somewhat differently from purified CK, possibly because of structural or conformational differences, with consequently altered recognition or affinity by receptors involved in removal. Quantification of enzyme clearance and delineation of mechanisms involved are important not only for improved enzymatic estimates of infarct size and damage in organs other than the heart but also to provide information needed for development of potential therapeutic approaches of enzyme deficiency states.

REFERENCES

1. Ashwell, G. and Morell, A. G. *Advances in Enzymology* (Ed. Alton Meister, Vol. 41, John Wiley and Sons, New York, 1974, pp. 99–128.
2. Askonas, B. A. The effect of thyroxine on creatine phosphokinase activity. *Nature*, **167,** 933–4 (1951).
3. Bangham, A. D. *Cell Membranes: Biochemistry, Cell Biology and Pathology* (Eds. Weissmann, G. and Clairborne, R.). H. P. Publishing Co., Inc., New York, 1975.
4. Batsakis, J. G. and Briere, R. O. *Interpretive Enzymology*, Charles C. Thomas, Springfield, 1967.
5. Boyd, J. W. The rates of disappearance of L-lactate dehydrogenase isoenzymes from plasma. *Biochimica et Biophysica Acta*, **132,** 221–31 (1967).
6. Cairns, J. A. and Klassen, G. Modification of acute myocardial infarction (AMI) by IV propranolol (P). *Circulation Supplement II*, **52,** II–107 (1975).
7. Clark, G. L., Roberts, R., and Sobel, B. E. The influence of creatine kinase (CK) transport in lymph on plasma CK curves after myocardial infarction. *Clinical Research*, **25,** 213A (1977).
8. Clark, G. L., Siegel, B. A., and Sobel, B. E. Qualitative external evaluation of regional cardiac lymph in intact dogs. *Physiologist*, **20,** No. 4, 17 (1977).
9. Doty, D. H., Bloor, C. M., and Sobel, B. E. Altered tissue lactic dehydrogenase activity after exercise in the rat. *Journal of Applied Physiology*, **30,** 548–51 (1971).
10. Feola, M. and Glick, G. Cardiac lymph flow and composition in acute myocardial ischemia in dogs. *American Journal of Physiology*, **229,** 44–8 (1975).
11. Fleisher, G. A. and Wakim, K. G. The fate of enzymes in body fluids—an experimental study. I. Disappearance rates of glutamic pyruvate transaminase under various conditions. *Journal of Laboratory and Clinical Medicine*, **61,** 76–85 (1963).
12. Fleisher, G. A. and Wakim, K. G. The fate of enzymes in body fluids—an experimental study. III. Disappearance rates of glutamic oxalacetic transaminase II under various conditions. *Journal of Laboratory and Clinical Medicine*, **61,** 98–106 (1963).
13. Griffiths, P. D. Serum enzymes in diseases of the thyroid gland. *Journal of Clinical Pathology*, **18,** 660–3 (1965).
14. Hermens, W. Th. *Handbook of Hemophilia* (Eds. Brinkhous, K. M. and Heinker, H. C.) American Elsevier Publishing Company, Inc., New York, 1975.

15. Hess, B. *Enzymes in Blood Plasma*, Academic Press, Inc., New York, 1963.
16. Karlsberg, R. P., Siegel, B. A., and Roberts, R. Effect of altered thyroid function on the disappearance of creatine kinase. *Clinical Research*, **25,** 297A (1977).
17. Maclean, D., Griffiths, P. D., and Emslie-Smith, D. Serum enzymes in relation to electrocardiographic changes in accidental hypothermia. *Lancet*, **2,** 1266–70 (1968).
18. Mahy, B. W. J. and Rowson, K. E. K. Isoenzyme specificity of impaired clearance in mice infected with Riley virus. *Science*, **149,** 756–9 (1965).
19. Mahy, B. W. J., Rowson, K. E. K., and Parr, C. W. Studies on the mechanism of action of Riley virus. IV. The reticuloendothelial system and impaired plasma enzyme clearance in infected mice. *Journal of Experimental Medicine*, **125,** 277–88 (1967).
20. Malmberg, P. Aspartate aminotransferase in dog heart lymph after myocardial infarction. *Scandinavian Journal of Clinical and Laboratory Investigation*, **30,** 153–8 (1972).
21. Markham, J., Karlsberg, R. P., Roberts, R., and Sobel, B. E. Mathematical characterization of kinetics of native and purified creatine kinase in plasma. In *Computers in Cardiology*, IEEE Computer Society, Long Beach, California, 1976.
22. Norris, R. M., Whitlock, R. M. L., Barratt-Boyes, C., and Small, C. W. Clinical measurement of myocardial infarction size. Modification of a method for the estimation of total creatine phosphokinase release after myocardial infarction. *Circulation*, **51,** 614–20 (1975).
23. Posen, S. Turnover of circulating enzymes. *Clinical Chemistry*, **16,** 71–83 (1970).
24. Qureshi, A. R. and Wilkinson, J. H. Inactivation *in vitro* of lactate dehydrogenase-5 in blood. *Annals of Clinical Biochemistry*, **14,** 48–52 (1977).
25. Qureshi, A. R. and Wilkinson, J. H. Rates of disappearance from plasma of enzymes labeled by coupling with a radioactive Iodo-ister. *Clinical Chemistry*, **22,** 1277–82 (1976).
26. Rapaport, E. The fractional disappearance rate of the separate isoenzymes of creatine phosphokinase in the dog. *Cardiovascular Research*, **9,** 473–7 (1975).
27. Roberts, R., Henry, P. D., and Sobel, B. E. An improved basis for enzymatic estimation of infarct size. *Circulation*, **52,** 743–54 (1975).
28. Roberts, R. and Painter, A. A. Radioimmunoassay for canine creatine kinase isoenzymes. *Biochimica et Biophysica Acta*, **480,** 521–6 (1977).
29. Roberts, R., Painter, A. A., and Sobel, B. E. The relation between clearance of creatine kinase enzyme activity and turnover of enzyme protein assessed by radioimmunoassay. *Clinical Research*, **25,** 249A (1977).
30. Roberts, R., Parker, C. W., and Sobel, B. E. Detection of acute myocardial infarction with a radioimmunoassay for MB creatine kinase. *Lancet*, **2,** 319–21 (1977).
31. Roberts, R. and Sobel, B. E. The effect of selected drugs and myocardial infarction on the disappearance of creatine kinase from the circulation in conscious dogs. *Cardiovascular Research*, **11,** 103–12 (1977).
32. Roberts, R., Sobel, B. E., and Parker, C. W. Radioimmunoassay for creatine kinase isoenzymes. *Science*, **194,** 855–7 (1976).
33. Schmidt, E. and Schmidt, F. W. Clinical enzymology. *FEBS Letters*, **62,** E62, (1976).
34. Shell, W. E., Kjekshus, J. K., and Sobel, B. E. Quantitative assessment of the extent of myocardial infarction in the conscious dog by means of analysis of serial changes in serum creatine phosphokinase activity. *Journal of Clinical Investigation*, **50,** 2614–25 (1971).
35. Sobel, B. E., and Braunwald, E. The cardiovascular system in thyrotoxicosis. In *The Thyroid*, 3rd edn., Harper and Row Publishers, Inc., New York (1971).
36. Sobel, B. E., Markham, J., Karlsberg, R. P., and Roberts, R. The nature of disappearance of creatine kinase from the circulation and its influence on enzymatic estimation of infarct size. *Circulation Research*, **41,** 836–43 (1977).
37. Sobel, B. E., Roberts, R., and Larson, K. B. Considerations in the use of bio-chemical markers of ischemic injury. *Circulation Research Supplement I*, **38,** I-99–I-106 (1976).
38. Sterling, K. Turnover rate of serum albumin in man as measured by ^{131}I tagged albumin. *Journal of Clinical Investigation*, **30,** 1228–37 (1951).
39. Strandjord, P. E., Thomas, K. E., and White, L. P. Studies on isocitric and lactic dehydrogenases in experimental myocardial infarction. *Journal of Clinical Investigation*, **38,** 2111–8 (1959).
40. Yasmineh, W. G., Pyle, R. B., Cohn, J. N., Nicoloff, D. M., Hanson, N. Q., and Steele, B. W. Serial serum creatine phosphokinase MB isoenzyme activity after myocardial infarction: Studies in the baboon and man. *Circulation*, **55,** 733–8 (1977).

CHAPTER 6

Tissue enzymes
A. F. Smith

INTRODUCTION	115
ENZYMES AS MARKERS OF TISSUE INJURY	115
ENZYMES IN TISSUES—DIFFERENTIAL TISSUE ENZYME DISTRIBUTION PATTERNS	117
Different Types of Cell Within an Organ	120
Location of Enzymes Within the Cell	120
Factors Affecting Tissue Enzyme Distribution Patterns	122
Source of Tissue	123
Technical Problems in Tissue Enzyme Analysis	123
Physiological Variations in Tissue Enzyme Activity	124
Effect of Disease on Tissue Enzymes	124
ENZYMES IN PLASMA	125
Normal Plasma Enzyme Activity	125
Causes of Increased Plasma Enzyme Activity	126
Relation Between Plasma and Tissue Enzyme Profiles	127
Selection of Enzyme Tests for Use in Diagnosis	129
REFERENCES	130

INTRODUCTION

The biochemist, when studying tissue enzymes, is principally interested in elucidating the mechanisms by which they act and in relating their properties and occurrence to their role in the overall metabolism of the cell. In the present context, however, enzymes are discussed mainly insofar as they may be of value as diagnostic tools. The presence of abnormal activities of enzymes in serum, or occasionally other fluids, is used to indicate whether or not tissue damage (or certain other changes) has occurred and to suggest which organ has been damaged. The kinetic, chemical, and physical properties of enzymes are not, in this diagnostic context, of primary importance, but are used analytically as aids to specific identification and quantitation. In this section, those properties of enzymes which render them particularly suitable as markers of cellular damage are discussed; some of the problems associated with their use for this purpose are also considered.

ENZYMES AS MARKERS OF TISSUE INJURY

There are numerous ways in which a tissue may be damaged by disease, but many of the cellular events which follow such damage show similar features. Therefore, although most of the events which will be described relate to ischaemia, which is the most common cause of myocardial injury, other sorts of damage (e.g. toxic) may be associated with similar changes.

In the early, reversible stages of ischaemia, the membrane pumps, which maintain the normal ionic gradient between the cell and its environment, fail and potassium and other ions leak out of the cell. Furthermore, overall disruption of normal metabolic interrelationships results in the subsequent accumulation of abnormal amounts of cellular metabolites, such as lactate and adenosine, which also leak out into the surrounding fluid. These changes which have been described in more detail in earlier chapters of this book, occur rapidly within minutes of the onset of ischaemia and, in theory, could allow very early detection of cellular damage. However, in practice, changes in serum concentration of ions or intermediary metabolites cannot be used in the intact organism to detect cellular injury since:

(i) Most of the substances are removed from the circulation very rapidly so that increases in serum levels are, even in otherwise favourable circumstances, only transitory.

(ii) The concentration gradient for these substances between cell and surrounding fluid is relatively shallow—often of the order of 10 or 100 to 1. Consequently, small amounts of damage would be undetectable.

(iii) The changes are not specific to any particular tissue.

The circumstances governing the release of macromolecules are different. They are released from the damaged cell at a later stage after the onset of injury, although the cell may not yet be irreversibly damaged. The increased efflux of macromolecules can presumably be attributed to a generalized increase in membrane permeability. Once they have been released into serum, however, most cellular proteins and probably other macromolecules, are removed from the circulation relatively slowly, having plasma half-lives of between a few hours and a few days. For a number of reasons, enzymes are a very suitable class of macromolecule to use as an index of tissue damage since:

(i) Many enzymes are present in the cytosol and, in general, cytosolic components tend to be released more quickly and more readily from damaged cells than other cellular components. Different circumstances attend the release of mitochondrial or membrane-bound enzymes: these differences may be of diagnostic use (see below).

(ii) High cell to plasma ratios for enzyme concentration (up to 100,000 to 1) are encountered. These high ratios mean that small amounts of tissue damage may be detectable by measuring serum enzyme concentration.

(iii) Simple assays are often available by which enzyme concentration can be determined by measuring catalytic activity. More recently, radioimmunoassay has also been used successfully.

(iv) Because of the functional differences between tissues they contain different amounts of various enzymes. Some enzymes are found almost exclusively in a single tissue. The release of such tissue-specific enzymes into plasma may allow localization of a disease process.

This last point is of fundamental importance in clinical enzymology and in the next section such differences in tissue enzyme patterns are discussed in more detail.

ENZYMES IN TISSUES—DIFFERENTIAL TISSUE ENZYME DISTRIBUTION PATTERNS

The approximate activity of several enzymes of diagnostic interest in human tissues is shown in Table 1 and Figure 1. Clearly the enzymes listed represent only a minute fraction of the total numbers present in cells. Some of the reasons why only a small selection of cellular enzymes have been found to be of diagnostic value will be considered later.

Certain functions, and the enzymes catalysing the corresponding metabolic pathways, are shared by almost all cells. The enzymes include those of the glycolytic pathway, the pentose shunt, the Krebs' cycle and of the mitochondrial electron transport chain. Although these enzymes may be present in the cell in fairly high concentrations, they are only of limited value as markers of cellular damage since they lack tissue specificity. This means that increases in serum activity of such widely distributed enzymes will not indicate which organ has been damaged—one of the most important features of an enzyme test.

Table 1. Activities of Some Enzymes of Clinical Importance in Human Tissues[4,7]

Tissue	GD	LD	Enzyme activity (iu/g tissue) MD	AST	ALT	CK
Liver	15	200	160	85	50	<10
Skeletal muscle	3	200	130	50	5	2000
Heart muscle	1	170	670	75	4	500
Renal cortex	3	160	150	15	3	20
CNS	<1	75	160	30	<1	200
Lung	<1	40	40	1	<1	10
Red blood cell	<1	50	40	1	<1	<10

GD—glutamate dehydrogenase.   CK—creatine kinase
ALT—alanine aminotransferase. AST—aspartate aminotransferase
MD—malate dehydrogenase.   LD—lactate dehydrogenase

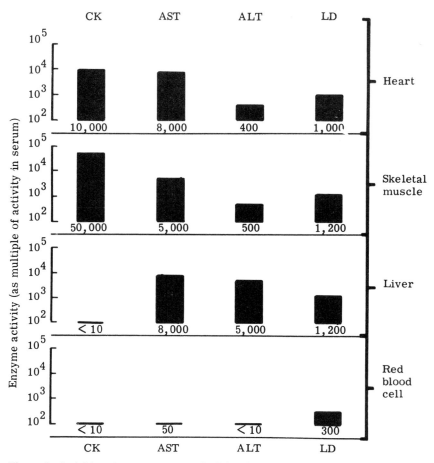

Figure 1. Activities of some enzymes of clinical importance in human tissues. Figures underneath each block represent the approximate cell/plasma ratio for the enzyme. Note the logarithmic scale on the ordinate

A minority of enzymes are tissue specific. This usually reflects a specialized function of the organ concerned. For example, enzymes of the urea cycle occur in liver but not elsewhere, so it is not surprising that an increase in serum ornithine carbamoyl transferase activity (one of the enzymes of the urea cycle) is a specific index of hepatocellular damage. Similarly, certain digestive enzymes, such as amylase, trypsin, and lipase, are confined to the pancreas and other organs closely related to the upper gastrointestinal tract. A rise in serum activity of these enzymes therefore provides strong evidence of pancreatic disease (although serum α-amylase activity may also rise in salivary gland disease and in certain other rather unusual circumstances).

In contrast, most of the enzymes used for clinical diagnosis are not tissue specific, but nevertheless do show wide variations in their prevalence in different organs. The predominant enzymes vary from tissue to tissue depending on the metabolic functions of the organ. Creatine kinase (CK), for example, is present in highest concentrations in skeletal muscle: this reflects the important role of CK in permitting creatine phosphate to act as a store of high-energy phosphate bonds which can readily be used to replete cellular ATP levels. Similarly alanine aminotransferase (ALT) is present in liver in much higher concentrations than in other tissues.

By considering combinations or profiles of these partially tissue-specific enzymes it is possible to improve specificity. To take a simple example: about equal activities of aspartate aminotransferase (AST) are present in the cytosol of heart, liver, and skeletal muscle, so measurement of AST activity in a tissue extract would not help distinguish these tissues from one another. However, the additional measurement of either CK or ALT, would readily allow distinction to be made between heart and skeletal muscle on the one hand and liver on the other (Figure 2).

It is important not to lay too much emphasis on relatively small differences between tissue enzyme profiles. Thus in the above example, although skeletal muscle characteristically contains four times the CK activity of heart, the use of profiles of enzyme activity might not always be able to distinguish the tissue reliably. This is due to the numerous factors which may

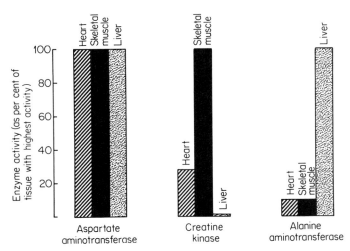

Figure 2. Differences in tissue activity of three of the enzymes used most commonly in clinical diagnosis

alter tissue enzyme activity. A more reliable method of distinguishing skeletal from cardiac muscle is to use isoenzyme analysis (see Chapters 7 and 10).

Most of the enzymes so far discussed are present in high activity in the cytosol. It has also been implicitly assumed that most of the enzymes released from a tissue originate from the predominant, usually specialized cell in that tissue. It is necessary, before considering the diagnostic implications of the differential distribution of enzymes in tissues, to discuss the occurrence of enzymes in the different cell types within a tissue and the intracellular distribution of enzymes.

Different Types of Cell Within an Organ

Most organs contain several different kinds of cell, although one cell type often predominates. Between cell types, the patterns of enzyme activity vary with the functions the cells perform. Thus the Kupffer cell has a different complement of enzymes from the hepatocyte. Most tissues contain fibrous and connective tissue components as well as more specialized cells but these supporting cells rarely cause diagnostic difficulty. This is partly because the supporting cells often contain relatively low concentrations of most of the enzymes used for clinical diagnosis and partly because such cells are often less affected by adverse conditions than the more specialized cells in a tissue.

The organs of major interest to the clinical enzymologist (liver, heart, muscle, etc.) are fairly homogeneous in the principal cell that they contain, so it is not possible to localize damage within the organ to a particular geographical or functional area. The kidney is an exception to this rule since the cells in different areas (glomerulus, proximal tubule, distal tubule, papilla) contain different concentrations of a number of enzymes. However, it has not proved possible to exploit these differences for diagnostic purposes.

Sometimes hyperplasia or neoplasia may cause an increase in the number of cells of a particular type in a tissue. The increased turnover of these cells may be reflected in an increase in plasma enzyme activity—of alkaline phosphatase in osteogenic sarcoma, for example. Similarly plasma acid phosphatase activity may be increased in Paget's disease and Gaucher's disease due to an increase in osteoclasts and Gaucher cells, respectively, in bone—both types of cell being relatively rich in acid phosphatase.

With a few exceptions, however, the presence of different cell types within a tissue rarely gives rise to diagnostic problems.

Location of Enzymes Within the Cell

Enzymes are distributed widely throughout the cell: they are present in the soluble cytoplasm, the nucleus, mitochondria, lysosomes, and other

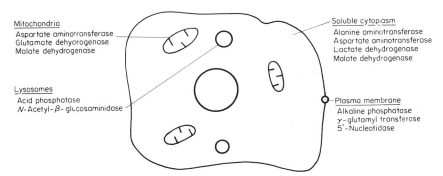

Figure 3. Location of selected enzymes, mostly of clinical importance, in the cell. The cell is illustrative only and has the enzyme characteristics of more than one tissue

intracellular organelles (Figure 3). The enzymes may be 'free' (i.e. not attached to any intracellular structure) or may be bound to membranes or other structures. In some cases, enzymes with similar catalytic activities may be present in more than one intracellular compartment—for example, there is malate dehydrogenase activity in both cytosol and mitochondria but the physical and chemical characteristics of the two enzymes are quite distinct from one another.

The majority of enzymes used diagnostically to demonstrate cellular damage are present mainly or entirely in the cytosol. Of the enzymes of primary importance in the detection of myocardial damage, all LD and about 90% of CK activity is present in the cytosol. Probably about half the AST activity of most cells is present in the cytosol, the remainder being present in mitochondria. In normal individuals, however, the serum AST arises from the soluble cytoplasm and not from mitochondria. The AST proteins from the two intracellular sites have different physical, biochemical, and kinetic properties and may be distinguished from one another fairly readily (see Chapter 7). These differences presumably reflect different functions that the enzymes perform in different parts of the cell—since the concentrations of substrates and products are likely to vary considerably within the different cellular subcompartments.

Mitochondria also contain a large number of enzymes, these include the Krebs' cycle enzymes and those catalysing the chain of electron transfer to molecular oxygen. Cytochrome oxidase is used extensively in research as a 'marker' for the presence of mitochondria in cell fractions but is not of value in diagnosis; the same applies to most, but not all, other mitochondrial enzymes. For example, glutamate dehydrogenase, which is also almost exclusively localized in mitochondria, has been found to be of value in the diagnosis of liver disease since its release from liver cells is particularly marked in patients with very extensive hepatocellular damage.[6] In general, the release of mitochondrial enzymes into serum implies a fairly severe form of cellular injury. However, the information provided is usually available from other indices of the severity of the disease and there are often technical difficulties in the measurement of mitochondrial enzymes, consequently they are rarely employed in routine diagnostic enzymology.

A microsomal component is also obtained by most fractionation techniques. A certain amount of confusion has resulted from this term. It should be emphasized that the 'microsome' is the product of an *in vitro* fractionation procedure and the resultant fraction contains a number of distinct subcellular structures. These include plasma membrane, outer mitochondrial membrane, Golgi complex, endoplasmic reticulum and, in some cases, sarcoplasmic reticulum. Therefore the enzyme actitivites exhibited by this fraction do not represent a single discrete intracellular organelle. The enzymes of clinical interest, particularly in liver cell homogenates, include alkaline phosphatase, 5′-nucleotidase, γ-glutamyl transferase, and leucine aminopeptidase (arylamidase). These enzymes are probably mainly located on plasma membranes and appear to be fairly firmly bound since solvents, such as butanol, or detergents may be required to extract them from tissue homogenates. Because these enzymes are so firmly bound to a cell component they are not released to any extent when the cell is damaged by anoxia. Serum levels rise in cholestatic liver disease, however, and such increases are mainly attributable to enzyme induction within the cell and regurgitation of enzyme into serum caused by the obstruction within the biliary tree. The above-mentioned enzymes are only present in heart in very small amounts and plasma membrane-bound enzymes are not of value in the diagnosis of myocardial disease except insofar as they may indicate the nature of secondary hepatic involvement resulting from circulatory or other changes.

The lysosomal enzyme acid phosphatase is of value in the diagnosis of prostatic carcinoma. This is due to the very high concentrations of acid phosphatase in prostate rather than to any inherent property of lysosomal enzymes in general. In animals, it has been shown that ischaemic damage to myocardium is associated with dissolution of lysosomes and release of lysosomal enzymes, such as N-acetyl-β-glucosaminidase, into the cytoplasm.[1] There is little infromation about lysosomal enzymes in the serum of patients with myocardial infarction.

Factors Affecting Tissue Enzyme Distribution Patterns

Although the distribution of many enzymes in normal human tissues is fairly well documented, there are sometimes discrepancies between the reported results. Most of these differences are attributable to the use of tissues from inappropriate sources or to analytical causes. In some cases, however, physiological or pathological factors may also cause variation in tissue enzyme activity.

Source of tissue

A number of sources have been used. Many original studies reported values obtained on animal tissues, which are readily available in adequate amounts and which can be obtained under optimal conditions. However, it is not valid to extrapolate results obtained for one species to other species, since there may be considerable differences between them. For example, uricase is present in most mammalian species but not in humans or higher apes. Of more direct relevance to cardiological diagnosis, in man about 30% of myocardial CK is present as a 'cardiospecific', or MB isoenzyme (see Chapter 7), whereas in the dog less than 5% of myocardial CK is present in a similar 'cardiospecific' form.

Autopsy material may also be unsuitable, as enzymes are inactivated fairly rapidly after death at rates depending on the lability of the enzyme and environmental conditions. The MB isoenzyme of CK tends to lose activity within a few hours of death, as do enzymes in tissues which readily undergo autolysis.

Samples from biopsies are better but may prove difficult to obtain or be too small for full analysis. There may also be ethical problems in obtaining biopsies from healthy tissue, for example normal myocardium.

Technical problems in tissue enzyme analysis

Difficulties arising from variation of results with different methods of tissue homogenization and extraction have already been mentioned. Purely analytical factors may also cause discrepant or misleading results, as there is likely to be interference with assay methods when extracts containing high concentrations of a large number of enzymes are being assayed. This is particularly so when assaying enzymes of relatively low substrate specificity or enzymes present in low activity, since interfering enzymes may remove substrates, activators, or inhibitors from the reaction mixture, thereby invalidating the assay.

In many other respects similar problems to those encountered with plasma enzyme analysis arise (Chapter 8). Thus some enzymes may require activation; for example, CK analysis requires the presence of sulphydryl groups in the reaction mixture if reliable results are to be obtained. In other instances more complex patterns of activation may be encountered: glutamate dehydrogenase only becomes maximally activated in the presence of both L-leucine and ADP. Some enzymes are very sensitive to the effects of heavy metal ions such as Ag^+ or Hg^+; chelating agents such as EDTA may have variable effects on enzyme activity, possibly attributable to the fact that they may remove both activators and inhibitors of the reaction. The above points illustrate the fact that it is often necessary to do a great deal of preliminary work before selecting suitable conditions for the assay of tissue enzyme activity.

Physiological variations in tissue enzyme activity

Altered physiological circumstances, due to changes in age, environment, or diet may also affect cellular enzyme activities. In the neonate, many enzyme systems still reflect the requirements of the foetal environment; for example, the enzyme responsible for bilirubin conjugation may be deficient at birth, especially in premature infants, and this may cause jaundice. In old age, tissue activities of many enzymes tend to fall below normal adult levels although some of this change may be more apparent than real reflecting only an increased proportion of fibrous elements in most tissues.

The effects of changes in environment have been studied more in bacteria than in man. In bacteria, many enzymes are induced by the presence of substrate and their synthesis may be repressed by the accumulation of products. Similar controls may operate in man; for example, the rate of formation of the rate-limiting enzyme in haem synthesis, δ-amino-laevulinate synthetase, is controlled by the concentration of the end-product, haem, in the cell. These mechanisms do not seem to play an important role in altering cellular activities of those enzymes which are used clinically to detect cellular damage. However, it is known that alcohol, certain drugs, and some hormones may induce synthesis of γ-glutamyl transferase in the liver, and as a result, there may also be a rise in the activity of the enzyme in serum.[3]

Genetic differences may account for variation between individuals. Cholinesterase and glucose-6-phosphate dehydrogenase have genetic variants of low enzyme activity which occur with varying frequency in different populations. However, no significant polymorphism of this type has been described for those enzymes currently used for the detection of myocardial damage.

Effect of disease on tissue enzymes

As well as acute ischaemic or toxic damage to cells, there are changes within a tissue which may occur more slowly or be less severe. Sometimes, as already mentioned, the proportions of various cell types in a tissue may be changed by disease. Thus there may be an increase in fibrous tissue in organs affected by ischaemia or an increase in inflammatory cells in infections. Obviously the tissue enzyme pattern will change towards that of the cell type which is increasing. In the heart, it has been suggested that the increase in LD_5 (see next chapter) which occurs in ischaemia and hypertrophy of the myocardium is due to the increased numbers of fibrous tissue cells which are present.[2] It is at least equally likely, however, that there has been a shift in the isoenzyme pattern within individual myocardial cells.

In the liver, cholestasis induces the synthesis of as much as a 10-fold increase in alkaline phosphatase concentration. Such dramatic increases in tissue enzyme concentration are not seen in skeletal or cardiac muscle.

Nevertheless, increased amounts of the so called 'cardiospecific' isoenzyme of CK may be present in skeletal muscle from patients with polymyositis or muscular dystrophy (see Chapter 7). Such changes may be of importance in the interpretation of serum enzyme results.

ENZYMES IN PLASMA

Normal Plasma Enzyme Activity

It is possible to distinguish several categories of enzyme present in plasma:

(i) Those fulfilling an essential role in the overall function of blood itself. These would include the enzymes responsible for blood clotting and fibrinolysis, and possibly other systems such as the renin/angiotensin system which are more difficult to classify. In a few cases the function of an enzyme secreted into plasma may not be certain, though it may be inferred. Thus cholinesterase may be concerned with the hydrolysis, in plasma, of substances which could interfere with neuromuscular function by inhibiting the acetyl cholinesterase present at nerve endings. Sometimes the enzymic activity of a serum protein may seem to be incidental to other functions: ceruloplasmin, a copper-carrying protein, can act as an oxidase for a variety of polyalcohols and polyamines, yet these functions do not appear to have any physiological role.

(ii) Those that are normally present in exocrine secretions. These are mainly the enzymes secreted by the gut mucosa or by exocrine organs related to the gut. Often the enzymes are secreted as inactive precursors, for example trypsinogen is the inactive precursor of trypsin.

(iii) Enzymes whose normal functions are exclusively intracellular. These enzymes, like those in the previous category, have no known function in blood and are, in general, those most used in clinical diagnosis as indices of various types of cellular damage. Their presence in the blood of normal individuals can be attributed to release from cells being replaced in the normal process of tissue wear-and-tear, cell turnover, or leakage of small amounts of enzyme from healthy cells.

(iv) Miscellaneous enzymes which do not fit readily into any of the above categories. Bone alkaline phosphatase, for example, is secreted by the osteoblast into the surrounding matrix.

In health, plasma enzyme activities remain relatively constant over long periods. This is the result of a steady-state equilibrium having been reached between:

(i) the rate of release of enzyme into blood;
(ii) the distribution of enzyme in the extracellular fluid; and
(iii) the rate of removal of enzyme from blood by excretion, inactivation, or degradation.

Causes of Increased Plasma Enzyme Activity

Any of the factors mentioned above which contribute to the normal steady-state condition may become disturbed by disease, thereby changing plasma enzyme activity. In practice, it is unusual for significant alterations in plasma enzyme activity to result from changes in their distribution in the extracellular fluid or from changes in the rate of their removal or inactivation. Increased plasma enzyme activity is usually caused by an increase in the rate of release from cells. This, in turn, may be caused by:

(i) Acute cell damage—such as that associated with ischaemia.

(ii) An increased rate of cell turnover—for example due to neoplasia. A similar effect is the increase in plasma alkaline phosphatase activity which occurs during periods of bone growth.

(iii) An increase in the intracellular concentration of an enzyme. Much of the increase in plasma γ-glutamyl transferase activity which occurs in patients being treated with barbiturates can be attributed to increased activity of the enzyme in liver cells.

(iv) Obstruction to the normal pathways of enzyme secretion. The increase in plasma amylase activity in pancreatic duct obstruction, although probably mainly due to associated cell damage, may be partly due to this factor. Similar considerations may partly account for the increase in plasma alkaline phosphatase activity which occurs in patients with cholestasis (although alkaline phosphatase is probably not normally actively secreted into bile).

After an episode of acute cell damage, not all intracellular enzymes are released at the same rate, nor to the same extent; as stated above, mitochondrial and membrane-bound enzymes are not released as readily as those present in the cytosol. In tissue perfusion experiments, it has also been shown that enzymes of low molecular weight are released more readily than larger molecules.[5] Enzymes which are unstable or are released slowly from damaged cells may, in fact, be inactivated or degraded by lysosomal enzymes before being released from the cell. Even after release from an injured cell, local inactivation of enzymes may occur before they reach the circulation.

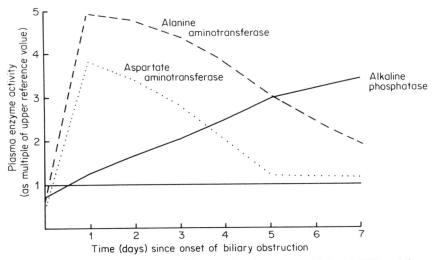

Figure 4. Appearance in plasma of the cytoplasmic enzymes, ALT and AST, and the plasma membrane enzyme, alkaline phosphatase, after an episode of biliary obstruction

Maximal plasma activities of enzymes normally present in the cytosol usually occur within a few hours of an episode of acute cell damage. The rate of increase in plasma activities of membrane-bound enzymes is usually much slower since the mechanisms causing the increase come into operation more slowly, and are not usually related to cell death. A typical course of events in a patient with cholestasis due to gallstones is shown in Figure 4. The rise in serum alkaline phosphatase activity occurs much more slowly than the rise in serum activity of the cytoplasmic enzymes AST and ALT. (This rise in aminotransferase activity is usually small since hepatocellular damage in patients with uncomplicated biliary obstruction due to gallstones is often minimal.)

Relation Between Plasma and Tissue Enzyme Profiles

For the reasons given above, even for enzymes present in the cytosol, the profile of enzymes entering blood is likely to differ from that present in the corresponding damaged organ. These differences become more marked after the enzymes have reached the blood stream, since they are excreted, inactivated, or transferred to another fluid compartment within the body at different rates (Table 2).

Redistribution within the extracellular fluid usually occurs fairly rapidly, smaller molecules leaving the intravascular compartment more readily than larger ones. After this phase, enzymes are removed from blood at an

Table 2. Plasma Half-lives and Molecular Weights of Some Enzymes used in Diagnosis (Approximate Values)

	Half-life (h)	Molecular weight
Alanine aminotransferase	50	110,000
Aspartate aminotransferase		
—cytoplasmic isoenzyme	20	120,000
—mitochondrial isoenzyme	6[a]	100,000
Creatine kinase		
—MM isoenzyme	20	80,000
—MB isoenzyme	10	80,000
Glutamate dehydrogenase	16	1,000,000
Lactate dehydrogenase		
—isoenzyme 1 (HHHH)	100	135,000
—isoenzyme 5 (MMMM)	10	135,000
Malate dehydrogenase		
—cytoplasmic isoenzyme	14	50,000
—mitochondrial isoenzyme	28	50,000

[a] Value obtained from animal experiments;[6] clinical data in humans suggests a longer half-life.

exponential rate which varies from enzyme to enzyme. Therefore, the plasma enzyme profile attributable to damage to an organ may change quite rapidly as those enzymes with longer half-lives (Table 2) in plasma become relatively more prominent. A patient in whom there had been an episode of acute hepatocellular damage might provide a hypothetical and oversimplified example. Shortly after the episode the relative serum activities of ALT, AST, and LD might resemble those in liver tissue (Figure 5) but after

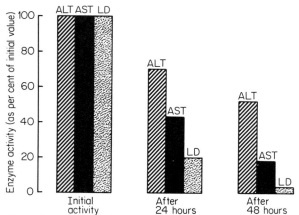

Figure 5. Example of how a plasma enzyme profile (similar to that in liver disease) might be expected to change with time. The rapid phase of distribution within the extra cellular fluid has been ignored and a single burst of enzyme release has been assumed for the sake of simplicity

2 days serum ALT activity remains raised whereas LD activity has returned to normal. These changing enzyme profiles attributable to the difference in plasma enzyme half-lives may be of value in diagnosis, since if such changes are interpreted correctly, they may provide evidence about the time at which tissue damage occurred.

In many patients, a further complication is also present if more than one organ has been damaged. This may be due to involvement in the primary disease process or may be secondary to disease principally affecting a single organ. For example, the liver may be damaged as a result of congestive cardiac failure in patients with myocardial infarction.

Selection of Enzyme Tests for Use in Diagnosis

In making this choice, the ability of each test to provide the answers to certain questions must be assessed. The first question 'Has tissue damage occurred?' might have the rider 'and how much damage?' added, since the answer is likely to have a bearing on both parts of the question. The basic requirement is for a sensitive test, so that small amounts of tissue damage may be detected. This will depend on:

(i) A high tissue/plasma enzyme activity ratio.
(ii) Ready release of enzyme after tissue damage.
(iii) Sufficient stability of enzyme activity in plasma for levels to remain raised for at least several hours.
(iv) The presence of relatively constant enzyme activity in healthy plasma. Obviously, if there are wide fluctuations in plasma enzyme activity in health, it will be more difficult to detect increases due to disease. This problem arises particularly with CK, since serum activity is affected by physiological factors such as exercise.

It may be more difficult to answer the second question 'Which organ has been affected?' since it is relatively uncommon for enzymes which are tissue specific to have the additional property of being sensitive indices of tissue damage. It has already been shown, however, that the relative distribution of enzymes in tissues of major clinical interest (Table 1) varies enough to allow a simple combination of enzyme tests to localize the site of damage in many patients, especially those in whom only a single organ has been affected. Unfortunately, patients often have damage to more than one organ and this may make the interpretation of results much more difficult. This is especially so in very ill patients, as the generalized disturbance tends to affect all tissues, with consequent release of enzymes from them. Similarly, after trauma or surgery, the extensive damage to muscle and other tissues may cause sufficient enzyme release to mask disease in organs not primarily affected by the trauma. This problem arises in the post-operative detection of myocardial infarction.

The third question is more practical—'Is it possible to measure activity of this enzyme reliably and conveniently?' Obviously, a suitable assay method must be available, and in most cases this will have to be relatively simple and cheap as well as being able to provide accurate and precise results. It is also necessary to ensure that the enzyme does not deteriorate under routine conditions of blood collection, transport, and storage.

The selection of enzymes to use, singly or in combination as a profile, in routine diagnosis is mainly based on clinical experience. If a group of enzyme tests to form a profile is required, then the enzymes will need to complement one another, and to be raised not only in the characteristic relationship to one another, but under different disease stimuli (if appropriate) and at different times in the course of the disease. By measuring plasma activities of a carefully selected combination of enzymes, useful information, additional to that obtainable from a single enzyme measurement, often becomes available:

(i) The enzyme profile may indicate the tissue which has been damaged.

(ii) The presence of relatively large amounts of mitochondrial enzymes suggests a more severe degree of cellular damage.

(iii) The pattern of enzyme results may indicate the likely stage which the disease has reached.

(iv) Unexpected patterns indicate the need to consider an alternative or additional diagnosis.

In clinical practice, enzyme profiles are used mainly in the diagnosis of liver and heart disease. Their restricted use may be attributed to a number of factors, but the principle reason is that in uncomplicated cases the result of a single enzyme test, taken in conjunction with clinical findings, is often sufficient to establish a diagnosis. In patients in whom there are complications, the use of enzyme profiles becomes less valuable as several organs are often affected by disease, and this may make interpretation difficult. The problems of lack of tissue specificity can, therefore, only partially be answered by the use of tissue profiles; isoenzyme analysis, discussed in the next chapter, is probably a more useful tool for this purpose.

REFERENCES

1. Hoffstein, S., Weissman, G., and Fox, A. L. Lysosomes in myocardial infarction: studies by means of cytochemistry and subcellular fractionation with observations on the effect of methyl prednisolone. *Circulation Supplement 1*, **53**, 34–40 (1976).
2. Revis, N. W., Thomson, R. Y., and Cameron A. J. V. Lactate dehydrogenase isoenzymes in the human hypertrophic heart. *Cardiovascular Research*, **11**, 172–6 (1977).

3. Rosalki, S. B. Gamma-glutamyl transpeptidase. *Advances in Clinical Chemistry*, **17,** 53–107 (1975).
4. Schmidt, E. and Schmidt, F. W. Enzym-muster menslicher Gewebe. *Klinische Wochenschrift*, **38,** 957–61 (1960).
5. Schmidt, E. and Schmidt, F. W. Release of enzymes from liver. *Nature*, **213,** 1125–6 (1967).
6. Schmidt, E., Schmidt, F. W., and Otto, P. Isoenzymes of malic dehydrogenase, glutamic oxaloacetic transaminase and lactic dehydrogenase in serum in disease of the liver. *Clinica Chimica Acta*, **15,** 283–9 (1967).
7. Smith, A. F. Separation of tissue and serum creatine kinase isoenzymes on polyacrylamide gel slabs. *Clinica Chimica Acta*, **39,** 351–9 (1972).

CHAPTER 7

Tissue isoenzymes
A. F. Smith and J. H. Wilkinson

INTRODUCTION	133
DEFINITION	133
STRUCTURAL BASIS OF ENZYME HETEROGENEITY	134
Hybrids of Two or More Polypepitide Chains	134
Structurally Unrelated, Genetically Independent Proteins	136
Genetic Variants (Allelozymes)	136
More Complicated Patterns of Heterogeneity	136
Polymerization of Enzyme Molecules	137
Formation of Complexes	137
ISOENZYMES OF IMPORTANCE IN CARDIOLOGY	138
Lactate Dehydrogenase Isoenzymes	138
Creatine Kinase Isoenzymes	139
Aspartate Aminotransferase Isoenzymes	141
CONCLUSION	142
REFERENCES	142

INTRODUCTION

It has long been known that certain enzymes occur in different forms in different tissues: the cholinesterase of the liver differs in its substrate specificity and its sensitivity to inhibitors from that associated with nervous tissue, and prostatic acid phosphatase may be distinguished from the erythrocytic enzyme by its marked inhibition in the presence of L-(+)-tartrate. More than 100 enzymes are known to occur in multiple forms which can be differentiated according to their physicochemical and catalytic properties. The heterogeneity of some enzymes, notably lactate dehydrogenase (LD) and creatine kinase (CK) is of considerable importance in the diagnosis of myocardial infarction and other cardiac conditions. This section is therefore devoted to a brief review of enzyme multiplicity, but for further information the reader is referred to specialist monographs.[7,18]

DEFINITION

The term *isozyme* (as opposed to isoenzyme) was first coined by Markert and Møller[8] in 1959 to describe the multiple forms of enzymes having similar catalytic properties, but for etymological reasons the term *isoenzyme* is generally preferred. The International Union of Biochemistry (I.U.B.) has approved both terms but has recommended that their use be restricted to those enzymes whose occurrence in multiple forms stems from genetic differences, and that the expression *multiple forms* be used as a more general term to include also those enzymes whose heterogeneity is due to association with variable amounts of charged moieties or to polymerization (see below). In clinical practice *isoenzymes* and *multiple forms* have been used interchangeably, but those most relevant to cardiac pathology, LD, CK, and aspartate aminotransferase (AST) are true isoenzymes as defined by the I.U.B.

STRUCTURAL BASIS OF ENZYME HETEROGENEITY

The more frequent ways in which different isoenzymes may arise are given below—with details of some of the enzymes of clinical interest as examples.

Hybrids of Two or More Polypeptide Chains

Lactate dehydrogenase provides the classical example of this type of isoenzyme heterogeneity. The LD of most human and animal tissues is separable by chromatography or electrophoresis into five components, the proportions of which vary from tissue to tissue. The enzyme molecule (mol. wt. 135,000 daltons) is a tetramer comprising four subunits, each with a mol. wt. of about 35,000 daltons. The subunits are of two types, H (heart) and M (muscle), which can undergo either homopolymerization to give HHHH or MMMM, or heteropolymerization to give the hybrid isoenzymes, HHHM, HHMM, and HMMM (Figure 1). The H subunit contains higher proportions of aspartate and glutamate than the M subunit which is relatively richer in arginine and methionine. Consequently HHHH has the greatest negative charge and MMMM is the most electropositive, and the five isoenzymes are readily separable by electrophoresis. Electrophoretically separated isoenzymes are numbered in order of their mobilities towards the anode, the fastest migrating HHHH being numbered LD_1 and the slowest (MMMM), LD_5.

Creatine kinase exhibits a similar type of heterogeneity, but in this case the enzyme molecule is a dimer comprising two types of subunit, M (muscle) and B (brain), and consequently only a single hybrid isoenzyme can be formed. As with LD, the proportion of each of the isoenzymes varies from tissue to tissue. On electrophoresis at pH 8.6 the MM isoenzyme remains near the point of application while the BB form migrates towards the anode with about the mobility of albumin (Figure 2). The MB isoenzyme has an intermediate mobility.

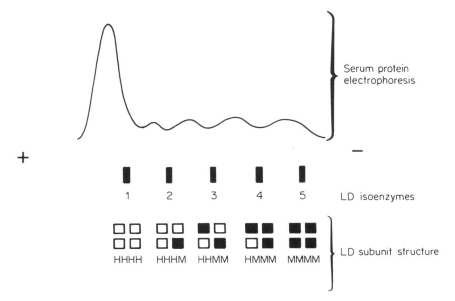

Figure 1. Isoenzymes of lactate dehydrogenase, showing polypeptide chain composition and electrophoretic mobility relative to serum proteins

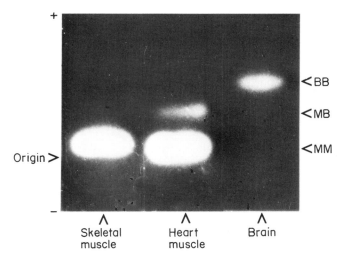

Figure 2. The separation on starch gel electrophoresis of creatine kinase isoenzymes in human brain, heart, and skeletal muscle (kindly donated by Dr. G. R. Doran)

A similar but more complex example of an enzyme which owes its heterogeneity to subunit combinations is provided by aldolase. Three quite distinct forms, A, B, and C predominate in skeletal muscle, liver, and brain respectively, but the A isoenzyme can produce sets of three hybrid isoenzymes with either the B or C form, thus suggesting that each isoenzyme is a tetramer.[1]

Structurally Unrelated, Genetically Independent Proteins

Aspartate aminotransferase occurs in cells in both the cytosol and the mitochondria. Studies of its amino acid composition and sequence, however, suggest that the two forms are quite different proteins.[9] It is not surprising, therefore, that they differ in their electrophoretic mobilities and other properties. Rather, it is remarkable that two dissimilar proteins should have similar catalytic activities.

Genetic Variants (Allelozymes)

The polymorphism shown by glucose-6-phosphate dehydrogenase in red cells provides a classic example of this type of isoenzyme variation caused by the presence of different alleles which may occupy the same gene locus. The isoenzyme variants, which are closely related to one another structurally, usually have slightly different electrophoretic mobilities; in some of the variants the enzyme protein has abnormally low enzymic activity. Patients with such isoenzymes of low enzyme activity may have haemolytic anaemia, such as that associated with the antimalarial drug primaquine or with favism. The atypical plasma cholinesterase variant associated with scoline apnoea provides another example of this type of isoenzyme variation.

More Complicated Patterns of Heterogeneity

Alkaline phosphatase offers an example of an enzyme which occurs in a different form in each tissue in which it is present. These differences are primarily due to differences in the primary structure of the enzyme molecule. In some cases the differences between the isoenzymes may be relatively small. Thus although alkaline phosphatase from liver differs from the isoenzyme from bone in stability to heating and in several other properties, antisera to one will generally cross-react with the other. In other instances the isoenzymes differ more markedly, the liver and intestinal isoenzymes, for example, do not seem to share so many common antigenic determinants as do the liver and bone isoenzymes. Differences between the isoenzymes, especially in electrophoretic mobility, can also be attributed to variation in the number of neuraminic acid residues attached to the molecule. When preparations of various tissues' alkaline phosphatase isoenzymes are treated with neuraminidase prior to electrophoresis, the anionic mobilities of all forms, except that from the intestinal mucosa, are significantly reduced.

Polymerization of Enzyme Molecules

After electrophoresis on media which have a molecular sieving effect, e.g. starch gel or polyacrylamide gel, cholinesterase preparations may show 5 to 7 distinct forms (although one of the forms predominates). If resubjected to electrophoresis immediately after elution from the gel, each form maintains its electrophoretic mobility, but after storage at room temperature for a few days, interconversion of the isoenzymes, especially to the predominant form, is apparent. This phenomenon has been attributed to reversible polymerization,[5] since this type of heterogeneity cannot be demonstrated on electrophoretic media which have no sieving effect.

Formation of Complexes

There are a number of special circumstances in which additional bands of enzyme activity, which are not present in tissues, appear in serum. Some of these bands are due to the occurrence of small conformational changes in the enzyme molecule or to changes, such as polymerization, which tend to occur under certain storage conditions. In other instances, the enzyme is present in serum in the form of a complex. For example, both alkaline phosphatase and LD may be present in serum bound to IgG or to IgA, AST may be bound to α_2-macroglobulin. In some patients with cholestatic liver disease, a different sort of complex is formed: alkaline phosphatase and other enzymes associated with the biliary canaliculus circulate in a particulate form—apparently each particle represents a fragment of the canalicular membrane. Because of their size, these aggregates remain at the origin on electrophoresis on media with a sieving effect.

Although the latter two categories of electrophoretic variants which show separate bands of enzyme activity cannot be regarded as true isoenzymes, it may at first be impossible to establish their true nature and the term isoenzyme may be used for want of a better description.

Of the enzymes so far cited as examples, CK, AST, and LD find widespread use in the diagnosis of heart disease, and isoenzyme studies add considerably to the value of CK and LD measurement. These enzymes are therefore considered in more detail below. Neither cholinesterase nor alkaline phosphatase is of primary importance in heart disease but they may be of value in the detection of secondary or coincidental damage in other tissues, e.g. hypotensive damage to the liver.

ISOENZYMES OF IMPORTANCE IN CARDIOLOGY

Lactate Dehydrogenase Isoenzymes

Different human tissues can broadly be classified into three groups according to whether their LD complements consist predominantly of the electrophoretically fast migrating isoenzymes, LD_1(HHHH) and LD_2(HHHM), the electrophoretically slow forms, LD_4(HMMM) and LD_5(MMMM), or those of intermediate mobility, mainly LD_3(HHMM). The cardiac enzyme, like that of the erythrocytes, brain, and kidney, belongs to the first group and is easily distinguished by electrophoresis from the liver and skeletal muscle enzymes which consist almost exclusively of the slow components (Figure 3). The main isoenzyme found in the leucocytes, adrenal, thyroid, and other tissues is LD_3.

It has been suggested that the different properties of the LD isoenzymes which occur in heart, muscle, and other tissues may be of functional significance. Thus the principal LD isoenzyme present in heart (HHHH) is maximally active in the presence of relatively low concentrations of pyruvate and is inhibited by excess pyruvate. On the other hand, the major isoenzyme present in most muscles (MMMM) requires a higher pyruvate concentration for maximal activity and is inhibited less by excess pyruvate. These kinetic properties seem to match the metabolic conditions in the two tissues. Heart, because it is not subject to sudden large variations in energy consumption, metabolizes carbohydrates and fatty acids at a relatively constant rate, with complete oxidation of metabolic intermediates such as pyruvate—therefore tissue concentrations of pyruvate and lactate remain low. Muscle, on the other hand, is subject to sudden bursts of activity in which the anaerobic metabolism of glycogen stores to form lactate is required to meet immediate energy requirements: in these circumstances, tissue pyruvate and lactate concentrations rise. In this context, it is of interest that some of the deeper-seated muscles, which are mainly concerned with postural control

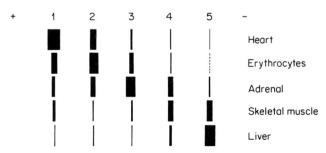

Figure 3. Distribution of lactate dehydrogenase isoenzymes in human tissues

and show a relatively constant state of tonic activity, have an isoenzyme pattern which tends to be similar to that of myocardium. As a generalization, therefore, it has been suggested that the presence of M- rather than H-type subunits represents an adaptive mechanism for the enhanced use of anaerobic glycolytic pathways.[2]

Although the teleological arguments presented above have been challenged, they are supported by results which show that the synthesis of M-type subunits of LD by tissue culture cells is partly regulated by the oxygen tension in the surrounding medium.[3] The isoenzymic pattern in the kidney may also be of relevance in this context, since the cortex which is well supplied with oxygen contains mainly isoenzymes rich in H subunits whereas the relatively poorly oxygenated renal papilla tends to contain LD isoenzymes comprising a higher proportion of M subunits.

Alterations in tissue isoenzyme patterns may also accompany disease. For example, analysis of human autopsy material has shown that patients with cardiac hypertrophy or coronary artery disease have a relative increase in those LD isoenzymes containing the M subunit.[10]

Changes in isoenzyme patterns may also occur in tissues which develop malignant change, or even in apparently normal tissue in the immediate vicinity of a neoplastic area.[6] Most neoplasms tend to show a relative increase in LD_5 activity. There are many examples of this type of change; in general, the isoenzyme pattern in such tumours tends to alter towards that prevailing in the tissue during foetal development although increases in LD_5 may merely reflect the increased glycolysis which occurs in tumour tissue.

Since LD is found in the cytosol of all human cells, an increase in the total serum enzyme activity is of little diagnostic specificity. However, the introduction of isoenzyme electrophoresis, or appropriate non-electrophoretic tests (Chapters 8, 9, 10) has greatly enhanced the value of this enzyme test, since the isoenzyme composition of the damaged tissue is partly reflected in the serum isoenzyme pattern. Thus after an episode of myocardial infarction, increased LD_1 activity is found in the serum, whereas in acute liver disease the serum LD_5 activity is disproportionately increased. The relative amounts of the two isoenzymes in plasma, however, may be greatly affected by the much shorter half-life of LD_5 than LD_1.[19] Although a predominant increase in serum LD_1 activity is also seen in megaloblastic and haemolytic anaemias, haematological investigations can usually clarify the diagnosis without difficulty.

Creatine Kinase Isoenzymes

The three CK isoenzymes present in the cytosol, MM, MB, and BB, occur in different proportions in different tissues in a manner analagous to the distribution of LD isoenzymes. In early foetal life, CK-BB predominates in

Table 1. Approximate Concentration of CK and its Cytoplasmic Isoenzymes in Human Tissues

iu g^{-1}	CK activity (iu g^{-1})	Isoenzyme composition		
		MM (%)	MB (%)	BB (%)
Skeletal muscle	2000	100	0	0
Cardiac muscle	500	70	30	0
Brain	200	0	0	100
Intestine	150	5	0	95
Bladder	150	5	5	90
Uterus	50	3	2	95
Thyroid[a]	30	15	0	85
Kidney[a]	20	10	0	90
Lung[a]	10	25	0	75
Prostate	10	3	2	95

[a] Conflict of reports about isoenzyme composition.

all tissues but during muscle development CK-BB is replaced, first by both CK-MB and CK-MM and finally, at about the sixth month of foetal development, by the MM isoenzyme alone. In the adult, three main patterns of CK isoenzyme distribution may be distinguished (Table 1).[4,13,17]

(i) Skeletal muscle. Tissue total CK concentration is very high—about 50,000 times than in normal serum. The MM isoenzyme predominates with MB and BB contributing less than 1% each to total CK activity.

(ii) Cardiac muscle. The concentration of CK is about one quarter of that in skeletal muscle and, in addition to the MM isoenzyme which predominates, between 25% and 35% of CK activity is attributable to the MB isoenzyme.

(iii) Brain. In the central nervous system total CK activity is about one half that in cardiac muscle, only the BB isoenzyme occurs. A number of other organs usually with less total CK activity than brain have a similar isoenzyme pattern, although small but variable amounts of MM and MB isoenzymes are also present. These tissues include intestine, bladder, uterus, thyroid, kidney, lung, and prostate.

Over 95% of normal serum CK activity is of the MM type.

The tissue distribution listed above suggests that the myocardium is the only tissue which contains significant amounts of CK-MB. If this were so, CK-MB would be an ideal tool for studying heart damage since diseases of other organs would not give rise to increased serum CK-MB activity. Unfortunately, the specificity of CK-MB for heart is not absolute and it is necessary to make some qualifications.

Firstly, there is evidence that some muscles, possibly those containing a high proportion of 'red' fibres, may have an isoenzyme pattern closer to that of cardiac than skeletal muscle.[11,13] However, the data about CK isoenzyme distribution in different muscles is conflicting, since other investigators have been unable to demonstrate CK-MB in any healthy skeletal muscles.[4,17]

Secondly, there is no doubt that the proportion of CK-MB in muscle rises considerably in muscular dystrophy—particularly of the Duchenne type.[15,16] Similar changes may occur in polymyositis, with 10% to 20% of muscle CK present in the MB form. It has been postulated that this change represents a reversion towards the more primitive foetal pattern, or alternatively that it is merely indicative of regenerating muscle fibres. The serum isoenzyme pattern also reflects these changes, so it is important that elevated serum CK-MB in these patients should not necessarily be attributed to the involvement of myocardium in the disease process.

Throughout this chapter, the general principle has been advanced that, where an organ is diseased or damaged, the isoenzymes appearing in the plasma will be of the type found in the damaged tissue. Creatine kinase offers a number of apparent contradictions to this principle. For example, although brain contains only CK-BB, cerebral infarction is characteristically associated with an increased serum CK activity which is due to the MM isoenzyme. The thyroidal CK is also of the BB type but increased CK activity in the serum of patients with hypothyroidism is due to the MM isoenzyme. The likely explanation of these findings is that there is concomitant muscle damage or change in muscle cell permeability in these disorders. In practice, it is very unusual to find serum CK-BB activity increased.

In addition to the isoenzymes which occur in the cytosol, mitochondrial and myofibrillar CK isoenzymes have also been described. However, since they have not been reported to occur in plasma in either health or disease, they will not be discussed further.

Aspartate Aminotransferase Isoenzymes

Those tissues which contain high concentrations of aspartate aminotransferase, such as heart, liver, and skeletal muscle, all have both cytoplasmic and mitochondrial forms of the isoenzyme. The mitochondrial isoenzyme is often present in higher concentrations within the cell but this is not reflected in the pattern in serum, since the circulating enzyme consists almost entirely of the cytosolic (anionic) form. Differences between the cytosolic forms from various tissues have not been demonstrated, so it is not possible to differentiate the cytosolic AST from liver, for example, from that from myocardium. The mitochondrial isoenzymes from different tissues also appear to be similar.

On *a priori* grounds it might be expected that the release of both mitochondrial and cytoplasmic AST would reflect cell necrosis, whereas lesser degrees of cell damage might be associated with release of the cytosolic, but not with mitochondrial, isoenzymes. In liver disease it has been possible to make this distinction since cases of acute hepatitis and acute toxic damage show relatively marked increases in mitochondrial AST compared with the much smaller increases, relative to cytoplasmic AST, seen in chronic liver disease.[12]

In myocardial infarction serum levels of the mitochondrial, as well as the cytoplasmic, isoenzyme of AST are raised (Chapter 9), but AST isoenzyme studies have not yet found widespread application as diagnostic aids. Nevertheless there is some evidence that mitochondrial enzymes, particularly AST, may provide a better index of the severity of the episode than cytoplasmic enzymes.[14]

CONCLUSION

The occurrence of isoenzymes of CK and, to a lesser extent, LD which are relatively cardiospecific is of great value in diagnostic cardiology. By the development of appropriate tests for these more specific isoenzymes, the diagnosis of myocardial damage can be made with more confidence than by the measurement of total enzyme activity alone.

REFERENCES

1. Baron, D. N., Buck, G. M., and Foxwell, C. J. Multiple forms of aldolase in mammalian brain, gonads, fetal tissue and muscle. *Advances in Enzyme Regulation*, **7**, 325–36 (1969).
2. Dawson, D. M., Goodfriend, T. L., and Kaplan, N. O. Lactic dehydrogenase: functions of the two types. *Science*, **143**, 929–33 (1967).
3. Goodfriend, T. L., Sokol, D. M., and Kaplan, N. O. Control of synthesis of lactic acid dehydrogenases. *Journal of Molecular Biology*, **15**, 18–31 (1966).
4. Jockers-Wretou, E. and Pfleiderer, G. Quantitation of creatine kinase isoenzymes in human tissues and sera by an immunological method. *Clinica Chimica Acta*, **58**, 223–32 (1975).
5. La Motta, R. V., McComb, R. B., Noll, C. R., Wetstone, H. J., and Reinfrank, R. F. Multiple forms of serum cholinesterase. *Archives of Biochemistry and Biophysics*, **124**, 299–305 (1968).
6. Latner, A. L., Turner, D. M., and Way, S. A. Enzyme and isoenzyme studies in preinvasive carcinoma of the cervix. *Lancet*, **ii**, 814–6 (1966).
7. Markert, C. L. (Ed.) *Proceedings of International Conference on Isoenzymes*, Vols. 1–4, Academic Press, New York and London, 1974–75.
8. Markert, C. L. and Møller, F. Multiple forms of enzymes: tissue, ontogenetic and species specific patterns. Proceedings of the National Academy of Sciences (USA), **45**, 753–63 (1959).
9. Martinez-Carrion, M. and Tiemeier, D. Mitochondrial glutamate-aspartate transaminase. 1. Structural comparison with supernatant isozyme. *Biochemistry*, **6**, 1715–22 (1967).
10. Revis, N. W., Thomson, R. Y., and Cameron, A. J. V. Lactate dehydrogenase isoenzymes in the human hypertrophic heart. *Cardiovascular Research*, **11**, 172–6 (1977).
11. Rosalki, S. B. Creatine phosphokinase isoenzymes. *Nature*, **207**, 414 (1965).

12. Schmidt, E., Schmidt, F. W., and Otto, P. Isoenzymes of malic dehydrogenase, glutamic oxaloacetic transaminase and lactic dehydrogenase in serum in disease of the liver. *Clinica Chimica Acta*, **15,** 283–9 (1967).
13. Smith, A. F. Separation of tissue and serum creatine kinase isoenzymes on polyacrylamide gel slabs. *Clinica Chimica Acta*, **39,** 351–9 (1972).
14. Smith, A. F., Wong, P. C.-P., and Oliver, M. F. Release of mitochondrial enzymes in acute myocardial infarction. *Journal of Molecular Medicine*, **2,** 265–9 (1977).
15. Somer, H., Dubowitz, V., and Donner, M. Creatine kinase isoenzymes in neuromuscular disease. *Journal of Neurological Sciences*, **29,** 129–36 (1976).
16. Takahashi, K., Shutta, K., Matsuo, B., Takai, T., Takao, H., and Imura, H. Serum creatine kinase isoenzymes in Duchenne muscular dystrophy. *Clinica Chimica Acta*, **75,** 435–42 (1977).
17. Tsung, S. H. Creatine kinase isoenzyme patterns in human tissue obtained at surgery. *Clinical Chemistry*, **22,** 173–5 (1976).
18. Wilkinson, J. H. *Isoenzymes.* 2nd ed., Chapman and Hall, London, 1970.
19. Wilkinson, J. H. and Qureshi, A. R. Catabolism of plasma enzymes, as studied with ^{125}I-labelled lactate dehydrogenase-1 in the rabbit. *Clinical Chemistry*, **22,** 1269–76 (1976).

CHAPTER 8

The measurement of enzymes

D. W. Moss

INTRODUCTION	145
PROGRESS OF ENZYMIC REACTIONS	147
EFFECT OF ENZYME CONCENTRATION ON RATE OF CATALYSED REACTION	147
The Enzyme–Substrate Complex	149
EFFECT OF SUBSTRATE CONCENTRATION ON RATE OF CATALYSED REACTION:	
THE MICHAELIS–MENTEN EQUATION	151
Consecutive Enzymic Reactions	158
Two-substrate Reactions	160
INHIBITION AND ACTIVATION OF ENZYME ACTIVITY	165
Inhibition of Enzyme Activity	165
Characteristics of Irreversible Inhibition	166
Kinetics of reversible inhibition	166
Competitive Inhibition	166
Non-competitive Inhibition	167
Uncompetitive Inhibition	168
Mixed Inhibition	169
Inhibition by reaction products	169
Activation of Enzymes	170
Coenzymes and Prosthetic Groups	172
EFFECT OF pH ON ENZYMIC REACTIONS	173
EFFECT OF TEMPERATURE ON ENZYMIC REACTIONS	174
METHODS OF MEASURING CHEMICAL CHANGES IN ENZYMIC REACTIONS	176
Automation of Enzyme Activity Measurements	181
Expression of Results of Enzyme Activity Measurements	182
Standardization and Quality Control of Enzyme Assays	183
IDENTIFICATION AND MEASUREMENT OF ISOENZYMES	187
Electrophoretic Separation of Isoenzymes	189
Ion Exchange Chromatography	191
Selective Inactivation in the Characterization of Isoenzymes	193
Catalytic Differences Between Isoenzymes	194
Immunochemical Methods of Isoenzyme Characterization	195
CONCLUSIONS	197
BIBLIOGRAPHY	198

INTRODUCTION

Enzymes are proteins with catalytic activity due to their powers of specific activation of their substrates. This definition indicates the particular advantages of enzymes in the detection of tissue damage. Because of their remarkable catalytic activity, a given number of enzyme molecules will convert an enormously greater number of substrate molecules to measurable products within a short space of time, so that the appearance of increased amounts of enzymes in the blood stream can be detected with great sensitivity, although the amount of enzyme protein released from damaged cells is negligible compared with the background level of non-enzymic proteins in blood. Since enzymes are specific in their actions, each type of enzyme catalysing only a unique chemical reaction or a limited range of related chemical reactions, a particular enzyme can selectively be identified and its activity measured. Therefore, amounts of enzymes released into the circulation in the investigation of heart disease can be estimated by measuring the catalytic activity of plasma or serum under defined conditions.

The only other analytical approach which appears able potentially to rival measurement of catalytic activity in sensitivity and specificity is the determination of enzymes as proteins by immunochemical procedures. In this approach, specificity is inherent in the reaction between the enzyme as antigen and its complementary antibody, while sensitivity is conferred by the ability to accumulate large counts of radioactivity in radioimmunoassay or similar techniques. The ability of immunochemical methods to distinguish between catalytically similar variants of the same enzyme (isoenzymes) has already found useful applications in clinical enzymology. The use of immunoassays to measure total amounts of enzyme has found little application so far, but offers the possibility of avoiding certain problems inherent in assays of catalytic activity; in particular, problems of comparability and standardization arising from the total dependence of the results of assays of catalytic activity on the precise conditions of measurement.

However, immunochemical methods themselves have their limitations. Assured supplies of specific antisera with reproducible properties are required, which may be expensive and for which suitable purified enzyme antigens are needed. Modern assays of enzyme activity are rapid and relatively inexpensive, and can be automated, whereas techniques such as radioimmunoassay may take longer for each measurement and may be less precise. Immunochemical and catalytic activity methods will probably provide results which are not necessarily completely correlated, since the former will measure all antigenically similar enzyme molecules, whether catalytically active or not. This may indeed prove to be an advantage in favour of immunoassay, but current clinical interpretations are based on experience of changes in enzymic activity in the plasma in disease so that, if the immunochemical methods were found to give substantially different results, new reference values and clinical experience would be required for their interpretation.

Since all enzymes are proteins they share many characteristic properties.

Therefore, in spite of the large number of different enzymes which are known to exist, the behaviour of enzymes can be described in general terms and principles to be applied in the design and performance of all enzyme assays can be formulated.

PROGRESS OF ENZYMIC REACTIONS

The progress of conversion of the substrate into products in the presence of an enzyme can be followed by measuring the decreasing concentration of the substrate or the increasing concentration of the products. Measurement of product formation is preferable where possible, since determination of the increase in concentration of a substance above an initially zero or low level is analytically more reliable than measurement of a fall from an initially high level. However, certain enzymes of importance in cardiology are usually measured in terms of a fall in substrate concentration.

Several distinct phases can be recognized as making up the progress curve of a typical enzymic reaction, although the relative lengths of the phases depend on the nature and conditions of the reaction (Figure 1). Immediately following the initiation of the reaction by the introduction of enzyme or substrate there is a period of induction before pronounced changes become established: this lag phase may last for only a few seconds or several minutes, depending to some extent on the experimental conditions. It is usually followed by a phase during which the concentrations of substrate and products change rectilinearly with respect to time; i.e. the rate of product formation is constant. This is the most useful phase from the point of view of enzyme activity determination and the rate of formation of product during this phase is usually referred to as the initial rate of reaction (v_0). As will be seen later, the rate of the reaction during this phase is essentially independent of the change in concentration of substrate; in other words, the reaction is effectively zero order with respect to substrate concentration.

The rectilinear phase of the progress curve may be very short or considerably prolonged, but it is followed by a progressive decline in the rate of reaction. This results from the operation of several factors, such as falling substrate concentration, increasing importance of the reverse reaction, accumulation of inhibitory products, and even progressive inactivation of the enzyme itself. Ultimately, the reaction ceases when equilibrium is reached or all the substrate has been consumed.

EFFECT OF ENZYME CONCENTRATION ON RATE OF CATALYSED REACTION

Addition of more enzyme to the reaction mixture increases the rate of the reaction catalysed by the enzyme. Provided that no other factor such as

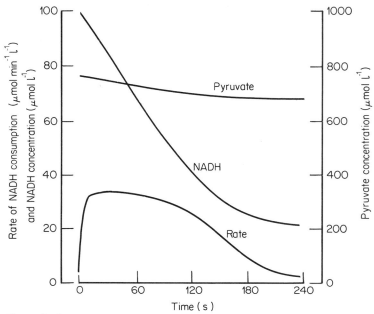

Figure 1. Changes in substrate concentrations and rate of reaction during an assay of lactate dehydrogenase activity at 37 °C in phosphate buffer with pyruvate and NADH as substrates. The reaction is followed by the fall in absorbance at 340 nm as NADH is oxidized to NAD^+. The rate of reaction rises rapidly to a maximum value, from which it declines only slightly until about half the NADH has been used up. During this phase of the reaction the rate is essentially zero order with respect to substrate concentration. At the point at which the rate falls below about 90% of its maximum value, NADH concentration is approximately $10 \times K_m$. K_m for NADH is of the order of 5×10^{-6} mol l^{-1}, whereas for pyruvate it is 9×10^{-5} mol l^{-1}. Thus an initial pyruvate concentration approximately 10 times that of NADH is used. (Concentrations are per litre of reaction mixture.)

substrate concentration is limiting, the rate of reaction is directly proportional to the concentration of active enzyme molecules. This is the basis of the quantitative determination of enzymes by measurement of reaction rates, and reaction conditions are selected to ensure that the observed reaction rate is proportional to enzyme concentration over as wide a range as possible.

The relationship between reaction rate and enzyme concentration is most clearly seen if the initial rate of reaction is measured, before factors which cause a fall in rate from its maximum value begin to operate: indeed, the value of v_0 theoretically increases without limit with increasing enzyme concentration, provided that sufficient substrate to saturate the enzyme is always present. In practice, a finite period of time is needed to accumulate a measurable amount of reaction products, so that, even in the best-designed methods, measurement of v_0 becomes impracticable above a certain limit of enzyme concentration.

The initial rate of reaction can be measured most reliably in methods in which the progress of reaction is monitored continuously, e.g. in a spectrophotometer, so that the appropriate portion of the progress curve can be identified. These 'kinetic' methods (as they have been called) are preferable to fixed-time methods, in which the total change determined after a predetermined period of incubation of enzyme with substrate is taken as a measure of the catalysed rate of reaction.

Provided that the progress curves given by the different enzyme samples all have the same shape during the fixed period of incubation, i.e. they are all described by equations of the same form, the different amounts of substrate transformed or products formed by the different samples during this time will be proportional to their relative enzyme activities. This is most obviously true if the several progress curves are all straight lines, and the reaction conditions are chosen to ensure that this is so for as wide a range of enzyme activity as possible. However, since the fixed-incubation approach does not allow the actual shape of each progress curve to be determined, the possibility remains that some samples will have progress curves which are non-rectilinear—perhaps because their enzyme activity is close to the useful limits of the method—and will therefore give rise to erroneous results. In order to provide as great a certainty as possible that all progress curves will be straight lines, the range of activity measurable by the method must be restricted. Samples with activities greater than this limit must be reanalysed, preferably with a reduced incubation time since dilution of enzyme-containing samples may result in a non-proportionate change in catalytic activity.

The ability to identify the reaction rate associated with the zero-order phase of the reaction for virtually all samples gives continuous-monitoring methods of enzyme assay a decisive advantage, and such methods should be used wherever possible. Nevertheless, reliable and reproducible measurements of enzyme activity can be made by fixed-time methods, provided that the limitations of this type of assay are appreciated. Improvements in photometry and other developments in methodology have allowed the duration of incubation in fixed-time methods to be shortened compared with older assays, with a corresponding increase in the measureable range of activity (Figure 2). Fixed-time assays can readily be adapted to some forms of automated analysers, particularly multichannel analysers.

The Enzyme–Substrate Complex

Catalysis by enzymes proceeds by way of a reversible reaction between enzyme and substrate to form an intermediate enzyme–substrate complex,

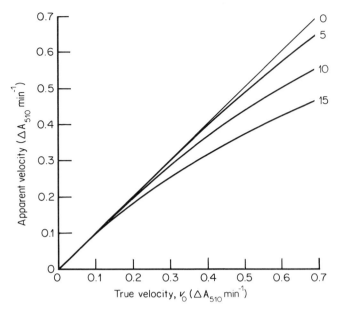

Figure 2. Apparent velocity of hydrolysis of phenyl phosphate by alkaline phosphatase measured over a fixed time interval, as a function of the true initial velocity, v_0. The curves 5, 10, and 15 correspond to incubation periods of 5, 10, and 15 min and show the increasing deviation of the apparent velocity from the initial rate as the incubation period is lengthened. Initial substrate concentration 4.75 mmol l^{-1} of reaction mixture in carbonate–bicarbonate buffer, pH 9.9 at 37 °C

which then breaks down into free enzyme and reaction products:

$$E + S \rightleftharpoons ES \rightarrow E + \text{products}$$

The rate of appearance of products at a given moment is proportional to the concentration of the ES complex. This explains the accelerating effect on the reaction rate of an increase in enzyme concentration: provided that an excess of substrate is present, the increased enzyme concentration results in an increase in the concentration of ES.

Experimental proof of the existence of an intermediate enzyme–substrate complex was first obtained for the reaction between peroxidase and hydrogen peroxide, and has been frequently confirmed for other enzymic reactions. Enzyme–substrate complexes consist of one molecule of enzyme combined with one, or at the most a few, molecules of substrate. Since enzyme molecules are usually large compared to molecules of their substrates, the observation of one-to-one stoichiometry is in keeping with the concept of a specialized substrate-binding region of the enzyme molecule (the active centre) and is at variance with theories of non-specific adsorption, as in catalysis at a surface.

The existence of a specialized substrate-binding region is also in accordance with the observed facts of enzyme specificity. All enzymes are specific; that is, there are limitations on the nature of the reaction which each one will catalyse. However, these limits vary widely from one enzyme to another. At one extreme are the enzymes of absolute specificity, which will catalyse only a single chemical change involving a single compound. Frequently, only one of a pair of isomeric compounds can serve as the enzyme's substrate. At the other extreme are enzymes which are group specific or reaction specific. These enzymes are able to catalyse a particular type of reaction provided only that the substrate contains the appropriate chemical bond or radical. Between these extremes are enzymes which can tolerate relatively minor variations in the structure of their substrates.

A knowledge of the limits of specificity of particular enzymes is important in designing assays for clinical use. Where enzymes of wide specificity are concerned, substrates can be chosen which react to give readily estimated products—substances with specific light-absorption properties, for example. The isoenzymic forms of an enzyme may act at different rates on substrate analogues, and this can be exploited in isoenzyme characterization and measurement, as in the case of lactate dehydrogenase. However, some of the enzymes which are of diagnostic importance in cardiology, such as creatine kinase, have absolute specificity, so that variation of the substrate to facilitate analysis is not possible.

EFFECT OF SUBSTRATE CONCENTRATION ON RATE OF CATALYSED REACTION: THE MICHAELIS–MENTEN EQUATION

As well as explaining the dependence of reaction rate on enzyme concentration under conditions in which excess substrate is present, the formation of an enzyme–substrate complex also accounts for certain characteristics of the relationship between reaction velocity and substrate concentration. If the enzyme concentration is fixed and the substrate concentration is varied, the rate of reaction is almost directly proportional to substrate concentration at low values of the latter, i.e. the reaction is first order with respect to substrate concentration. At high substrate concentrations, on the other hand, variation in substrate concentration has no effect on rate, and the reaction is zero order with respect to substrate concentration (Figure 3).

At low substrate concentrations, addition of more molecules of substrate increases the concentration of the enzyme–substrate complex and consequently the rate of reaction. When substrate concentration is high, all the enzyme molecules are combined with substrate so that further addition of substrate molecules can produce no increase in the concentration of the

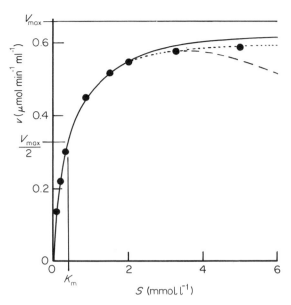

Figure 3. Relationship of rate of reaction, v, to substrate concentration, s, for the hydrolysis of α-naphthyl phosphate by alkaline phosphatase from human liver in carbonate–bicarbonate buffer, pH 9.9 at 37 °C. The full curve corresponds to the Michaelis–Menten equation, $v = V_{max} \cdot s/(s + K_m)$, for values of V_{max} of 0.66 μmol min^{-1} ml^{-1} and K_m of 0.4 mmol l^{-1}. The experimentally determined points (●) fit the theoretical curve closely at lower substrate concentrations, but begin to deviate from it at higher values of /s (dotted curve). In some cases, the phenomenon of inhibition by excess substrate is observed (broken line)

enzyme–substrate complex. Under these conditions, an increased rate of reaction can only be achieved by addition of more enzyme.

The idea that the enzyme is saturated with substrate at high concentrations of the latter was introduced by Henri in 1902, and in 1913 Michaelis and Menten extended his theory to derive a general equation relating the catalysed rate of reaction, v, and substrate concentration, s, for a given concentration of enzyme:

$$v = \frac{V_{max} \cdot s}{s + K_m} \quad (1)$$

In their derivation, Michaelis and Menten made certain assumptions, notably that an equilibrium is set up between enzyme and substrate molecules on the one hand, and the molecules of the enzyme–substrate complex on the other, and that the equilibrium is not disturbed by the breakdown of the complex into free enzyme and products. The term K_m in the equation is then the equilibrium constant of the reversible reaction between enzyme and substrate. The term V_{max} represents the limiting velocity of reaction which is only reached at infinitely high substrate concentrations.

It is now known that the enzyme and substrate are not in equilibrium in many enzymic reactions, but that a steady state exists. A rate equation derived from steady-state kinetics has the same form as the Michaelis–Menten equation, but the meaning of K_m is different—instead of being an equilibrium constant it contains additional rate constants.

Experimental data for the dependence of reaction rate on substrate concentration usually agree closely with the hyperbolic form predicted by the rate equation derived from either equilibrium or steady-state assumptions (Figure 3). However, since the interpretation of the meaning of K_m depends on assumptions which cannot be tested in a simple experiment of this nature, the symbol K_m is reserved for the experimentally determined constant. If a definite kinetic meaning is attached to the constant, other symbols are chosen, such as K_s if an equilibrium constant is implied.

By substituting $v = V_{max}/2$ in the Michaelis–Menten equation s becomes equal to K_m. The value of K_m is therefore obtained by determining the substrate concentration at which v equals half the maximum velocity obtained at high substrate concentrations. Extrapolation of the hyperbolic velocity–substrate curve to the limiting value of v is not reliable. Therefore, a transformation of the equation is usually used to derive V_{max} and K_m. The most popular of these transformations is usually attributed to Lineweaver and Burk: since from the Michaelis–Menten equation,

$$\frac{1}{v} = \frac{K_m}{V_{max}} \cdot \frac{1}{s} + \frac{1}{V_{max}} \qquad (2)$$

a plot of $1/v$ against $1/s$ is a straight line of slope K_m/V_{max} with intercepts $-1/K_m$ on the abscissa and $1/V_{max}$ on the ordinate (Figure 4).

The double-reciprocal plot of $1/v$ against $1/s$ has been criticized on the grounds that values of v obtained at low substrate concentration, where experimental error is greatest, have a disproportionate effect on the slope and intercepts of the straight line fitted to the experimental points. However, the individual points can be given weights appropriate to their reliability, and the plot has the advantage that the effects of various factors on both K_m and V_{max} can readily be seen from the resulting changes in the respective intercepts. This is particularly useful in studies with inhibitors, for example.

The value of K_m is usually independent of the amount of enzyme used for its determination. However, it is generally affected by alterations in other conditions; particularly in pH, but also in such factors as ionic strength or the nature of the buffer solution. Therefore, when the determination of K_m

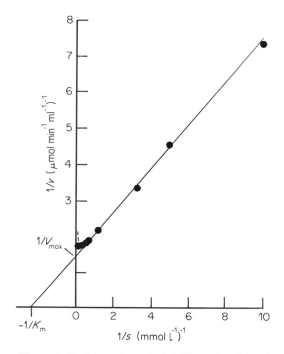

Figure 4. Double-reciprocal plot $1/v$ against $1/s$, of the data shown in Figure 3. At low values of $1/s$ (high substrate concentration) the points deviate from the straight line. The broken line shows the effect of inhibition by excess substrate on this type of plot

is used to aid in enzyme characterization, e.g. in distinguishing between isoenzymic forms of an enzyme, the conditions of measurement must be standardized.

Determination of the dependence of the catalysed rate of reaction on substrate concentration is an essential part of the process of defining conditions for the assay of enzyme activity, not only to ensure that a substrate concentration is chosen which is adequate to maintain zero-order kinetics during the assay, even when the enzyme concentration is high, but also to detect deviations from the typical hyperbolic relationship. Since the maximum velocity is, in theory, reached only at infinitely high substrate concentrations, an effective limiting value of v is reached which is below the theoretical maximum. Nevertheless, the observed reaction velocity usually becomes essentially independent of s when the latter exceeds about 10 times K_m. When $s = 10 \times K_m$, v is approximately 91% of the calculated V_{max} if the theoretical relationship holds. Practical considerations such as the solubility or cost of the substrate may limit the choice of substrate concentration.

In some enzymic reactions the observed velocity reaches a maximum then falls as substrate concentration is increased (Figures 3 and 4). The phenomenon of inhibition by excess substrate is probably due in many cases to formation of catalytically ineffective enzyme–substrate complexes in which more than one substrate molecule combines with the active centre of the enzyme. Competition between the first substrate and a second substrate involved in a two-substrate reaction may account for some examples of inhibition by excess substrate; e.g. water is the second reactant in hydrolytic reactions, and its effective concentration may be reduced at high substrate concentrations. More rarely, an enzyme may be found to be activated by increasing concentrations of its substrate, i.e. v may be greater at high values of s than predicted by the Michaelis–Menten equation.

Instead of the usual hyperbolic dependence of reaction velocity on substrate concentration, some enzymes show a sigmoid relationship between these variables (Figure 5). Thus the increase in v for a given increase in s is lower than that expected from the Michaelis–Menten equation at low values of s, but greater than predicted at higher values of s. The nearly flat portion of the curve at low substrate concentrations essentially constitutes a threshold of substrate concentration, below which v is low and is little affected by changes in s. Around a critical concentration of substrate the enzyme responds markedly to changes in s, switching from very low to

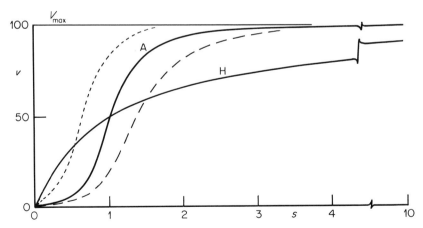

Figure 5. Dependence of v upon s for an allosteric enzyme (curve A) compared with that for an enzyme showing typical Michaelis–Menten kinetics (H). Both have similar V_{max} values, but for H, velocity is greater at low substrate concentrations. Above a threshold value of s of about 0.6 mmol l^{-1}, however, velocity approaches V_{max} much more rapidly in the case of A as substrate concentration increases. Activators of A reduce the sigmoidicity of curve A (dotted curve); i.e. they lower the substrate concentration threshold, whereas inhibitors (broken curve) increase the sigmoidicity and raise the threshold value

pronounced activity. Not surprisingly, therefore, enzymes which occupy rate-limiting positions in metabolic reaction sequences are often found to exhibit sigmoid kinetics.

In general, the existence of a sigmoid curve for v against s implies that more than one molecule of substrate binds to each molecule of enzyme and that binding of the second molecule is affected by the presence of the substrate molecule already bound, and so on until all the substrate-combining sites are filled. When the attachment of succeeding molecules is facilitated by those bound earlier, the interaction between them is described by the term cooperativity.

For an enzyme with n binding sites per molecule for the substrate, if equilibrium is rapidly attained between the free enzyme and the enzyme with all the sites filled,

$$E + nS \rightleftharpoons ES_n$$

For the reaction to reach equilibrium the degree of cooperativity must be high, i.e. each bound molecule of substrate increases the affinity of the remaining sites so greatly that partly filled enzyme molecules have only a transient existence. The rate of the catalysed reaction is then virtually entirely accounted for by the breakdown of the ES_n complex, and the reaction velocity is given by the equation:

$$v = \frac{V_{max} \cdot (s)^n}{K + (s)^n} \qquad (3)$$

in which K is the equilibrium constant for the reversible reaction between enzyme and substrate.

The rate of reaction thus depends on the nth power of the substrate concentration. When $n = 1$, the equation reduces to the Michaelis–Menten equation, but when n is greater than 1, plots of v against s take on an increasingly sigmoid appearance. In order to determine the value of n, the log of $v/(V-v)$ may be plotted against log s to obtain a straight line of slope n and intercept on the ordinate at $-\log K$, since

$$\log \frac{v}{V-v} = n \log s - \log K \qquad (4)$$

Plots of this type are usually referred to as Hill plots, because of the similarity of the equation on which they are based to that proposed by Hill to describe the oxygen saturation curve of haemoglobin.

Values of n of greater than one indicate the existence of cooperative effects in the binding of successive substrate molecules, and values of less than one that the ligands behave anticooperatively, i.e. that binding of the first molecule reduces affinity for further binding. If n is found to be unity it does not necessarily imply that only one molecule of substrate is bound to each molecule of enzyme, but that binding of the first molecule of substrate neither helps nor hinders the attachment of any further substrate molecules to the same molecule of enzyme.

The degree of sigmoidicity of the curve relating v and s may be altered by the presence of an appropriate modifier. An inhibitor tends to make the curve more sigmoid, increasing the concentration of substrate needed to produce a significant increase in v. On the other hand, an activator may bring the curve to a more hyperbolic shape (Figure 5). These modifiers combine with the enzyme at sites distinct from the substrate-binding site and have been termed allosteric modifiers to emphasize their supposed effects on the conformation of the enzyme molecules. The existence of different binding sites, with its implication of different specificity requirements for substrate and modifiers, has obvious advantages for the operation of a regulatory enzyme in a metabolic sequence. The modifier need not resemble the substrate of the enzyme in structure, as a competitive inhibitor would have to do, for example, and so could be the product of a later enzyme in the same sequence or even a molecule produced in a different sequence of reactions. Allosteric modifiers thus offer increased possibilities for control of metabolism.

Although the modifier need not be an analogue of the substrate, the site at which the modifier binds to the enzyme has its own specificity requirements. Therefore, analogues of the modifier may be found which compete with the modifier itself for its own binding site, so preventing the activation or inhibition which follows combination between the enzyme and the true modifier. In some cases the effect of the modifier may be abolished without affecting the catalytic activity of the enzyme; e.g. if some allosteric enzymes are heated briefly the shape of the curve of v against s may change from sigmoid to hyperbolic, without a fall in V_{max}. This provides further evidence for the existence of separate binding sites for substrate and modifier.

The separate but identical multiple binding sites for substrate, and the different binding sites for activators or inhibitors could arise by repeated sequences of amino acids in the primary structure of a single polypeptide chain. More easily envisaged is a set of protein subunits, called monomers or protomers, which associate with each other to form the active enzyme oligomer and each of which carries a substrate-binding site, as well as activator and/or inhibitor-binding sites. All the enzymes exhibiting cooperativity which have so far been investigated have been found to be oligomeric and most of the models proposed for the action of allosteric enzymes have assumed the existence of this quaternary level of protein structure (Figure 6).

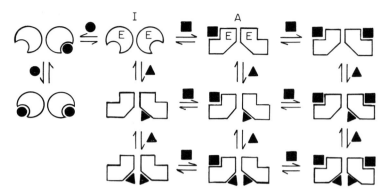

Figure 6. Diagram of the effects of activators and inhibitors on an allosteric enzyme, based on the model proposed by Monod, Wyman, and Changeux. The enzyme protomers (E) can exist in two forms, I and A, in equilibrium with each other. Inhibitors (small circles) stabilize the enzyme in the form (circles) which does not bind substrate molecules. Binding of substrate (squares) or an activator (triangles) stabilizes the enzyme in the form to which substrate binds readily, displacing the equilibrium between the two forms of the enzyme in favour of the substrate-binding conformation, so that further substrate molecules are bound more readily

Consecutive Enzymic Reactions

All reactions catalysed by enzymes are in theory reversible, the enzyme catalysing the forward and backward reactions equally. However, in practice, the reaction is usually found to be more rapid in one direction than the other, so that an equilibrium is reached in which the product of either the forward or the backward reaction predominates, sometimes so markedly that the reaction is virtually irreversible. The equilibrium point is determined by the respective free-energy changes in the two reactions.

If the product of the reaction in one direction is removed as it is formed, e.g. because it is the substrate of a second enzyme present in the reaction mixture, the equilibrium of the first enzymic process will be displaced so that the reaction may go to completion in that direction. Reaction sequences in which the product of one enzyme-catalysed reaction becomes the substrate of the next enzyme and so on, often through many stages, until conversion of the original metabolite to a final product has been accomplished, are characteristic of living processes. In the laboratory, also, two or three or more enzymic reactions may be linked together in this way in order to provide a means of measuring the activity of the first enzyme or the concentration of the initial substrate in the chain.

For such a series of linked reactions, a Michaelis–Menten curve can be constructed for each enzyme taking part. The maximum velocity (V_{max}) of each stage in the overall process will depend on the amount of the particular enzyme catalysing that stage which is present in the system. Therefore, however much substrate is introduced into the system at the beginning of the sequence, the rate of appearance of end-products will not increase beyond the limit set by the lowest of the several values of V_{max}. The rate of flux through the system will only increase if the value of V_{max} for the rate-limiting enzyme is raised, either by provision of more enzyme (e.g. by *de novo* synthesis in living tissue, or by adding more enzyme in the laboratory), or by a modification of the existing enzyme molecules to render them more effective catalytically. However, if the concentration of the first substrate and the rates of the succeeding reactions are such that none of the enzymes is saturated with its substrate, an increase in the concentration of the primary substrate will result in a greater rate of appearance of products.

When a secondary enzyme-catalysed reaction, known as an indicator reaction, is used to determine the activity of a different enzyme, e.g. in the determination of aspartate aminotransferase activity in serum by reduction of the oxaloacetate fomed in the aminotransferase reaction to malate in the presence of malate dehydrogenase and NADH, it is essential that the primary reaction should be the rate-limiting step. The activity of the indicator enzyme must be such as to ensure the virtually instantaneous removal of the product of the first reaction, to prevent reversal of the first reaction becoming significant. Although the measured enzyme is typically acting under conditions of saturation with respect to its substrate, the concentration of the substrate of the indicator enzyme (i.e. the product of the first reaction) remains in the region of the Michaelis–Menten curve in which v is directly proportional to s, so that the rate of reaction catalysed by the indicator enzyme is directly proportional to the rate of product formation in the first reaction. A transition period occurs after the start of the first reaction to allow the concentration of its product to reach a steady state.

Since the rate of the second reaction depends on the activity of the indicator enzyme as well as on the concentration of its substrate (i.e. the product of the primary reaction), the duration of the lag period is reduced by increasing the concentration of the indicator enzyme, and by specifying a low steady-state concentration of the product of the primary reaction.

The rate of the indicator reaction, v_i, is related to substrate concentration and therefore to the product concentration, p, by the Michaelis–Menten equation

$$v_i = \frac{V^i_{max} \cdot p}{p + K^i_m} \tag{5}$$

in which V^i_{max} and K^i_m are the maximum velocity and Michaelis constant of the indicator enzyme. For the rate of the indicator reaction to match that of the primary reaction, v_i must equal the limiting velocity of the primary reaction, v_t, which the assay system is expected to measure. Therefore, the activity of indicator enzyme needed is given by

$$v_t = \frac{V^i_{max} \cdot p}{p + K^i_m} \tag{6}$$

or, rearranged,

$$V^i_{max} = v_t\left(1 + \frac{K^i_m}{p}\right) \tag{7}$$

For example, in a coupled assay designed to measure up to 300 μmol min^{-1} l^{-1} of aspartate aminotransferase activity in serum, corresponding to 25 μmol min^{-1} l.$^{-1}$ when diluted in the assay mixture, if a steady-state concentration of oxaloacetate of 1 μmol l^{-1} is chosen for p, the appropriate activity of the indicator enzyme malate dehydrogenase (K^i_m approximately 15 μmol oxaloacetate l^{-1}) would be

$$V^i_{max} = 25(1 + \tfrac{15}{1}) = 400 \ \mu\text{mol min}^{-1}\,\text{l}^{-1} \text{ of assay mixture.}$$

Thus although the ratio of activities of the indicator and primary enzymes will vary from one assay method to another depending on the range of activity which it is desired to measure, the Michaelis constant of the indicator enzyme, and the lag period which is considered acceptable, the catalytic concentration of the indicator enzyme in the reaction mixture must be much greater than that of the enzyme being determined.

Two-substrate Reactions

Most reactions catalysed by enzymes are of the type:

$$A + B \rightleftharpoons C + D$$

When one of the substrates is water (i.e. when the process is one of hydrolysis) with the reaction taking place in aqueous solution, only a fraction of the total number of water molecules present participates in the reaction. The small change in the concentration of water has no effect on the rate of reaction. A similar situation prevails in some indicator reactions, such as that involving malate dehydrogenase discussed above. In this case the second substrate, NADH, is present in excess and changes in its concentration during the reaction have a negligible effect on reaction rate. Quite apart from their applications as indicator reactions, oxidation–reduction processes in which dehydrogenases transfer hydrogen from donor molecules to specific coenzymes, the latter acting as second substrates, are of great importance in biochemistry.

If such a reaction proceeds by way of intermediate enzyme–substrate complexes, so that

$$E + A \rightleftharpoons EA$$

followed by

$$EA + B \rightleftharpoons EAB \rightarrow C + D$$

and if A and B combine with separate sites on the enzyme molecule, the rate of the reaction from left to right is given by

$$v = \frac{V_{max} \cdot ab}{ab + bK_m^a + aK_m^b + K_s^a K_m^b} \qquad (8)$$

K_m^a and K_m^b are the Michaelis constants for the two substrates and a and b are their concentrations. K_s^a is the equilibrium constant for the reversible reaction between the enzyme and A. If the equation is rearranged into the double reciprocal form,

$$\frac{1}{v} = \frac{1}{a}\left(\frac{K_m^a}{V_{max}} + \frac{K_m^b K_s^a}{b V_{max}}\right) + \frac{1}{V_{max}}\left(1 + \frac{K_m^b}{b}\right) \qquad (9)$$

it can be seen that a plot of $1/v$, against $1/a$ will give a straight line, but both the slope of the line and its intercept on the ordinate are affected by b, the concentration of the second substrate (Figure 7). Similarly, a plot of $1/v$ against $1/b$ is rectilinear, but with the slope and intercept dependent on a. If, however, the concentration of one substrate (e.g. b) is infinitely great, the rate equation reduces to the usual form for one-substrate reactions:

$$v = \frac{V_{max} \cdot a}{a + K_m^a} \qquad (10)$$

Values of K_m and V_{max} for each substrate can be derived from experiments in which the concentration of the first substrate is held constant at saturating levels while the concentration of the second substrate is varied, and vice versa. There is of course no reason why the K_m values for the two substrates should be the same, or even similar; e.g. pyruvate and NADH, the two-substrate pair in the reaction catalysed by lactate dehydrogenase of beef heart, have K_m values of 2×10^{-5} mol l^{-1} and 3×10^{-6} mol l^{-1}, respectively.

It may be difficult in practice to ensure that a saturating concentration of the first substrate is reached or maintained while the concentration of the second substrate is varied. In such cases it is preferable to construct several double-reciprocal plots at known, fixed concentrations of the first substrate then, by replotting the slopes and intercepts given by these plots, to derive the values of K_m and V_{max} corresponding to infinite concentration of the first substrate. However, this procedure is usually unnecessary for dehydrogenase reactions.

It has been assumed in the foregoing discussion that not only does the binding of one substrate not influence the binding of the other, but also that

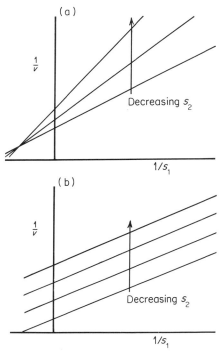

Figure 7. Double-reciprocal plots of $1/v$ against $1/s_1$ for two-substrate reactions, showing the effect of falling concentration of the second substrate, s_2. (a) In a dehydrogenase reaction in which a ternary complex is formed, and (b) in a ping-pong bi-bi reaction mechanism (e.g. aminotransferase) in which no ternary complex is formed

the Michaelis–Menten assumptions with regard to the establishment of equilibria are satisfied. If these last assumptions are justified, the order of combination of substrates with the enzyme has no effect on the linearity of $1/v$ against $1/s$ plots. However, a random order of combination gives non-linear plots in the steady-state kinetic treatment.

Equilibrium conditions are likely to be set up when the complex between the enzyme and both substrates breaks down to products relatively slowly compared with its rate of formation. If one substrate combines with the enzyme much more rapidly than the other, the order of combination becomes effectively non-random. Again, in some reactions no ternary complex, EAB, may be formed, since the binding of the first substrate is followed by release of the first products before the second substrate is bound.

This 'ping-pong bi-bi' type of mechanism occurs in reactions catalysed by aminotransferases.

The relationship between reaction velocity and the concentrations of the two substrates in ping-pong bi-bi reactions reduces to the form

$$v = \frac{V_{max} \cdot ab}{ab + bK_m^a + aK_m^b} \tag{11}$$

The reciprocals of v and a are related by the equation

$$\frac{1}{v} = \frac{1}{a} \cdot \frac{K_m^a}{V_{max}} + \frac{1}{V_{max}} \left(1 + \frac{K_m^b}{b}\right) \tag{12}$$

so that a plot of $1/v$ against $1/a$ is unchanged in slope by variation in the concentration of b, but the intercept on the ordinate, and therefore the value of V_{max}, changes as b is varied (Figure 7). Similar equations describe the variation of V_{max} with a when $1/v$ is plotted as a function of $1/b$.

The selection of reaction conditions for the measurement of enzymic activity involving two substrates can be approached empirically by varying the concentration of the first substrate while keeping the concentration of the second substrate constant, until maximum activity is reached. The process is then repeated with the concentration of the first substrate held at the value thus determined while the concentration of the second substrate is varied. The results have been published of many attempts to define optimal substrate concentrations in this way for the measurement of clinically important enzymes, such as the aminotransferases. However, the recommended concentrations frequently differ, and indeed there is no reason why this experimental approach should give a single solution to the problem of optimization of substrate concentrations in two-substrate reactions.

Inspection of the rate equation for a ping-pong bi-bi reaction shows that a particular observed reaction velocity, expressed as a fixed fraction of the theoretical V_{max} value, will be given by an infinite number of pairs of values of the substrate concentrations a and b. If the two Michaelis constants of the reaction, K_m^a and K_m^b, are known, a plot of the concentration of the first substrate, a, against that of the second substrate, b, for all values of v which are the same fraction of V_{max} will be a hyperbola (Figure 8).

In practice the choice of substrate concentrations is limited by such considerations as the solubility of the substrates, the viscosity and light-absorbance of concentrated solutions, and the relative costs of the reagents. These factors have been given different degrees of emphasis by the authors of various recommended methods for enzymes of diagnostic importance. Differences in other chosen reaction conditions, such as pH and temperature, also contribute to the different results obtained when alternative methods are used to measure a given enzyme activity and the effect of these

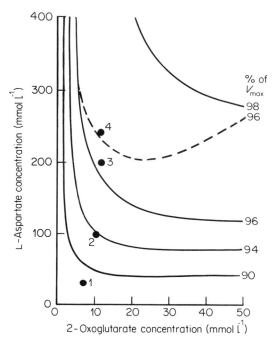

Figure 8. Relationship between calculated reaction velocity (expressed as a percentage of the theoretical V_{max}) and concentration of the two substrates, L-aspartate and 2-oxoglutarate, for human heart aspartate aminotransferase. The solid curves make no allowance for inhibition by 2-oxoglutarate; when this is taken into account, the broken line (showing the effect on the 96% V_{max} curve as an example) is obtained. The points (●) show substrate concentrations (1) in the original method of Karmen (1955), (2) recommended by a U.K. Working Party on Enzyme Standardization (1972), (3) according to German (1972) and Scandinavian (1974) Enzyme Committees, and (4) provisionally recommended by the Expert Panel on Enzymes of IFCC (1977). Values for K_m (aspartate), K_m (2-oxoglutarate), and K_i (2-oxoglutarate) of 4.6, 0.18, and 44 mmol l^{-1}, determined under the conditions provisionally recommended by the Expert Panel on Enzymes, have been used for calculation (personal communication from H. U. Bergmeyer). The values of these constants may be slightly different under other conditions, e.g. when phosphate rather than Tris buffer is used

differences is usually more significant than that of relatively small variations in substrate concentrations from one method to another.

The rate equation for a two-substrate reaction is modified if one or both substrates inhibit the enzyme at high concentrations. For example, aspartate aminotransferase is inhibited by excess concentrations of the substrate 2-oxoglutarate and this inhibition is competitive with respect to L-aspartate. Therefore, to maintain a given reaction velocity at high 2-oxoglutarate concentrations, the concentration of L-aspartate has to be increased above the value needed at lower concentrations of 2-oxoglutarate. This has the effect of altering the shape of the line linking pairs of substrate concentrations which correspond to a given reaction velocity from a hyperbola to a parabola (Figure 8).

INHIBITION AND ACTIVATION OF ENZYME ACTIVITY

The rates of enzymic reactions are often found to be affected by changes in the concentrations of substances other than the enzyme or substrate. These modifiers may be activators, i.e. they increase the rate of reaction, or their presence may inhibit the enzyme, reducing the reaction rate. Activators and inhibitors are usually small molecules, or even ions. They vary in specificity from modifiers which exert similar effects on a wide range of different enzymic reactions at one extreme, to substances which affect only a single reaction at the other. Reagents such as strong acids or multivalent anions and cations, which denature or precipitate proteins, destroy enzyme activity and so may be regarded as extreme examples of non-specific enzyme inhibitors. However, these effects, which depend on the properties of enzymes as proteins rather than as catalysts, are not usually included in discussions of enzyme inhibition although they have obvious practical implications in the treatment and storage of specimens in which enzyme activity is to be measured.

Some phenomena of enzyme activation or inhibition are due to interaction between the modifier and a non-enzymic component of the reaction system, such as the substrate. In most cases, however, the modifier combines with the enzyme itself, in a manner analogous to the combination of enzyme and substrate.

Inhibition of Enzyme Activity

Inhibitors are divisible into reversible and irreversible types. Reversible inhibition implies that the activity of the enzyme is fully restored when the inhibitor is removed from the system in which the enzyme acts by some physical separative process such as dialysis, gel filtration, or chromatography. An irreversible inhibitor, on the other hand, combines covalently with the enzyme so that physical methods are ineffective in separating the two. Organophosphorus compounds are extremely potent inhibitors of esterases, including acetylcholinesterase. The enzyme breaks one of the bonds in the inhibitor, but part of the molecule is left bound to the active centre of the enzyme, preventing further activity. In some cases enzymes which have combined with irreversible inhibitors can be reactivated by a chemical reaction which removes the blocking group: e.g. the phosphoryl enzymes formed with organophosphorus compounds can sometimes be reactivated by treatment with oximes or hydroxamic acids.

Characteristics of irreversible inhibition

An irreversible inhibitor is not in equilibrium with the enzyme. Its effect is progressive with time, becoming complete if the amount of inhibitor present exceeds the total amount of enzyme. The rate of the reaction between enzyme and inhibitor is expressed as the fraction of the enzyme activity which is inhibited in a fixed time by a given concentration of inhibitor. The velocity constant of the reaction of the inhibitor with the enzyme is a measure of the effectiveness of the inhibitor.

When the inhibitor is added to the enzyme in the presence of its substrate, combination of the enzyme and inhibitor may be delayed because a proportion of the enzyme molecules will be combined with the substrate and therefore protected from reacting with the inhibitor. However, as the substrate molecules are broken down the active centres become available for combination with the inhibitor. Thus inhibition will eventually become complete even though an excess of substrate may initially be present, compared with the inhibitor concentration. Furthermore, addition of more substrate is ineffective in reversing the inhibition, in contrast to its effect on reversible competitive inhibition discussed later.

Kinetics of reversible inhibition

Reversible inhibition is characterized by the existence of an equilibrium between enzyme, E, and inhibitor, I:

$$E + I \rightleftharpoons EI$$

The equilibrium constant of the reaction K_i, is a measure of the affinity of the inhibitor for the enzyme, just as the value of the substrate constant, K_s, reflects the affinity of the enzyme for its substrate.

Competitive inhibition. The inhibitor is usually a structural analogue of the substrate and binds to the enzyme at the substrate-binding site but, because it is not identical with the substrate, breakdown into products does not take place. When the process of inhibition is fully competitive the enzyme can be combined with either the substrate or the inhibitor, but not with both simultaneously. Two equilibria are therefore possible:

$$E + S \rightleftharpoons ES \rightarrow E + \text{products}$$

and

$$E + I \rightleftharpoons EI$$

An equation can be derived which relates the observed reaction velocity to the concentrations of substrate, s, and inhibitor, i:

$$v = \frac{V_{max} \cdot s}{s + K_m\left(1 + \dfrac{i}{K_i}\right)} \tag{13}$$

This is the Michaelis–Menten equation, but with K_m modified by a term including the inhibitor concentration and inhibitor constant. V_{max} is unaltered. Therefore, curves of v against s in the presence and absence of inhibitor reach the same limiting value at high substrate concentrations, but, when the inhibitor is present, K_m is apparently greater. Plots of $1/v$ against $1/s$ with and without inhibitor cut the ordinate at the same point, but have different slopes and intercepts on the abscissa (Figure 9).

Non-competitive inhibition. A non-competitive inhibitor is usually unlike the substrate in structure. It is assumed to bind at a site on the enzyme molecule distinct from the substrate-binding site; thus there is no competition between inhibitor and substrate and a ternary enzyme–inhibitor–substrate complex can form. Attachment of the inhibitor to the enzyme does not alter the affinity of the enzyme for its substrate, but the enzyme–inhibitor–substrate complex does not breakdown to give products. The following equilibria are therefore possible:

$$E + S \rightleftharpoons ES \rightarrow E + \text{products}$$
$$E + I \rightleftharpoons EI$$
$$EI + S \rightleftharpoons EIS \quad \text{and} \quad ES + I \rightleftharpoons EIS$$

Since S and I do not interfere with the binding of each other, the rate constants for the combination of S with EI are the same as those for the binding of S with E. Similarly, the same rate constants apply to the reaction of I with ES as with E.

Derivation of an expression for v as for the competitive case gives the following:

$$v = \frac{V_{max} \cdot s}{(K_m + s)\left(1 + \dfrac{i}{K_i}\right)} \tag{14}$$

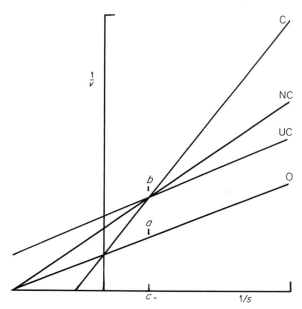

Figure 9. Effects of different types of inhibitors on the double-reciprocal plot of $1/v$ against $1/s$. Each of the inhibitors has been assumed to reduce the activity of the enzyme by the same amount, represented by the change in $1/v$ from a to b at a substrate concentration of c. Line 0 is the plot for enzyme without inhibitor, C with a competitive inhibitor, NC with a non-competitive inhibitor, and UC with an uncompetitive inhibitor

The effect of a non-competitive inhibitor is therefore to divide V_{max} by a factor of $(1+i/K_i)$. Plots of v against s have the same shape when the inhibitor is present as when it is omitted, but do not reach the uninhibited value of V_{max}, however great the substrate concentration. Double-reciprocal plots of $1/v$ against $1/s$ are altered in both slope and intercept on the ordinate by the presence of the inhibitor (Figure 9).

Uncompetitive inhibition. In this rather unusual type of inhibition, parallel lines are obtained when plots of $1/v$ against $1/s$ with and without the inhibitor are compared (Figure 9), i.e. the slope remains constant but the intercept on the ordinate is altered by the presence of the inhibitor.

Uncompetitive inhibition is rare in single-substrate reactions: an example of importance in clinical enzymology is the inhibition of intestinal and placental alkaline phosphatases by L-phenylalanine. It can occur in such reactions when the inhibitor binds to the ES complex, but not to the free enzyme. A ternary complex can form, but only after the enzyme has

THE MEASUREMENT OF ENZYMES

combined with the substrate:

$$E + S \rightleftharpoons ES \rightarrow \text{products}$$
$$ES + I \rightleftharpoons ESI$$

The rate equation is

$$v = \frac{\dfrac{V_{\max} \cdot s}{\left(1 + \dfrac{i}{K_i}\right)}}{\dfrac{K_m}{\left(1 + \dfrac{i}{K_i}\right)} + s} \tag{15}$$

$V_{\max}$ and K_m are both divided by $(1 + (i/K_i))$ when the inhibitor is present, compared with the reaction without inhibitor.

Uncompetitive inhibition is more common in two-substrate reactions. For an enzymic process in which A and B are converted to products C and D which follows the route

$$E + A \rightleftharpoons EA \xrightleftharpoons{+B} EAB \rightleftharpoons C + D$$

an inhibitor (I) which is similar in structure to B can compete with B for its binding site on EA. Therefore, at constant concentrations of A, plots of $1/v$ against $1/b$ with and without the inhibitor will be characteristic of competitive inhibition. With the same inhibitor but with a varied and b fixed, however, uncompetitive inhibition results because I binds to the enzyme–substrate complex EA, as in the single–substrate case of uncompetitive inhibition just described.

Because in two-substrate reactions the uncompetitive inhibitor of the first substrate is a competitive inhibitor of the second substrate and therefore related to the latter in structure, it is easier to postulate possible inhibitors and to test them than is the case with single-substrate reactions, where there is usually no obvious structural relationship between the substrate and an uncompetitive inhibitor.

Mixed inhibition. Plots of $1/v$ against $1/s$ for the inhibited and uninhibited reactions are often found not to fit any of the three patterns described above. Instead, the lines intersect to the left of the ordinate but above the abscissa. These cases are usually interpreted as implying that the inhibitor interferes with both the binding of the substrate to the enzyme and its subsequent breakdown.

Inhibition by reaction products

Accumulation of reaction products can contribute in two ways to the fall in velocity of an enzymic reaction with time. First, the increasing concentration of products will tend to drive the reaction backwards, if it is freely reversible. Second, a product may itself be an inhibitor of the forward reaction so that, even if the reaction is not readily reversible, it proceeds against a rising concentration of inhibitor. A familiar example is the release of the competitive inhibitor, inorganic phosphate, by the action of alkaline phosphatase on its substrates. In this case both organic phosphates and inorganic phosphate bind to the active centre of the enzyme with similar affinities; i.e. K_m and K_i are of the same order of magnitude.

An increasing concentration of an inhibitory reaction product is a frequent cause of non-linearity of reaction progress curves during fixed-time methods of enzyme assay. For example, oxaloacetate produced by the action of aspartate aminotransferase inhibits the enzyme, particularly the mitochondrial isoenzyme. Continuous-monitoring methods allow the rate of reaction to be determined before inhibition by the reaction products becomes significant. The inhibitory product may also be removed as it is formed by a coupled enzymic reaction, as in the determination of aspartate aminotransferase with malate dehydrogenase as the indicator enzyme.

Known inhibitors of an enzymic reaction other than the products of the reaction itself must be excluded from the assay system for that enzyme; e.g. certain buffer ions may inhibit particular enzymes, so that the choice of buffer solutions is restricted. However, selective enzyme inhibition can also be used to increase the specificity of the assay system for the enzyme being measured. For example, the presence in serum specimens of the enzyme adenylate kinase, which converts ADP to AMP and ATP, interferes with the measurement of creatine kinase activity. Addition of high concentrations of AMP to the assay system reduces this interference by inhibiting adenylate kinase, although AMP also inhibits creatine kinase, but to a lesser extent. Selective enzyme inhibition is also a useful technique in the characterization of isoenzymes.

Activation of Enzymes

Activators are considered to increase the rates of enzyme-catalysed reactions by promoting formation of the most active state of the enzyme itself or of other reactants such as the substrate. This generalization probably covers a wide variety of mechanisms of activation, some of which are discussed here.

Enzyme activity depends on the presence of specific chemical groups in the active centre and these groups may be altered by environmental factors, with a resulting loss of catalytic activity. Reagents which are able to reverse such changes reactivate the enzyme. Many enzymes, such as creatine kinase, depend on the presence of reduced sulphydryl (—SH) groups for their activity, so that oxidation of these groups inactivates the enzyme. Addition of compounds containing reduced sulphydryl groups to the enzyme solution or the assay mixture reactivates creatine kinase, and this procedure is now followed routinely in methods for the measurement of this enzyme in serum.

Many enzymes contain metal ions as an integral part of their structures, e.g. zinc in alkaline phosphatase and carboxypeptidase A. The function of the metal may be to stabilize tertiary and quaternary protein structure, and removal of the metal ions by treatment with concentrated EDTA solution is often accompanied by conformational changes with inactivation of the enzyme. The enzyme can often be reactivated by dialysis against a solution of the appropriate metal ion. This process may take some time, because rearrangement of the polypeptide chains into the active conformation is not instantaneous.

The metal ion component of many enzymes appears to play a direct part in catalysis, possibly in addition to any structural role they may fulfil. A metal ion may function in catalysis by providing an electropositive centre in the enzyme with which negatively charged groups in the substrate can form coordinate links.

When the activator ion is an essential part of the functional enzyme molecule, whether as a purely structural element or with an additional catalytic role, it is usually incorporated quite firmly into the enzyme molecule. It is not necessary, therefore, to add the activator to reaction mixtures and excess of the ion may in fact have an inhibitory effect. Activators which are not part of the molecule produce their effects when added to enzyme–substrate reaction mixtures.

For some enzymes combination of the apparent substrate with a metal ion may be necessary before full, or even any, catalytic activity is observed. In these cases the true substrate of the enzyme is the metal–substrate complex, and it is the concentration of this complex which influences the rate of reaction. In many reactions catalysed by ATPases and other phosphate-transferring enzymes, e.g. creating kinase, there seems to be obligatory formation of complexes between the substrate and magnesium ions. In these and other cases the metal ion may act as a bridge between the substrate and the enzyme, or it may alter the configuration of the substrate, or it may neutralize ionic charges in the substrate which would otherwise hinder the approach of the substrate to the active centre.

Some enzymes possess activator-binding sites which are separate from the site at which the substrate is bound. The effects of combination with the activator are transmitted through the protein to produce an effect on the substrate-binding site, either facilitating binding (i.e. affecting K_m), or increasing the rate of breakdown of the ES complex into products (i.e. increasing V_{max}). Both K_m and V_{max} may be affected. Cooperative effects, in which the binding of one molecule of a ligand facilitates the binding of additional molecules, are characteristic of the allosteric enzymes already discussed.

When v is plotted as a function of a, the concentration of a reversibly combining activator, a hyperbola is usually obtained at a fixed value of s, from which a value for K_A can be derived, representing the value of a at which v is half the maximum velocity observed at high concentrations of the activator. The simplest interpretation of K_A is that it is the dissociation constant of the equilibrium between E and A. However, this is only true when the combination of enzyme and activator is independent of the reaction between E and S; i.e. the same value for K_A is obtained at all concentrations of the substrate. If the free enzyme and the enzyme–substrate complex have different affinities for the activator the value for K_A varies with s. The kinetics are further complicated when the activator and substrate combine with each other.

Activation of salivary and pancreatic amylases by chloride ions (one of the rare examples of activation by anions as distinct from cations) probably involves a reversible combination of chloride with the enzyme. Addition of 5 mmol of chloride l^{-1} increases amylase activity almost threefold, at the same time shifting the pH optimum from 6.5 to 7.0. The chloride ion may combine with a positively charged group in the enzyme and change the ionization constant of a group important in catalysis. However, other anions such as bromide or iodide are less effective activators of amylase, so that some degree of specificity is involved in the process of activation.

Coenzymes and Prosthetic Groups

An absolute distinction between coenzymes and activators is difficult to draw. Coenzymes are usually of more complex constitution than activators, although smaller molecules than the enzyme proteins themselves. Some compounds which are classed as coenzymes (e.g. the dinucleotides NAD and NADP) are specific substrates in two-substrate reactions and their effect on the rate of reaction follows the Michaelis–Menten pattern of dependence on substrate concentration.

A number of coenzymes are more or less permanently bound to the enzyme molecules, where they form part of the active centre and undergo cycles of chemical change during the reaction. An example of these prosthetic groups, as they are termed, is pyridoxal phosphate, a component of aspartate and alanine aminotransferases. The active holo-enzyme results from the combination of the inactive apo-enzyme with the prosthetic group. Prosthetic groups, like activators with a structural role, do not usually have to be added to elicit full catalytic activity of a sample of the enzyme unless previous treatment has caused the prosthetic group to be lost from some enzyme molecules. However, both normal and pathological serum samples contain appreciable amounts of apo-aminotransferases, which can be converted to the active holo-enzymes by a suitable period of incubation with pyridoxal phosphate. It is not clear whether the presence of the apo-enzymes results from the loss of the prosthetic group during release of enzymes from their cells of origin, or whether the enzymes in tissues are incompletely saturated with pyridoxal phosphate.

The effect of incubation with pyridoxal phosphate is to increase the aspartate aminotransferase activity of serum specimens by an average of about 50%, but the effect on alanine aminotransferase levels is considerably less. There is considerable variation from sample to sample in the degree of activation and also, in the case of aspartate aminotransferase, between different diseases. For example, a greater degree of activation is usually seen in sera in which the elevated enzyme activity is due to myocardial infarction than when the cause of elevation is liver disease, but this seems to be of little diagnostic significance.

The differences between sera in activation of the aminotransferases by pyridoxal phosphate show that the relative proportions of apo- and holo-enzymes can vary, as well as the total concentrations of enzyme molecules. In order to ensure that assay methods measure only differences in the total enzyme content of specimens, therefore, the most recent descriptions of methods for aminotransferase assay specify conditions for activation of apo-enzymes which may be present. This principle applies also to other known cofactors in all enzyme assays.

EFFECT OF pH ON ENZYMIC REACTIONS

When pH is varied, the velocity of reaction in the presence of a constant amount of enzyme is typically greatest over a relatively narrow range of pH. The effect of pH changes on v is reversible, except after exposure to extremes of pH at which denaturation of the enzyme may occur. There is a great deal of variation between enzymes with regard to the pH at which activity is at a maximum and in the width of the zone of pH over which the enzyme is active. Some of the effects of pH on v may be due to changes in ionization of the substrate, rather than of the enzyme.

Since enzymes are proteins, they possess a large number of chemical groups which are capable of existing in different ionic forms. Of these groups, those on the outside of the molecule will respond to pH changes in the aqueous environment by changes in ionization. The relative proportions of the ionized and un-ionized forms of each group are given by the equation

$$\text{pH} = \text{p}K_a + \log \frac{\text{(concentration of ionized form)}}{\text{(concentration of un-ionized form)}} \qquad (16)$$

in which $\text{p}K_a$ is the negative logarithm of the ionization constant of the group concerned.

The existence of a fairly narrow pH range of optimum activity for most enzymes suggests that one particular ionic form of the enzyme molecule, of the many that potentially can exist, is the catalytically active one. Changes in the ionization of groups remote from the active centre of the enzyme would not be expected to affect the rate of reaction greatly, although affecting other properties of the enzyme molecule, such as solubility or electrophoretic mobility. When plots of the variation of velocity with substrate concentration are made at different pH values for a given amount of an enzyme, it is usually found that the Michaelis constant, as well as the maximum velocity, is markedly dependent on pH. Thus both binding of the substrate by the enzyme and the rate of breakdown of the enzyme–substrate complex are affected by changes in pH. The probable identity of groups present in the active centre has been inferred from a study of pH–activity relationships for a number of enzymes.

It is usual to choose the pH of maximum velocity for the assay of a particular enzyme; not only does this increase the sensitivity of the measurement, but it also reduces errors caused by slight differences in pH from one measurement to another, since the variation of activity with pH is at a minimum in the vicinity of the optimum pH. When K_m has a marked dependence on pH, i.e. when the curves of v against s vary in shape with pH, the observed pH optimum depends on substrate concentration. Therefore, the optimum pH for enzyme assay must be determined or redetermined after the appropriate substrate concentration has been selected.

The buffering capacity present in the reaction system must be sufficient to control the pH both of the reactants and of the serum sample, which may itself exert a powerful buffering action in assays in which a high ratio of sample volume to total reaction volume is used. Since buffers have their maximum buffering capacity close to their pK_a values, as far as possible a buffer system should be chosen for which pK_a is close to the desired pH of the assay. Interaction between certain buffer ions and other components of the assay system (e.g. activating metal ions) may eliminate these buffers from consideration. Although substrates such as amino acids in aminotransferase reactions have buffering properties, their concentrations are usually too low to affect the control of pH in the complete assay system. Exceptions are peptides and amino alcohols which act as amino acid or phosphate acceptors in some transferase reactions at concentrations high enough to provide most or all of the necessary buffering capacity.

EFFECT OF TEMPERATURE ON ENZYMIC REACTIONS

The rate of a reaction catalysed by an enzyme increase with increasing temperature. Although there are significant variations from one enzyme to another, the rate approximately doubles, on average, for each 10 °C rise in temperature; i.e. Q_{10} is of the order of 2. A difference of 1 °C between two measurements of enzyme activity thus causes a difference of about 10% in observed rates, emphasizing the need for close control of temperature in all such measurements.

The initial rate of reaction measured instantaneously goes on increasing with rising temperature, at least in theory. In practice, however, a finite time is needed in all methods to allow the components of the reaction mixture, including the enzyme solution, to reach temperature equilibrium and to permit the formation of a measurable amount of product. During this period the enzyme is undergoing thermal inactivation and denaturation. This process, which is one of the factors contributing to non-linearity of reaction progress curves, has a very large temperature coefficient for most enzymes and thus becomes virtually instantaneous at temperatures in the region of 60–70 °C.

The counteracting effects of the increased rate of the catalysed reaction and more rapid enzyme inactivation as the temperature is raised, account for the existence of an apparent 'optimum temperature' for enzyme activity often referred to in the older literature of enzymology. The apparent optimum temperature is dependent on the time taken to make measurements of activity. With older fixed-time methods of assay, requiring lengthy periods of incubation of enzyme and substrate, enzyme inactivation takes effect at lower temperatures and the phenomenon is more easily seen.

Thermal inactivation of enzymes is influenced by other factors as well as duration of exposure to a particular temperature. These include the presence of substrate and its concentration, and the pH, nature, and ionic strength of the buffer. The presence of other proteins, as in serum samples, may help to stabilize the enzyme. Differences in rates of thermal inactivation are an important basis for the characterization of isoenzymes.

The rate of a chemical reaction is related to the absolute temperature by the empirical equation of Arrhenius:

$$2.303 \frac{d(\log k)}{dt} = \frac{E}{RT^2} \qquad (17)$$

in which k is the rate constant for the reaction, T the absolute temperature, and R the gas constant (1.987 cal deg^{-1} mol^{-1}). E is also a constant for the particular reaction, called the energy of activation. Integrating the Arrhenius equation,

$$\log k = \frac{-E}{2.303 RT} + \text{constant} \qquad (18)$$

Since the maximum velocity is given by $k \times$ (enzyme concentration), a plot of $\log V_{max}$ against $1/T$ is linear, of slope $-E/2.303R$. The energy of activation of the reaction can thus be calculated. Comparison of the values of E

obtained for an enzymic reaction in the presence and absence of the catalyst (if the reaction proceeds at a measurable rate without the enzyme) shows that the energy of activation is much lower when the enzyme is present. Therefore, the enzyme increases the rate of reaction by lowering the energy barrier which otherwise separates the reactants and the products, although it is not yet clear how this lowering is brought about.

The choice of temperature for the assay of enzymes of clinical importance has been the subject of extensive debate. Temperatures close to the ambient temperature, e.g. 25 or 30 °C, are more suitable in manual or semi-automated continuous-monitoring enzyme assays, in which temperature fluctuations are magnified and equilibration periods are prolonged during the necessary repeated openings of the cuvette compartment of the photometer by the existence of a large difference between the ambient and measurement temperatures. This objection to the use of higher temperatures does not usually apply when automated methods of assay are in use in which samples are processed within a controlled environment, nor in the analysis of batches of specimens by fixed-time methods, and 37 °C is often chosen in these circumstances.

The greater rate of reaction at 37 °C compared with 25 or 30 °C increases the sensitivity of assay, shortening the time required for measurement, and this favours the use of the higher temperature. However, enzyme inactivation and deterioration of substrates and coenzymes are also more rapid. Whatever temperature is chosen, precise temperature control during the enzymic reaction is essential, preferably within ±0.1 °C and certainly to within ±0.2 °C. Cooling to the surrounding air is more rapid at 37 °C than at lower temperatures and this is frequently regarded as an advantage of this particular temperature. However, cycles of passive heat loss alternating with periods of heating are unlikely to meet tightening specifications for temperature control, making positive cooling increasingly necessary.

Differences in assay temperature constitute a major cause of lack of agreement between results for a particular enzyme reported from different laboratories. Because of the diversity of methods and temperatures in use, reports of results of enzyme assays should include appropriate reference ranges.

METHODS OF MEASURING CHEMICAL CHANGES IN ENZYMIC REACTIONS

The amount of substrate transformed into products during an enzyme-catalysed reaction is almost invariably measured by some form of photometric analysis. The reaction may be accompanied by a change in the absorbance characteristics of some component of the assay system, either in the visible or ultraviolet spectrum, making possible continuous monitoring in a recording spectrophotometer. One or more additional reactions may be necessary to convert a product of the primary enzymic reaction to a photometrically detectable form. If the indicator reactions can proceed simultaneously with the primary reaction without interfering with it, continuous monitoring is still possible. However, in some cases the process of colour development inhibits the enzyme being measured, or the latter, together with other proteins, interferes with colorimetry so that deproteinization is necessary. In these instances, only fixed-time assay methods are practicable.

The absorbance (A) of light of a particular wavelength by a solution is given by $A = \log(1/T)$ where T is the transmission, i.e. the ratio of emergent to incident light intensity. T is usually expressed as a percentage, so that $A = \log(100/T)$ or $2 - \log T$. A is related to the concentration, c, of the absorbing solute by Beer's Law and to the path-length, l, of the solution by Lambert's Law, so that

$$A = kcl \qquad (19)$$

in which k is a constant. When c is expressed in mol l^{-1} and l is 1 cm, k is the molar absorptivity, ε, of the solute at the wavelength of the incident light and under the defined conditions of pH, temperature, and solvent. Since A is directly proportional to c, photometers, which actually measure the ratio of transmitted and incident light, are usually provided with meters with logarithmic scales reading in absorbance units. An output signal converted to log T and therefore proportional to concentration changes is needed for the direct recording of reaction progress curves.

The Beer–Lambert law holds in practice only over a defined range of concentration for a given solute. However, developments in photometry have reduced the importance of various instrumental factors which themselves can cause non-rectilinearity of curves relating absorbance to concentration, so that it is not uncommon to find with modern instruments that the law holds over a greater range of concentration than appeared to be the case with older equipment. Nevertheless, the usable range of concentration should be determined for each method and type of photometer.

In photometric analysis in the visible spectrum, simple colorimeters with photoelectric cell detectors and wavelength selection by filters are often used. The day-to-day stability and reproducibility of these instruments are often not great. Moreover, the stoichiometry of the chromogenic reactions may be uncertain and the final coloured product poorly characterized. It is usual in colorimetry to compare the absorbance of the unknown solution with that of a similarly treated standard solution of the product being determined.

The introduction of prism- or diffraction-grating spectrophotometers capable of isolating a narrow beam of monochromatic light in the ultraviolet or visible spectrum and with stable and sensitive photomultipliers as detectors has greatly improved the reproducibility of photometric measurements. Consequently, it has become customary to make use of the known molar absorptivity of well-defined reaction products when calculating changes in their concentrations from measurements made with spectrophotometers. However, this procedure can lead to serious errors if the assumptions on which it is based are not justified.

The observed absorbance of a solution of an absorbing solute varies with the bandwidth of the beam of light used to measure it, lower readings being obtained as the bandwidth is increased (Figure 10). The error is greater when the solute has a narrow absorption peak. Inaccuracies in the wavelength setting of the spectrophotometer, or the choice of a filter with a wavelength of maximum transmission which is not close to the wavelength of maximum absorption, so that readings are being taken on the side of the absorption peak, magnify errors due to the use of wide bandwidths. Also, the spectrum of the compound being measured and overlapping spectra of interfering substances may be incompletely resolved at wide bandwidths. Published values for molar absorptivities apply only to measurements made with narrow bandwidths in correctly calibrated spectrophotometers or

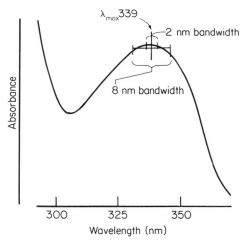

Figure 10. Ultraviolet absorption spectrum of NADH recorded in a narrow bandwith spectrophotometer. The absorbance reading at a bandwidth of 8 nm is approximately 98% of that obtained with a 2 nm bandwidth. The error of the absorbance measurement becomes much greater if the wavelength setting of the instrument does not correspond with the λ_{max} of 339 nm

spectral-line photometers. Therefore, before these values are used to calculate enzyme activities, their applicability to the instruments with which measurements are made should be verified.

In designing spectrophotometers, much effort is devoted to ensuring that all the light picked up by the detector has indeed been transmitted through the whole length of the cuvette containing the sample, by minimizing stray light and ensuring that the incident beam is not scattered by impinging on the cuvette walls. Wide slit widths may increase light scattering and also make it more probable that stray light will reach the detector. The errors produced by stray light are greater when the intensity of the transmitted beam is low, i.e. when absorbance is high, and when wide slit widths are needed to achieve adequate sensitivity. When the measurement of enzyme activity depends on the fall of absorbance from an initially high level, as in the determination of aspartate aminotransferase activity, the capacity of the photometer reliably to measure high initial absorbances limits the choice of reaction conditions, such as the concentrations of NADH and 2-oxoglutarate, and the ratio of serum volume to total volume.

High absorbances can be measured with greater precision in double-beam than in single-beam spectrophotometers because, in the former, a strongly absorbing blank solution can be placed in the reference beam to attenuate it to the same low level of intensity as the beam transmitted through the sample. The detector is therefore required to compare two signals of approximately equal strengths. However, as the level of light in both beams is low, the signal-to-noise ratio is low also, limiting the accuracy of measurement of high absorbances in double-beam instruments.

The use of double-beam spectrophotometers allows a correction for blank reactions to be made simultaneously with the recording of the absorbance changes in the test solution, whereas in single-beam instruments a separate blank recording may be required. A distinct but related problem is that of side reactions other than the one being determined which take place in the incomplete reaction mixture, i.e. before the reaction under study is initiated by addition of one of its specific substrates. A familiar example is the reduction of pyruvate in serum samples, with a corresponding oxidation of NADH, which takes place in reaction mixtures used for aspartate aminotransferase estimations before the aminotransferase reaction is triggered by addition of 2-oxoglutarate. It is usual to allow these sample-dependent changes to go to completion by an appropriate pre-incubation period before the enzyme estimation proper is started, although if the true reagent-blank reaction (e.g. any reagent-dependent aminotransferase reaction due to contamination of the indicator enzyme with aminotransferase) is negligible, a simultaneous correction can be applied in a double-beam instrument by means of a sample blank placed in the reference beam.

The most widely used values for molar absorptivities in measuring enzyme activities are those for the reduced coenzymes NADH and NADPH. For both these compounds molar absorptivities of $6.22 \times 10^3 \, l \, mol^{-1} \, cm^{-1}$ have been in general use for nearly 30 years. However, it is now apparent that a number of variables affect the molar absorptivities of the coenzymes. As the temperature increases, the wavelength of maximum absorbance decreases from about 340 nm at 0 °C to about 338.5 nm at 38 °C, with λ_{max} for NADPH being about 0.5 nm greater than for NADH. The absorption peaks also broaden with increasing temperature, and absorbance measured at a fixed wavelength (e.g. 340 nm) falls as the temperature is increased. Molar absorptivities increase as pH rises, whereas the values are lower in solutions of higher ionic strengths.

These recent observations indicate that, if quoted values for molar absorptivities of NADH or NADPH are used for calculation of enzyme activities, not only must the photometer employed be of similar performance characteristics to the instrument used to derive the original values, but also due regard must be given to the conditions of temperature and pH under which the cited values were obtained. Molar absorptivities of $6.3 \times 10^3 \, l \, mol^{-1} \, cm^{-1}$ have been suggested for both NADH and NADPH when measurements are made at 340 nm. With these values, variations in temperature or pH of the magnitude usually existing between different assay methods in clinical enzymology should not introduce errors due to changes in molar absorptivity of more than 1 or 2% (Bergmeyer, 1977).

Measurement of absorption of light is sometimes replaced by fluorimetry as a means of determining the amount of chemical change accompanying an enzymic reaction. Measurement of fluorescence instead of absorbance offers the advantages of greater sensitivity—usually of the order of 10^2 to 10^4 times that of absorptiometry—and the existence of an additional criterion, the fluorescence emission spectrum, by which the compound being measured can be distinguished from other compounds with similar absorption spectra.

The intensity of fluorescence emission (F) of solutions with low light absorbance is given by the expression

$$F = I_0(2.3\varepsilon c d) \cdot \phi \tag{20}$$

in which I_0 is the intensity of the exciting light beam and ε, c, and d have the same significance as in absorptiometry. The term ϕ is the quantum efficiency of fluorescence of the fluorescent compound. Thus the intensity of fluorescence varies with fluctuations in the intensity of the exciting beam and, since high-energy light sources which are often unstable are needed to produce measurable fluorescence, these fluctuations can cause poor reproducibility of fluorescence measurements unless instrumental design provides for their correction. If the intensity of the exciting light is kept constant as its wavelength is varied, and if the quantum efficiency of fluorescence is independent of wavelength as seems generally to be the case, F varies in the same way as ε with the exciting wavelength. In other words, the fluorescence excitation spectrum is the same as the absorption spectrum, not surprisingly since only light which is absorbed is capable of exciting fluorescence. Various instrumental factors such as the variation of the quantum intensity of the light source with wavelength distort the observed excitation spectrum so that correction for these factors is necessary before the identity of the two spectra becomes apparent.

As the concentration of a fluorescent solute increases beyond a certain level, the intensity of fluorescence no longer increases and may decrease. This effect, known as concentration quenching, is due to loss of energy by molecules in the excited state, as a result of collisions between them, without emission of light. It is essential, therefore, to ensure that quantitative measurements are made within the range in which fluorescence is proportional to concentration. Fluorescence is reduced at higher temperatures and, like absorbance, is usually dependent on pH. There is no absolute measure of fluorescence corresponding to molar absorptivity in absorptiometry because of the dependence of fluorescence emission on the intensity of the exciting beam, so that standardization is always required in fluorimetry. Permanent arbitrary standards consisting of fluorescent substances in solid solution in plastic rods or prisms are often used.

Absorptiometric methods are usually of adequate sensitivity for the assay of enzyme activities in serum. However, the greater sensitivity of fluorimetry may be valuable in the measurement of individual isoenzyme fractions.

Automation of Enzyme Activity Measurements

The increasing demand for enzyme estimations in diagnosis and treatment has been met by the introduction of automated methods of analysis, as has the growing demand for other types of analyses. Two separate lines of development of automatic enzyme analysers can be distinguished (Moss, 1977).

The first of these is the mechanization of fixed-time methods of analysis, e.g. with the Technicon Corporation continuous-flow 'AutoAnalyzer'. Provided that sound fixed-time methods are chosen, such automated methods perform at least as well and often better than their manual counterparts, because of the improved reproducibility of measurement of sample and reagent volumes generally associated with automated analysis. However, automated fixed-time procedures share the same limitations as manual methods of this type, namely, that they are valid only over a defined range of enzyme activity and that the shapes of individual progress curves cannot be monitored. Because chemical reactions do not go to completion in the 'AutoAnalyzer', calibration with known standard solutions is always necessary. The precise times at which reagent streams mingle to initiate and terminate enzymic reactions may be difficult to determine exactly; therefore, calibration with serum or other samples of known enzyme activity (determined for example, by a spectrophotometric method) is preferable to the use of standard solutions of the reaction product.

The second approach to automated enzyme analysis is represented by the many instruments now available which are designed to mechanize the successive stages of sample and reagent volume measurement, pre-incubation and temperature control, initiation of the enzymic reaction, monitoring of absorbance changes, and calculation of enzyme activity, which together constitute an enzyme measurement by the continuous-monitoring procedure. These automatic enzyme analysers are capable of providing satisfactory analytical data with greatly increased productivity. However, in their design, as in the design of all such apparatus, choices and compromises may have been made in the interests of simplicity, reliability, or reduced capital cost which introduce into the analytical process limitations which were not present in the original manual method, of which the operator should be aware. For example, the infinite number of points represented by the analogue signal of changing absorbance with time drawn on a recorder chart may be replaced by three, or even two, digitized readings during the progress of the reaction. This simplification of the reaction-monitoring process may result in failure to detect non-rectilinear progress curves. Again, the mode of operation of the automatic analyser may require the enzymic reaction to be initiated by the addition of serum to the reaction mixture, so that pre-incubation to allow sample-dependent side reactions to be completed is impossible.

Expression of Results of Enzyme Activity Measurements

Whatever the method or instrumentation used to determine enzymic activity, the result of such a determination takes the form of an amount of chemical change produced in a defined period of time. The units of amount and time were often arbitrarily chosen in older enzyme assay methods, giving rise to a series of eponymous enzyme units such as the Karmen unit of aspartate aminotransferase activity, which corresponds to a change in absorbance of 0.001 per minute in the reaction mixture under defined conditions. A uniform system of expressing enzyme activities, the International Unit (iu), was proposed by the Enzyme Commission of IUB and has been widely adopted. This corresponds to the transformation of 1 μmol of substrate in 1 minute under the conditions of measurement. However, the International Unit, like the eponymous units which it replaces, is method dependent and a particular result expressed in these terms cannot be interpreted without a knowledge of the method to which it refers: results obtained by different procedures for measuring the activity of a particular enzyme are not comparable, even if all the results are expressed in International Units, and even if the differences between the methods are relatively minor ones.

It is usually necessary in clinical enzymology to express results as the concentration of enzyme activity in the original sample, usually serum, so that a suitable volume term must be added, e.g. 1 ml in the case of Karmen Units. The definition of the International Unit itself contains no volume term and concentrations of enzyme activity are usually expressed as iu l^{-1}.

Changes in the methods of expressing enzyme activity to conform with SI practice are currently under discussion, and will probably result in the eventual adoption of the 'katal' (the catalytic activity of any enzyme which produces the transformation of 1 mol of substrate per second under defined conditions), one iu being approximately equal to 16.67 nanokatals (abbreviated to nkat). The volume term for concentrations remains the litre. Expression of enzyme activities in katals, as with other units, does not remove the need to specify the conditions under which the results were obtained.

Standardization and Quality Control of Enzyme Assays

The proliferation of methods for measurement of the activities of enzymes of diagnostic importance, together with the multitude of different units of activity and reference ranges to which this diversity of methods inevitably gives rise, constitute a serious barrier to the interpretation of data obtained in different laboratories, whether included in the records of individual patients or reported in the scientific literature. Two approaches to standardization of enzyme assays are possible (Moss, 1971). The first is to provide enzyme preparations of stated catalytic activity which can be used as calibration standards, in the way in which solutions of accurately known concentration are used to standardize determinations of such substances as glucose. However, this approach requires the use of enzyme preparations of assured stability: because of the susceptibility of enzymes to denaturation, sufficiently stable enzyme standards are not yet available, although research into methods of stabilizing enzyme molecules may rectify this in the future.

The second approach is to standardize the conditions of assay, and this has been the objective of various working-parties and committees. The specific aims of these groups have not been identical, leading to some variation in recommendations, ranging from methods intended for use in small groups of neighbouring laboratories and therefore reflecting the practice prevailing in those laboratories, to rigorously specified reference methods intended to provide a criterion of analytical performance against which routine methods can be judged. Between these extremes are the methods proposed by national or international enzyme committees which

Table 1. Reaction Conditions Recommended for the Assay of the Three Enzymes of Greatest Diagnostic Value in Cardiology. All concentrations refer to the complete reaction mixtures containing the sample. (a) Committee on Enzymes of the Scandinavian Society for Clinical Chemistry and Clinical Physiology (1974), (b) Expert Panel on Enzymes, International Federation of Clinical Chemistry (1977), (c) provisional recommendations agreed between societies for clinical chemistry in the German Federal Republic, France, and Scandinavia (1977).

Lactate dehydrogenase (L-lactate: NAD—oxidoreductase, EC 1.1.1.27)

$$\text{pyruvate} + \text{NADH} + \text{H}^+ \rightleftharpoons \text{L-lactate} + \text{NAD}^+$$

(a) Substrate concentrations: pyruvate 1.2 mmol l^{-1}, NADH 0.15 mmol l^{-1}
    pH 7.4; buffer: Tris 50 mmol l^{-1}, EDTA 5 mmol l^{-1}
    Temperature: 37.0 °C
    Sample volume: total volume ratio 1:45

Consumption of NADH is recorded continuously by fall in absorbance at 340 nm.

Aspartate aminotransferase (L-aspartate: 2-oxoglutarate aminotransferase, EC 2.6.1.1; also called aspartate transaminase or glutamic-oxaloacetic transaminase)

$$\left.\begin{array}{c}\text{L-aspartate}\\+\\\text{2-oxoglutarate}\end{array}\right\} \rightleftharpoons \left\{\begin{array}{c}\text{glutamate}\\+\\\text{oxaloacetate}\\+\\\text{NADH}+\text{H}^+\end{array}\right\} \xrightarrow{\text{malate dehydrogenase}} \left\{\begin{array}{c}\text{malate}\\+\\\text{NAD}^+\end{array}\right.$$

(a) Substrate concentrations: L-aspartate 200 mmol l^{-1}, 2-oxoglutarate 12 mol l^{-1}, NADH 0.15 mmol l^{-1}
    Indicator enzyme: malate dehydrogenase 600 iu l^{-1}
    pH 7.7; buffer: Tris 20 mmol l^{-1}, EDTA 5 mmol l^{-1}
    Temperature: 37.0 °C
    Sample volume: total volume ratio 1:8.33

(b) Substrate concentrations: L-aspartate 240 mmol l^{-1}, 2-oxoglutarate 12 mmol l^{-1}, NADH 0.18 mmol l^{-1}
    Indicator enzyme: malate dehydrogenase 420 iu l^{-1}
    Activator: pyridoxal-5-phosphate 0.10 mmol l^{-1}
    pH 7.8; buffer: Tris 80 mmol l^{-1}
    Temperature: 30 ± 0.05 °C
    Sample volume: total volume ratio 1:12

In both (a) and (b) consumption of NADH is recorded continuously at 340 nm. Both include lactate dehydrogenase to ensure reduction of pyruvate in the sample. Method (b) specifies both sample- and reagent-blank corrections.

Creatine kinase (ATP: creatine phosphotransferase, EC 2.7.3.2)

$$\left.\begin{array}{c}\text{creatine phosphate}\\+\\\text{ADP}\end{array}\right\} \rightleftharpoons \left\{\begin{array}{c}\text{creatine}\\+\\\text{ATP}\\+\\\text{D-glucose}\end{array}\right\} \xrightarrow{\text{hexokinase}} \left\{\begin{array}{c}\text{ADP}\\+\\\text{glucose-6-phosphate}\\+\\\text{NADP}^+\end{array}\right\} \xrightarrow{\text{glucose-6-phosphate dehydrogenase}} \left\{\begin{array}{c}\text{glucono-lactone-6-phosphate}\\+\\\text{NADPH}+\text{H}^+\end{array}\right.$$

Table 1. (*Continued*)

(c) Substrate concentrations: creatine phosphate 30 mmol l^{-1}, ADP 2 mmol l^{-1}, D-glucose 20 mmol l^{-1}, NADP 2 mmol l^{-1}
Auxiliary enzyme: hexokinase 2500 iu l^{-1}
Indicator enzyme: glucose-6-phosphate dehydrogenase 1500 iu l^{-1}
Activators: Mg^{2+} 10 mmol l^{-1}, N-acetylcysteine 20 mmol l^{-1}
Inhibitors of adenylate kinase: AMP 5 mmol l^{-1}, diadenosine pentaphosphate 0.01 mmol l^{-1}
pH 6.7 at 25 °C; buffer: imidazole-acetate 100 mmol l^{-1}
Temperature: 25, 30, or 37 °C according to national recommendations
Sample volume: total volume ratio 1:25

Production of NADPH is recorded continuously by increase in absorbance at 340 nm. Addition of EDTA 1.5–2.0 mmol l^{-1} is now recommended.

---

are designed to combine analytical reliability with applicability to current standards of instrumentation and skill in well-equipped laboratories. Examples of this approach are the methods recommended by the Scandinavian Committee on Enzymes (Table 1).

Many laboratories now perceive the advantages of being able to refer their enzyme results to a particular set of recommended methods, to provide at least a partial solution to the problems of enzyme standardization, and methods such as those defined by the Scandinavian Committee on Enzymes are being used increasingly. These recommended methods are frequently referred to as 'optimized methods', to reflect the attempts made in their formulation to select a value for each variable such that enzyme activity approaches the maximum velocity attainable for a given sample of enzyme, since under these conditions small changes in factors other than the amount of active enzyme present will have minimum effects on the measured rate of reaction. Usually the reaction conditions finally chosen give a reaction velocity rather less than the maximum ultimately attainable because of the need to take into account practical considerations such as the limited solubility of some reactants, or the light absorbance of concentrated solutions. These compromises are reflected in the description 'optimized' rather than 'maximized'. Optimized methods are typically able to deal with greater ranges of enzyme activity than methods with less adequate reaction conditions, making the need for special treatment of high-activity samples, such as preliminary dilution, less frequently necessary. Non-proportionate changes of activity as a result of dilution of enzyme samples is sometimes encountered, notably with creatine kinase, for which an apparent activation on dilution is reported from time to time. Several explanations have been offered in the case of creatine kinase, including enzyme reactivation or dilution of an inhibitor in serum, but the phenomenon seems to be inconstant and is said to be absent when modern assay methods are used.

Although the selection of reliable methods is an important step towards consistent and reproducible performance of enzyme assays, even the best methods are ineffective unless sufficient attention is paid to reagent quality. The possibility that impure or defective reagents may find their way into the laboratory, or that reagents may deteriorate due to overlong or unsuitable storage, must be kept in mind. Consistency of reagents can be checked by analysing samples with both old and new reagents. If discrepancies are found, the faulty reagent can be identified by a process of elimination.

The application of quality-control procedures to ensure that satisfactory analytical performance of enzyme assays is maintained on a day-to-day basis is also complicated by the tendency of enzyme preparations to undergo denaturation with loss of activity. This makes it difficult to distinguish between poor analytical performance and denaturation as possible causes of a low result obtained for a control sample introduced into a batch of analyses. Lyophilized preparations containing various enzymes are available from commercial sources and these have a useful function in quality control, as have serum pools prepared in the laboratory, assayed for enzyme activity and stored at $-20\,°C$ in small portions for daily use. However, the use of both types of control materials is attended with uncertainties for the reasons mentioned above. Furthermore, it is often difficult to reproduce the values for activity stated by the manufacturers for freeze-dried sera, even when the same methods are apparently in use, so that recalibration in the laboratory is necessary.

A further useful procedure in the control of enzyme analysis is to carry forward from one batch to the next one or two samples, so that the successive results can be compared: some loss of activity usually takes place between the first and second analyses, but for many enzymes this is less than 10 per cent with suitable interim storage. When large batches of specimens (50 to 100 or more) drawn from a constant type of population are routinely analysed, comparison of the daily batch-means provides a useful additional means of quality control in enzymology and one which makes no assumptions about the stability of enzymes in control materials. However, it may be difficult to decide how much day-to-day variation in batch mean arises from analytical variability and how much from changes in the composition of the population from which the samples come.

In spite of these various difficulties, systematic application of quality-control programmes is essential in enzymic analysis, as in other forms of clinical analysis, if the results obtained are to be useful in diagnosis and treatment. Reproducibility of results of enzyme assays on a day-to-day basis is usually of the order of $\pm 10\%$ coefficient of variation for activities within the normal range and $\pm 5\%$ coefficient of variation for elevated activities with well-controlled, modern analytical techniques.

IDENTIFICATION AND MEASUREMENT OF ISOENZYMES

Isoenzymes are proteins with similar catalytic properties but which differ in structure. The differences in structure arise at the level of the gene, either as a result of the existence of multiple loci coding for a particular enzyme, or because of allelic variation at a single locus. Some families of isoenzymes are due to a combination of both types of genetic multiplicity. In the case of multimeric enzymes, the polypeptides of different structures determined by different loci or alleles may aggregate to form hybrid enzyme molecules consisting of a mixture of unlike subunits. Thus a further series of heteropolymeric isoenzymes is formed. A familiar example is that of lactate dehydrogenase. Two loci determine the synthesis of two distinct subunits of this enzyme in most human tissues. However, the catalytically active enzyme molecule is a tetramer and, besides the two homopolymers each consisting of four identical subunits, random association of unlike subunits gives rise to three further heteropolymers (Figure 11). When the

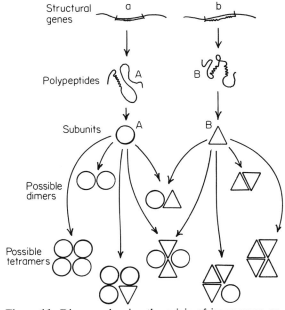

Figure 11. Diagram showing the origin of isoenzymes, assuming the existence of two distinct gene loci. When the active enzymes are polymers containing more than one subunit, hybrid isoenzymes consisting of mixtures of different subunits may be formed. One such isoenzyme can be formed in the case of a dimeric enzyme such as creatine kinase, and three if the enzyme is a tetramer (e.g. lactate dehydrogenase). In both cases two homopolymeric isoenzymes can also exist

active enzyme is a dimer, as is creatine kinase, the existence of two distinct subunits can generate three isoenzymes.

Isoenzymic variation due to the existence of multiple alleles provides the molecular basis for many inherited metabolic diseases. However, the isoenzymes which are of importance in diagnostic enzymology are the products of multiple gene loci. Modified forms of enzyme molecules can also be formed as a result of the operation of various processes, *in vivo* or *in vitro*, after biosynthesis of the constituent polypeptides has taken place. Such postgenetic modifications are of little diagnostic significance, except in so far as they may confuse the interpretation of patterns of true isoenzymes.

Where multiple gene loci exist which determine the structure of a particular enzyme, the different loci are not usually expressed to equal extents in all tissues. The rates of degradation of the separate gene products may also differ between tissues. Thus the contribution of the various isoenzymes to the total enzyme activity may vary quite widely from one tissue to another. Release of enzymes into the circulation as a result of tissue damage causes the distribution of isoenzymes in the plasma to approach that of the damaged tissue. The increased organ specificity thus conferred on enzyme tests constitutes an important aspect of the diagnostic use of isoenzyme studies. Of equal importance, however, is the fact that the isoenzyme pattern of a sample of serum or plasma may be abnormal, even though the total enzyme activity in question may be within normal limits, thus increasing the sensitivity of enzyme tests in diagnosis.

Since isoenzymes are, by definition, genetically distinct proteins, they differ in primary structure (i.e. in amino acid sequence) and consequently in higher levels of protein structure also. The differences between isoenzymes are therefore ultimately to be defined by structural studies of the individual protein molecules. However, in order to make the advantages of isoenzyme characterization available for clinical purposes, methods are needed which can distinguish qualitatively or quantitatively between isoenzymes within the limits imposed by the nature of clinical samples and the time available for analysis. Several techniques are of established value in these respects.

Differences in structure between the isoenzymic forms of an enzyme may be very small, amounting perhaps to only a single amino acid residue in the case of isoenzymes determined by allelic genes. Where an amino acid with an ionizable side chain replaces a non-ionizable or differently ionizable residue, the isoenzymes differ in net molecular charge. Only about half of the constituent amino acids of proteins have ionizable side chains; nevertheless, the primary structure determines the overall conformation of the protein molecule, and thus the amino acid side chains in contact with the aqueous environment, so that the majority of amino acid substitutions probably result in a change in net molecular charge, even if ionizable amino acid residues are not directly involved. Consequently, methods of protein separation which depend on differences in net molecular charge, such as electrophoresis or ion exchange chromatography, have a particularly important place in isoenzyme methodology.

Electrophoretic Separation of Isoenzymes

Zone electrophoresis on various supporting media combines the advantages of high resolution with applicability to small volumes of sample, when suitably sensitive methods are used to detect and to measure the separated enzyme zones. Certain supporting media, notably starch and polyacrylamide gels, have pores which approach in size the molecular dimensions of protein molecules, so that an element of segregation according to size is introduced into the resolution of protein mixtures on these media. The members of the lactate dehydrogenase isoenzyme system do not differ markedly from each other in molecular size, nor do the isoenzymes of creatine kinase. The use of gel media does not significantly affect the electrophoretic separation of either set of isoenzymes, therefore, and similar patterns are obtained with, for example, starch gels and with non-sieving supports such as cellulose acetate film. However, the restricted diffusion of protein molecules and reaction products in the gels tends to preserve discrete enzyme zones during staining of enzyme zones after electrophoresis. Against this advantage must be set the greater technical difficulty of preparing and handling gels compared with films, the relative impermeability of some gels to enzyme-location reagents, and difficulties which may be encountered in densitometric scanning of stained gels.

The methods used to locate isoenzyme zones after electrophoresis are usually adaptations of those used in histochemistry to demonstrate enzyme activities in tissue sections by formation of a coloured precipitate. For example, in locating zones of lactate dehydrogenase activity, oxidation of the substrate, lactate, is coupled through several stages to the reduction of a tetrazolium salt:

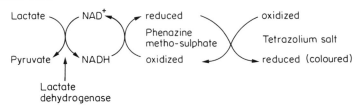

After a suitable interval the reaction is stopped and the coloured bands are fixed, e.g. by treating the strip with dilute acetic acid. With suitable modifications, this sequence of reactions can be applied to the location of other enzymes which directly, or indirectly through coupled reactions, effect the reduction of NAD^+ or $NADP^+$.

When the purpose of isoenzyme separation is to distinguish between possible alternative sources of a raised enzyme activity—e.g. between heart and liver as sources of lactate dehydrogenase in serum—inspection of the relative intensity of the coloured isoenzyme zones may provide sufficient information. However, in order to exploit the potential diagnostic sensitivity of isoenzyme separation, quantitative determination of the relative proportions of the different fractions is necessary. For example, changes in the relative proportions of isoenzymes 1 and 2 of lactate dehydrogenase, or an increase in the normally very low activity of the MB isoenzyme of creatine kinase, are sensitive indicators of myocardial damage which require quantitative methods for their demonstration.

Lactate dehydrogenase isoenzymes are well separated by zone electrophoresis, so that densitometric scanning of stained electrophoretic strips gives a good estimate of the activities of the individual isoenzymes, provided that the total activity applied to the strip and the staining conditions are such that the intensity of the colour is proportional to enzymic activity. The possible errors inherent in fixed-time staining methods can be avoided by repeatedly scanning the strip at 340 nm in a densitometer sensitive to this wavelength at timed intervals after treating the strip with a buffered substrate–coenzyme solution. The rate of reaction corresponding to each zone is then derived from successive readings. Although potentially accurate, this method requires appropriate apparatus and is time consuming.

Similar considerations apply to the location and quantitation of zones of creatine kinase activity after electrophoresis. As with lactate dehydrogenase, here also separation of the isoenzymes is good, but the need to detect and measure low activities of the MB isoenzyme calls for the use of sensitive techniques, such as measurement of the fluorescence of reduced NADP in a suitable fluorescence scanner, or in a fluorimeter by eluting the reduced NADP produced by the action of the isoenzyme zones *in situ* in the gel support in which electrophoresis has been carried out.

Elution of electrophoretically separated isoenzyme zones, followed by measurement of the enzymic activities of the eluates, provides an alternative to densitometric methods of isoenzyme quantitation and one which has been applied in the case of creatine kinase. This approach is capable of giving accurate estimates of isoenzyme activities and avoids some of the difficulties associated with densitometry. However, the small sample volumes which can be applied to some media, such as cellulose acetate, require the use of sensitive methods of assay of activity in eluates, while recoveries from other media, such as starch gel, are low or variable.

The qualitative separation of isoenzymes by zone electrophoresis is not in general affected by changes in composition of the serum itself. Occasionally, association between isoenzymes and other proteins *in vivo* may result in the appearance of zones with unusual electrophoretic mobility, as may rare genetic mutations affecting the locus determining the structure of a particular isoenzyme, or modifications in isoenzyme structure accompanying malignant transformation. Dissociation and regrouping of subunits during storage of serum samples, or the selective inactivation or alteration of certain isoenzymes, have been reported to change the electrophoretic patterns of both lactate dehydrogenase and creatine kinase isoenzymes. While these possibilities must be kept in mind, isoenzyme patterns determined by electrophoresis are in the main characteristic and reproducible. Consequently, a considerable advantage of electrophoresis is the high degree of certainty which it brings to the identification of specific isoenzymes.

Ion Exchange Chromatography

This technique also makes use of differences in net molecular charge at a given pH to separate isoenzymes. A typical ion exchange material is DEAE-cellulose, in which ionizable diethylaminoethyl groups are attached to an inert cellulose matrix. At alkaline pH, the groups are positively charged and negatively charged protein molecules are attracted to them. Attachment of the protein molecules to the exchanger is reversed either by a progressive change in pH which reduces the ionic charges causing attachment, or by a progressive increase in the concentration of a negatively charged ion which competes with the protein molecules for the ion exchange groups.

When a mixture of isoenzymes, e.g. of lactate dehydrogenase, is eluted from a column of DEAE-cellulose by a gradient of increasing chloride concentration, the least anionic isoenzyme (e.g. isoenzyme 5 of lactate dehydrogenase) appears first in the eluate, followed successively by increasingly anionic forms. Thus the separation obtained is similar to that seen on zone electrophoresis, and the individual isoenzymes can be characterized in terms of the chloride concentration (or pH, if a pH gradient is used) at which they are eluted. Ion exchange chromatography is not in general as highly resolving of closely similar proteins as zone electrophoresis, but relatively large amounts of proteins can be separated with good recoveries of enzymic activity, so that the method is of great value in enzyme purification.

The need to measure the enzymic activity of multiple fractions of the total eluate from an ion exchange column in order to locate and quantitate the isoenzyme peaks makes the method laborious, while elution itself may take several hours. When the isoenzymes to be separated are quite distinct in their elution characteristics, stepwise elution or batch processes may be used. In the first technique, the ion exchange column (usually small) is washed with a given volume of eluant of constant composition. The total eluate is collected for measurement of enzymic activity. The column is then washed with a further fixed volume of eluant of a composition chosen to elute a second isoenzyme and the whole eluate is collected. Further washings may be interposed between the two elution steps. In batch processes, the sample and successive eluants are shaken with solid ion exchange material in which the inert support may be beads of glass or macroporous resin.

The simplicity of stepwise ion exchange methods makes them suitable for routine analysis of serum samples and they have been widely applied to the quantitation of creatine kinase isoenzymes (Mercer, 1974). However, these methods are susceptible to error if all the activity of a particular isoenzyme is not eluted by the appropriate eluant, or by succeeding washes, so that a fraction of its activity is eluted with a second isoenzyme, the activity of which is consequently overestimated. Errors of this type are particularly likely to affect estimates of the second isoenzyme when the isoenzyme eluted in the first step is present in large excess in the original mixture compared with the second isoenzyme. This is the case in stepwise elution methods for the determination of the MB isoenzyme of creatine kinase, when failure to remove the MM isoenzyme completely causes overestimation of the MB form. Strong binding or denaturation of an isoenzyme on the ion exchange material may cause its underestimation. The dilution of the isoenzymes in the total volume of eluate plus washings may make direct estimation of enzyme activity without preliminary concentration difficult. Increasing the volume of sample applied to the column to counteract the dilution effect may also increase the possibility of carry-over of one isoenzyme into a succeeding fraction.

Other forms of chromatography which have been applied to fractionation of isoenzyme mixtures include high-pressure liquid chromatography, which combines properties of high resolution with rapidity, and affinity chromatography. The latter makes use of differences between isoenzymes in their affinities for a specific ligand which is attached to an inert insoluble support used as the stationary phase in a chromatography column or in a batch technique. Both these techniques have so far been little used in routine analysis of clinical specimens. As already mentioned, related isoenzymes are usually of similar sizes so that exclusion chromatography (gel filtration) is of little application in separating them, though an important preparative technique.

Probably the most highly resolving of all protein separation methods which depend on differences in ionization characteristics is isoelectric focusing, in which protein molecules migrate through a stabilized pH gradient under the influence of a potential difference applied across it, and concentrate as discrete zones at the pH regions corresponding to their isoelectric points. However, the time required to reach equilibrium is lengthy, enzymes are susceptible to inactivation at their isoelectric pH values, and interaction with the carrier ampholytes used to produce the pH gradient may take place.

Selective Inactivation in the Characterization of Isoenzymes

The relatively small changes in structure which distinguish one isoenzyme from another may result in quite marked differences in stability. Thus comparison of rates of inactivation under controlled conditions has become an important technique in isoenzyme characterization. Exposure to elevated temperatures or to concentrated solutions of urea are frequently chosen as denaturing agents. Rates of enzyme inactivation by both these agents are critically dependent on the conditions of the experiment which must therefore be strictly controlled if reliable comparisons between samples are to be made. For example, temperature coefficients of inactivation by heat are of the order of 10–100 or more, so that variations in temperature of fractions of a degree markedly alter rates of inactivation. For heat inactivation studies, therefore, large, well-stirred water-baths with high-performance thermostats are needed to minimize temperature fluctuations and the duration of inactivation must be accurately timed. When urea is used as the inactivating agent, freshly prepared solutions must be used since cyanate forms on storage of such solutions and may itself act as an enzyme inhibitor.

The differences in stability between isoenzymes 1 and 5 of lactate dehydrogenase are marked; e.g. after heating at 60 °C for 60 minutes, less than 30% of the initial activity survives in serum samples in which isoenzyme 5 predominates, compared with more than 60% when isoenzyme 1 is the major component. This test is therefore widely used to distinguish between liver and heart as possible sources of a raised serum lactate dehydrogenase activity. However, the heat stability of lactate dehydrogenase in sera with normal total activities ranges between 30 and 60%. Because of this, a considerable elevation of activity due to entry of isoenzyme 1 into the circulation is needed to produce an abnormal heat stability in a specimen from a patient whose original lactate dehydrogenase heat stability was towards the lower end of the normal range. The method is therefore insensitive to small changes in the isoenzyme composition of serum. Differences in heat stability between isoenzymes 1 and 2 of lactate dehydrogenase are much less marked than those between isoenzymes 1 and 5 and, although the greater stability of isoenzyme 1 compared with isoenzyme 2 can be demonstrated under suitable conditions, determinations of heat stability are not sufficiently sensitive to demonstrate an increase in the activity of isoenzyme 1 relative to that of isoenzyme 2.

Catalytic Differences between Isoenzymes

Although the isoenzymic forms of a particular enzyme catalyse the same chemical reaction, they are rarely completely identical in catalytic properties: e.g. isoenzymes may not have identical Michaelis constants for their substrates, their pH optima may differ, inhibitors may produce quantitatively different effects with various isoenzymes, and isoenzymes may differ in their relative rates of action on alternative substrates in those instances in which the substrate specificity of the enzyme allows the structure of the substrate to be varied. The extent of the variations between isoenzymes in some or all of these respects is considerable but, when significant differences do exist, they can be made the basis of methods of identification and measurement of particular isoenzymes. Under the most favourable circumstances, this approach to isoenzyme measurement has the advantage that only slight changes in the usual method of measuring enzyme activity may be needed to provide information on the isoenzyme composition of a sample. On the other hand, all such methods depend on the fundamental assumption that the properties of the respective isoenzymes remain constant in every sample of serum or plasma in which they occur. Differences in catalytic properties have been used in isoenzyme studies of both lactate dehydrogenase and creatine kinase.

The isoenzymes of lactate dehydrogenase will oxidize 2-hydroxybutyric acid, the next higher homologue of lactic acid, and the rate at which this substrate is converted is greater for the electrophoretically more anodal isoenzymes. Therefore, measurement of the rate at which 2-hydroxybutyrate is oxidized (or, more usually, at which 2-oxobutyrate is reduced) reflects the extent to which the distribution of lactate dehydrogenase isoenzymes in serum is shifted towards a more anodal pattern (Elliott, Jepson, and Wilkinson, 1962). Activity towards the alternative substrate is usually termed 'α-hydroxybutyrate dehydrogenase (HBD) activity', although the activity is a property of lactate dehydrogenase isoenzymes, and often the sensitivity of the method to changes in isoenzyme distribution is increased by expressing the results as a ratio of the respective activities with the two substrates, pyruvate and 2-oxobutyrate. Lactate dehydrogenase isoenzymes also differ in the rates at which they utilize synthetic analogues of their second substrate, NAD.

Although the use of alternative substrates is analytically very convenient and yields quantitative data, differences in substrate specificity between the most similar pairs of isoenzymes are rather small, so that the method, like heat inactivation, is relatively insensitive to changes in the relative proportions of, for example, isoenzymes 1 and 2 of lactate dehydrogenase. Like heat inactivation, the tissue specificity of the method is seen to best advantage in deciding between heart and liver as possible sources of a raised total serum lactate dehydrogenase activity. However, determination of 'HBD' activity is more sensitive than measurement of total lactate dehydrogenase activity in serum among the enzyme tests which are useful in the investigation of myocardial infarction; thus, it offers some of the increased sensitivity associated with the clinical use of isoenzymes, though to a rather smaller degree than the quantitative determinations of individual isoenzymes.

With lactate dehydrogenase and also with other isoenzyme systems, differences in catalytic properties between isoenzymes may be most apparent when conditions such as substrate concentration are suboptimal. This may affect precision, since the activity being measured is lowered, and may require the use of methods of high sensitivity. The isoenzymes of creatine kinase differ in their Michaelis constants and consequently in the ratios of their activities measured at low and high substrate concentrations. A method of determining MB-creatine kinase has been based on these differences but, because of the very low activities observed at low substrate concentrations, a highly sensitive assay based on emission of light in a coupled ATP-dependent luciferase reaction has to be employed (Witteveen, Sobel, and DeLuca, 1974).

Selective inhibition is a useful technique in analysing mixtures of certain isoenzymes. Isoenzymes 1 and 5 of lactate dehydrogenase are inhibited to different extents by low concentrations of sulphite, oxamate, or oxalate, intermediate degrees of inhibition being shown by the hybrid isoenzymes. Selective activation of creatine kinase isoenzymes with various sulphydryl reagents has been observed under certain conditions and a method based on this observation has been advocated for the determination of the MB isoenzyme of creatine kinase in serum. However, the consistency and reproducibility of the differences between creatine kinase isoenzymes in their activation characteristics have been questioned.

Immunochemical Methods of Isoenzyme Characterization

Different isoenzymes or their constituent polypeptide subunits are usually sufficiently distinct in structure to be recognized as antigenically distinct when injected into animals of species other than that from which they originate. Thus antisera can be prepared which are specific for the particular isoenzymic form of the enzyme used as the antigen. The use of isoenzyme-specific antisera brings to isoenzyme characterization the great specificity associated with the antigen–antibody reaction: an antiserum prepared against isoenzyme 1 of human lactate dehydrogenase does not cross-react with isoenzyme 5, and *vice versa*. Similarly, an antiserum to human MM creatine kinase does not cross-react with the BB isoenzyme, nor an anti-BB antiserum with the MM dimer. However, since the specific antigens are the polypeptide subunits which make up the active isoenzyme molecules, although no cross-reaction is observed with homopolymers of the non-immunizing antigen, heteropolymers (i.e. hybrid isoenzymes) cross-react to various degrees, reflecting their subunit compositions. For example, the MB isoenzyme reacts with antisera raised against either the MM or BB isoenzymes, but to a lesser extent than that with which the BB or MM dimers themselves respectively react with such antisera.

The availability of specific antisera to isoenzymes opens up a wide range of possible methods of measurement. Many of the immunochemical methods which have been applied to isoenzyme analysis have continued to make use of the catalytic activity of the isoenzymes, either as the means of quantitative measurement (e.g. by determination of residual activity after reaction with antiserum), or as a means of locating the enzyme–antigen precipitate, if the reaction with antibody does not inhibit activity, in immunodiffusion or 'rocket' electroimmunoassays. Radioimmunoassays, in which isoenzyme labelled with a radioactive tracer competes with unlabelled isoenzyme for antibody-binding sites, have also been applied to isoenzyme measurement, and these methods do not depend on the catalytic activity of the isoenzyme being determined.

Among the isoenzymes of diagnostic value in cardiology, immunochemical methods have been explored most actively in the determination of the MB isoenzyme of creatine kinase in serum. Immunoprecipitation has been used, in which residual creatine kinase activity is measured after precipitation of the MM and BB forms with appropriate antisera (Jockers-Wretou and Pfleiderer, 1975). Each antiserum precipitates part of the MB activity, but the proportion of MB isoenzyme in the mixture can be calculated from the two residual activities and the total creatine kinase activity. An alternative procedure employs an anti-MM antiserum which inhibits almost completely the activity of M subunits of creatine kinase, whether present as MM dimers or as MB hybrid isoenzymes (Gerhardt *et al.* 1977). Thus residual activity reflects the proportion of MB creatine kinase in the isoenzyme mixture, provided that no BB dimer is present. Correction is necessary for the slight excess of M subunit activity which is not inhibited as well as for adenylate kinase activity incompletely suppressed by inhibitors of this enzyme and which may become significant in comparison with the low activities of the MB isoenzyme remaining after immunoinhibition.

A radioimmunoassay based on the use of MB creatine kinase labelled with ^{125}I and an anti-BB antiserum has been described (see Chapter 10) which is capable of detecting picomolar amounts of the MB isoenzyme in the presence of a large excess of the MM isoenzyme by competitive displacement (Roberts, Sobel, and Parker, 1976). This method is independent of catalytic activity and therefore will measure both active and inactive enzyme molecules, provided that the process of inactivation of the enzyme in plasma does not destroy its antigenic determinants. Since catalytically active enzyme molecules may represent only a fraction of those released from myocardial cells and surviving in the circulation, the ability to measure total amounts of enzyme molecules may help to improve estimates of infarct size. However, similar estimates of enzyme release were obtained by both immunochemical and catalytic activity measurements in a number of patients.

Both the antibody inhibition method and the radioimmunoassay procedure for the determination of MB creatine kinase assume that the BB dimer is absent from the serum samples, since this isoenzyme would be measured together with the MB form. This assumption is probably true for uncomplicated myocardial infarction. However, BB creatine kinase occurs in several tissues besides brain and a number of reports have appeared of the presence of this isoenzyme in serum in a variety of conditions. Although many of these conditions are unlikely to confuse the investigation of suspected myocardial infarction, one recent report indicates that BB creatine kinase may be present transiently in the sera of patients during coronary-artery bypass grafting (Vladutiu et al. 1977). If BB creatine kinase is found to occur regularly under these circumstances or in other conditions in which diagnosis of myocardial infarction is the objective of measuring creatine kinase isoenzymes, modification of these two methods to increase their specificity for MB creating kinase will be needed.

CONCLUSIONS

Many different methods for the identification and measurement of isoenzymes in serum are available, based both on structural differences—in net molecular charge, in stability, and in antigenicity, and on differences in catalytic behaviour—in relative substrate specificity, Michaelis constants, and response to specific inhibitors. The choice of a particular method depends on a number of factors; amongst them, the resources of skill and equipment available, and the purpose for which the investigations are intended. If differentiation between possible alternative tissue sources of a raised enzyme activity is required, qualitative or semiquantitative methods will probably be adequate. For the detection of minimal myocardial damage, however, methods capable of measuring small changes in the activity of MB creatine kinase, or changes in the relative activities of isoenzymes 1 and 2 of lactate dehydrogenase are needed.

Whatever methods of isoenzyme characterization are chosen, quality control and strict adherence to specified procedures are as essential as in other branches of diagnostic enzymology (Rosalki, 1974).

BIBLIOGRAPHY

Bergmeyer, H. U. New values for molar absorption coefficients of NADH and NADPH for use in routine laboratories. *Zeitschrift für Klinische Chemie und Klinische Biochemie*, **13**, 507–8 (1977).

Bergmeyer, H. U. and Gawehn, K. *Principles of Enzymatic Analysis*, Weinheim, Verlag Chemie, 1977.

Bergmeyer, H. U., Scheibe, P., and Wahlefeld, A. W. *Optimization Methods for Aspartate Aminotransferase and Alanine Aminotransferase*. *Clinical Chemistry*, **24**, 58–73 (1978).

Committee on Enzymes of the Scandinavian Society for Clinical Chemistry and Clinical Physiology. Recommended methods for the determination of four enzymes in blood. *Scandinavian Journal of Clinical and Laboratory Investigation*, **33**, 291–306 (1974).

Elliott, B. A., Jepson, E. M., and Wilkinson, J. H. Serum 'α-hydroxybutyrate dehydrogenase'. A new test with improved specificity for myocardial lesions. *Clinical Science*, **23**, 305–16 (1962).

Expert Panel on Enzymes, International Federation of Clinical Chemistry. Provisional recommendations on IFCC methods for the measurement of catalytic concentrations of enzymes, Parts 2 and 3, *Clinica Chimica Acta*, **70**, F19 (1976); **80**, F21 (1977).

Gerhardt, W., Ljungdahl, L., Börjesson, J., Hofvendahl, S., and Hedenäs, B. Creatine kinase B-subunit activity in human serum. 1. Development of an immunoinhibition method for routine determination of S-creatine kinase B-subunit activity. *Clinica Chimica Acta*, **78**, 43–53 (1977).

German Society for Clinical Chemistry. Standard method for the determination of creatine kinase activity. *Zeitschrift für Klinische Chemie und Klinische Biochemie*, **15**, 255–60 (1977).

Jockers-Wretou, E. and Pfleiderer, G. Quantitation of creatine kinase isoenzymes in human tissues and sera by an immunological method. *Clinica Chimica Acta*, **58**, 223–32 (1975).

London, J. W., Shaw, L. M., Fetterolf, D., and Garfinkel, D. A systematic approach to enzyme assay optimization, illustrated by aminotransferase assays. *Clinical Chemistry*, **21**, 1939–52 (1975).

Mercer, D. W. Separation of tissue and serum creatine kinase isoenzymes by ion-exchange column chromatography. *Clinical Chemistry*, **20**, 36–42 (1974).

Moss, D. W. Accuracy, precision and quality-control of enzyme assays. *Journal of Clinical Pathology*, **24**, Supplement (Association of Clinical Pathologists) 4, 22–30 (1971).

Moss, D. W. The relative merits and applicability of kinetic and fixed-incubation methods of enzyme assay in clinical enzymology. *Clinical Chemistry*, **18**, 1449–54 (1972).

Moss, D. W. Automatic enzyme analyzers. *Advances in Clinical Chemistry*, **19**, 1–56 (1977).

Moss, D. W. and Butterworth, P. J. *Enzymology and Medicine*, Pitman Medical, London, 1974.

Rand, R. N. Practical spectrophotometric standards. *Clinical Chemistry*, **15**, 839–63 (1969).

Roberts, R., Sobel, B. E., and Parker, C. W. Radioimmunoassay for creatine kinase isoenzymes. *Science*, **194**, 855–7 (1976).

Rosalki, S. B. Standardization of isoenzyme assays, with special reference to lactate dehydrogenase isoenzyme electrophoresis. *Clinical Biochemistry*, **7**, 29–40 (1974).

Vladutiu, A. O., Schachner, A., Schaeffer, P. A., Schimert, G., Lajos, T. Z., Lee, A. B., and Siegel, J. H. Detection of creatine kinase BB isoenzyme in sera of patients undergoing aortocoronary bypass surgery. *Clinica Chimica Acta*, **75**, 467–73 (1977).

Wilkinson, J. H. *Isoenzymes*, 2nd edn., Chapman and Hall, London, 1970.

Witteveen, S.A.G.J., Sobel, B. E., and DeLuca, M. Kinetic properties of the isoenzymes of human creatine phosphokinase. *Proceedings of the National Academy of Sciences (USA)*, **71**, 1384–7 (1974).

CHAPTER 9

Enzymes and routine diagnosis
A. F. Smith

INTRODUCTION	200
THE IDEAL ENZYME TEST—CLINICAL REQUIREMENTS	201
Sensitivity	202
Specificity	202
Time Course of Enzyme Rise after Myocardial Infarction	202
REFERENCE VALUES (THE 'NORMAL RANGE')	204
ENZYMES COMMONLY USED FOR THE DIAGNOSIS OF MYOCARDIAL INFARCTION	206
Aspartate Aminotransferase	208
Method of Assay	208
Sample Collection and Storage	209
Creatine Kinase	209
Methods of Assay	210
Sample Collection and Storage	211
Creatine Kinase Isoenzymes	212
Electrophoretic Methods	212
Ion Exchange Methods	213
Immunological Methods	213
Differential Activation	214
Lactate Dehydrogenase	214
Methods of Assay	215
Total Lactate Dehydrogenase	215
Heat-stable Lactate Dehydrogenase	217
Urea-stable Lactate Dehydrogenase	218
Hydroxybutyrate Dehydrogenase	218
Selection of Methods	219
Sample Collection and Storage	220
Lactate Dehydrogenase Isoenzymes	220
ACUTE MYOCARDIAL INFARCTION—THE UNCOMPLICATED CASE	221
Serum Enzyme Activity and Prognosis	224
Relationship between the Different Serum Enzymes	225
Optimal Times for Obtaining Serum Samples	225
PROBLEMS OF SPECIFICITY AND SENSITIVITY	226
Use of Isoenzyme Measurements	226
Creatine Kinase Isoenzymes	226
Lactate Dehydrogenase Isoenzymes	228
Serum Enzyme Tests Other Than Aspartate Aminotransferase, Creatine Kinase, and Lactate Dehydrogenase	228

Enzymes Present in Myocardium	228
Enzymes Originating From Tissues Other Than Myocardium	230
Distinction between Myocardial Infarction and Ischaemia	231
Angina	231
Acute Coronary Insufficiency	231
EFFECTS OF MYOCARDIAL COMPLICATIONS WHICH MAY ARISE IN PATIENTS WITH INFARCTS	232
Extension of the Myocardial Infarct	232
Arrhythmias	233
Direct Current Countershock	233
Closed Chest Cardiac Massage	233
CHEST AND CARDIOVASCULAR DISORDERS OTHER THAN MYOCARDIAL INFARCTION	234
Myocarditis and Pericarditis	234
Dissecting Aneurysm of the Aorta	234
Tachycardia	234
Cardiac Failure	234
Pulmonary Embolism and Infarction	235
Pneumonia	236
CIRCUMSTANCES WHERE THE DIAGNOSIS OF MYOCARDIAL INFARCTION MAY PRESENT SPECIAL DIFFICULTIES	236
Operative Procedures	236
Cardiac Catheterization	236
Surgical Operations	236
Heart Operations	237
Accidental Trauma	237
Comatose Patients	237
Cerebrovascular Accidents	238
Drug Overdosage	238
Hypothermia	238
FACTORS COMMONLY CAUSING MISINTERPRETATION OF ENZYME RESULTS	238
Intramuscular Injections	238
Alcohol	239
Exercise	239
RELATIONSHIP BETWEEN THE ELECTROCARDIOGRAM AND ENZYME TESTS	240
CONCLUSION: SELECTION OF ENZYME TESTS FOR DIAGNOSIS OF MYOCARDIAL INFARCTION	241
REFERENCES	242

INTRODUCTION

In 1954, it was first reported, in patients with acute myocardial infarction, that serum aspartate aminotransferase (AST) activity became raised soon after the onset of symptoms, reached a peak at about 24 hours and thereafter returned to normal over the next few days.[42] Shortly afterwards, similar changes in serum lactate dehydrogenase (LD) activity were described. Since relatively straightforward methods for the analysis of the activity of these enzymes in serum also became available at about the same time,[34] there was only a short delay before both AST and LD were being used extensively to aid the routine diagnosis of acute myocardial infarction.

Many other enzymes were subsequently shown to be released from damaged myocardial tissue, but most of them were shown to have no diagnostic advantages over AST and LD. However, in 1960, raised serum creatine kinase (CK) activity was demonstrated in patients with myocardial infarction.[16] In spite of difficulties in developing a reliable method for serum CK assay, the differences in its organ specificity from those of AST and LD rendered CK a useful additional diagnostic tool. In particular, CK is absent from liver and red blood cells, tissues from which release of AST or LD can give rise to difficulties in the interpretation of serum enzyme results.

By the 1960s enough published evidence was available to suggest that serum AST, CK, and LD activity were raised in over 95% of patients with acute myocardial infarction.[2,28,74,78] It was possible to conclude, therefore, that enzyme tests were sufficiently sensitive for routine diagnostic purposes. However, the major problem still remains: serum enzyme tests are insufficiently specific, since it is often not possible to identify with confidence the organ from which an enzyme has been released into serum.

Difficulties in interpretation have also arisen due to differences in enzyme methodology and the units in which enzymes are reported. Currently, both international and national bodies are trying to achieve as much standardization of enzyme methodology as possible, so that comparable results for enzyme tests may be obtained in different laboratories.[7,14]

Serum enzyme tests are used in cardiology for purposes other than the diagnosis of myocardial infarction. However, this chapter will be concerned with the use of enzyme tests for the routine clinical diagnosis of myocardial infarction. In particular, practical aspects relating to current methodology, the selection of tests and the interpretation of results will be considered.

THE IDEAL ENZYME TEST—CLINICAL REQUIREMENTS

Primarily, the clinician requires to know whether or not a patient has had a myocardial infarct; other variables, such as the size of the infarct, may also be of clinical importance. Therefore, a satisfactory enzyme test will have the following characteristics:

1. Serum enzyme activity will be abnormal in all patients who have had an acute myocardial infarct. In other words, the test should be sensitive.
2. Serum enzyme activity will be normal in all patients who have not had acute myocardial infarcts: namely, the test should be specific.
3. Serum enzyme activity should become raised shortly after the onset of the episode, to allow early diagnosis, and should remain raised sufficiently long so that the rise will not be missed.

Unfortunately, no enzyme test completely fulfils these criteria. Furthermore, clinicians, who have to make decisions based on laboratory results, may not appreciate the ways in which most enzyme tests fail to behave in such an 'ideal' fashion. In this section, the above requirements are discussed in more detail.

Sensitivity

The clinical sensitivity of a test may be defined as the ability of that test to classify as positive those patients who have the disease in question—myocardial infarction in the present context. Clinical sensitivity should be distinguished from analytical sensitivity: the ability of an analytical method to detect small quantities of a measured component. Clinical sensitivity may be expressed mathematically:

$$\frac{\text{number of diseased patients with positive test}}{\text{total number of diseased patients tested}} \times 100$$

Thus a test which is 100% sensitive will be abnormal in all patients with the disease in question.

Specificity

The clinical specificity of a test may be defined as the ability of the test to classify as negative those who do not have the disease in question. As with sensitivity, the clinical specificity must be distinguished from the analytical specificity of a test, which relates to the ability of a method to determine solely the component it purports to measure. In mathematical terms, the clinical specificity of a test for myocardial infarction (MI) would be expressed:

$$\frac{\text{number of non-MI patients with negative test}}{\text{total number of non-MI patients tested}} \times 100$$

Thus a test which is 100% specific for myocardial infarction will always give normal results, or results which are not characteristic of myocardial infarction, in patients who have not had myocardial infarcts.

Time Course of Enzyme Rise after Myocardial Infarction

The period during which a serum enzyme remains raised after myocardial infarction will depend mainly on the rate of release of the enzyme from damaged tissue and on the rate of its removal from the intravascular space. These rates vary from enzyme to enzyme.

Selection of the most appropriate enzyme test may depend on the circumstances in which the patient is seen. In Coronary Care Units, to which most patients are usually admitted within 6 hours of the onset of symptoms, enzymes whose serum activity rises rapidly may be required if early confirmation of the diagnosis is desired. Similarly, the detection of episodes of reinfarction is easiest with enzymes whose serum activity has returned to normal relatively rapidly, so that a well-defined secondary rise is apparent. However, some patients are not seen until several days after the episode of chest pain. In these circumstances, an enzyme of which the serum activity remains elevated for a relatively long period is required.

Evaluations of serum enzyme tests have not always ensured that specimens are obtained at the correct time. This is the case with most investigations which are usually quoted as finding a low incidence of abnormal enzyme results in patients with myocardial infarction. Some reports, for example, have included patients seen, for the first time, 4 days after the acute episode, by which time serum AST and CK activity might have been expected to have returned to normal in an appreciable percentage of patients[78,9C] (Table 1, page 224).

Clearly, the ideal test for myocardial infarction would be 100% sensitive and 100% specific, providing samples were taken at the correct time. No test, enzyme or other, fulfils these criteria, although the e.c.g. is usually regarded as being 100% specific—if correctly interpreted.

In practice, sensitivity is much less of a problem than specificity, since nearly all patients with myocardial infarcts have raised serum enzyme activities. Those who do not show raised serum enzymes are likely to have had only small infarcts and to have a relatively good prognosis.[12]

The importance of lack of specificity is more difficult to assess. Firstly, it is impossible to obtain accurate figures for the specificity of a test, as it is rarely possible to exclude a diagnosis of myocardial infarction with certainty. Secondly, the figures obtained are dependent on the population being sampled; for example, it is likely that a general medical ward would admit more patients with raised serum enzyme activity due to causes other than myocardial infarction than a Coronary Care Unit.

Between 10% and 30% of patients admitted to a Coronary Care Unit who have *not* had myocardial infarcts are found to have raised serum CK, AST, or LD activity.[74,76] The incidence in general hospital wards is probably of the same order.[24] If AST, CK, and LD are considered as a group, and due attention is paid to the relative size of the increase of each enzyme and to the timing of peak activity, the specificity improves to over 95%, since non-myocardial factors which may affect one of the enzymes may not affect the others.

Alternatively, specificity may be improved by measuring 'heart-specific' enzymes such as the MB isoenzyme of CK (CK-MB) or the most anodal of the LD isoenzymes (LD_1).

The clinical considerations which have been discussed provide the most cogent evidence on which to select an enzyme test as suitable for routine use, or to reject it as unsuitable. It would be unrealistic, however, to assume that analytical factors may not also be of major importance. The following analytical properties of an enzyme test would need to be taken into account:

1. accuracy and precision;
2. simplicity and rapidity of performance;
3. suitability for partial or complete automation;
4. availability of stable and pure reagents;
5. cost of reagents and necessary capital equipment.

REFERENCE VALUES (THE 'NORMAL RANGE')

In health, the activity of most serum enzymes fluctuates within fairly wide limits. For example, one study has shown such intra-individual fluctuations for serum AST activity to have a coefficient of variation of about 25%.[92] There are additional, even larger, variations between individuals. These within-individual and between-individual variations account for the wide span of most 'normal ranges'.

However, there are semantic difficulties associated with the use of the word 'normal'.[80] Firstly, individuals with serum enzyme results outside certain arbitrary 'normal' limits may be labelled as 'abnormal'—a word with potentially misleading and emotional overtones. Secondly, in statistical jargon, 'normal' describes the shape of a frequency distribution curve. For these reasons, the term 'reference values' is now preferred. The set of values may be obtained either from a single individual or from a group, but in either event the source of the data used to obtain the reference values must be stated precisely.

There are numerous practical difficulties in defining reference ranges for most serum enzymes; insufficient attention to these problems has been paid in the past, consequently the results of several of the earlier studies on the relative value of serum enzyme tests in myocardial infarction must be called into doubt. It was common practice, for example, to obtain reference ranges by analysing sera from a group of healthy young adults, usually male. The mean and standard deviation (SD) were calculated and the 'normal' range defined as mean ± 2 SD, namely that portion of a Gaussian distribution which includes about 95% of the population results. Unfortunately, this approach was frequently invalid since:

1. The reference range often related to a different population (age, sex, mobility, etc.) from the one from which patients' samples were obtained.
2. The population sampled was often far too small (see below).

3. The assumption that the results were distributed in a Gaussian manner could not be justified.

Serum CK exemplifies some of the difficulties that may be encountered. The distribution of serum CK activity in a group of healthy adult males aged between 20 and 60 years, is shown in Figure 1. The values are skewed to the right and, of the two overlays shown, the distribution is much closer to log-Gaussian than Gaussian. If, because of too small a sample size or lack of appreciation of the problem, the results were incorrectly assumed to be distributed in a Gaussian manner, the resulting 'upper limit of normal' (mean ± 2 SD) is 115 iu l^{-1}. On the other hand, if the distribution is assumed to be log-Gaussian, the 'upper limit of normal' or upper reference value is 135 iu l^{-1}. In fact, there is no intrinsic reason why the results should follow any particular mathematical distribution, so it may be more valid, providing an adequately large sample is available, to use non-parametric tests, which make no inherent assumptions about the distribution of the data.[63]

'Normal' serum CK activity is also lower in women than in men, lower in in-patients than in out-patients, and lower in the morning than in the

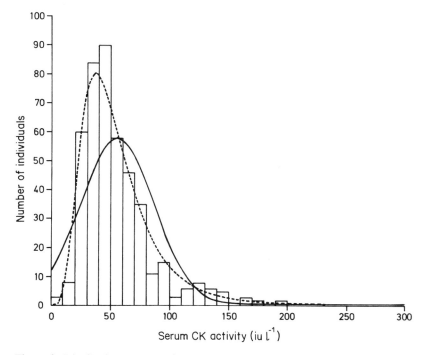

Figure 1. Distribution of serum CK activity in a group of healthy adult males. —— Gaussian curve overlay, – – – log-Gaussian curve overlay to fit observed data

evening.[27] Some of these differences may be attributed to release of CK from muscle during exercise (see below), and they can cause much difficulty in the interpretation of serum CK results close to the upper reference value.

It is not so difficult to define reference ranges for serum AST and LD activity. Although both are slightly skewed to the right, this is less marked than with CK.

The activity of the MB isoenzyme of CK (CK-MB) in serum is normally very low. This presents an additional problem in determining reference ranges, since some methods are unable to detect CK-MB in normal serum, while other methods may be both inaccurate and imprecise at these low levels of activity. The problem is likely to remain until methods capable of determining the amounts of CK-MB activity in normal serum with accuracy and precision are developed. Meantime, most measurements of low activities of CK-MB should probably be regarded as only semiquantitative.

ENZYMES COMMONLY USED FOR THE DIAGNOSIS OF MYOCARDIAL INFARCTION

In this section, the properties and tissue distribution of AST, CK, and LD are briefly recapitulated. Assay methods are discussed in relation to individual enzymes, but certain general points regarding enzyme assay in routine diagnosis may conveniently be made initially.

1. As recommended by the International Federation of Clinical Chemistry (IFCC), continuous monitoring ('kinetic') methods are suggested for routine use.[7] This is practicable in nearly all laboratories, since suitable automatic or semi-automatic instruments for measuring absorbance change are commercially available. Some laboratories, for reasons of convenience, ease of specimen handling and data processing use continuous-flow methods (usually AutoAnalyzer II, SMA 12 or SMAC, Technicon Instrument Co. Ltd., Tarrytown, New York) for enzyme analyses. In general, the results using these methods correlate well with those using discrete analysis but there may be difficulties with standardization and the range over which enzyme activity can be measured is often rather limited.

2. The recommended methods are all NAD or NADP linked; absorbance is usually monitored at 340 nm but some laboratories may prefer to use the mercury line at 334 nm. Measurements at 365 nm are less satisfactory, as absorbance changes are smaller and a narrow bandpass spectrophotometer, or the use of a mercury lamp as light source, is required if the published molar absorbtivity of NADH at this wavelength is to be valid.[94]

3. In accordance with IFCC recommendations, 30 °C is suggested as the temperature of choice for enzyme assay. Nevertheless, it is unlikely that laboratories will abandon overnight their established practice of using 25 °C

or 37 °C, since the change would entail a change in reference ranges, as the activity of most serum enzymes is about 50% greater at 37 °C than 30 °C. Such changes are likely to cause confusion to those interpreting the results, and the clinician may need convincing of the necessity of making alterations in analytical methods or units of reporting unless they can be shown to improve diagnostic efficiency.

4. For the reasons given above, enzyme methods are unlikely to become uniform for some time. Because of the diversity of reaction conditions, it would be misleading to try and give precise reference ranges for the enzymes described, since general, rather than detailed, recommendations will be given when reaction conditions are discussed.

5. The complexity of some of the reaction mixtures, as well as the instability of reagents have led to the development of numerous commercially available 'kits', in which the reaction mixtures are usually freeze dried, and the user only has to reconstitute, add serum, and measure absorbance change. Although the kits tend to be fairly expensive, they are very suitable for laboratories with small workloads or where the reagents are especially complex, as with CK methods. The reagent concentrations and specified reaction conditions vary from kit to kit. When originally introduced, kits often used suboptimal reaction conditions since such kits were based on 'state of the art' technology. Later, when advances in technology demonstrated that the suboptimal reaction conditions were causing low results compared with newer methods, the manufacturers were often, with good reason, unwilling to change the formulation of their kits since laboratories were accustomed to the numerical values obtained with the existing ones. At present, therefore there are kits on the market which use 'original' reagent concentrations and which tend to give low results, and there are others which use 'optimized' reaction conditions. These 'optimized' kits generally employ higher substrate concentrations, and therefore sustain higher enzyme activity, and often the reaction conditions are close to those of nationally or internationally recommended methods. In selecting a kit for routine use, only ones in which the reaction conditions are specified and seem to be fairly close to those of recommended methods should be selected.

In the interests of uniformity it is suggested that, where IFCC recommended or nationally recommended methods are available, these should be adopted where possible, especially if a test is being introduced or changed for other reasons. Although there may be detailed differences between IFCC and national recommendations, these are unlikely, in practice, to give rise to significant differences in measured enzyme activity.

Apart from the difficulties in reaching agreement between any large number of people, often with strong opinions of their own, there are dangers in advocating too strongly universal adoption of a single nationally or internationally recommended method. The principal danger is that, by giving an 'official' stamp of approval to certain reaction conditions, further improvements in methods will be delayed or prevented.

Aspartate Aminotransferase

This is present in many tissues, notably liver, myocardium, skeletal muscle, and kidney. Two principal isoenzymes have been described, one cytoplasmic, the other mitochondrial. Although the tissue concentration of the mitochondrial isoenzyme is often greater than that of the cytoplasmic isoenzyme, the latter predominates in serum, both in health and disease: normally less than 10% of total serum AST activity can be attributed to the mitochondrial isoenzyme.

In addition to myocardial disease, serum AST activity is raised in some liver and muscle diseases and occasionally in disorders affecting kidney, pancreas, or other tissues.

Method of assay

Most laboratories now use a variant of the original Karmen method,[34] in which the production of oxaloacetate is coupled with NADH oxidation:

$$\text{L-aspartate} + \text{2-oxoglutarate} \xrightleftharpoons{\text{AST}} \text{oxaloacetate} + \text{L-glutamate}$$

$$\text{oxaloacetate} + \text{NADH} + \text{H}^+ \xrightarrow[\text{dehydrogenase}]{\text{malate}} \text{L-malate} + \text{NAD}^+$$

Serum is incubated with a mixture of aspartate, NADH, and malate dehydrogenase in Tris/HCl buffer pH 8.0. The reaction is then started by the addition of oxoglutarate; absorbance change at 340 nm is monitored. The main problem that may occur with this analysis is that NADH may be converted to NAD^+ by enzymic reactions other than the reaction catalysed by AST. In particular, lactate dehydrogenase may react with NADH and the endogenous pyruvate present in serum, causing consumption of NADH before the addition of oxoglutarate, and possibly during the measured reaction as well. To overcome this effect, LD is usually added in excess to the pre-incubation reaction mixture, so that the side reaction can go to completion before the addition of oxoglutarate. Furthermore, the assay should be repeated if the initial absorbance is too low—indicating a low initial concentration of NADH.

It has recently been shown that inclusion of pyridoxal-5-phosphate in the reaction mixture enhances serum AST activity by a variable amount.[64] Some sera, particularly those from patients with liver disease, show only small amounts of activation, often between 0 and 20%, whereas other sera, particularly those from patients with myocardial infarction may show increases in activity of up to 70%.[62] The IFCC recommended method includes pyridoxal-5-phosphate in the reaction mixture.[22] As with the problem of reaction temperature, routine laboratories seem likely to delay the adoption of these recommendations since the differential activation of sera from patients with different types of disease will require a reappraisal of the modified test.

Continuous-flow methods are used extensively for serum AST assay, often employing the same coupled reaction sequence to measure activity. Accuracy and precision at low levels of enzyme activity tend to be less good than with discrete analytical methods.

Sample collection and storage

These present few special problems. Although AST is present in red cells, small amounts of haemolysis do not invalidate the test to the same extent that LD assays are affected. Serum AST is relatively stable for several days if stored at −20 °C or +4 °C.

Creatine Kinase

This enzyme is present in very high concentration in skeletal muscle, in lower concentration in cardiac muscle (about 25% of the activity present in skeletal muscle), and is also present in brain, thyroid, kidney, gut, and other tissues. Three isoenzymes are described in the soluble cytoplasm of the cell, each isoenzyme being a dimer—conventionally designated MM, MB, or BB. Skeletal muscle contains mainly the MM isoenzyme, with most authors finding negligible quantities of the MB and BB isoenzymes (less than 1% of each).[58,65,85] Brain contains the BB isoenzyme, whereas the myocardium contains about 70% MM and 30% MB. The important feature is that the MB isoenzyme is usually considered to be heart specific. By and large this is true, but there are exceptions.[75] Serum normally contains mainly the MM isoenzyme, with less than 5% of total CK in the form of the MB isoenzyme.[50,54]

Increased serum CK activity is almost invariably due to release from skeletal or cardiac muscle. Since CK is present in high concentration in skeletal muscle and the body contains such a large mass of muscle, a relatively minor increase in the rate of release of muscle CK will cause an increased serum CK activity. This probably accounts for the increased serum CK activity seen in physiological conditions, such as after exercise, and in disorders which are not thought of as primarily having any effect on skeletal muscle,[27] such as coma from acute poisoning with hypnotics or severe generalized disease of any type.

Methods of assay

Numerous methods of CK measurement have been described in the past, testament to the unsatisfactory nature of many of them. Most laboratories now use a method in which the ATP produced in the 'reverse' direction is coupled, through hexokinase (HK) and glucose-6-phosphate dehydrogenase (G-6-PD) with reduction of NADP:[59,68]

$$\text{creatine phosphate} + \text{ADP} \xrightleftharpoons{\text{CK}} \text{ATP} + \text{creatine}$$

$$\text{ATP} + \text{glucose} \xrightleftharpoons{\text{HK}} \text{glucose-6-phosphate} + \text{ADP}$$

$$\text{glucose-6-phosphate} + \text{NADP}^+ \xrightleftharpoons{\text{G-6-PD}} \text{NADPH} + \text{6-phosphogluconate} + \text{H}^+$$

Serum is pre-incubated with a reaction mixture containing ADP, glucose, NADP, HK, G-6-PD, magnesium ions, a sulphydryl agent, and an inhibitor of adenylate kinase; the reaction is then started with creatine phosphate. The magnesium ions are required as a cofactor and the sulphydryl agent reactivates CK (see below). Diadenosine pentaphosphate or AMP or both are included to inhibit a 'blank' reaction due to adenylate kinase activity;[82] adenylate kinase may also be inhibited by fluoride. Imidazole, pH 6.7, has been suggested as the most appropriate buffer.[81] The absorbance change is monitored by 340 nm, preferably for 5 minutes or more.

Certain points are of practical and theoretical importance:

1. The sulphydryl groups of CK become oxidized, or otherwise altered, during storage of serum rendering the enzyme inactive (see below). The process can be prevented by storage in the presence of a sulphydryl compound, and by the inclusion of a sulphydryl compound in the reaction mixture. Early methods, before 1965, often did not employ sulphydryl compounds and produced low and inconstant CK activities. It is not possible to reach valid conclusions based on the results of such methods. *N*-acetylcysteine has been recommended as the most appropriate sulphydryl compound to reactivate CK.[81]

2. Rectilinear reaction rates are only obtained after an initial 'lag' phase which may last for between 1 and 15 minutes. To keep the lag period to a minimum, CK should be reactivated with sulphydryl compounds before the reaction is started and fairly high concentrations of HK, G-6-PD, and NADP are required in the reaction mixture.[81]

3. Dilution of serum enhances CK activity.[26] There is no doubt that many of the earlier methods of CK assay were associated with apparent activation of CK activity when the specimen was diluted. At the time, this was attributed mainly to the presence of inhibitors of CK in serum. However, the methods now being recommended[81] do not show this dilution effect until activities more than 10 to 20 times the upper reference value for healthy

adults are being measured. It seems likely that most of the dilution effect was caused by purely methodological factors—some of these attributable to the rather complex coupled assay system. The main factors contributing to the apparent dilution effect were:

1. The use of suboptimal creatine phosphate and NADP concentrations (for reasons of economy).
2. The presence of relatively long 'lag' phases due to delayed activation of CK by sulphydryl compounds. The association of this factor and the previous one caused some substrate depletion and product accumulation before a steady-state reaction rate was achieved. This effect would be more marked with high-activity specimens.
3. The use of inadequate amounts of the coupling enzymes glucose-6-phosphate dehydrogenase and hexokinase. As is shown on page 159, this will also cause a long 'lag' phase and the presence of relatively high concentrations of intermediate reaction products, such as ATP and glucose-6-phosphate which may inhibit the overall reaction sequence.

Therefore, if more recent methods of CK assay are used and if the minimum volume of serum, compatible with the degree of precision required, is present in the reaction mixture, very few samples should need diluting.

In practice, methods of CK analysis are less convenient than those for AST and LD, since the reagents tend to be expensive and unstable. The rate of analysis of specimens also tends to be low, as the reaction has to be followed for longer, due to the 'lag' period. Most laboratories use commercially available reagent kits.

Continuous-flow methods are also used. Although they are usually also based on the 'reverse' reaction, creatine formation is measured directly by a colorimetric reaction with diacetyl and orcinol.

Sample collection and storage

No special precautions for sample collection and transport to the laboratory are necessary. Serum is usually preferred to plasma, but heparinized blood is satisfactory. Minor degrees of haemolysis do not affect enzyme activity. During storage, CK activity declines rapidly due to oxidation of sulphydryl groups on the enzyme. Most of the activity is readily restored by 15 to 30 minutes pre-incubation with a sulphydryl reagent prior to assay, either by addition of mercaptoethanol to serum (to a concentration of 5 to 10 mmol l^{-1}) or by pre-incubation of serum with assay mixture, containing a sulphydryl agent, before the addition of starting reagent (creatine phosphate). However, for optimal preservation of all three CK isoenzymes, mercaptoethanol should be added to serum prior to storage. For short periods of preservation, the storage temperature is not critical, for longer periods −20 °C is probably advisable.

Creatine Kinase Isoenzymes

There was considerable delay between the time when it was originally recognized that CK-MB might be a specific index of myocardial damage and the introduction of the test into routine use.[87] This delay was largely due to difficulties in developing accurate methods for measuring CK-MB activity in serum.

Even now there are many methods being used in routine Clinical Chemistry laboratories, their very number attesting to the fact that none of the techniques can be regarded as ideal. Most of the methods are supplied as 'kits' which are commercially available; for most laboratories, use of such kits is probably advisable unless the workload is very large or considerable experience with CK isoenzyme analysis has been obtained previously. Electrophoretic, ion exchange, selective activation, and immunological methods are all available commercially, although there may be several kits using a particular method.

Electrophoretic methods

These methods mostly use agarose or cellulose acetate as support medium.[52] Starch and polyacrylamide gel have also been used but the methods are more laborious and do not offer any significant compensating advantages.[75] After electrophoretic separation of the isoenzymes, the gel is usually overlaid with a reagent mixture similar to that used for measuring total CK activity. After incubation for 15 minutes to 1 hour, bands of fluorescence, due to NADPH, appear at the site of CK activity. Many of the kits are designed primarily for semiquantitative or qualitative use and a visual appraisal of the bands of CK activity under u.v. light is all that is required. However, some methods, especially those using agarose gel, are suitable for quantitation which is usually performed by densitometric scanning of the gel for NADPH fluorescence. The more accurate alternative, whereby NADPH is quantitated after elution from the gel,[58] is probably too tedious for routine use. The electrophoretic methods are simple, fairly rapid and relatively inexpensive. Furthermore, they tend to be reliable since the enzyme bands can be *seen*, so it is unlikely that small changes in reagent concentrations or other experimental conditions will lead to seriously erroneous results. Nevertheless, overheating of the gel must be avoided, since this may cause inactivation of CK, especially the BB and MB isoenzymes. Electrophoretic methods, however, are relatively insensitive, being unable to detect CK-MB activities less than 5–10 iu l.$^{-1}$; therefore they cannot detect CK-MB activities within, or just above, the reference range. The lack of sensitivity is coupled with relatively poor precision of the methods which rely on densitometric scanning. A significant number of laboratories are unwilling to use electrophoretic methods because of these disadvantages.

Ion exchange methods

Short disposable columns packed with DEAE-cellulose, DEAE 'Sephadex' (Pharmacia, Uppsala, Sweden), or other suitable ion exchange medium are usually used.[50,52,54] After application of diluted serum to the top of the column, the MM, MB, and BB isoenzymes are eluted sequentially either by a continuous ionic gradient or with two or three changes of buffer of increasing ionic strength. The successive fractions so obtained are assayed for CK activity. A large number of variants of this method have been published, some, which are available commercially, supply the pre-packed columns and buffer solutions ready made-up. The more satisfactory variants are capable of greater sensitivity than electrophoretic methods. Fairly good precision is obtainable but some methods are inaccurate as there is carry-over of one fraction into the next, e.g. the MM isoenzyme may be found in the MB fraction. Another source of inaccuracy in some methods is that only two fractions are collected (one containing the MM isoenzyme, the other both the MB and BB isoenzymes) on the assumption that the BB isoenzyme is rarely found in serum: this assumption may not be justified. The ion exchange methods are, therefore, satisfactory if used meticulously and if a method or kit of proven performance is selected. Erroneous results are likely to arise if the methods are not used with care since small alterations in separation conditions may cause potentially dangerous and misleading carry-over of the MM isoenzyme into the MB fraction thus causing a gross overestimation of serum CK-MB activity.

Immunological methods

These have been introduced more recently. They rely on the antigenic differences between the B and the M polypeptide chains: antisera raised against the MM isoenzyme show little cross-reaction with the BB isoenzyme, and vice versa. Two main types of immunological method have been used— immunoinhibition and radioimmunoassay. Most experience has been obtained with immunoinhibition methods which are available commercially. A simple and commonly used commercial kit employs an anti-MM antibody to inhibit MM isoenzyme activity: any residual activity is attributable to CK containing B subunits (i.e. CK-BB and CK-MB). The method is very simple and quick (taking about 15 minutes to perform) but antisera may vary in their characteristics and some have been reported to show cross-reaction with B subunits and variable inhibition of the MB isoenzyme. The methods also assume that uninhibited activity is due to MB isoenzyme alone, on the premise that serum contains no BB isoenzyme—which is not always true. The methods using radioimmunoassay are probably more sensitive but are more time consuming and are also dependent on the specificity of the antisera.

Differential activation

The differential activation of serum CK activity by different sulphydryl compounds is used by another method, for which the reagents are commercially available. The method is based on the assumption that whereas dithiothreitol will activate both CK-MB and CK-MM activity, glutathione will activate only CK-MM. The difference in measured activity in the presence of the two activators is, therefore, taken to represent CK-MB activity. There are practical and theoretical disadvantages to this method, although it is technically simple. First, any method which relies on measurement of a relatively small difference between two results, is likely to lack precision—especially when CK-MB activity is low. Second, the basic premise that CK-MM is fully activated, and CK-MB not activated, by glutathione is open to question since many investigators have been unable to repeat the original findings.

It is clear that no method is ideal. Laboratory factors must be matched with the clinical requirements such as the required sensitivity and specificity of the assay. Commercial 'kits' will usually be the most convenient choice: the simplest of these are undoubtedly based on immunoinhibition of CK-MM. Electrophoretic methods are simple and reliable but not sensitive whereas ion exchange methods are sensitive but not always reliable.

Precautions necessary for specimen transport and storage have already been considered in the section on CK methods.

Lactate Dehydrogenase (LD)

This enzyme is widely distributed in the soluble cytoplasm of most tissues. Although all five isoenzymes are present in all the tissues, the relative proportions in each tissue vary considerably.[91] In heart, red cells and red cell precursors, the anodal isoenzymes LD_1 and LD_2 predominate. In liver and skeletal muscles the cathodal enzymes LD_5 and, to a lesser extent, LD_4 predominate. In the kidney, the isoenzyme composition depends on the part of the nephron from which the enzyme is derived. Spleen, lung, leucocytes, and certain other tissues tend to have an intermediate isoenzyme pattern, often with LD_3 present in the highest concentration. In normal serum, the anodal isoenzymes predominate, with LD_2 present in highest concentration, then LD_1, LD_3, LD_4, and LD_5 in decreasing order of activity.

In practice, LD is most often released into serum at increased rates in diseases of heart, liver, skeletal muscle, kidney or red cells, or in neoplastic disease. The resulting pattern of LD isoenzymes in serum reflects the pattern present in the diseased organ, superimposed on the normal serum pattern. It should be borne in mind, however, that the serum isoenzyme pattern also reflects the rate of *removal* of the isoenzymes from the circulation. Therefore the serum pattern may cause a misleadingly low estimate of the amount of LD_5 being released from damaged tissue, since LD_5 has a plasma half-life of about 10 hours, whereas that of LD_1 is about 4 days.

Methods of assay

The major discussion of methodology will relate to serum total LD activity. However, there are a number of methods which measure preferentially the anodal isoenzymes of LD—sometimes such methods are, rather misleadingly, termed *'heart-specific'* LD assays. These methods will also be described in outline since many laboratories use them in place of, or in addition to, total LD assay.

Total LD

The reaction catalysed by LD:

$$\text{lactate} + NAD^+ \rightleftharpoons \text{pyruvate} + NADH + H^+$$

is freely reversible and may conveniently be monitored in either direction.

Pyruvate to lactate ($P \rightarrow L$) methods are probably the most widely used, especially in Europe. Serum is usually pre-incubated with buffered NADH and the reaction started by the addition of pyruvate. Selection of appropriate reaction conditions may not be easy, since the kinetic characteristics of the isoenzymes differ from one another with regard to substrate requirements, pH optima, etc. However, substrate concentration is most critical, the pH, type of buffer, and NADH concentration being less important; most methods use phosphate or Tris buffer at pH 7.2 or pH 7.4 and an NADH concentration between 0.12 and 0.20 mmol $l.^{-1}$.[23]

The main problem lies in the selection of the most appropriate pyruvate concentration. At a pyruvate concentration of about 0.6 mmol $l.^{-1}$, which is optimal for LD_1 (Figure 2), the activity of LD_5 is submaximal, whereas if the pyruvate concentration is raised to 2.5 or 3.0 mmol $l.^{-1}$, optimal for LD_5, the excess pyruvate inhibits LD_1. Since most normal sera contain mainly the anodal isoenzymes, LD_1, LD_2, and LD_3 routine methods have usually used

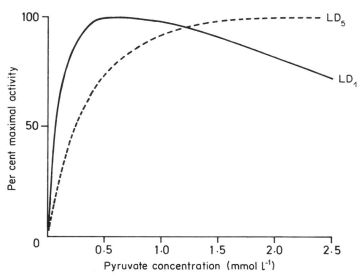

Figure 2. Effect of pyruvate concentration on LD activity.

pyruvate concentrations between 0.6 and 1.0 mmol l.$^{-1}$. Although one author has advocated a pyruvate concentration of 2.5 mmol l.$^{-1}$ (for an assay at 37 °C),[46] this has been shown to cause inhibition of the LD in sera from patients with myocardial infarction—as might have been expected (Figure 2).[86] The optimal concentration, allowing the best compromise between the requirements of LD_1 and LD_5, is about 1.2 mmol l.$^{-1}$.[14,23]

Inaccurate results may be caused by instability of pyruvate or by the development of inhibitors in NADH preparations.[6] At room temperature or +4 °C, pyruvate solutions deteriorate rapidly, so this reagent should always be prepared immediately before use or stored at −20 °C. Inhibitors of LD may develop in NADH solutions stored at any temperature, or even in the solid form if allowed to get damp. It is advisable to prepare NADH solutions immediately before use, preferably from a freshly opened, previously sealed vial.

Lactate to pyruvate (L → P) methods are also used extensively, especially in the United States. There is variation in the pH, type of buffer, and substrate concentrations used by various methods. This may cause confusion, since the variables are interdependent and the activities in different buffers vary considerably; for example, activities in 2-amino-2-methyl-1-propanol buffer are 50 to 100% higher than corresponding activities in pyrophosphate buffer.[10]

The choice between L → P and P → L methods will mainly be based on previous experience and convenience, since results by the two methods have been shown to correlate well, although activities are greater using the P → L method.[23] The reagents for L → P methods are more stable, but appropriate precautions should eliminate any problems with pyruvate or NADH reagents. The variation in results associated with lack of uniformity of pH, buffer type, and substrate concentration associated with the L → P methods may incline laboratories to select methods using pyruvate as substrate.

The discrete methods just discussed, performed on a reaction rate analyser or centrifugal analyser may be regarded as the methods of choice. Nevertheless, many laboratories employ continuous-flow methods, which usually use lactate as substrate. The results with these methods have been found to correlate well with those using discrete analysis, and the precision is usually quite good. However, the presence of several isoenzymes makes secondary standardization particularly liable to give rise to difficulties, especially if the potential dangers are not recognized.

Heat-stable LD

The anodal isoenzymes of LD are relatively stable to heating at 60 °C for 1 hour, or for shorter periods at 65 °C (Figure 3).[44,79] A number of methods have been described—these differ mainly in details of the heating procedure.

In one method, serum is heated in a small tube in a water bath kept at 60 °C for 1 hour.[9] The serum is cooled, and then assayed for LD activity.

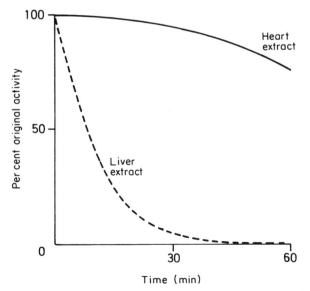

Figure 3. Effect of incubation of enzyme at 60 °C on LD activity.

Since the heated serum contains mainly LD_1 and LD_2, the pyruvate concentration used in the assay should be fairly low (about $0.6\,\mathrm{mmol\,l.^{-1}}$). Methods using lactate as substrate may also be used.

Urea-stable LD

If the reaction mixture used for LD assay contains $2.0\,\mathrm{mol\,l.^{-1}}$ urea, the cathodal isoenzymes are almost completely inhibited (Figure 4). The relationship between urea concentration, pyruvate concentration and optimal differentiation between LD_1 and LD_5 is complex and depends on the temperature and pH of the reaction mixture.[29,37] For example, in the presence of $2.0\,\mathrm{mol\,l.^{-1}}$ urea the optimal pyruvate concentration for LD_1 becomes about $3.0\,\mathrm{mmol\,l.^{-1}}$, instead of $0.6\,\mathrm{mmol\,l.^{-1}}$ in the absence of urea.[9]

A number of methods have been described; these differ in details of reagent concentrations and will, therefore, give numerically different results from one another.

'Hydroxybutyrate dehydrogenase' (HBD)

In this test, pyruvate is replaced by 2-oxobutyrate as substrate for LD. Since LD_1 and LD_2 have relatively higher affinities for 2-oxobutyrate than

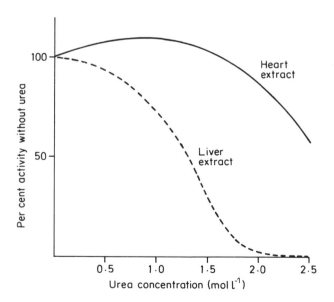

Figure 4. Effect of inclusion of urea in the reaction mixture on LD activity.

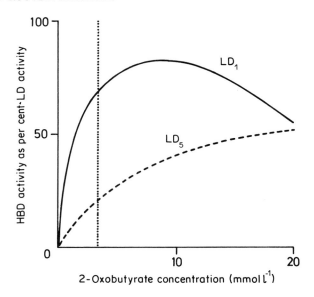

Figure 5. Effect of oxobutyrate concentration on HBD activity (expressed as a percentage of the activity obtained using 1.2 mmol/l.$^{-1}$ pyruvate as substrate instead of oxobutyrate). The vertical dotted line represents the oxobutyrate concentration originally employed to effect maximal separation between LD_1 and LD_5.

have LD_4 and LD_5 (Figure 5), serum HBD activity tends to reflect mainly the presence of the anodal LD isoenzymes. It is apparent that optimal differentiation between LD_1 and LD_5 (Figure 5) occurs at low substrate concentrations. A low substrate concentration was originally chosen for this reason.[69] However, other authors, in adapting the method for use at 37 °C have 'optimized' reaction conditions, by increasing the substrate concentration from 3.3 mmol l.$^{-1}$ to 15 mmol l.$^{-1}$, although it would be expected that this would remove most of the differences between the reaction rates sustained by LD_1 and LD_5.[14,20] In other words, adoption of an analytical optimal method can cause a loss of the clinical discriminating value of the test—although this is poor even when the original assay conditions are used.

Selection of methods

For those laboratories using LD assays mainly for the diagnosis of heart disease, one of the 'heart-specific' LD assays is likely to prove more useful than total LD assay. Of the 'heart-specific' assays, heat-stable LD has been shown to be the most sensitive and specific of the tests although only marginally better than urea-stable LD.[5,9] In general, HBD is much less effective, having little advantage in sensitivity or specificity over total LD. In spite of these findings, HBD has remained the most widely used of the tests, probably as a result of convenience and for historical reasons rather than positive decision. However, its retention as a routine test does not seem to be justified since it has few compensating advantages to offset its relative lack of specificity. Use of heat-stable LD is rather time consuming, as it has an additional heating stage; this also tends to cause poor precision. On the other hand, urea-stable LD results may be affected if the period of pre-incubation of sample with urea is not kept constant, since there is a continuing very slow inactivation of LD. On balance, urea-stable LD probably offers the best compromise between analytical convenience and clinical utility.

Whichever method is selected, it should be remembered that none are truly 'heart specific', since there are other tissues, such as red cells, which have an isoenzyme pattern almost identical to that of myocardium. Furthermore, relatively large increases in serum activity of HBD, but not usually urea-stable LD or heat-stable LD, may be present in patients with increases in LD_5.

Sample collection and storage

Insufficient attention to details of collection, transport, separation, and storage of specimens may cause inaccurate results. Serum is preferable to plasma but heparinized plasma may be used providing the conditions of separation ensure adequate removal of platelets; presence of platelets in plasma may cause spuriously high results. Haemolysis invalidates the test, since red cells also contain high concentrations of LD. Therefore, care must be taken at the time of collection and during transport and separation, to ensure the minimum of trauma to the blood. The cathodal isoenzymes of LD may be unstable if stored at +4 °C or −20 °C.[40] Although storage at room temperature has been recommended, this carries high risk of bacterial growth in the specimen and would entail special handling arrangements. It would seem advisable to assay specimens on the day they are received in the laboratory, although this advice is more relevant for total LD assays than 'heart-specific' assays.

Lactate Dehydrogenase Isoenzymes[91]

Electrophoretic techniques are nearly always used to quantitate the individual LD isoenzymes in serum. The isoenzymes can be separated on cellulose acetate, agar, agarose, polyacrylamide gel, or starch gel; they are then 'stained' by an appropriate reaction. The method used to demonstrate the isoenzymes has usually involved coupling the reduction of NAD^+ with that of a tetrazolium salt—usually nitro blue tetrazolium ((NBT). Methylphenozium methosulphate (phenazine methosulphate or PMS) is required as an intermediate carrier:

```
Lactate      NAD⁺      Reduced      NBT
                         PMS
        ⤫          ⤫          ⤫
Pyruvate     NADH      PMS       Insoluble coloured
                                     formazan
```

When reduced, the NBT forms an insoluble dark-blue formazan which is deposited at the site of LD activity.

In practice, agarose is probably the best separating medium. After electrophoresis, the agarose gel is overlaid with a reaction mixture containing lactate, NAD^+, PMS, and NBT, often, but not always, incorporated into a second agarose gel to minimize diffusion of the bands of enzyme activity which form after incubation. These isoenzyme bands may then be quantitated by densitometry if required.

Alternatively, the PMS and NBT may be omitted and the LD isoenzymes then appear as areas of fluorescence. These may also be quantitated by using an appropriate fluorescence densitometer.

The results yielded by both these methods must be regarded as semiquantitative, since inaccuracies may arise at all stages of the analysis. Even before analysis, LD_5 may become partially inactivated during storage under conditions which will not affect the other isoenzymes. During electrophoresis, LD_5 tends to become inactivated if there is any overheating of the gel. The staining conditions are also likely to favour one group of LD isoenzymes at the expense of the others. For example, the use of high lactate concentrations to 'drive' the reaction tends to favour LD_5, since LD_1 and LD_2 are inhibited by excess lactate. In addition, areas of high LD activity tend to become substrate depleted, causing underestimation of prominent bands.

In spite of these drawbacks, it is often possible to select conditions, especially if NADH fluorescence is being quantitated, such that reasonably precise results are obtained.

ACUTE MYOCARDIAL INFARCTION—
THE UNCOMPLICATED CASE

In the 'classical' case of myocardial infarction it is usual to assume that the onset of acute symptoms, usually cardiac pain, coincides with the formation of an occlusive thrombus in one of the coronary arteries. This thrombus is assumed to be the initiator of a chain of events which shortly leads to infarction of the segment of myocardium supplied by the affected vessel.

However, there are patients in whom no recent thrombus is detectable in the coronary arteries at death; instead, there is severe narrowing of all three coronary vessels and the infarcted areas are diffuse rather than being confined to that part supplied by a single coronary artery. It is possible that these different pathological varieties of infarction may be associated with different patterns of enzyme release. Such variations in the pathophysiology of the disease may be responsible for some of the deviations from the 'typical' enzyme findings to be described.

It is convenient to adopt the usual convention, which times the myocardial infarct from the onset of severe continuous chest pain, or other symptoms attributable to infarction.

If serum enzyme activity is measured at frequent intervals after infarction and plotted against time, the resulting curve has several important features. Firstly, there is always a 'lag' phase of several hours, before any rise in serum enzymes is detectable. Secondly, serum enzyme activity becomes abnormal and rises fairly sharply to a peak. Thirdly, enzyme activity returns to normal at a rather variable rate, depending on the enzyme.

The lag phase is of variable duration, but always lasts at least 3 hours. Generally, the duration of the lag phase will depend on the rate of release of the enzyme from the damaged myocardium, its rate of local destruction and the extent of dilution of the enzyme in the plasma compartment. In practice, serum CK-MB activity usually becomes raised between 3 and 6 hours after infarction, serum AST and total CK become raised between 4 and 8 hours, and serum LD activity becomes abnormal an hour or so after AST and CK (Figure 6). By 12 to 16 hours almost all cases show increased activity of the four enzymes mentioned (Figure 7).

The practical importance of the lag phase lies in the fact that, by taking specimens during this phase, it is often possible to obtain baseline 'pre-infarct' serum enzyme activities. Subsequent increases in enzyme activity can then be shown to be associated with the event causing the acute onset of symptoms, rather than being attributable to some other cause which has been present for some time and may be unrelated to the acute disorder. This may not be applicable where there is doubt about the time of onset of symptoms or where admission to hospital is delayed, but is of particular value where episodes of chest pain occur in patients who are already in hospital and from whom blood can therefore be obtained almost immediately.

Peak enzyme activity usually occurs after a relatively short period during which serum enzyme activity has been rising rapidly. This rising phase represents a period when the rate of enzyme release exceeds the rate of destruction or removal, whereas at the peak the two rates are equal.

Peak serum CK-MB activity usually occurs between 16 and 24 hours after infarction. Serum CK and AST activity reach their peak slightly later,

ENZYMES AND ROUTINE DIAGNOSIS 223

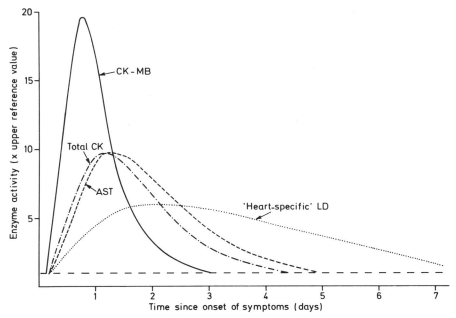

Figure 6. Changes in serum enzyme activity after myocardial infarction.

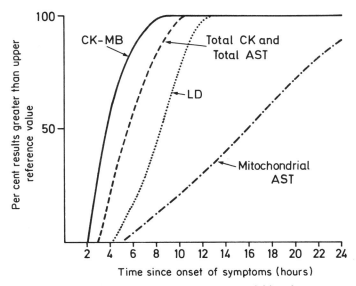

Figure 7. Incidence of abnormal serum enzyme activities after myocardial infarction.

Table 1. Serum Enzyme Activities after Acute Myocardial Infarction

Enzyme	'Lag' period (h)	Peak activity (h)	Duration of rise (days)	Relative size of peak ($\times$ upper ref. value)
Total CK	4–10	20–30	3–6	10
CK–MB	3–8	16–24	1–4	20
AST	4–10	20–30	3–6	10
LD	6–12	30–60	7–14	6

usually at 20 to 30 hours, whereas serum LD activity is more variable, usually being maximum at between 30 and 60 hours after infarction (Table 1).

In practice, it is unusual, unless sampling is undertaken at very frequent intervals, to obtain blood at a time when the enzyme activities are maximum. Nevertheless this is usually of little significance as the peaks are fairly broad, except for CK-MB which usually has a sharper peak.

In patients admitted to the Royal Infirmary, Edinburgh, the average magnitude of the peak activities for the commonly used myocardial enzymes (Table 1) decreases in the order CK-MB, AST, CK, LD. These values were obtained from any unselected group of patients admitted to a Coronary Care Unit; they are slightly atypical in that most authors consider that total CK activity shows a greater rise than serum AST activity. Such discrepancies are probably due to differences in reference ranges. Between patients there may be great variation, with barely detectable increases in serum enzyme activity in some patients and 40- or 50-fold increases, in relation to the upper reference value, in other cases.

Duration of raised serum enzyme activity depends mainly on the rate of destruction and removal of the enzyme, the length of the period during which enzyme was released from the myocardium, and the magnitude of the peak enzyme activity. The theoretical aspects of this are covered elsewhere in this book but in practice it is important to know when to expect raised serum enzyme activities since this will affect the interpretation of the results.

Serum CK-MB activity returns to normal most rapidly; 30–40% of patients have normal activities by 48 hours. Serum CK and AST activities usually remain raised for between 3 and 6 days, and serum LD activity for between 7 and 14 days (Table 1).

Serum Enzyme Activity and Prognosis

Peak serum enzyme activity correlates fairly well with the mortality rate.[12] For example, in patients who survive for 24 hours or more, the mortality in the first 10 days is 5% or less if peak serum AST or CK activity is less than 3 times the upper reference value, whereas the mortality rate is over 30% in patients with peak serum CK or AST activity more than 12 times the upper reference value.[74] Peak serum enzyme activity, therefore, acts as a crude index of infarct size, a variable which can probably be determined more reliably by more sophisticated analysis of serum enzyme changes (see Chapters 11 to 15). For routine purposes, however, mathematical analysis of enzyme decay curves is unlikely to justify the extra work involved.

Even in patients surviving a month after infarction, the subsequent prognosis is poorer in those with markedly raised enzyme activities.[83]

Relationship between the Different Serum Enzymes

The magnitudes, relative to the upper reference value, of the various peak serum enzyme activities correlate well with one another. This relationship has considerable diagnostic value, since large increases in the activity of one of the enzymes should, in the uncomplicated case and at the appropriate time, be accompanied by increases of comparable magnitude in the other enzymes. If this is not so, an additional or alternative diagnosis should be considered. For example, if serum AST activity is raised to 5 times the upper reference value 24 hours after an episode of chest pain, whereas serum CK activity remains normal, then it is extremely unlikely that the raised AST activity can be attributed to acute myocardial infarction; liver disease, or even analytical error, should be considered as more likely alternatives.

Similarly, although patients may differ in the time at which peak enzyme activity occurs, in a particular patient the peaks of the serum enzymes tend to occur at the appropriate time relative to one another, either earlier or later than the times shown in Table 1.

Optimal Times for Obtaining Serum Samples

It is apparent from the foregoing discussion that the successful demonstration of a rise in enzyme activity will depend, to a certain extent, on the time at which samples are taken. Probably, the commonest errors in interpretation arise from failure to take into account the timing of the blood sample in relation to the onset of symptoms.

In most cases, a compromise is required between the maximum possible amount of information, only obtainable by frequent serial sampling, and priorities of convenience and cost. In patients in whom the electrocardiogram shows changes diagnostic of myocardial infarction and the history is typical, it could be argued that no measurements of enzyme activity are required. On the other hand, there are patients with small infarcts in whom the increase in serum enzyme activity is transient and who, therefore, require blood to be taken at several times if the increase is not to be missed.

A reasonable compromise is to measure serum enzyme activity on admission to hospital (i.e. usually less than 6 hours after the onset of symptoms) and at approximately 24 and 48 hours. Such a regime usually allows an approximately 'baseline' activity to be obtained and also provides specimens at times when CK, AST, and LD are all fairly close to their peak activities. In the absence of persisting symptoms, there is little to be gained by continuing to measure enzyme activities in samples taken after 48 hours, since increases in serum enzyme activity, if they are going to occur, will almost invariably have done so by this time.

PROBLEMS OF SPECIFICITY AND SENSITIVITY

Use of Isoenzyme Measurements

In most cases of myocardial infarction the diagnosis is obvious and the enzyme changes are easy to interpret. Discussion of the selection of the most appropriate tests to perform is best deferred until some of the major causes of the lack of enzyme specificity have been considered. It is in the minority of patients who are atypical, or in whom the diagnosis is complicated by the presence of other diseases, that the major diagnostic challenge lies. In these patients isoenzyme measurements may be especially valuable.

CK isoenzymes

Fairly recently, a great deal of attention has been directed towards measurement of CK-MB activity in the serum of patients with myocardial infarction.[38,76,89] It has been shown that, in the majority of disorders in which serum total CK activity is raised due to the release of enzyme from skeletal muscle, there is no accompanying rise in serum CK-MB activity. Many of these disorders are discussed in the ensuing pages. Since, in CK-MB, we have the most cardiac-specific test for myocardial damage presently available, the possible ways in which the test could mislead should be considered.

Tissue studies, which have been performed in both animals and man, have shown considerable variation in the distribution of CK-MB in the same tissue but different species. For example, dog myocardium contains less than 5% CK-MB (in relation to total CK) whereas human myocardium contains about 30% CK-MB. Therefore, it is not possible to draw conclusions about CK-MB distribution in man from animal studies.

Isoenzyme distribution studies in man have usually been either incomplete or have used unsatisfactory methods of quantitation. Autopsy material is unsatisfactory since CK-MB is unstable after death. Early investigations probably overestimated CK-MB concentration in skeletal muscle, since most subsequent studies have found no CK-MB in skeletal muscle,[58,65,85] in contrast to the 10% to 30% figures previously reported.[25] However, there are other reports which, while agreeing that most skeletal muscle contain no CK-MB, suggest that certain muscles (e.g. diaphragm), may contain significant amounts of CK-MB.[75]

Skeletal muscle from patients with muscular dystrophy has also been shown to contain greater than normal amounts of CK-MB. This has been attributed to a failure of development from the foetal state, in which CK-BB and CK-MB predominate.[72]

In sera from patients, there is also evidence that CK-MB activity may not always be attributable to heart disease. In most patients with muscular dystrophy both CK-MM and CK-MB activity in serum are raised, with CK-MB often comprising about 20% of total CK activity.[77] Similar serum patterns may be seen in patients with polymyositis. There is not likely to be any problem in diagnosis in these patients with overt muscle disease, unless it is hoped to use isoenzyme tests to indicate whether myocardium is involved in the dystrophic process. However, there are asymptomatic patients with subclinical muscle disease who have increased serum CK-MB as well as CK-MM activity; this has been demonstrated in two families affected by malignant hyperpyrexia.[95] Therefore, it would be unwise to assume that the presence of CK-MB in serum necessarily indicates myocardial damage. More experience is needed before the circumstances which may cause misleading results are fully delineated.

Difficulties of a different nature may arise with patients who are first seen more than 1 or 2 days after the acute episode of myocardial infarction, since serum CK-MB activity returns to normal very rapidly. In patients seen more than 2 days after infarction LD isoenzyme studies are more likely to be of value than CK isoenzyme studies.

Quantitative measurements of serum CK-MB may not always be required, since the sensitivity and specificity of the test are such that a qualitative result will often provide all the necessary information. Using electrophoretic methods, qualitative analysis is less time consuming, since only a visual assessment of the electrophoretic strip under ultraviolet light is required, thereby eliminating the need for expensive fluorometric scanning equipment. However, there are occasions when quantitative results may provide useful additional information:

1. When evaluating borderline results. For example, if CK-MB shows small increases in the serum of patients with angina, then quantitative evaluation may be required to distinguish these small increases from the larger ones present in patients with myocardial infarction. Similar considerations apply to the increases in serum CK-MB activity which occur after cardiac surgery (see below).

2. When it is necessary to demonstrate *changes* in serum CK-MB activity, for example in patients in whom it is suspected that there has been extension of an established infarct.

It is also true that given the possible choice between being provided with a quantitative and a semiquantitative result, most cardiologists would probably prefer a numerical result.

LD isoenzymes

The advantages of using a 'heart-specific' LD test have already been discussed. Clearly, even greater specificity may be achieved by individual measurements of the activity of all five isoenzymes. The main features seen in myocardial infarction are: firstly, a predominance of the anodal isoenzymes of LD and, secondly, a reversal of the LD_1/LD_2 ratio so that LD_1 activity becomes greater than that of LD_2. These findings are not absolutely characteristic of myocardial infarction since they may also be present in patients with megaloblastic or haemolytic anaemia or in specimens which have become haemolysed after collection. There may also be less common circumstances where this isoenzyme pattern is seen.

Lactate dehydrogenase isoenzyme studies are not so valuable in the diagnosis of myocardial infarction as CK isoenzyme measurements. In most laboratories the extra time and expense involved in measurement are probably not justified, although some authors find the combination of LD and CK isoenzyme results particularly useful.[89]

Serum Enzyme Tests other than AST, CK, and LD

A rather conventional approach has been adopted in this chapter, in that consideration has been confined to CK, AST, LD, and their isoenzymes. This is mainly because there is little or no convincing evidence of the superiority of other enzyme tests. However, brief mention will be made of some of the other enzymes which originate from damaged myocardium, and some which do not originate from myocardium but which may be increased in the serum of patients after myocardial infarction.

Enzymes present in myocardium

It is convenient to discuss separately enzymes which are present in different cell compartments since the behaviour of mitochondrial enzymes is different from that of cytoplasmic enzymes.

Cytoplasmic enzymes include aldolase, adenylate kinase (myokinase) glyceraldehyde phosphate dehydrogenase, malate dehydrogenase and phosphohexose isomerase. The pattern of elevation of all these enzymes after myocardial infarction is roughly similar to that of serum AST and CK, although the relative height of peak activity tends to be slightly smaller. None of the enzymes is specific to myocardium and little or no additional information is provided by their measurement. Alanine aminotransferase (ALT) also comes in this category but is best considered in relation to enzymes originating mainly from tissues other than myocardium.

There is little information about the release of *mitochondrial enzymes* from damaged myocardium. Most information is available about mitochondrial AST (m-AST) although some of the evidence is conflicting.

After myocardial infarction, m-AST activity in serum usually begins to rise at between 8 and 24 hours (Figure 7), reaches a peak at about 48 hours and thereafter slowly returns to normal over the next few days (Figure 8).[8] Although the overall pattern of release appears to be similar to that of cytoplasmic enzymes, particularly LD, the initial lag phase is longer, presumably because the mitochondrial enzymes lie more deeply within the cell and may be more firmly bound to cell structures. On average the relative magnitude of the rise in serum enzyme activity is rather less than that for total AST or LD, and the increases in m-AST correlate rather poorly with those of the cytoplasmic enzymes. Increased m-AST activity is present in over 90% of patients with myocardial infarction.

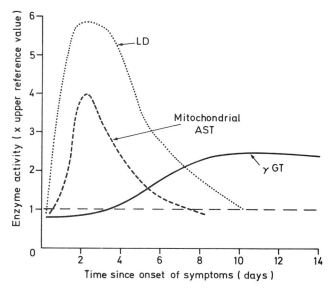

Figure 8. Changes in serum enzyme activity after myocardial infarction.

Similar changes in serum mitochondrial malate dehydrogenase activity occur, but glutamate dehydrogenase which is also a mitochondrial enzyme, shows only inconstant changes which are probably related to liver damage rather than to myocardial damage.

Mitochondrial enzyme measurements are not used for the routine diagnosis of myocardial infarction. However, the differences in release pattern, although minor, and the poor correlation with changes in cytoplasmic enzymes suggest that measurement of mitochondrial enzyme activity may provide additional information not available from cytoplasmic enzyme measurements. One might speculate that mitochondrial enzymes are only released from cells which are irreversibly damaged, whereas it is known that lesser degrees of injury are capable of resulting in cytoplasmic enzyme release, for example the release of creatine kinase from muscle after exercise (see below). These differences could be of importance in determining the amount of myocardial necrosis that has occurred.

Enzymes originating from tissues other than myocardium

A number of enzymes, usually used for the diagnosis of liver disease merit consideration here since their measurement may facilitate interpretation of 'cardiac' enzyme tests.

Ornithine carbamoyl transferase is liver specific and appears in serum whenever there is hepatocellular injury. Serum levels are not raised in myocardial infarction unless there is, additionally, hepatocellular involvement. Technical difficulties have limited the widespread use of this test.

Alanine aminotransferase (ALT) is also present in high activity in the cytosol of liver cells and is often regarded as being specific for liver. This is not so, since heart tissue contains moderate activities of ALT (Chapter 6): over 80% of patients with myocardial infarction show slight increases in serum ALT activity, rarely exceeding 3 times the upper reference value. The relatively early peak activity, at about 30 hours, strongly suggests a myocardial origin for the enzyme, since the other liver-specific enzymes become raised much later and not necessarily in the same patients. The major value of serum ALT in the diagnosis of myocardial infarction is as an indicator of the likely origin of a raised serum AST activity, since some patients with suspected infarction may have liver damage due, for example, to heart failure, alcohol excess, or viral infection. In liver disease, serum AST and ALT tend to parallel one another, whereas in myocardial infarction serum AST activity is always much higher than serum ALT activity, at least during the first 2 days.

γ-Glutamyl transferase (γGT) is also present in liver (and other tissues). Like alkaline phosphatase, it is attached to plasma membranes and is only released into serum to a limited extent after hepatocellular injury (Chapter

6). Although γGT is not present in myocardium, significant increases in serum γGT activity occur after myocardial infarction in slightly over 50% of patients. The rise in serum γGT activity does not usually begin until about the 5th day after myocardial infarction, peak activities occur about the 10th day and enzyme levels return to normal after 20 to 30 days.[1,30] Similar but smaller increases in serum alkaline phosphatase activity may also occur.[21]

The origin of the raised serum γGT activity is uncertain. Some consider that the γGT concentration increases in myocardial tissue undergoing repair and that this is released into serum. There is little evidence to support this, since attempts to demonstrate γGT in organizing myocardium have proved unsuccessful. It seems more likely that the γGT arises from liver, although there is little relationship between serum γGT activity and other indices of hepatocellular damage, either clinical or biochemical.

Although the cause of the increased serum γGT activity after myocardial infarction may remain uncertain, the enzyme has little or no part to play in the routine diagnosis of heart disease.

Distinction between Myocardial Infarction and Ischaemia

Angina

It is usually assumed, more as a matter of definition than one on which factual evidence is available, that serum enzymes do not rise following attacks of angina. If the chest pain, of presumed cardiac origin, is associated with rises in serum enzymes, the patient is usually considered to have had a small infarct.

In fact, it seems rather unlikely that this dogma is always correct. Most of the enzyme tests used previously were probably insufficiently sensitive or not specific enough to detect very small increases in the serum activity of myocardial enzymes. However, since ischaemia, short of cell necrosis, may cause rises in serum total CK after exercise, it seems reasonable to suppose that similar but smaller increases in serum enzymes may follow reversible ischaemic damage to cardiac muscle. Such increases may only be detectable by methods which can measure serum CK-MB activity accurately and precisely. The conclusions of some authors that, in patients with heart disease, a rise in serum enzymes can be equated with myocardial necrosis is probably not justified although in practice only serum CK-MB is likely to cause real difficulties in interpretation.

Acute coronary insufficiency

This is essentially a clinical diagnosis the pathological basis of which remains in doubt. Although, by definition, patients do not show electrocardiographic evidence of recent infarction, this evidence may appear later as an infarct develops. In other patients the chest pain eases fairly rapidly.

In most studies, up to half the patients given this diagnosis on purely clinical and electrocardiographic grounds have serum enzyme changes.[28,74] This is usually taken as presumptive evidence that infarction has occurred. However, it is extremely difficult to establish that the enzyme changes could not have been caused by a prolonged period of myocardial ischaemia.

The enzyme changes in the patients who progress, often slowly, to definite infarction may also be rather atypical, since serum enzymes may show only small increases in activity or even, in rare instances, remain normal.[74]

As with angina, it is probably reasonable to categorize these patients as having had a myocardial infarct or not on the basis of the serum enzyme findings, but it should be recognized that this is an expedient which is based on insubstantial evidence.

EFFECT OF MYOCARDIAL COMPLICATIONS WHICH MAY ARISE IN PATIENTS WITH INFARCTS

Extension of the Myocardial Infarct

Although the patterns of enzyme release vary from patient to patient, three basic forms can be distinguished:[47]

1. Patients showing 'typical' uncomplicated release curves. Enzymes are liberated from damaged myocardium for less than 30 hours; in patients coming to autopsy, the microscopic changes in the necrotic myocardium are homogeneous. The enzyme changes in these patients have already been outlined.

2. Patients in whom the peak activities of serum enzymes occur later, and, indeed, may be less evident as distinct peaks than in the previous category. Quantitative analysis of the serum activity curves suggests that enzyme release occurs for longer than 30 hours. It seems likely that fresh areas of infarction continue to develop over a variable, but extended period; autopsy data support this conclusion since the microscopic findings are more heterogeneous than those in patients in the previous group.

Patients in this category sometimes present diagnostic difficulty, since the relatively slow release of enzyme over an extended period may cause serum enzyme findings to be essentially negative.

3. Patients who have two separate and distinct episodes of acute myocardial infarction.

In these patients, who develop further chest pains while in hospital and who are suspected of having had a further episode of acute infarction, diagnosis should usually present little difficulty. A blood sample taken immediately after the onset of symptoms will establish a new baseline, and a further sample taken about 24 hours later will detect any secondary rise in serum enzyme activity attributable to a second infarct. In these circumstances, CK-MB measurement is particularly useful since serum activity is unlikely to be affected by other conditions which may complicate myocardial infarction, such as heart failure or pulmonary embolism. It is also the serum enzyme most likely to have returned to normal after the original episode.

The patients in the last two categories are important since therapeutic measures might be able to prevent continuing myocardial damage. However, since the patterns of enzyme release reflect myocardial events after their occurrence, enzyme tests are not of particular value in predicting in which patients extension of the infarct is likely to occur.

Arrhythmias

During the first few hours after a myocardial infarct, ventricular fibrillation is very common and accounts for the majority of deaths which occur before arrival in hospital. In hospital, however, ventricular fibrillation becomes much less common, other arrhythmias predominating. There is no evidence that the arrhythmias which commonly complicate myocardial infarction have any significant causal effect on serum enzyme activity. On the contrary, the increased incidence of arrhythmias in patients with large increases in serum enzymes is probably because these patients have had larger infarcts.[66] It is the therapeutic measures which are taken to correct the arrhythmias which cause changes in serum enzyme activity.

Direct Current Countershock

Evidence obtained from patients who have not had myocardial infarcts suggests that small increases in serum CK activity, rarely to more than twice the upper reference value, occur in about 20% of patients having d.c. countershock to correct an arrhythmia.[36] If the measure has to be applied several times, the incidence of abnormal enzyme activities will rise. Serum CK-MB activity remains normal, so that CK probably arises from skeletal, rather than cardiac, muscle.[19]

Closed Chest Cardiac Massage

This is commonly associated with extensive trauma to the anterior chest. Therefore, it is not surprising that increases in serum CK are frequent and may be of considerable magnitude. Serum CK-MB usually remains normal or only slightly raised.

CHEST AND CARDIOVASCULAR DISORDERS OTHER THAN MYOCARDIAL INFARCTION

Myocarditis and Pericarditis

Myocarditis may be due to a number of causes, either primary (e.g. viral) or secondary to generalized disease such as acute rheumatic fever. Whatever the cause, serum enzymes will rise in most patients in whom the myocardial inflammation is active.[13,57] The size of the enzyme increase is usually less than that seen in myocardial infarction and will, in general, reflect the amount of myocardial damage.

In pericarditis, serum enzyme activities are usually normal but may be raised if there is significant underlying myocardial damage.

In both myocarditis and pericarditis, serum enzyme and isoenzyme patterns will be identical to those seen in myocardial infarction. However, clinical problems in diagnosis are uncommon.

Dissecting Aneurysm of the Aorta

The symptoms of acute dissecting aneurysm of the ascending aorta may closely resemble those of acute myocardial infarction. Serum enzyme changes are relatively poorly documented, possibly due to the rapidly fatal outcome in so many cases. Most patients show normal levels of serum enzyme activity, but there are several reports of rises in serum CK, AST, and LD activity, either alone or in combination. It is usually possible to attribute the rise to coexisting myocardial or hepatocellular damage.

Tachycardia

As has already been stated, most forms of arrhythmia do not cause increased serum enzyme activity. However, there are cases in which small increases in serum CK, AST, and LD activity have been observed following periods of tachycardia lasting several hours.[28,74] After reversion to normal rhythm there has been no evidence of myocardial infarction on the electrocardiogram.

In patients with tachycardia of longer duration, the incidence of abnormal enzyme results may be 50% or more; such patients often have increases in liver-specific enzymes, indicating that the liver is the likely source of the raised serum enzymes.[70]

Cardiac failure

Patients suspected of having had myocardial infarcts not infrequently have cardiac failure caused by pre-existing heart disease. Although mild congestive cardiac failure is only rarely accompanied by any increases in serum enzyme activity, more severe cardiac failure may cause moderate increases in serum AST, and sometimes LD_5, activity.[4] Serum CK and LD_1 activity remain normal. Problems in diagnosis could occur in those laboratories which rely solely on AST and total LD for the diagnosis of myocardial infarction. However, even in these circumstances a rise in serum ALT activity of about the same magnitude as that of AST suggests the presence of hepatocellular damage.

Pulmonary Embolism and Infarction

Pulmonary embolism, with or without infarction, is a common disorder which is often difficult to diagnose clinically. For example, one study has shown that less than 50% of patients dying from pulmonary infarction were correctly diagnosed before death, one of the commonest misdiagnoses being myocardial infarction.[51] More recently, diagnosis has probably improved since more accurate diagnostic procedures, such as lung scanning and arteriography, are available. However the need for simple tests for pulmonary embolism remains, since many patients continue to die without a correct diagnosis having been made.[53]

There has been controversy about the value of enzyme tests in making a diagnosis of pulmonary embolism, thereby distinguishing it from pneumonia or myocardial infarction. Some published series find the incidence of abnormal serum LD activity in patients with pulmonary embolism to be over 90%,[3] but in more recent investigations the incidence has been between 50% and 80%.[39,71] It has been proposed that the following triad indicates a diagnosis of pulmonary infarction: raised serum LD activity, raised serum bilirubin, and normal serum AST activity.[88] Although it is true that serum AST activity is raised in only about 20% of patients with pulmonary infarction, most investigators have found this triad to be of little discriminatory value.

The isoenzyme pattern of the raised serum LD in pulmonary infarction does not correspond to that present in lung tissue. Two different types of pattern may be present in serum.[39,84] In some patients, LD_1 predominates, the LD activity tending to rise relatively soon, less than 4 days, after the episode. Increased serum enzyme activity in these patients is presumably due to haemolysis of red cells trapped in the infarcted area of lung. Other patients show a predominant increase in LD_5, the enzyme usually appearing in serum more than 4 days after infarction. These patients often also show increased serum ALT, AST, and sometimes ornithine carbamoyl transferase activity, almost certainly due to hepatocellular damage.

The increased CK activity occasionally described in the serum of patients with pulmonary infarction is unlikely to have arisen from lung tissue, which contains little CK.[60] The rises are more likely to be due to intramuscular injections or associated disease.[39]

One must conclude that serum enzyme tests, although often abnormal in patients with pulmonary infarction, are of little value in making a positive diagnosis. The differentiation between pulmonary and myocardial infarction normally presents few problems to the clinician; where there is doubt, serum enzymes may help since the findings are different in the two conditions.

Pneumonia

Small rises in enzyme activity occasionally occur but these are of little value in diagnosis.[33,60]

CIRCUMSTANCES WHERE THE DIAGNOSIS OF MYOCARDIAL INFARCTION MAY PRESENT SPECIAL DIFFICULTIES

Usually, patients with myocardial infarction present a fairly straight-forward history indicating the diagnosis. There are, however, patients in whom myocardial infarction may supervene on an already serious condition. Since enzyme changes are common in all disorders in which there is tissue damage, the detection of myocardial necrosis in these patients may be difficult.

Operative Procedures

Cardiac catheterization

Cardiac catheterization with or without coronary angiography, is associated with an increase in serum CK activity in up to 100% of patients, the percentage depending on the centre from which the results have been reported.[11,67] Serum AST, LD, and CK-MB activity are usually normal or show negligible increases. The increases in serum CK activity are rarely due to damage to cardiac muscle, but may be caused by the intramuscular injection of the drug used to pre-medicate the patient, since there are usually no increases in patients given oral pre-medication.[11]

Surgical operations

Post-operative myocardial infarction is a relatively common problem which may be difficult to diagnose In unselected surgical patients, the incidence of post-operative myocardial infarction is about 2% but this proportion rises to between 10% and 20% in patients over 50 years old with hypertension or other forms of heart disease.[17,48]

After surgery, a variable proportion of patients show increases in serum CK, AST, and LD activity.[31,61] Serum CK-MB activity remains normal.[15,76]

In most patients, the increase in serum CK activity is greater than in serum AST or LD activity, presumably because the increases are due to damage to skeletal muscle. In general, the extent of the rise in activity will depend on the amount of muscle trauma associated with the operation. Small operations, such as breast biopsy, are not accompanied by increased serum CK activity, whereas large increases follow major surgical operations. However, the anaesthetic may also be important since suxamethonium, especially if used in conjunction with halothane, may cause considerable rises in serum CK activity even in patients who have had only minor eye operations.[32]

Heart operations

These may, in addition, be accompanied by a relatively transient rise in serum CK-MB activity, with normal activity usually being regained within 24 hours.[15] However, the proportion of total CK present in the MB form is less after surgical operations on the heart than after myocardial infarction, presumably because only a relatively small amount of the enzyme arises from damaged myocardium.[76] In many patients, the post-operative electrocardiogram is abnormal after cardiac, especially bypass, operations and may be difficult to interpret. In this group, those patients in whom there is a large and prolonged increase in serum CK-MB activity or in whom the LD_1/LD_2 ratio is greater than 1, should be suspected of having sustained post-operative myocardial necrosis.[15]

In conclusion, total CK, AST, and LD are of only limited value in the diagnosis of post-operative myocardial infarction. Measurement of serum CK-MB activity, on the other hand, is a valuable diagnostic test in these circumstances, even in patients who have had operations on the heart.

Accidental trauma

Sometimes the hypotension associated with blood loss accompanying an accident may precipitate a myocardial infarct. The enzyme changes in patients who have been injured in accidents are poorly documented but, in general, large increases in serum activity may be expected if there has been significant tissue trauma.[35] As with post-operative patients, serum CK-MB activity seems the test most likely to be of value.

Comatose Patients

Most patients who have been unconscious for more than a few hours develop increased serum enzyme activities. This may often be due to the primary cause of the coma, but it is unlikely that this is always so.

Cerebrovascular accidents

These accidents are associated with increased serum CK activities in 50% to 70% of cases.[18] The degree of enzyme rise is variable but very large increases may be seen, sometimes up to 20 times the upper reference value. It is not possible to correlate the degree of enzyme rise with the clinical state of the patient. The CK is of skeletal muscle origin.

Drug overdosage

Drug overdosage is almost invariably associated with increases in serum enzyme activity if the patient becomes deeply unconscious. Often the increases in serum enzymes can be attributed to specific organ damage, for example to the liver or kidney, but this is unlikely to be the case in certain patients who have taken overdoses of hypnotics. In patients with barbiturate or methaqualone overdosage, with no evidence of liver or kidney damage, increases in serum CK, AST, and LD activity occur, often to very high levels, especially of CK.[93] The enzymes probably arise from skeletal muscle, although the cause of their release is not clear; serum CK-MB activity is normal.[76]

Hypothermia

Hypothermia is often encountered in old people who may have had cerebrovascular accidents or become immobilized due to injury. Large increases in serum CK activity are seen in over 90% of cases; these are accompanied by slightly smaller increases in LD and AST.[45]

There are a large number of other conditions such as diabetic ketoacidosis and severe renal disease which may be accompanied by impairment of consciousness. Serum CK, AST, and LD may be raised in any of these disorders, although the pattern of rise will vary and depend on the primary disease. The practical point is that enzyme tests may be misleading if used to try to detect evidence of myocardial damage in these patients. The only enzyme test of discriminatory value in these circumstances is CK-MB.

FACTORS COMMONLY CAUSING MISINTERPRETATION OF ENZYME RESULTS

Intramuscular Injections

There is little doubt that many of the increases in serum CK activity, which in the past have been attributed to specific disease processes, have instead been due to intramuscular injections.[49] For example, after intramuscular injections of chlorpromazine or diazepam, about 50% of patients show a significant increase in serum CK activity.[41] Whereas single injections may cause as much as a 10-fold increase in serum CK activity, repeated injections or the development of clinical signs of cellulitis at the injection site may be associated with much larger increases. Serum CK activity usually remains raised for 2 or 3 days. Serum AST and LD activity are scarcely affected. The degree of enzyme rise also depends on the drug being injected, since injections of physiological saline do not cause increased serum CK activity.

Since many patients with suspected myocardial infarction receive injections of pain-killing drugs such as morphine, any increase in serum CK that is not accompanied by rises in serum AST and LD activity in these patients should be appraised critically. In some Coronary Care Units, the use of intramuscular injections is avoided, most drugs being given intravenously. Serum CK-MB isoenzyme measurements should prove valuable if there is doubt about the diagnosis, since CK-MB is not released into serum after intramuscular injections.[35]

Alcohol

If taken in excess, alcohol may damage both liver, causing increased serum LD_5 and AST activity, and skeletal muscle, causing increased serum CK activity. Depending on the location of the hospital to which the patient is admitted, between 1% and 10% of patients admitted with a provisional diagnosis of myocardial infarction will have recently consumed an excess of alcohol. A proportion of these patients, probably between 20% and 60%, will have raised serum CK or AST, or both, often to very high activities.[43,55]

Exercise

After periods of severe, unaccustomed or prolonged exercise, there may be increases in serum enzyme activity. Usually, peak activities occur 8 to 24 hours after the exercise. The size of the increase depends on a number of factors:

1. The concentration of enzyme in muscle. Rises in serum CK activity are much greater than those of AST or LD. Serum CK-MB activity remains normal.
2. The duration and severity of exercise. Whereas short periods of exercise (less than 30 minutes) usually result in relatively small increases in serum CK activity, longer periods provide a much more effective stimulus.[56] For example, a 50 mile London to Brighton walk resulted in 12- to 70-fold increases in serum CK activity.[27]
3. Physical fitness. Physical training reduces the tendency for serum

enzyme activities to rise after exercise.[56] Similarly, trained athletes show smaller increases than untrained people taking the same amount of exercise.[73]

Problems in diagnosis arise only in patients in whom severe exercise is thought to have precipitated a myocardial infarct. Misinterpretation of an increase in serum CK activity should not occur if CK-MB isoenzyme measurements are available. There should also be little or no accompanying increase in serum 'heart-specific' LD activity.

RELATIONSHIP BETWEEN THE ELECTROCARDIOGRAM AND ENZYME TESTS

In routine clinical practice, the two principal aids to the diagnosis of acute myocardial infarction are the electrocardiogram (e.c.g.) and serum enzyme tests. It is of interest to compare the roles of the two types of test.

Most patients with acute myocardial infarction show characteristic e.c.g. changes in the Q wave, the ST segment, and the T wave:

Abnormal Q waves (more than 25% of the R wave and lasting 0.04 second or more) appear at a variable time (hours to days) after the onset of infarction. The Q waves usually persist indefinitely.

ST segment elevation occurs in leads over the area of infarction. This is usually the first e.c.g. abnormality, occurring within minutes of infarction. The changes are relatively transient, lasting from a few hours to several days.

T wave changes. Some patients show transient early, 'giant' upright T waves in the leads overlying the infarct. Later, after 1 or 2 days, more constant changes occur, namely symmetrical T wave inversion maximum at about 2 weeks. The inverted T waves may persist indefinitely but often return to normal.

In general, the abnormal Q waves provide the best positive evidence of myocardial infarction since ST segment and T wave changes may occur in ischaemia and a variety of other disorders. The chance of the e.c.g. providing definitive evidence of infarction will depend on the timing of the e.c.g. recordings in relationship to the clinical course of the disorder.

In patients who are admitted within the first few hours of the onset of symptoms, it is likely that ST segment changes, but not Q waves, will be present on the e.c.g. These changes are not, as stated above, diagnostic of infarction and may be caused by ischaemia. In many cases, the early presumptive diagnosis of myocardial infarction will be made on the basis of ST segment changes *supported by* a rise in serum enzymes.

Over the next 48 hours, during which Q waves are likely to develop, the enzyme changes are also most likely to yield strong presumptive evidence as to whether or not infarction has occurred. However, there are a minority of cases in whom the e.c.g. may not show changes diagnostic of infarction because:

1. Q waves do not develop in patients with subendocardial infarction;
2. the patient already has Q waves caused by a previous infarct;
3. the e.c.g. pattern is obscured by left bundle branch block or certain other abnormalities.

In patients seen for the first time 2 days to 2 weeks after the infarction, most will have abnormal e.c.g.s still showing progressive changes, such as T wave inversion. The enzyme findings are likely to be of less value.

Therefore, although the e.c.g. provides specific evidence of myocardial infarction in most patients, there are three main circumstances where enzyme tests provide additional valuable, or even crucial, information. The first, as mentioned, is when it is required to make a relatively early diagnosis and ST segment changes, but not Q waves, are present. The second, is when the e.c.g. does not develop diagnostic changes at any stage after infarction. The third, is when an attempt at the estimation of infarct size is being made. In general, the e.c.g. does not provide much useful evidence about the size of an infarct, although experimental studies on the use of ST segment 'mapping' in patients with anterior infarcts are in progress in some centres. Usually, the clinical findings and routine enzyme results will give a fairly good indication of infarct size.

Since the e.c.g. may be normal in patients with myocardial infarcts, it may be impossible to *exclude* a diagnosis of myocardial infarction in a patient with some other cause for his symptoms. In such a patient, the presence of normal serum enzyme activities provides useful confirmatory evidence that infarction has not occurred.

CONCLUSION: SELECTION OF ENZYME TESTS FOR DIAGNOSIS OF MYOCARDIAL INFARCTION

The routine clinical chemistry department will be faced with the choice of enzyme tests to offer the clinical staff. The eventual choice will depend on both clinical and analytical factors. Furthermore, different clinical circumstances may dictate a variety of alternative requirements. For example, if it is required to exclude patients who have not had an infarct, possibly to free a bed in a Coronary Care Unit, then a sensitive test is needed so that no patients with infarcts are discharged. In such circumstances, a certain lack of specificity would be acceptable since no harm would be likely to arise to a patient who had not had a myocardial infarct but who remained in a Coronary Care Unit. On the other hand, if a positive diagnosis is required, most problems arise from the lack of specificity of enzyme tests.

Measurement of serum AST and 'heart-specific' LD on samples taken on admission and at 24 and 48 hours will suffice to make a positive diagnosis in nearly all cases. Serum CK may be measured in place of, or in addition to, AST, but CK is no more specific for myocardium than AST and has been shown to be raised in a large variety of disorders.[55] If serum AST is thought to have arisen from liver, measurement of serum ALT activity may prove a useful additional test to confirm this.

LD isoenzyme studies have been used in some centres with success. The diagnostic pattern that seems to be fairly characteristic of myocardial infarction is a reversal of the normal LD_1/LD_2 ratio to values greater than 1.0.[89]

CK-MB isoenzyme measurement. The availability of practicable methods of analysis has greatly increased the use of these measurements in routine laboratories. The combined advantages of great specificity and sensitivity probably render this enzyme the most useful single test for the diagnosis of myocardial damage. In post-operative patients and also in patients with tissue trauma or damage due to other causes, CK-MB is the only test which is likely to confirm a diagnosis of myocardial infarction. Similarly, in patients who have had intramuscular injections or who are admitted after periods of exercise which may have caused muscle CK release, MB isoenzyme measurements may be valuable.

Many laboratories will be unable, for reasons of cost and additional workload, to perform isoenzyme measurements, particularly of CK-MB, in all patients suspected of having had myocardial infarct. However, where possible the test should be available to resolve problems in patients in whom the other diagnostic features are equivocal. It is worth stressing that, providing specimens are obtained at the appropriate time and results are correctly interpreted, the vast majority of myocardial infarcts can be diagnosed using clinical and electrocardiographic data combined with measurement of serum CK or AST, or both, and 'heart-specific' LD activity.

REFERENCES

1. Agostoni, A., Ideo, G., and Stabilini, R. Serum gamma-glutamyl transpeptidase activity in myocardial infarction. *British Heart Journal*, **27**, 688 (1965).
2. Agress, C. M. Evaluation of the transaminase test. *American Journal of Cardiology*, **3**, 74 (1959).
3. Amador, E. and Potchen, E. J. Serum lactic dehydrogenase activity and radioactive lung scanning in the diagnosis of pulmonary embolism. *Annals of Internal Medicine*, **65**, 1247 (1966).
4. Auvinen, S. Evaluation of serum enzyme tests in the diagnosis of acute myocardial infarction. *Acta Medica Scandinavica Supplement*, 539 (1974).
5. Auvinen, S. and **Konttinen**, A. The diagnostic value of serum LDH isoenzymes and heat-stable and urea-stable LDH measurements. *Acta Medica Scandinavica*, **189**, 191 (1971).

6. Berry, A. J., Lott, J. A., and Grannis, G. F. NADH preparations as they affect reliability of serum lactate dehydrogenase determinations. *Clinical Chemistry*, **19,** 1255 (1973).
7. Bowers, G. N., Bergmeyer, H. U., and Moss, D. W. Provisional recommendation (1974) on IFCC methods for the measurement of catalytic concentration of enzymes. *Clinica Chimica Acta*, **61,** F11 (1975).
8. Boyde, T. R. C. Serum levels of the mitochondrial isoenzyme of aspartate aminotransferase in myocardial infarction and muscular dystrophy. *Enzymologia Biologica et Clinica*, **9,** 385 (1968).
9. Brydon, W. G. and Smith, A. F. An appraisal of routine methods for the determination of the anodal isoenzymes of lactate dehydrogenase. *Clinica Chimica Acta*, **43,** 361 (1973).
10. Buhl, S. N., Jackson, K. Y., Lubinski, R., and Vanderlinde, R. E. A search for the best buffer to use in assaying human lactate dehydrogenase with the lactate-to-pyruvate reaction. *Clinical Chemistry*, **22,** 1872 (1976).
11. Chahine, R. A., Eber, L. M., and Kattus, A. A. Interpretation of the serum enzyme changes following cardiac catheterisation and coronary angiography. *American Heart Journal*, **87,** 170 (1974).
12. Chapman, B. L. Correlation of mortality rate and serum enzymes in myocardial infarction. *British Heart Journal*, **33,** 643 (1971).
13. Choresis, C. and Leonidas, J. Serum transaminases in diphtheritic myocarditis. Their relation to electrocardiographic findings. *Acta Paediatrica*, **51,** 293 (1962).
14. The Committee on Enzymes of the Scandinavian Society for Clinical Chemistry and Clinical Physiology. Recommended methods for the determination of four enzymes in blood. *Scandinavian Journal of Clinical and Laboratory Investigation*, **33,** 291 (1974).
15. Dixon, S. H., Limbird, L. E., Roe, C. R., Wagner, G. S., Oldham, H. N., and Sabiston, D. C. Recognition of postoperative acute myocardial infarction. Application of isoenzyme techniques. *Circulation Supplement 3*, **47** and **48,** 137 (1973).
16. Dreyfus, J. C., Shapira, G., Resnais, J., and Scebat, L. La créatine-kinase sérique dans le diagnostic de l'infarctus myocardique. *Revue français d'Etudes Cliniques et Biologiques*, **5,** 386 (1960).
17. Driscoll, A. C., Hobika, J. H., Etsten, B. E., and Proger, S. Clinically unrecognised myocardial infarction following surgery. *New England Journal of Medicine*, **264,** 633 (1961).
18. Dubo, H., Park, D. C., Pennington, R. J. T., Kalbag, R. M., and Walton, J. N. Serum-creatine-kinase in cases of stroke, head injury and meningitis. *Lancet*, **2,** 743 (1967).
19. Ehsani, A., Ewy, G. A., and Sobel, B. E. Effects of electrical counter-shock on serum creatine phosphokinase (CPK) isoenzyme activity. *American Journal of Cardiology*, **37,** 12 (1976).
20. Ellis, G. and Goldberg, D. M. Serum α-hydroxybutyrate dehydrogenase activity. *American Journal of Clinical Pathology*, **56,** 627 (1971).
21. Ewen, L. M. and Griffiths, J. Patterns of enzyme activity following myocardial infarction and ischemia. *American Journal of Clinical Pathology*, **56,** 614 (1971).
22. Expert Panel on Enzymes of IFCC. Provisional recommendation on an IFCC method for the measurement of catalytic concentrations of enzymes. Part 2. Aspartate aminotransferase. Measurement of the catalytic concentration in human serum. *Clinica Chimica Acta*, **70,** F19 (1976).
23. Gay, R. J., McComb, R. B., and Bowers, G. N. Optimum reaction conditions for human lactate dehydrogenase isoenzymes as they affect total lactate dehydrogenase activity. *Clinical Chemistry*, **14,** 740 (1968).
24. Goldberg, D. M. and Winfield, D. A. Diagnostic accuracy of serum enzyme assays for myocardial infarction in a general hospital population. *British Heart Journal*, **34,** 597 (1972).
25. Goto, I., Nagamine, M., and Katsuki, S. Creatine phosphokinase isoenzymes in muscles. Human fetus and patients. *Archives of Neurology*, **20,** 422 (1969).
26. Graig, F. A., Smith, J. C., and Foldes, F. F. Effect of dilution on the activity of serum creatine phosphokinase. *Clinica Chimica Acta*, **15,** 107 (1966).
27. Griffiths, P. D. Serum levels of ATP: creatine phosphotransferase (creatine kinase). The normal range and effect of muscular activity. *Clinica Chimica Acta*, **13,** 413 (1966).

28. Griffiths, P. D. ATP: creatine phosphotransferase in the diagnosis of acute chest pain. *British Heart Journal*, **28**, 199 (1966).
29. Hanson, N. Q. and Freier, E. F. Stability in urea as a measure of cardiac lactate dehydrogenase on the centrifugal analyzer. *Clinical Chemistry*, **20**, 769 (1974).
30. Hedworth-Whitty, R. B., Whitfield, J. B., and Richardson, R. W. Serum γ-glutamyl transpeptidase activity in myocardial ischaemia. *British Heart Journal*, **29**, 432 (1967).
31. Hobson, R. W., Conant, C., Mahoney, W. D., and Baugh, J. H. Serum creatine phosphokinase: analysis of postoperative changes. *American Journal of Surgery*, **124**, 625 (1972).
32. Innes, R. K. R. and Strømme, J. H. Rise in serum creatine phosphokinase associated with agents used in anaesthesia. *British Journal of Anaesthesia*, **45**, 185 (1973).
33. Irani, F. A. and Newhouse, M. T. Serum lactic dehydrogenase in pneumonia. *Southern Medical Journal*, **65**, 858, 874 (1972).
34. Karmen, A. A note on the spectrophotometric assay of glutamic-oxalacetic transaminase activity in human blood serum. *Journal of Clinical Investigation*, **34**, 131 (1955).
35. Klein, M. S., Shell, W. E., and Sobel, B. E. Serum creatine phosphokinase (CPK) isoenzymes after intramuscular injections, surgery and myocardial infarction. *Cardiovascular Research*, **7**, 412 (1973).
36. Konttinen, A., Hupli, V., Louhija, A., and Hartel, G. Origin of elevated serum enzyme activities after direct current counter-shock. *New England Journal of Medicine*, **281**, 231 (1969).
37. Konttinen, A. and Lindy, S. Assay of cardiac lactate dehydrogenase isoenzymes by means of urea. *Acta Medica Scandinavica*, **181**, 513 (1967).
38. Konttinen, A. and Somer, H. Determination of serum creatine kinase isoenzymes in myocardial infarction. *American Journal of Cardiology*, **29**, 817 (1972).
39. Konttinen, A., Somer, H., and Auvinen, S. Serum enzymes in acute pulmonary embolism: Extrapulmonary sources in acute pulmonary embolism. *Archives of Internal Medicine*, **133**, 243 (1974).
40. Kreutzer, H. H. and Fennis, W. H. S. Lactic dehydrogenase isoenzymes in blood serum after storage at different temperatures. *Clinica Chimica Acta*, **9**, 64 (1964).
41. Küster, J. Increased creatine-kinase concentrations after intramuscular injection of 'Diazepam'. (A contribution to the differential diagnosis of myocardial infarction.) *German Medicine*, **2**, 154 (1972).
42. La Due, J. S., Wroblewski, F., and Karmen, A. Serum glutamic oxalacetic transaminase in human acute transmural myocardial infarction. *Science*, **120**, 497 (1954).
43. Lafair, J. S. and Myerson, R. Alcoholic myopathy. With special reference to the significance of creatine phosphokinase. *Archives of Internal Medicine*, **122**, 417 (1968).
44. Latner, A. L. and Skillen, A. W. The heat stability index of lactate dehydrogenase in cardiac infarction. *Proceedings of the Association of Clinical Biochemists*, **2**, 100 (1963).
45. Maclean, D., Griffiths, P. D., and Emslie-Smith, D. Serum enzymes in relation to electrocardiographic changes in accidental hypothermia. *Lancet*, **2**, 1266 (1968).
46. McQueen, M. J. Optimal assay for LDH and αHBD at 37 °C. *Annals of Clinical Biochemistry*, **9**, 21 (1972).
47. Mathey, D., Bleifeld W., Buss, H., and Hanrath, P. Creatine kinase release in acute myocardial infarction: correlation with clinical, electrocardiographic and pathologic findings. *British Heart Journal*, **37**, 1161 (1975).
48. Mauney, F. M., Ebert, P. A., and Sabiston, D. C. Post-operative myocardial infarction: A study of the predisposing factors, diagnosis and mortality in a high risk group of surgical patients. *Annals of Surgery*, **172**, 497 (1970).
49. Meltzer, H. Y., Mrozak, S., and Boyer, M. Effect of intramuscular infections on serum creatine phosphokinase activity. *American Journal of Medical Sciences*, **259**, 42 (1970).
50. Mercer, D. W. and Varat, M. A. Detection of cardiac-specific creatine kinase isoenzyme in sera with normal or slightly increased total creatine kinase activity. *Clinical Chemistry*, **21**, 1088 (1975).
51. Miller, R. and Berry, J. B. Pulmonary infarction: a frequently missed diagnosis. *American Journal of Medical Sciences*, **222**, 197 (1951).
52. Morin, L. G. Evaluation of current methods for creatine kinase isoenzyme fractionation. *Clinical Chemistry*, **23**, 205 (1977).

53. Morrell, M. T. and Dunuik, M. S. The post-mortem incidence of pulmonary embolism in a hospital population. *British Journal of Surgery*, **55,** 347 (1968).
54. Nealon, D. A. and Henderson, A. R. Separation of creatine kinase isoenzymes in serum by ion-exchange column chromatography (Mercer's method, modified to increase sensitivity). *Clinical Chemistry*, **21,** 392 (1975).
55. Nevins, M. A., Saran, M., Bright, M., and Lyon, L. J. Pitfalls in interpreting serum creatine phosphokinase activity. *Journal of the American Medical Association*, **224,** 1382 (1973).
56. Nuttall, F. Q. and Jones, B. Creatine kinase and glutamic oxalacetic transaminase activity in serum: Kinetics of change with exercise and effect of physical conditioning. *Journal of Laboratory and Clinical Medicine*, **71,** 847 (1968).
57. Nydick, I., Tang, J., Stollerman, G. H., Wroblewski, F., and La Due, J. S. The influence of rheumatic fever on serum concentrations of the enzyme glutamic oxalacetic transaminase. *Circulation*, **12,** 795 (1955).
58. Ogunro, E. A., Hearse, D. J., and Shillingford, J. P. Creatine kinase isoenzymes: their separation and quantitation. *Cardiovascular Research*, **11,** 94 (1977).
59. Oliver, I. T. A spectrophotometric method for the determination of creatine phosphokinase and myokinase. *Biochemical Journal*, **61,** 116 (1955).
60. Perkoff, G. T. Demonstration of creatine phosphokinase in human lung tissue. *Archives of Internal Medicine*, **122,** 326 (1968).
61. Person, D. A. and Judge, R. D. Effect of operation on serum transaminase levels. *Archives of Surgery*, **77,** 892 (1958).
62. Ratnaike, S. and Moss, D. W. Apoenzyme of aspartate aminotransferase in serum in health and disease. *Clinica Chimica Acta*, **74,** 281 (1977).
63. Reed, A. H., Henry, R. J., and Mason, W. B. Influence of statistical method used on the resulting estimate of normal range. *Clinical Chemistry*, **17,** 275 (1971).
64. Rej, R., Fasce, C. F., and Vanderlinde, R. E. Increased aspartate aminotransferase activity of serum after *in vitro* supplementation with pyridoxal phosphate. *Clinical Chemistry*, **19,** 92 (1973).
65. Roberts, R., Henry, P. D., and Sobel, B. E. An improved basis for enzymatic estimation of infarct size. *Circulation*, **52,** 743 (1975).
66. Roberts, R., Husain, A., Ambos, H. D., Oliver, G. C., Cox, J. R., and Sobel, B. E. Relation between infarct size and ventricular arrhythmia. *British Heart Journal*, **37,** 1169 (1975).
67. Roberts, R., Ludbrook, P. A., Weiss, E. S., and Sobel, B. E. Serum CPK isoenzymes after cardiac catheterisation. *British Heart Journal*, **37,** 1144 (1975).
68. Rosalki, S. B. An improved procedure for creatine phosphokinase determination. *Journal of Laboratory and Clinical Medicine*, **69,** 696 (1967).
69. Rosalki, S. B. and Wilkinson, J. H. Reduction of α-ketobutyrate by human serum. *Nature*, **188,** 1110 (1960).
70. Runde, I. and Dale, J. Serum enzymes in acute tachycardia. *Acta Medica Scandinavica*, **179,** 535 (1966).
71. Sasahara, A. A., Cannilla, J. E., Morse, R. L., Sidd, J. J., and Tremblay, G. M. Clinical and physiologic studies in pulmonary thromboembolism. *American Journal of Cardiology*, **20,** 10 (1967).
72. Shapira, F., Dreyfus, J.-C., and Allard, D. Les isozymes de la créatine kinase et de l'aldolase du muscle foetal et pathologique. *Clinica Chimica Acta*, **20,** 439 (1968).
73. Shapiro, Y., Magazanik, A., Sohan, E., and Reich, C. B. Serum enzyme changes in untrained subjects following a prolonged march. *Canadian Journal of Physiology and Pharmacology*, **51,** 271 (1973).
74. Smith, A. F. Diagnostic value of serum-creatine-kinase in a Coronary Care Unit. *Lancet*, **2,** 178 (1967).
75. Smith, A. F. Separation of tissue and serum creatine kinase isoenzymes on polyacrylamide gel slabs. *Clinica Chimica Acta*, **39,** 351 (1972).
76. Smith, A. F., Radford, D., Wong, C. P., and Oliver, M. F. Creatine kinase MB isoenzyme studies in diagnosis of myocardial infarction. *British Heart Journal*, **38,** 225 (1976).
77. Somer, H., Donner, M., Murros, J., and Konttinen, A. A serum isoenzyme study in muscular dystrophy. *Archives of Neurology*, **29,** 343 (1973).

78. Sørensen, N. S. Creatine phosphokinase in the diagnosis of myocardial infarction. *Acta Medica Scandinavica*, **174,** 725 (1963).
79. Strandjord, P. E., Clayson, K. J., and Freier, E. F. Heat-stable lactate dehydrogenase in the diagnosis of myocardial infarction. *Journal of the American Medical Association*, **182,** 1099 (1962).
80. Sunderman, F. W. Current concepts of 'normal values', 'reference values', and discrimination values' in clinical chemistry. *Clinical Chemistry*, **21,** 1873 (1975).
81. Szasz, G., Gruber, W. and Bernt, E. Creatine kinase in serum: 1. Determination of optimum reaction conditions. *Clinical Chemistry*, **22,** 650 (1976).
82. Szasz, G., Gerhardt, W., Gruber, W., and Bernt, E. Creatine kinase in serum: 2. Interference of adenylate kinase with the assay. *Clinical Chemistry*, **22,** 1806 (1976).
83. Thygesen, K., Nielsen, B. L., and Nielsen, J. S. Enzymes and long-term survival after first myocardial infarction. *Acta Medica Scandinavica*, **199,** 75 (1976).
84. Trujillo, N. P., Nutter, D., and Evans, J. M. The isoenzymes of lactic dehydrogenase. II. Pulmonary embolism, liver disease, the postoperative state and other medical conditions. *Archives of Internal Medicine*, **119,** 333 (1967).
85. Tsung, S. H. Creatine kinase isoenzyme patterns in human tissue obtained at surgery. *Clinical Chemistry*, **22,** 173 (1976).
86. Tuckerman, J. F. and Henderson, A. R. Some observations on serum lactate dehydrogenase (pyruvate → lactate) assays optimised at 37°. *Clinica Chimica Acta*, **49,** 241 (1973).
87. Van der Veen, K. J. and Willebrands, A. F. Isoenzymes of creatine phosphokinase in tissue extracts and in normal and pathological sera. *Clinica Chimica Acta*, **13,** 312 (1966).
88. Wacker, W. E. C., Rosenthal, M., Snodgrass, P. J., and Amador, E. A triad for the diagnosis of pulmonary embolism and infarction. *Journal of the American Medical Association*, **178,** 8 (1961).
89. Wagner, G. S., Roe, C. R., Limbird, L. E., Rosati, R. A., and Wallace, A. G. The importance of the identification of the myocardial-specific isoenzyme of creatine phosphokinase (MB form) in the diagnosis of acute myocardial infarction. *Circulation*, **47,** 263 (1973).
90. White, L. P. Serum enzymes. 1. Serum lactic dehydrogenase in myocardial infarction. *New England Journal of Medicine*, **255,** 984 (1956).
91. Wilkinson, J. H. *Isoenzymes*, 2nd edn. Chapman and Hall Ltd., London, 1970.
92. Winkel, P., Statland, B. E., and Bokelund, H. Factors contributing to intra-individual variation of serum constituents: 5. Short term day-to-day and within-hour variation of serum constituents in health subjects. *Clinical Chemistry*, **20,** 1520 (1974).
93. Wright, N., Clarkson, A. R., Brown, S. S., and Fuster, V. Effects of poisoning on serum enzyme activities, coagulation and fibrinolysis. *British Medical Journal*, **3,** 347 (1971).
94. Ziegenhorn, J., Senn, M., and Bücher, T. Molar absorptivities of β-NADH and β-NADPH. *Clinical Chemistry*, **22,** 151 (1976).
95. Zsigmond, E. K., Starkweather, W. H., Duboff, G. S., and Flynn, K. A. Abnormal creatine-phosphokinase isoenzyme pattern in families with malignant hyperpyrexia. *Anaesthesia and Analgesia*, **51,** 827 (1972).

CHAPTER 10

Radioimmunoassay of creatine kinase isoenzymes
R. Roberts and B. E. Sobel

INTRODUCTION	247
DEVELOPMENT OF A CK ISOENZYME RADIOIMMUNOASSAY (RIA)	248
SENSITIVITY AND SPECIFICITY	251
CLINICAL APPLICATIONS	253
TECHNICAL CONSIDERATIONS	254
FUTURE DIRECTIONS	254
CONCLUDING COMMENTS	255
REFERENCES	255

INTRODUCTION

As discussed in Chapters 8 and 9, quantitative isoenzyme assays have been developed based on detection of enzymatic activity. However, additional approaches were required not only to improve precision and sensitivity but also to permit detection of isoenzyme protein, whether or not it was enzymatically active. For this reason, a radioimmunoassay (RIA) was developed for detection and quantification of plasma MB creatine kinase (CK).[17] With conventional, enzymatically based quantitative assays, individual isoenzymes of CK in plasma can be reliably detected with a sensitivity of 5 to 10 iu l.$^{-1}$.[18] However, plasma CK-MB is present in normal subjects at a lower level, approximately 1 iu l.$^{-1}$. Furthermore, since CK-MB represents only 15 to 20% of total myocardial CK activity and since its disappearance from blood is faster than that of the MM isoenzyme,[12,13] plasma CK-MB after myocardial infarction rarely exceeds 15% of total plasma CK activity. The RIA can detect as little as 1×10^{-3} iu l.$^{-1}$ reproducibly and therefore can be employed to quantify CK-MB in plasma from normal subjects and to detect even modest elevations associated with infarction.

The RIA was developed in part also because of its potential for diagnostic applicability in conditions other than myocardial infarction. Since the BB isoenzyme is abundant in brain and in the gastrointestinal tract, elevations of CK-BB in plasma might be a useful diagnostic marker for diseases involving these organs. However, until recently, elevated plasma CK-BB activity had been paradoxically undetectable despite obvious insults affecting these tissues.[7] Henderson et al.[7] utilizing a column chromatographic method have shown minimal elevations of plasma CK-BB after cerebral injury. However, even peak levels of plasma CK-BB are rarely within the sensitivity of available enzymatically based quantitative assays. Consistent detection of these modest elevations requires more sensitive methods. The low level of peak plasma CK-BB activity after cerebral injury is probably related to several factors including: (1) lability of BB and probable extensive local denaturation; (2) rapid disappearance of BB from plasma;[12] (3) possible impedance to transport of CK-BB into blood by the blood–brain barrier; and (4) rapid denaturation of BB in unprotected plasma samples. Fortunately, radioimmunoassay of CK-BB permits reliable detection of elevated plasma CK-BB after cerebral injury providing diagnostic and potentially prognostic information.[1]

Radioimmunoassay provides particular advantages for elucidating the metabolism of plasma enzymes (see Chapter 5). As pointed out by Posen,[11] very little is known regarding turnover of plasma enzymes. It is not yet clear whether disappearance of enzyme activity from the systemic blood circulation is due to inactivation, denaturation, or removal of enzyme protein in part because generally available assays detect only enzyme activity rather than the concentration of enzyme protein. The consistent delay in release of enzyme into blood after tissue injury probably depends in part on the time required for breakdown of cell membranes. However, the possibility that release of enzyme which is enzymatically inactive occurs early, during this apparent lag phase, has not been excluded. The potential value of characterizing enzyme turnover with the use of RIA is promising.

DEVELOPMENT OF A CK ISOENZYME RADIOIMMUNOASSAY (RIA)

Although the first successful isoenzyme radioimmunoassay for the diagnosis of myocardial infarction detected CK isoenzymes,[17] the approach is applicable to radioimmunoassays for other plasma enzymes as well. Details of the method have been reviewed recently.[15–17] The basic requirements for radioimmunoassay are listed in Table 1. In the CK isoenzyme RIA, human

Table 1. Requirements for a Radioimmunoassay

1. Isolation and purification of antigen
2. Induction and harvesting of antibody specific for the substance to be assayed
3. Utilization of antibody with high affinity
4. Labelling of antigen with high specific radioactivity
5. Separation of bound from free antigen

CK isoenzymes, used as antigens, were isolated from human heart (MM and MB) and brain (BB). CK isoenzymes were isolated from tissue homogenates by repetitive ethanol extraction followed by batch adsorption and column chromatography.[3] Preparations of BB (60 iu mg^{-1}) and MM (500 iu mg^{-1}) were shown by SDS gel electrophoresis and polyacrylamide electrophoresis to be 90% pure with respect to protein and entirely pure with respect to other CK isoenzymes.[3] To develop antibodies, 1 mg of CK-MM or CK-BB was injected subcutaneously into rabbits followed by 0.2 mg every 10 days for a total of three doses with subsequent booster injections of 0.2 mg monthly. Several species have proven suitable for antibody production including goats, sheep, guinea-pigs, and mice. In general, and in the case of the CK radioimminoassay, antibody production is elicited with injections of antigen of 0.1–1.0 mg kg^{-1} of body weight followed by 1/10th of the dose in booster injections. The smaller the amount of protein used for immunization, the higher the affinity of the antibodies. In the case of the CK antiserum, Ouchterlony plates demonstrated that MM antiserum cross-reacts with CK-MM and CK-MB but not with CK-BB. Conversely, BB antiserum cross-reacts with CK-BB and CK-MB but not with CK-MM. Thus it appeared that the antibodies harvested were specific for the M and B subunits respectively. Based on the rationale that CK-BB is not normally present in plasma and is not released after myocardial infarction, we assumed that antiserum to the B subunit would be virtually specific for detecting plasma CK-MB and could therefore be used to develop a competitive displacement assay.

Although antibodies to CK isoenzymes have been available for a long time[6,8] and have been used in immunoinhibition assays,[8] radioimmunoassays have not been possible in part because of changes in enzyme structure of the antigen as reflected by loss of enzyme activity associated with radioactive labelling due to oxidation and dissociation of subunits. Radioactive labelling of antigen with high specific radioactivity for radioimmunoassays, required to provide needed sensitivity, necessitates use of an isotope which emits gamma radiation. For this reason, labelling is generally performed by iodination with either ^{125}I or ^{131}I. Radioactive labelling with a β-emitter such as ^{14}C or ^{3}H would not provide sufficient sensitivity for an RIA as illustrated in the following example. If one considers one molecule of insulin, which contains 243 carbon atoms, labelling of all the carbon atoms with ^{14}C would give rise to only 1/10th the number of disintegrations/minute that would occur if labelling of only one amino acid had been performed with ^{131}I.

Iodination of protein by conventional methods utilizes chloramine-T or lactoperoxidase. Both transfer iodine to the protein by means of oxidation. With the chloramine-T method, hyperchloric acid is produced to generate free iodine, a potent oxidizing agent. Attempts to label CK isoenzymes with

^{125}I and the chloramine-T method resulted in low specific radioactivity, low binding, and marked loss of enzyme activity of the CK isoenzyme antigen.[17] In 1973 Bolton and Hunter[2] developed a method for iodinating proteins by acylation rather than oxidation. Utilizing the chloramine-T method, ^{125}I was first incorporated into an ester N-succinimidyl-3-(4-hydroxphenyl-propionate). Subsequently, following dialysis for 24 to 48 hours, the ester with attached ^{125}I was transferred to amino groups on the protein molecule to be labelled by acylation, avoiding exposure of the protein to oxidation. The Bolton and Hunter method permits iodination of all exposed amino acids as opposed to the chloramine-T or lactoperoxidase techniques which iodinate only tyrosine and histidine.

With the Bolton and Hunter method, we were able to iodinate CK isoenzymes with high specific radioactivity (0.012 μCi μg^{-1} of protein) and with less than 5% loss of enzyme activity under conditions in which 5 mCi of ^{125}I mg^{-1} CK isoenzyme protein were utilized. It is possible to obtain much higher specific radioactivity (30–40 μCi μg^{-1}) if one uses smaller amounts of protein. Iodination of non-enzymatically active protein with substitution of one iodine atom per molecule usually does not alter the affinity of an antigen for its specific antibody. However, in our experience in utilizing this approach with an enzymatically active protein, this degree of substitution is likely to be associated with decreased affinity of the antigen for binding the antibody, requiring much longer incubation periods in the ultimate radioimmunoassay. Antigen with a specific radioactivity of 0.01 μCi mg^{-1} provides adequate sensitivity of a radioimmunoassay for plasma enzymes and has the advantage of not impairing binding affinity and therefore permitting short incubations during the radioimmunoassay procedure.

Iodination with ^{125}I as opposed to ^{131}I was selected for several reasons. The short half-life of ^{131}I (8 days) might seem to be preferable despite the short shelf-life. However, ^{131}I is produced as a mixture of 80% ^{127}I and 20% ^{131}I. The ^{127}I has a half-life of about 45 days. Counting efficiency for ^{131}I is only 40% in contrast to 80% for ^{125}I. Radiation from ^{125}I is less penetrating and thus less dangerous when utilized on a routine basis.

The final step in the development of a radioimmunoassay is selection of a method for separating antigen–antibody complexes from unbound radioactively labelled antigen. We chose a salt precipitation method utilizing a saturated ammonium sulphate solution.[4] Ammonium sulphate (40–50% saturated) precipitates large molecules such as antibodies but not smaller molecules of 100,000 daltons or less. For the CK isoenzyme RIA, labelled CK antigen was incubated with antiserum. The antigen–antibody pellet was separated from the supernatant fraction by centrifugation and counted for ^{125}I radioactivity to provide a quantitative measure of the amount of antigen bound to antibody.

Initial attempts to separate antibody-bound ^{125}I-CK from free ^{125}I-CK antigen were compromised by erratic precipitation of unbound ^{125}I-CK, presumably due to dissociation of the enzyme antigen into subunits with subsequent polymerization into complexes that were large enough to precipitate with ammonium sulphate. Incubating the antigen and antibody in a medium containing 1.6 mol l.$^{-1}$ Tris buffer and 10 mmol l.$^{-1}$ mercaptoethanol stabilized the CK molecules and prevented the problem. Albumin and rabbit gamma globin were included to prevent non-specific binding to the tube. Plastic tubes were found to be preferable to glass because they were less prone to bind protein.

SENSITIVITY AND SPECIFICITY

The sensitivity and specificity of BB antiserum for the B subunit is illustrated in Figure 1. Antiserum in a dilution of 1:150 precipitated about 95% of the ^{125}I-labelled BB and about 55% of the ^{125}I-MB but did not exhibit any specific binding to ^{125}I-MM. Furthermore, there was no specific binding of ^{125}I-MM to the BB antiserum and no inhibition of ^{125}I-BB binding to BB antiserum even when unlabelled MM was present in 25,000-fold excess over CK-BB or CK-MB. With MM antiserum, cross-reactivity

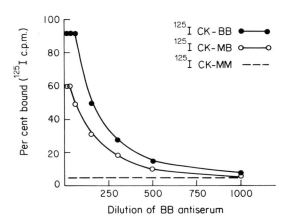

Figure 1. Shown here is the specificity of BB antiserum for the B subunit. Maximum binding to CK-BB is about 92% at a titre of about 1 in 100. Maximum binding to CK-MB is about 60%. The decreased binding with decreasing concentrations of antiserum reflects the dependence of the binding on the specific antibody. There was no specific binding to CK-MM, being similar to that of background counts

was evident with CK-MM and CK-MB but not with CK-BB. Normal rabbit serum produced only 4–5% non-specific binding. Thus binding of the ^{125}I-labelled antigen was immunologically specific with no cross-reactivity between the MM and BB systems.

Utilizing BB antiserum, we then developed a competitive displacement assay for plasma CK-MB. A typical inhibition curve is illustrated in Figure 2. As can be seen, the curve is quite steep over the range of antigen concentration from 1.0 to 100 nanograms representing 0.016–1.6 miu ml^{-1} of enzyme activity and the coefficient of variation in triplicate determinations was less than 5%. Thus the radioimmunoassay exhibited more than 1000-fold increased sensitivity compared to enzymatically based conventional quantitative assays.[9] As shown in Figure 3, representing results from analysis of serial samples from a patient with acute myocardial infarction, results with radioimmunoassay of plasma samples from patients with this disorder corresponded closely to those obtained with a batch adsorption, glass-bead technique previously developed and standardized in our laboratory.[5]

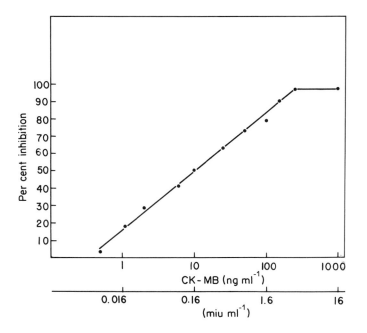

Figure 2. Shown here is a typical CK-MB inhibition curve illustrating a sensitivity of about 1 ng or 0.016 miu ml^{-1}

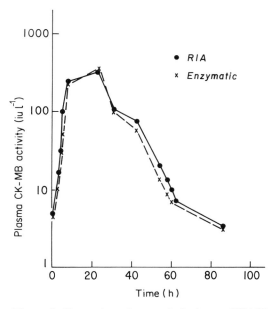

Figure 3. Shown here is a typical plasma CK-MB time–activity curve in a patient after myocardial infarction comparing values obtained with the radioimmunoassay with that obtained by an assay based on enzymatic activity. As can be determined from the two curves, the values are virtually identical. The conversion from enzyme immunoprotein concentration to enzyme activity must be verified for each system since the two may not change in parallel under some conditions

CLINICAL APPLICATIONS

Based on results with the CK isoenzyme RIA, normal plasma CK-MB activity averages 1 iu l.$^{-1}$ with a range of 0–5 iu l.$^{-1}$.[9,14] In general, with the RIA one can detect a onefold increase in plasma CK-MB within 4 hours of chest pain in patients developing myocardial infarction in contrast to the 4- to 5-fold increase frequently required for reliable detection with enzymatically based assays. Accordingly, the RIA permits earlier detection of myocardial infarction.

Patients with angina pectoris exhibit no increase in plasma CK-MB activity detectable with the radioimmunoassay. Neither do patients with marked increases in total CK activity resulting from intramuscular injections or non-cardiac surgery.

In preliminary studies utilizing RIA for detection of plasma CK-BB, elevations have been recognized in the range of 2–10 iu l.$^{-1}$ after cerebral vascular injury. Elevations appear within 4–6 hours after the initial injury and peak approximately 6–12 hours after the insult. Because of rapid disappearance, plasma BB is generally no longer detectable within 24 hours.[12]

TECHNICAL CONSIDERATIONS

In spite of the apparent similarity of MM from skeletal muscle and from heart, MM antiserum developed to heart MM has a different binding affinity for skeletal MM. Thus the source of antigen is important and must be taken into account.

Since enzymatically active proteins are unstable during storage, particularly when radioactively labelled, precautions are required.[10] Thus it is important that the labelled antigen preparation used to obtain a standard inhibition curve be characterized carefully for both specific enzyme and specific radioactivity. If considerable loss of enzymatic activity has occurred in the labelled antigen, it may be necessary to remove inactive protein by either ethanol precipitation or chromatographic separation. If the chromatographic approach is used, gel should be equilibrated in buffer containing 20 mmol l.$^{-1}$ mercaptoethanol to minimize instability of antigen during chromatography. Similarly, labelled CK antigen can be stored at −20 °C for up to 3 weeks if it is protected with mercaptoethanol. Nevertheless, on occasions ^{125}I may separate from the antigen requiring removal prior to use of the preparation in an RIA. Each new batch of antiserum must be characterized with respect to binding affinity and specificity.

The succinimidyl ester used to label antigen has a shelf-life of years, but if hydration occurs, labelling will be impaired. Antiserum can generally be stored for years at −28 to −70 °C. However, with long-term storage there may be an occasional change in the conformational structure of the molecule resulting in alterations in its binding affinity.

FUTURE DIRECTIONS

Since the radioimmunoassay for CK-MB is capable of detecting modest elevations in plasma CK-MB within 4 hours after the onset of chest pain in patients with myocardial infarction, it offers potential as a rapid screening test for this disorder. However, presently 6-hour incubation periods are generally required. Solid-phase absorbent techniques which permit almost instantaneous separation of bound from free antigen should ultimately obviate this delay. Development of radioimmunoassays for isoenzymes of other enzymes should not only improve diagnostic sensitivity and specificity for several clinical conditions, but also provide a research tool for the study of their metabolism.

CONCLUDING COMMENTS

The first radioimmunoassay for the diagnosis of myocardial infarction based on detection of plasma isoenzymes was developed for CK-MB. Based on experience with this prototype procedure, requirements for development of radioimmunoassays and potential difficulties likely to be encountered in interpreting and performing such assays are considered. The radioimmunoassay for CK isoenzymes is reviewed particularly with respect to its high sensitivity and capacity for prompt detection of myocardial infarction compared to conventionally available assays. Potential applicability of the RIA as a research tool to evaluate mechanisms involved in removal of enzymes is discussed since clarification of mechanisms of enzyme removal is important not only for estimating infarct size but also for designing therapy for enzyme deficiency states.

REFERENCES

1. Bell, R. D., Rosenberg, R., Ting, R. C., Mukherjee, A., Stone, M. J., and Willerson, J. T. Determination of creatine kinase BB isoenzyme by radioimmunoassay in cerebrospinal fluid and sera from patients with neurological disease. *Clinical Research*, **25**, 208A (1977).
2. Bolton, A. E. and Hunter, W. M. The labelling of proteins to high specific radioactivities by conjungation to a ^{125}I-containing acylating agent (application to the R-I-A). *Biochemistry Journal*, **133**, 529–38 (1973).
3. Carlson, E., Roberts, R., and Sobel, B. E. Preparation of individual CPK isoenzymes from myocardium and brain. *Journal of Molecular and Cellular Cardiology*, **8**, 159–67 (1976).
4. Farr, R. A. A quantitative immunochemical measure of the primary interaction between I*BSA and antibody. *Journal of Infectious Disease*, **103**, 239–162 (1958).
5. Henry, P. D., Roberts, R., and Sobel, B. E. Rapid separation of plasma creatine kinase isoenzymes by batch adsorption with glass beads. *Clinical Chemistry*, **21**, 844–9 (1975).
6. Jockers-Wretou, E., Grabert, K., and Pfleiderer, G. Quantitative immunological determination of creatine kinase isoenzymes in serum. *Zeitschrift für Klinische Chemie*, **13**, 85–92 (1975).
7. Nealon, D. A. and Henderson, A. R. Separation of creatine kinase isoenzymes in serum by ion-exchange chromatography (Mercer's method modified to increase sensitivity). *Clinical Chemistry*, **21**, 392–8 (1975).
8. Neumeier, D., Knedel, M., Wurzburg, U., Hennrich, N., and Lang, H. Immunologischer nachweis von creatinkinase-MB in serum bein myokardinfarkt. *Klinische Wochenschrift*, **53**, 329–37 (1975).
9. Painter, A., Sobel, B. E., and Roberts, R. Prompt detection of myocardial infarction by radioimmunoassay of MB creatine kinase. *American Journal of Cardiology*, **39**, 317 (1977).
10. Parker, C. W. *Radioimmunoassay of Biologically Active Compounds*, Prentice-Hall, Englewood Cliffs, 1976.
11. Posen, S. Turnover of circulating enzymes. *Clinical Chemistry*, **16**, 71–83 (1970).
12. Rapaport, E. The fractional disappearance rate of the separate isoenzymes of creatine phosphokinase in the dog. *Cardiovascular Research*, **9**, 473–7 (1975).
13. Roberts, R., Henry, P. D., and Sobel, B. E. An improved basis for enzymatic estimation of infarct size. *Circulation*, **52**, 743–54 (1975).

14. Roberts, R., Painter, A. A., and Sobel, B. E. Earlier diagnosis of acute myocardial infarction with a radioimmunoassay for MB creatine kinase. *Clinical Research*, **25,** 511A (1977).
15. Roberts, R., Parker, C. W., Sobel, B. E., and Painter, A. A. Radioimmunoassay of MB ('myocardial') CPK. *Clinical Research*, **24,** 238A (1976).
16. Roberts, R., Parker, C. W., Sobel, B. E., and Painter, A. A. Radioimmunoassay of human MB CPK. *Circulation Supplement II*, **54,** II-152 (1976).
17. Roberts, R., Sobel, B. E., and Parker, C. W. Radioimmunoassay for creatine kinase isoenzymes. *Science*, **194,** 855–7 (1976).
18. Smith, A. F., Radford, D., Wong, C. P., and Oliver, M. F. Creatine kinase MB isoenzyme studies in diagnosis of myocardial infarction. *British Heart Journal*, **38,** 225–32 (1976).

CHAPTER 11

Enzymatic estimation of infarct size

B. E. Sobel, J. K. Kjekshus, and R. Roberts

INTRODUCTION	257
RELATIONSHIPS BETWEEN DEPLETION OF MYOCARDIAL CK AND IRREVERSIBLE ISCHAEMIC INJURY	259
ESTIMATION OF INFARCT SIZE BASED ON ANALYSIS OF PLASMA CK TIME–ACTIVITY CURVES	261
INITIAL APPROXIMATIONS, REFINEMENTS AND EVOLUTION OF MODELS USED TO ANALYSE PLASMA CK TIME–ACTIVITY CURVES	263
Experimental Studies	265
Clinical Studies	266
FACTORS INFLUENCING ENZYMATIC ESTIMATION OF INFARCT SIZE, TECHNICAL CONSIDERATIONS	268
Parameters in the Model	269
CK Disappearance Rate (k_d)	269
CK Distribution Volume and the Ratio CK_R/CK_D	272
PHYSIOLOGICALLY BASED MODELS	275
ESTIMATION OF INFARCT SIZE BASED ON PLASMA CK-MB TIME–ACTIVITY CURVES	277
STRENGTHS AND LIMITATIONS OF ENZYMATIC ESTIMATION OF INFARCT SIZE	281
FUTURE DIRECTIONS	282
REFERENCES	285
APPENDIX	287

INTRODUCTION

Recently, the concepts that acute myocardial infarction is a dynamic process susceptible to favourable therapeutic modification and that the overall extent of infarction is an important determinant of prognosis, have been explored vigorously. These hypotheses stimulated development of new methods for evaluation of the severity of myocardial damage designed to overcome deficiencies in traditional techniques with limited applicability to intact patients. Analysis of plasma time–activity curves of circulating enzymes is one such approach.[2,28,39,40,43]

Although infarction has been defined traditionally on the basis of morphological criteria, often such criteria cannot be employed in intact animals or patients. Since morphological criteria of necrosis evolve slowly, they may not provide an accurate index of the extent of injury when survival after infarction is relatively short. Electrophysiological criteria including ST segment elevation in epicardial or praecordial recordings may be useful in evaluating directional changes reflecting the severity of regional ischaemia in an individual case,[24] but absolute estimation of the extent of infarction based on analysis of the electrocardiogram is difficult. Radionuclide imaging techniques, useful in detecting injured myocardium, provide limited quantification because of superimposition of overlapping regions of myocardium on two-dimensional displays, poor resolution, and limited contrast.[52] Although promising techniques such as positron emission transaxial tomography may overcome many of these difficulties,[51,54] calibration of results with independent criteria of the overall extent of infarction will probably be necessary.

In part because of extensive information available for many years characterizing changes in plasma enzyme activity accompanying myocardial infarction,[49] it seemed likely that analysis of serial changes would be particularly useful in assessing the extent of ischaemic injury.

Characterization of behaviour of biochemical constituents and markers in the circulation has of course been explored vigorously in many disciplines. Elucidation of the kinetics of coagulation factors—knowledge of which is of crucial importance in implementing therapy for patients with deficiency states—is one prominent example. The entire field of pharmacokinetics is another. Numerous investigators have recognized the probable relationship between loss of a biochemical constituent from an organ undergoing injury and the nature of its plasma time–activity curve. However, direct, experimental verification of quantitative relationships between the two has been elusive because of concomitant synthesis and loss of the marker locally within the organ undergoing injury; contributions to circulating levels of the marker by marker released from other organs; incomplete information regarding local degradation of the marker, its volumes of distribution, and its modes of transport into the circulation; lack of definition of factors regulating removal or denaturation of the marker once it has been distributed into vascular and extravascular compartments; and difficulties in experimental design such as avoidance of the use of anaesthetized acute animal preparations because of the influences of anaesthesia and the response to trauma on behaviour of the marker under investigation. In this selective review, the focus will be primarily on studies in which experimental verification of presumed relationships has been undertaken.

For assessment of myocardial ischaemic injury, markers such as creatine kinase (CK) seem particularly advantageous, since contrary to the case with lower molecular weight species such as myoglobin, CK in plasma is not cleared by the kidney. Accordingly, prevailing levels of plasma CK activity after myocardial infarction should be independent of renal blood flow. In addition, in contrast to many other enzymes, CK in the heart is confined virtually exclusively to myocardial components rather than other cellular elements such as connective tissue, marginating white cells, and other participants in the inflammatory exudate, or vascular endothelium. The importance of this consideration is underscored by the observation that after experimental myocardial infarction total activity of enzymes, such as lactate dehydrogenase (LD), in extracts of the heart frequently exhibit greater activity than the amount of activity present in the heart prior to infarction— presumably because of substantial contributions to activity from elements other than cardiac muscle, such as cells participating in the inflammatory response to infarction (unpublished observations and reference [16]). For this reason, it would be difficult to design experiments in which results of analysis of plasma LD curves could be compared in a quantitative way to LD depletion from the heart, since total LD depletion is not directly related to the overall extent of myocardial necrosis.

In our initial approach to the problem of estimating the extent of infarction (referred to as 'infarct size' for convenience in this selective review) from plasma enzyme time–activity curves, it was first necessary to determine whether depletion of the selected enzyme marker from the heart undergoing ischaemic injury was directly related to the extent of infarction assessed with independent criteria. Elucidation of this relationship is necessary to determine whether release of the marker *in the setting of ischaemic injury* is a manifestation of cell death or of reversible injury. Differentiation between the two possibilities is particularly important since several pharmacological and experimental manipulations, such as perfusion of myocardium with calcium-free solutions, can alter membranes resulting in leakage of markers out of non-ischaemic cells, despite the absence of irreversible injury. Only if release of the marker from the heart into the circulation in the setting of ischaemia is directly related to cell death, and only if depletion of the marker from myocardium is proportional to infarct size, can one hope to estimate infarct size from analysis of time–activity curves of the marker in plasma. Naturally, proper interpretation of such curves is dependent upon the characterization and consideration of several additional factors influencing transport of the marker, distribution, and its removal from the circulation as well as the influence of haemodynamic, pharmacologic, and physiologic perturbations on these processes.

RELATIONSHIPS BETWEEN DEPLETION OF MYOCARDIAL CK AND IRREVERSIBLE ISCHAEMIC INJURY

Under carefully defined conditions, depletion of myocardial CK activity associated with myocardial infarction correlates with infarct size estimated independently by techniques such as assay of the myocardial distribution of radioactively labelled microspheres or morphometric assessment of necrosis.[16,21] In rabbits subjected to coronary occlusion, the mass of infarction estimated by weighing excised regions of the left ventricle correlates with overall myocardial CK depletion. Twenty-four hours after coronary occlusion in open chest dogs, CK depletion correlates with the intensity of ischaemic injury sustained initially, reflected by the magnitude of ST segment elevations from corresponding sites in epicardial recordings obtained 15 minutes after the occlusion. Recent observations indicate that CK depletion from hearts of experimental animals correlates with microscopic, histochemical, and ultrastructural criteria of infarction, depressed free fatty acid accumulation, and altered attenuation of transmitted ultrasound. Analogous results have been obtained in patients with histochemical studies demonstrating close correlations between the overall extent of infarction and the estimated amount of enzyme lost from the heart.[3,19,23,54]

It has been recognized for some time that the percentage increase in serum transaminase activity correlates with infarct size determined morphologically in anaesthetized dogs. Unfortunately however, analysis of plasma transaminase time–activity curves cannot be verified with reference to depletion of transaminase from the heart because of potential contributions from non-myocardial elements. On the other hand, based on comparison of myocardial CK depletion, measured directly, to plasma CK time–activity curves, it appeared that the cumulative release of CK into blood was proportional to CK depletion, in turn proportional to infarct size measured with independent criteria.[39]

If myocardial CK depletion is indeed a valid index of infarct size, release of CK activity from the ischaemic myocardium must be directly related to cell death rather than to reversible injury. Compilation of findings from a large number of clinical studies does suggest that elevation of plasma enzymes derived from the heart in patients with myocardial ischaemia is a reflection of irreversible injury manifested by other independent criteria of infarction rather than of coronary insufficiency and angina pectoris alone.[50] Recent observations, in which scintigraphic criteria have been used to detect infarction, support this interpretation.

To determine whether experimentally induced release of CK is tantamount to cell death in the face of an ischaemic insult, we performed a recent series of studies in conscious and open chest dogs.[1] Coronary occlusion was induced by constriction of an exteriorized occluder and maintained for 10 minutes to 48 hours at which time perfusion was restored or the animal was sacrificed. When plasma CK and CK-MB activity was assayed in samples obtained serially for 24 hours and results were compared to microscopic changes in hearts from the same animals examined 48 hours after coronary occlusion, close concordance was demonstrated between the presence of myocardial necrosis detectable morphologically and release of

CK-MB (an isoenzyme relatively specific to myocardium) into the circulation. In this study, blood samples were obtained from the external jugular vein at 30-minute intervals for 6 hours beginning 30 minutes prior to occlusion and hourly for at least 18 hours thereafter in all dogs studied. An equivalent volume of normal saline was given as replacement in order to maintain blood volume. Since enzyme release may not be evident for several hours after the onset of ischaemia, sampling was continued for a minimum of 24 hours in each case. Forty-eight hours after occlusion, hearts were examined for gross and microscopic evidence of necrosis. This interval was selected to provide sufficient time for evolution of conventional morphological criteria of irreversible injury.

Among sham operated control dogs, CK-MB activity did not increase in plasma despite the fact that total plasma CK increased, presumably as a result of surgical trauma to skeletal muscle. No electrocardiographic changes indicative of infarction occurred, nor were sustained ventricular dysrrhythmias encountered. An example of enzyme changes under these conditions is shown in Figure 1.

Among all animals subjected to myocardial ischaemia induced by coronary occlusion of selected duration, microscopic evidence of necrosis was apparent whenever the occlusion was of sufficient duration to lead to elevation of plasma CK-MB activity exceeding 100% of baseline. In contrast, no necrosis was detectable in the hearts of any of the animals with transitory coronary occlusion maintained for less than 30 minutes, an interval of ischaemia that was not associated with elevation of plasma CK-MB activity. Analysis of data from all animals demonstrated a strong correlation between the presence of elevation of plasma CK-MB activity and histological criteria of irreversible injury. Since necrosis was associated with definite elevation of plasma CK-MB activity in every case, these observations support the view that elevation of plasma CK-MB activity induced by myocardial ischaemia is a reflection of cell death rather than of reversible injury.

Although enzymes are released from viable cells in some conditions unrelated to ischaemia, such as perfusion of isolated hearts with calcium-free media or exercise in the intact organism (leading to release of enzyme from skeletal muscle), the bulk of clinical and experimental data suggests that enzyme release from myocardium resulting from ischaemia occurs only when the insult is associated with irreversible injury.

ESTIMATION OF INFARCT SIZE BASED ON ANALYSIS OF PLASMA CK TIME–ACTIVITY CURVES

After correlations between myocardial CK depletion and infarct size had been characterized, relations were examined between plasma CK changes

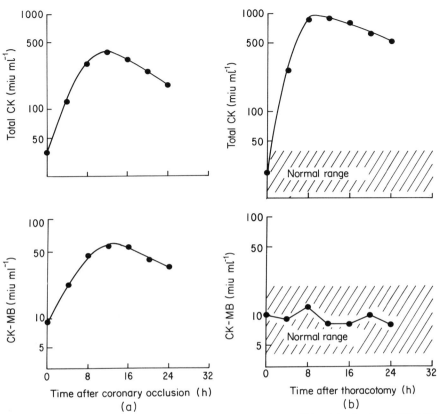

Figure 1. (a) A representative example of changes in plasma CK activity (top) and CK-MB activity (bottom) after coronary occlusion in a conscious dog. As can be seen, MB activity increases, peaking approximately 16 hours after coronary occlusion and the MB time–activity curve parallels the total CK time–activity curve. [Reproduced, with permission, from *Circulation*, **54**, 187–93 (1976).] (b) A representative example of changes in plasma CK (top) and CK-MB (bottom) after thoracotomy in an anaesthetized dog. The changes in total CK activity resemble those associated with acute myocardial infarction as exemplified in Figure 1(a), but CK-MB activity does not increase in a parallel fashion nor does it exceed the upper limit of the normal range. The changes in total plasma CK appear to reflect release of enzyme from organs other than the heart associated with trauma sustained during the course of thoracotomy. [Reproduced, with permission, from *Circulation*, **54**, 187–93 (1976).]

and myocardial CK depletion in conscious dogs subjected to experimental coronary occlusion.[39,40] The use of conscious animals was based on the need for avoiding spurious estimates of myocardial CK release into plasma reflecting release of enzyme from non-cardiac sources. CK released from the heart into the circulation was estimated from simple mathematical analyses derived from the concept that the rate of change of plasma CK activity reflected two competing phenomena: release of CK from the heart and disappearance of CK from the blood according to first-order kinetics. In animals with haemodynamically uncomplicated infarction, CK lost from the heart, measured directly, correlated well with CK released into blood estimated from analysis of the plasma CK time–activity curves ($r > 0.9$). These findings were not surprising in view of earlier observations by others indicating that marked elevations of activities of several enzymes in plasma and serum were associated with extensive infarction.

INITIAL APPROXIMATIONS, REFINEMENTS, AND EVOLUTION OF MODELS USED TO ANALYSE PLASMA CK TIME–ACTIVITY CURVES

Before examining the reproducibility of enzymatic estimates of myocardial CK depletion, let us first consider the initially formulated model in some detail, some potential pitfalls in both experimental evaluation of the model and its use in the clinical setting, and some recent progress in refining enzymatic estimates of infarct size.[31,45-47]

In our initial formulation, serial changes in plasma CK activity were assumed to reflect release from irreversibly injured myocardium into blood and simultaneous disappearance from the circulation. Under a variety of experimental conditions, the distribution volume of intravenously injected, purified canine myocardial CK appeared to conform to plasma volume when estimates were based on the use of an assumed, monoexponential disappearance function. Furthermore, since the log of CK activity declined linearly with correlation coefficients exceeding 0.95 in numerous experiments, the disappearance of enzyme activity could be crudely approximated by a first-order function. Instantaneous changes in plasma CK activity could then be expressed in terms of the rate of its appearance in plasma ($f(t)$) and a disappearance function, $k_d E$ where k_d equals fractional disappearance rate of CK from plasma: $dE/dt = f(t) - k_d E$.

With the use of an estimate of the CK distribution volume and an estimate of k_d based on results of experiments in which purified enzyme was injected into conscious animals intravenously, the cumulative amount of CK activity that would have appeared in blood had no disappearance occurred after experimentally induced myocardial infarction was calculated in a series of conscious dogs. In animals with haemodynamically uncomplicated infarction, the proportion of depleted myocardial CK, measured directly, appearing in blood was quite constant. Accordingly, infarct size could be estimated by analysis of serial changes in plasma CK activity with this model and results expressed as CK-gram-equivalents (with one gram-equivalent defined as the amount of damaged myocardium accounting for CK release equivalent to that from one gram of myocardium undergoing homogeneous necrosis).

Unfortunately, myocardial CK depletion and CK distribution volume cannot be measured directly in patients. Accordingly, values used in calculations of cumulative CK released after myocardial infarction in man have been based on corresponding values in the canine model. Bleifeld has examined the hearts of 17 consecutive patients dying after myocardial infarction with sufficient data for analysis of plasma CK time–activity curves.[3] It was found that morphometric estimations of infarct size correlated closely with those based on analysis of plasma CK time–activity curves ante-mortem ($r = 0.89$). These results support the view that enzymatic estimates of infarct size are concordant with morphological estimates despite the fact that several parameters used in calculations in patients are extrapolated from the canine model.

Although it has been known for some time that the MB isoenzyme of CK is a relatively specific marker of myocardium, progress in developing enzymatic estimates based on MB time–activity curves in plasma was hampered until recently by the lack of a quantitative technique for assaying the isoenzyme (see Chapters 8, 9, and 10). Qualitative approaches, such as electrophoresis and fluorometric scanning, are not suitable for this purpose for several reasons including asymmetric distribution of isoenzymes along the axis of scanning; limitation of diffusion of substrate into the supporting medium; a loss of the visualizing dye from the supporting medium; non-linearity of reaction rate associated with one or more isoenzymes because of the disparate activities of each in the sample; and variation in affinity of individual isoenzymes for specific reagents in the incubation medium.[33] When a quantitative assay from CK-MB activity became available, it was utilized for improved estimates of infarct size. Since canine myocardium contains less than 2% CK-MB activity, compared to total myocardial CK, verification of estimates of infarct size based on MB plasma time–activity curves is difficult based on correlation with depletion of myocardial MB because of the large contribution of non-CK-MB to total myocardial CK. Accordingly, we examined the correlation between infarct size estimated from plasma MB curves and from total CK curves in patients with haemodynamically uncomplicated infarctions.[28] Parameters in the two sets of calculations differed because the disappearance rate of the MB isoenzyme is more rapid than that of total CK and because the amount of MB in one gram of human myocardium represents only 12–15% of total CK activity. Nevertheless, estimates of infarct size based on MB correlated closely with those based on total plasma CK ($r = 0.97$ (Figure 2)) in patients selected to minimize the likelihood that non-cardiac CK release might distort estimates based on total plasma CK.

Despite obvious imperfections, enzymatic estimates of infarct size from

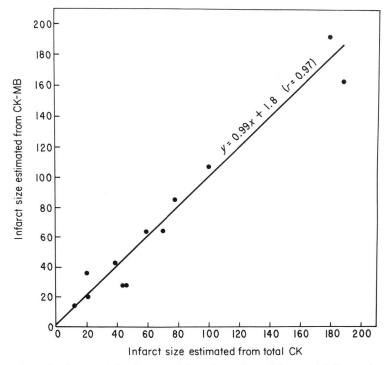

Figure 2. The correlation between infarct size estimated from serial changes in plasma CK-MB activity (ordinate) and infarct size estimated from serial changes in total plasma CK activity (abscissa). These results were obtained in 12 patients with acute myocardial infarction uncomplicated by cardiogenic shock or marked pulmonary oedema. [Reproduced by permission of Grune and Stratton from *Progress in Cardiovascular Diseases*, **18**, 405–20 (1976).]

several centres have correlated with biochemical and morphological analyses of the hearts of experimental animals, morbidity and mortality in patients, and histochemical assessment of necrosis among patients dying with acute myocardial infarction. Improved estimates have been obtained with the use of CK-MB, and further improvements can be anticipated from characterization of mechanisms underlying kinetics of CK in the circulation. Consideration of some of these results, and others already available, is useful in evaluating the applicability of enzymatic estimation of infarct size in the experimental and clinical setting.[3,14,18,22,25,28,38,55]

Experimental Studies

Among 22 conscious dogs with coronary occlusion, agreement between myocardial CK depletion and cumulated CK released from the heart calculated from plasma CK time–activity curves was close. Results fit the regression line: CK released (iu) = $272 \times g$ infarct $- 318$, $r = 0.96$, $n = 22$. The standard deviation of the slope of this line was 18.[39] Thus there was good correspondence between the percentage of infarct calculated from plasma values and infarct size estimated from myocardial CK depletion over a wide range of infarct size. Comparable correlations between estimates of infarct size based on evaluation of plasma CK time–activity curves and those based on myocardial CK depletion measured directly have been obtained in other species, such as baboons,[55] in whom the organ distribution of CK-MB is similar to that in man, permitting improved estimates based on MB rather than total CK activity. Even when myocardial infarction involves the presence of experimentally induced reperfusion, the correlation between CK appearance in blood and CK depleted from myocardium remains close ($r > 0.9$) although the slope of the function relating the two appears to change.[19] Failure to observe a consistent relationship between these two parameters by some investigators[36] appears to be due in part to several factors including: (i) failure to subtract high baseline values from observed values prior to calculating cumulative CK released from plasma CK-time–activity curves; (ii) study of experimental animal preparations in a non-steady state, reflected by the presence of high baseline values; (iii) sampling by repetitive venepuncture with potential marked contamination of samples by CK liberated from locally traumatized tissue; (iv) failure to exclude enzyme from non-cardiac sources from values utilized in the calculations; and (v) inadequate control of assay conditions to avoid spuriously low estimates of CK activity in samples with high activity because of dilution effects on results in most assay systems.[46] When these and other precautions are not observed, marked variance in plasma CK values occurs and results become uninterpretable.

Clinical Studies

Recently, enzymatic estimates of infarct size have been utilized in patients to examine relationships between infarct size and ventricular performance, electrical instability of the heart, altered ventricular compliance, morbidity, and mortality after the onset of myocardial infarction as well as exercise tolerance during the late recovery period.[3,4,7,25,29,38,41,43]

In patients with myocardial infarction, enzymatically estimated infarct size correlates with the frequency of ventricular dysrrhythmia during the first 10 hours after hospital admission. Not only the number of ventricular premature complexes but also the number of episodes of couplets and runs of ventricular tachycardia appears to be directly related to enzymatically estimated infarct size.

Impairment of left ventricular performance assessed with radioisotope angiocardiography in patients studied 24 to 48 hours after hospital admission correlated with enzymatically estimated infarct size[17] as did a reduction in ejection fraction assessed by contrast medium left ventricular angiocardiography in studies by Rackley and his associates.[38] Mathey and his coworkers have observed a correlation between the impairment of haemodynamic performance and infarct size.[13] Patients without haemodynamic dysfunction had small infarcts (averaging 17 CK-g-equiv.) and a mortality in hospital of only 5%. In contrast, patients with moderate increase in left ventricular filling pressure and slight decreases in cardiac output have larger infarcts (averaging 42 CK-g-equiv.) with a mortality of 21%. Among those patients with marked impairment of left ventricular function with a high pulmonary artery end-diastolic pressure (exceeding 20 mmHg), average infarct size was 99 CK-g-equiv. and mortality was 60%.[3,13]

Norris and his colleagues have previously demonstrated that the presence or absence of pulmonary venous congestion is a powerful clinical discriminator distinguishing survivors from non-survivors after infarction. With the use of a modified technique for enzymatic estimation of cumulated CK release, they have observed a correlation between pulmonary venous hypertension detectable radiographically and enzymatically estimated infarct size.[25] In our initial study of the correlation of enzymatically estimated infarct size and prognosis in a series of 35 patients, among the 12 patients with infarct size exceeding 65 CK-g-equiv., 8 died within 6 months of infarction.[43] The 4 survivors manifested marked functional impairment. Only 1 of 21 patients with apparently small infarcts died within 6 months. The average infarct size in survivors in Class I and Class II (New York Heart Association clinical classification) was less than 50% as large as infarct size in the other patients studied. In a later series of 69 consecutively admitted patients whose plasma CK values were increasing at the time of admission and in whom myocardial infarction was documented conventionally with patients with shock excluded, mean infarct size among survivors was significantly less than that among non-survivors (22 ± 2.5 versus 66 ± 6.9 CK-g-equiv, m^{-2} body surface area (mean $\pm$ SEM)). These observations indicate that enzymatically estimated infarct size appears to reflect a biologically significant parameter.[41]

Recently, Rackley and his associates have observed close correlations between enzymatically estimated infarct size and the mass of the left ventricle exhibiting dyskinesis detectable by left ventricular angiocardiography in patients with acute myocardial infarction. When patients with previous, remote myocardial infarction were excluded, the correlation coefficients between infarct size estimated enzymatically and the mass of the left ventricle exhibiting dyskinesis exceeded 0.9, as did the correlation coefficient of the relation between decreased ventricular compliance or increased pulmonary artery end-diastolic pressure and enzymatically estimated infarct size.[38] Bleifeld and his associates have observed similar correlations between haemodynamic impairment and enzymatic estimation of infarct size in patients with initial infarctions and have also observed a close correlation between infarct size estimated enzymatically ante-mortem and infarct size estimated histochemically by analysis of serial sections of the heart at post-mortem with a coefficient correlation of 0.89.[3] Recovery of exercise tolerance in the convalescent period appears to be directly related to enzymatically estimated infarct size judging from a recent report of patients undergoing a controlled rehabilitation programme.[4] These observations indicate that despite their obvious limitations, enzymatic estimates of infarct size provide insight into one important determinant of impairment of cardiac electrical instability, ventricular performance, compliance, morbidity, and early mortality associated with infarction—namely, the mass of tissue undergoing irreversible injury.

FACTORS INFLUENCING ENZYMATIC ESTIMATION OF INFARCT SIZE, TECHNICAL CONSIDERATIONS

Although analysis of plasma CK time–activity curves has been considered from a theoretical point of view, insufficient attention has been devoted to assurance of data of high quality.[46] One important potential source of error is sampling technique. Because the CK content in skeletal muscle exceeds 3000 iu g^{-1} in contrast to plasma CK activity in the range of 0.1–1.5 iu ml^{-1}, even minor trauma to muscle near the sampling site can distort results markedly. For this reason, plasma CK time–activity curves should not be assessed after surgery in experimental animals until values have returned to control levels, repetitive venepuncture should be avoided, and plasma samples must be collected after thorough washout of dead space in catheters. Failure to avoid these pitfalls leads to results incompatible with typical kinetics of CK in the circulation.

Because assays of CK activity are dependent on temperature, pH, dilution, inhibitors, cofactors, and competing enzymes in the sample, spurious results may occur unless adequate controls are employed. For example, with glutathione as a thiol donor, falsely low values for apparent CK activity may be obtained when glutathione reductase is present in the sample—a common occurrence after myocardial infarction. Dilution of samples should be sufficient to avoid saturation of coupling enzymes in the assay system, but dilution should not be so extreme that the sensitivity of the detection system is taxed. Standard curves are needed to avoid unrecognized error introduced by dilution because of effects on inhibitors or conformational changes in the CK molecules themselves. Unfortunately, control of some of these factors is particularly difficult in studies with CK-MB, a phenomenon that may account for undue reliance on techniques such as electrophoresis with visual or fluorometric scanning that provide only semiquantitative information.

Since enzymatic estimates of infarct size have been expressed in terms of the amount of CK lost from 1 gram of myocardium undergoing homogeneous infarction, the value of 1 CK-g-equiv. should be appropriate for the assay of plasma samples utilized in a particular study. In our initial studies in conscious dogs, this value was 800 iu g^{-1} in contrast to the substantially higher values of 1515 iu g^{-1} in the dog and 1200 iu g^{-1} in man obtainable with currently available assay systems that optimize CK activity and inhibit myokinase more effectively. Obviously, gross overestimation of infarct size would occur if calculations based on parameters determined with the first assay system were utilized with results of determinations of plasma CK activity assayed with a system yielding substantially higher results.[46]

Parameters in the Model

Several parameters utilized in the model initially formulated for enzymatic estimation of infarct size merit particular consideration including k_d (the instantaneous fractional CK disappearance rate), the distribution volume of enzyme released from the heart, and the ratio of CK released to CK depleted from myocardium (CK_R/CK_D).[31,37,45,46,48]

CK disappearance rate (k_d)

Although early disappearance of purified myocardial CK injected intravenously in conscious dogs deviates from first-order kinetics, and although different CK isoenzymes exhibit disparate disappearance rates,[26] in the intial model it was assumed for convenience that a first-order rate constant could be used as a crude approximation of k_d. Both, average rates obtained from fitting single exponentials to the tail of plasma CK time–activity curves and individualized estimates of k_d have been used. However, release of even small amounts of CK into blood during the interval in which total CK activity is decreasing after infarction may invalidate individualized estimates. Regardless of what method is used to estimate a parameter value for k_d, the same approach should be used in calculations of infarct size and the determination of the proportionality constant relating CK-g-equiv. calculated from plasma CK time–activity curves to grams of infarction. This is the case because the proportionality constant comprises a derived value for CK_R/CK_D and therefore contains k_d in its denominator. On the other hand, calculations of infarct size include k_d in the numerator. Thus a systematic error in the estimate of k_d appearing in both terms will be offset.[45]

Since marked variation in the rate of disappearance of CK within the interval during which an individual estimate of infarct size is obtained could distort the estimate substantially, this variable was assessed in conscious dogs with haemodynamic and pharmacological perturbations simulating those associated with myocardial infarction.[28,31,46,48] In conscious dogs without haemodynamic or physiological perturbations, enzyme activity and radioactivity of canine myocardial CK radioactively labelled with ^{14}C-formaldehyde decreased in a generally parallel fashion suggesting that the decline of enzyme activity was due to removal of intact enzyme molecules from the circulation. The distribution volume of CK calculated from the dilution principle approximated plasma volume within a range of 10%. In keeping with the well-recognized wide variation of CK disappearance rate from animal to animal or patient to patient, in these studies k_d varied within a range of 0.004 to 0.007 min^{-1}. The absence of enzyme activity or radioactivity in acid-precipitable material in urine confirmed the expectation, based on the molecular weight of CK, that renal excretion of intact enzyme molecules did not contribute appreciably to disappearance of CK from the circulation. In contrast to the variation of k_d from animal to animal, results of recent studies indicate that k_d remains remarkably constant in the same animal on successive occasions from day to day. Thus when repetitive injections of CK were made on three successive days in the same conscious animal, estimates of k_d were close in each case (Figure 3).

To evaluate effects of haemodynamic perturbations on the disappearance rate of intravenously injected canine myocardial CK in conscious dogs, interventions were induced simulating those seen in patients with acute myocardial infarction.[28] Animals were instrumented at least 1 week prior to each study with devices such as inflatable balloon cuff catheters around the inferior vena cava used to reduce cardiac output, occlusive devices around renal, hepatic, or coeliac arteries for subsequent use in diminishing renal or hapatic perfusion, epicardial pacing wires sutured to the right ventricle for control of heart rate, and electromagnetic flow probes around the aorta for assessment of cardiac output. After the animals had recovered completely from surgery required for implantation of the instrumentation, radioactively labelled myocardial CK was injected intravenously for 4 hours before implementation of the selected intervention, and blood samples were obtained through inlying venous catheters at 30-minute intervals. The rate of CK disappearance was assessed during the 4-hour interval prior to and the 4-hour interval following implementation and maintenance of each intervention. In additional experiments, effects of alteration of reticuloendothelial system activity were evaluated by inducing blockade with intravenous injection of zymosan, 10 mg kg^{-1}.

The ratio of CK disappearance before and after haemodynamic interventions, including profound reduction of cardiac output, doubling of heart rate, interruption of renal perfusion, and reduction of hepatic perfusion by more than 50%, varied by an average of less than 10%. In contrast, blockade of

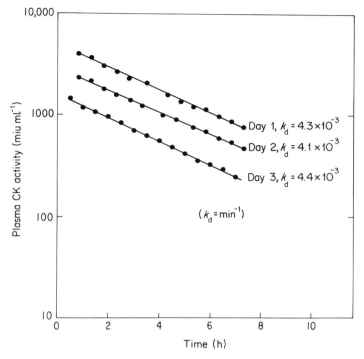

Figure 3. The consistency of the rate of disappearance of canine myocardial CK injected intravenously in the same recipient dog on three consecutive days. After each injection CK activity in plasma was assayed serially. As can be seen, the rate of disappearance (reflected by the slope of the semilog plot of serial values) did not vary substantially on the three occasions. Similar results were obtained in each of five dogs. [Reproduced, with permission, from *Cardiovascular Research*, **11**, 103–12 (1977).]

the reticuloendothelial system or administration of anaesthesia led to a marked reduction in the rate of CK disappearance. These results indicated that profound haemodynamic perturbations simulating those likely to occur in patients with complicated myocardial infarction do not substantially alter k_d and hence are unlikely to influence markedly enzymatic estimates of infarct size by altering enzyme disappearance. In additional experiments in which radioactively labelled CK was injected intravenously in conscious dogs subsequently subjected to acute myocardial infarction, the disappearance rate of radioactive counts remained constant during the course of the experiment, varying by less than 5%. Since the fractional disappearance rate of counts should continue to reflect the true fractional disappearance rate of enzyme even when non-labelled CK is being added to the blood pool due to continuing release from the heart (because of the principle of conservation of tracer), these observations indicate that k_d is not markedly affected by the process of infarction itself.

Parallel observations in patients, obtained by Norris,[25] indicate that CK disappearance remains remarkably constant within the same patient in association with repeated episodes of infarction. These data in experimental animals and patients support the view that CK disappearance rates remain relatively constant during evolution of infarction despite the variance in disappearance rate from animal to animal or patient to patient.

As noted previously, enzymatic estimation of infarct size depends on the value selected for k_d in an interesting fashion, since the proportionality constant (K, see Appendix) reflecting the amount of CK recovered in blood to that lost from myocardium includes a parameter in its denominator that is derived experimentally and contains k_d as one term. Thus an error in the estimate of k_d used to calculate infarct size (see Appendix) will be offset by the error in k_d appearing implicitly in the denominator of this proportionality constant as long as the same approach is used in estimating k_d in studies in which K is determined and in those in which calculations of infarct size are made.[45]

Since k_d remains quite constant within an individual experimental animal or patient from day to day and regardless of haemodynamic or pharmacological perturbations, it may be possible to reduce the error in estimation of infarct size dependent on the estimate of k_d well below that apparent from consideration of the variance of k_d from animal to animal or patient to patient alone, by estimating k_d in individual cases with the use of tracer techniques.

CK distribution volume and the ratio CK_R/CK_D

CK distribution volume has been determined by the dilution principle after intravenous injection of partially purified canine myocardial CK in conscious dogs. In early studies from several laboratories, high estimates of distribution volume resulted probably because of partial denaturation of the exogenous CK occurring during its isolation. When improved preparative techniques were utilized, estimates of distribution volume based on a one-compartment model approximated plasma volume. These revised estimates of distribution volume do not necessitate alteration in estimates of infarct size published previously since the proportion of CK recovered in blood is a parameter obtained empirically and based on an assumed value for distribution volume. On the other hand, a revision in the estimate of distribution volume does lead to a change in the estimate of the ratio of CK_R/CK_D. It is for this reason that the value delineated in the Appendix differs from that published in 1971.[25,39]

The term 'CK_R/CK_D', i.e. the ratio of CK lost from the heart that appears in blood, has been found experimentally to average 0.15 ± 0.016 (SEM) (based on a distribution volume equal to plasma volume). The relatively low proportion of CK reaching blood has not been well explained until recently. Local inactivation of enzyme within necrotic myocardium might be one factor responsible. Another has been suggested in recently completed studies. It appears likely that transport of enzyme in cardiac lymph is an important route of egress of CK from the heart into the circulation. Since lymph flow is so low (of the order of 1.5–3.0 ml h^{-1} $(100$ g$)^{-1}$ myocardium in anaesthetized dogs), transport of enzyme in lymph is likely to result in prolonged exposure. We have recently shown that CK, in contrast to several other enzymes, undergoes rapid inactivation when exposed to lymph *in vitro* and *in situ*. Although inactivation can be retarded by addition of buffer, it appears to play a major role in influencing plasma CK time–activity curves after myocardial infarction. For example, the observed ratio of CK_R/CK_D in animals subjected to coronary occlusion followed by occlusion of the lymphatics draining the heart was substantially less than the ratio in animals with coronary occlusion alone even though the disappearance rate of enzymes that did reach the systemic blood circulation was comparable in the two groups, and even though the amount of CK depleted from the heart was also comparable. Although alterations in lymphatic flow during infarction have not been characterized thoroughly, it appears likely that variation due to intramyocardial oedema, changes in venous pressure, regional myocardial compliance, and heart rate may alter the duration of exposure of CK liberated from necrotic tissue to lymph *in vivo* and accordingly the proportion of depleted myocardial CK ultimately appearing in the systemic blood circulation.[6] Characterization of lymphatic efflux from the heart may therefore be useful in refining enzymatic estimates of infarct size, by providing estimates of a parameter that could be employed in individual estimates derived from analyses developed with the use of physiologically based mathematical models.

One other factor that may contribute to the relatively low ratio of CK_R/CK_D recognized previously is the probability that CK released from the heart is distributed in at least one extravascular compartment. Thus during the development of plasma CK time–activity curves after myocardial infarction, net elimination of enzyme from the blood pool may be slowed because of resupply from the extravascular space. This gives rise to estimates of k_d based on a one-compartment model that are too low. Coupled with the larger distribution volume compatible with a two-compartment model, the analysis obtained would account for recovery of approximately twice as much CK depleted from the heart judging from a given plasma CK time–activity curve as that recognized previously.[31]

Mathematical treatment of data obtained from conscious experimental animals has recently helped to clarify these phenomena. Time–activity curves of plasma CK activity after intravenous injection of purified enzyme, crude extracts of myocardium, or plasma with high CK activity obtained previously from the same animal conform much more closely to double-, rather than single-exponential functions with standard deviations to the fit reduced by an average of approximately 50%.[45] Thus distribution in at least one extravascular compartment is probable. Accordingly, it seems likely that resupply of the vascular pool with enzyme initially distributed in an extravascular space proceeds simultaneously with removal (or inactivation) of enzyme in the vascular pool. Under these conditions, the observed elimination rate of CK from blood is not a direct reflection of the true, instantaneous disappearance rate, because loss of activity is partially offset by resupply of the vascular pool. Improved estimates of the true enzyme disappearance rate can be obtained by curve fitting plasma CK curves after intravenous injection of enzyme and obtaining parameters needed to solve the series of differential equations characterizing a two-compartment model entailing extravascular as well as vascular distribution. True enzyme disappearance rates (k_d) based on estimates made in this fashion appear to be more than 2-fold greater than observed elimination rates. Higher values of k_d would not alter previous estimates of infarct size because k_d enters into estimation of another derived parameter—CK_R/CK_D. However, higher values would account for substantially more recovery of CK lost from the heart.[45]

Under some conditions, variation in rates of exchange between the vascular and extravascular compartments or alterations in the relative volumes of distribution attributable to each could influence enzymatic estimates of infarct size if k_d were estimated from observed elimination rates in an individual case. Even if average values of k_d are used, the absolute amount of enzyme released from the heart would differ if the distribution volume deviated markedly from the average. Changes in rates of exchange between compartments might be anticipated in association with profound haemodynamic compromise, hypovolaemia, sepsis, or any other phenomenon altering permeability of lymphatic or vascular tissue. Accordingly, continued progress in refining enzymatic estimates of infarct size may be facilitated by development and verification of direct means to assess distribution volumes and true enzyme disappearance rates in each individual case with the use of tracers.

The two-compartment model is important for another reason as well. Because of resupply of the vascular space from an extravascular pool and from contained release of enzyme from the heart relatively late after the onset of infarction (documented experimentally[45]), k_d *cannot* be estimated accurately in any individual case simply by analysis of the declining portion of the plasma time–activity curve after infarction. Accordingly, use of average values determined independently or individual values obtained by analysis of tracer kinetics is necessary.

Naturally, in utilizing estimates of infarct size based on analysis of plasma CK time–activity curves for evaluating interventions designed to protect ischaemic myocardium, one must determine whether or not the intervention itself modifies parameters in the model—particularly the CK_R/CK_D ratio. The measured CK_R/CK_D ratio has remained quite consistent under a variety of experimental conditions with conscious dogs subjected to coronary occlusion in whom the ratio can be determined by direct assay of myocardial CK depletion at the end of each experiment and at selected intervals during the course of the study. The rate of washout of enzyme from the heart will not influence results as long as the CK_R/CK_D ratio remains constant, since the mathematical model utilized employs integration of changes throughout the entire interval of the experiment. Although we have observed some variation in the CK_R/CK_D ratio in conscious dogs subjected to coronary occlusion coupled with interventions such as administration of propranolol or acceleration of heart rate by ventricular pacing, the ratio has consistently been observed to average 0.15 ± 0.016 (SEM) under these conditions. On the other hand, after experimentally induced reperfusion, the ratio may increase in a systematic fashion which can be incorporated in a revised estimate of parameters used to calculate infarct size in animals subjected to this particular experimental intervention.[19]

PHYSIOLOGICALLY BASED MODELS

Although correlations between plasma CK time–activity curves and infarct size have provided a framework for estimating infarct size on an empirical basis, improved estimates may result from utilization of physiologically rather than empirically based models. Physiologically based models (see also Chapter 12) are those utilizing parameters representing processes governing release, transport, distribution, and removal of CK from the heart and circulation rather than empirical algorithms for analysis of plasma CK time–activity curves such as the log-normal function.[20,44–47] Empirical and physiologically based models differ in the same way as projections of stock market prices based on purely 'technical' considerations, i.e. the observed behaviour of the market in the past, differs from that based on analysis of forces present in the economy and acting at a specific time to influence the behaviour of the market as a whole. The estimated weight to be attributed to specific economic forces and assessments of the extent to which each is operating at any given time are parameters analogous to corresponding factors related to physiological processes influencing plasma CK curves. Although development of physiologically based models has not yet supplanted empirical approaches to enzymatic estimation of infarct size, some insight has been gained regarding physiological processes influencing and underlying the kinetics of CK in the circulation after myocardial infarction. For example, in a recent study the elimination rate of circulating CK after infarction was found to be substantially less than the rate of elimination of partially purified canine myocardial CK injected intravenously in conscious dogs.[31] Part of this difference was found to be due to a modest amount of continuing release of enzyme from the ischaemic heart relatively late during the evolution of infarction blunting the rate of decline of CK activity from blood. Since the amount of CK in myocardium is of the order of 2000 iu g^{-1}, and since peak values of CK activity in blood after infarction are generally of the order of 1–2 iu ml^{-1}, even modest continuing release could markedly diminish the observed decline of activity in blood. On the other hand, some of the disparity between elimination rates under the two conditions (after intravenous injection in contrast to after myocardial infarction) appeared to reflect biochemical differences between native CK circulating in blood after infarction and CK concentrated or purified from either plasma or myocardium. Thus injection of purified CK intravenously led to a much faster elimination rate than that seen after injection of an equal volume of plasma with high CK activity (induced by myocardial infarction in the donor animal) into the same recipient conscious dog.[45]

In addition to these contributions to the relatively slow rate of elimination of CK after myocardial infarction, as noted previously recent results indicate that CK released from the heart is distributed in at least one extravascular compartment with continuing exchange with blood. Thus observed elimination of CK from the systemic blood circulation appears to be substantially slower than the true, instantaneous fractional CK disappearance rate (k_d) because of resupply of the vascular from the extravascular pool. Since a double-exponential function provides parameter estimates for a system of differential equations describing a two-compartment model, and since it provides considerably closer fits to the data than those obtained with a single-exponential function, CK, like several other proteins, appears to be distributed in at least one extravascular pool.[45] Estimates of k_d, the true fractional disappearance rate, based on the two-compartment model are higher than those obtained from a one-compartment model described by a single-exponential function in which k_d is assumed to be identical to the observed rate of elimination of the enzyme from blood. Although use of the higher estimate would not change relative estimates of infarct size it would account for a substantially larger proportion of CK released from the heart than that accounted for previously.[45,46]

One other outcome of recent efforts to develop and evaluate physiologically based models of CK release, transport, and removal has been the implication of an important role for transport of enzyme in lymph.[5] It has been recognized for some time that activities of several enzymes increase markedly in lymph draining the heart after myocardial infarction, sometimes exceeding activities of the corresponding enzymes in blood. Although cardiac lymph flow is low, even after infarction—probably of the order of

$1-3 \text{ ml h}^{-1} (100 \text{ g})^{-1}$—these observations suggest that an appreciable amount of enzyme may reach the systemic circulation via transport in lymph. In addition, since exposure of enzyme to lymph might be quite prolonged because of the low flow, inactivation of enzyme in lymph may contribute to the relatively small amount of CK appearing in blood compared to that lost from the heart itself. In evaluating these possibilities we have observed marked inactivation of CK via lymph *in vitro*, attributable to a substance of low molecular weight, not chelated by EDTA, associated with inactivation of titratable reduced thiol groups, and associated with non-enzymatically mediated proteolysis of CK molecules. In addition, interruption of efflux of cardiac lymph *in vivo*, implemented several hours after the onset of coronary occlusion in conscious dogs, reduced the proportion of CK appearing in blood, compared to that depleted from the heart, without modifying the overall extent of apparent myocardial infarction or CK depletion from the heart. Cardiac lymphatic occlusion had no effect on the rate of elimination of CK from blood once it had reached the systemic blood circulation. These findings indicate that transport and inactivation of CK in lymph may influence plasma CK time–activity curves after myocardial infarction substantially by decreasing the CK_R/CK_D ratio when lymph flow is markedly reduced or by increasing it when flow is markedly augmented. Recent progress in external assessment of cardiac lymphatic flow in semiquantitative terms in intact animals suggests that it may soon be possible to refine enzymatic estimates of infarct size further by incorporation of a parameter reflecting the contribution of regional lymphatic flow in individual estimates.[6]

ESTIMATION OF INFARCT SIZE BASED ON PLASMA CK-MB TIME–ACTIVITY CURVES

Estimates of infarct size would of course be spuriously influenced by CK released from tissues other than the heart. Based on recent observations confirming the impression that the MB isoenzyme of CK is a remarkably specific marker of heart muscle, infarct size has been estimated on the basis of serial changes in CK-MB in plasma rather than from changes in total CK activity alone.[28] Unfortunately, this approach was not practical until quantitative, in contrast to semiquantitative, electrophoretic methods were developed.[30,33,35] Several years ago, with a quantitative, fluorometric-kinetic procedure for assay of CK-MB not requiring scanning of electrophoretic media, the quantitative distribution of isoenzymes of CK in human and canine tissues was evaluated.[27] Subsequently, with the use of this assay technique as a standard, more convenient quantitative procedures were developed including batch adsorption of individual isoenzymes with glycophase DEAE glass beads facilitating extension of these observations with a larger number of samples.[15] When the distribution of CK-MB was evaluated in extracts of human tissue obtained at surgery, myocardium was found to be virtually the only tissue surveyed containing appreciable amounts of the CK-MB isoenzyme. In contrast, in dogs, several tissues besides the heart—notably the gastrointestinal tract—contained substantial amounts of CK-MB.

In keeping with the proportion of CK in human myocardium represented by the MB isoenzyme (approximately 12–15%), peak plasma CK-MB activity after myocardial infarction averages approximately 12% of total CK in the same sample.[15,27,28,53] After haemodynamically uncomplicated myocardial infarction, the time–activity curve of CK-MB generally parallels that of total CK, although the MB isoenzyme exhibits a more rapid rate of elimination from blood.[28] When endogenous non-cardiac enzyme is released into blood after intramuscular injections, the serial changes in total CK after myocardial infarction are distorted but the CK-MB curve continues to reflect the characteristic changes indicative of an uncomplicated infarct. A similar lack of distortion of the CK-MB curve is evident in patients with non-cardiogenic shock with persistent plasma total CK elevations that do not reflect persistent myocardial necrosis.[12]

The specificity of elevated CK-MB activity in plasma as an index of myocardial infarction has been examined extensively. Among a large number of patients undergoing non-cardiac surgical procedures including thoracic, abdominal, orthopaedic, genitourinary, and gastrointestinal tract surgery, none exhibited plasma CK-MB elevations in any of serial samples obtained prior to surgery and for 24 hours subsequently.[32] Similar results were seen among more than 50 patients undergoing cardiac catheterization in which none exhibited elevated plasma CK-MB.[34] Patients in both of these categories exhibited marked elevations of total plasma CK activity because of trauma to tissues other than the heart associated with these procedures.

Since electrical cardioversion may be required in patients with evolving acute myocardial infarction, studies were undertaken to determine the extent to which this procedure alters plasma MB time–activity curves.[9] Thus total and plasma CK-MB activity levels were measured serially in 30 patients without myocardial infarction treated with direct current electrical countershock and in 17 patients with acute myocardial infarction. In addition, serial determinations of total and CK-MB were performed in closed chest anaesthetized dogs subjected to 10 repetitive countershocks at 15 second intervals with a delivered energy of 240 Watt-seconds per shock. In all cases, MB in plasma was assayed quantitatively. Less than 0.004 iu ml^{-1} of CK-MB occurred in plasma of normal human subjects. In contrast, patients with myocardial infarction with total plasma CK levels comparable to those seen after electrical cardioversion exhibited a substantial rise in plasma CK-MB activity with peak values averaging 0.039 ± 0.006 (SEM).

Fifteen of the 30 patients without infarction treated with electrical countershock had elevated total CK activity peaking within 4 hours. Among them, elevation of CK-MB accounted for the overall rise. In 2 patients, trivial elevations of CK-MB were observed, insufficient to interfere with quantitative estimation of infarct size by analysis of CK-MB time–activity curves. Among the dogs given massive shocks, myocardial necrosis was demonstrable in 6, each of whom had elevated plasma CK-MB activity of the order of 0.05 iu ml^{-1}. Results of these studies indicate that although electrical countershock, when used in doses exceeding those delivered in most clinical settings, can produce myocardial injury in dogs associated with elevation of CK-MB activity in plasma, countershock as conventionally employed in patients does not generally produce myocardial damage or elevation of plasma CK-MB activity sufficient to interfere with quantitative estimation of infarct size based on analysis of plasma CK-MB time–activity curves.

In view of the apparent specificity and sensitivity of increased plasma CK-MB as an index of myocardial infarction, infarct size has been estimated enzymatically from serial changes in plasma MB activity in patients with myocardial infarction.[28] Patients selected were those without haemodynamically complicated infarction, in order to permit comparison of results with enzymatic estimates based on analysis of total plasma CK activity. Estimated infarct size based on analysis of total CK ranged from 12 to 187 CK-g-equiv. The relation between estimates of infarct size based on MB and those based on total CK conformed to a regression line with a slope of 0.99, an intercept of 1.8, and a standard deviation of the slope of 0.07. The correlation coefficient of the line was 0.97. Thus despite the markedly different disappearance rates of MB and total CK, estimates of infarct size based on analysis of serial changes in plasma total CK and CK-MB correlated closely. Although enzymatic estimates of infarct size were verified initially in dogs based on analysis of plasma total CK changes by measuring myocardial CK depletion directly, an analogous procedure is difficult with respect to the CK-MB isoenzyme. Unfortunately, the proportion of CK-MB in dog myocardium is so low (less than 2%) (Figure 4) that assessment of CK-MB disappearance from the heart is prone to a large percentage error because of the high background contributed by isoenzymes other than the MB form and because of the inevitable variation of CK-MB from heart to heart. It was for this reason that one approach to evaluation of enzymatic estimation of infarct size based on analysis of CK-MB activity was comparison of results with this isoenzyme to those with total CK in patients with uncomplicated myocardial infarction.

The assumption that changes in total plasma CK conform to either a one- or a two-compartment model is a simplification used to permit convenient treatment of data. It neglects the possibility that the amount or rate of release of individual CK isoenzymes from the heart undergoing necrosis may

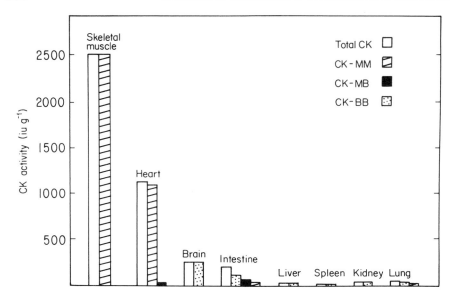

Figure 4. The relative contribution of activity associated with each isoenzyme of CK (MM, MB, and BB) to total CK activity in extracts of canine tissue. As can be seen, the proportion of MB in myocardial extracts is less than 2%, and in contrast to the case in man, MB contributes substantially to total CK activity in extracts of the gastrointestinal tract. [Reproduced, with permission, from *Circulation*, **52**, 743–54 (1975).]

differ depending on the intracellular locus of each species and its characteristic rate of inactivation in body fluids. The assumption also neglects documented differences in fractional rates of disappearance from the circulation of different CK isoenzymes. In dogs, the overwhelming predominance of the MM form in both myocardium and blood may minimize these difficulties. In patients, the use of CK-MB time–activity curves in contrast to total CK time–activity curves achieves the same purpose.

Use of any model for enzymatic estimation of infarct size is not applicable if substantial release of non-cardiac CK occurs and if its presence is influential in results obtained with the model. With experimental preparations such as circumflex coronary occlusion in the dog, a model frequently evoking hypoperfusion of the periphery or shock, or with the use of anaesthetized open chest animals, non-cardiac CK release is substantial and not excluded from calculations of infarct size estimated from plasma CK curves. To avoid this pitfall in patients, plasma CK-MB time–activity curves can be utilized since results obtained appear to be independent of release of non-cardiac CK.

STRENGTHS AND LIMITATIONS OF ENZYMATIC ESTIMATION OF INFARCT SIZE

In general, results obtained with the use of enzymatic estimates of infarct size appear to support the following:

(1) Myocardial CK depletion in experimental animals depends on the distribution and extent of infarction and is reflected by cumulative CK appearing in blood estimated from analysis of plasma CK time–activity curves. In patients, enzymatic estimates based on analysis of plasma CK time–activity curves indicate that infarct size is an important determinant of ventricular dysrrhythmia early after myocardial infarction, peak blood levels of free fatty acids (L. H. Opie, personal communication), impairment of ventricular compliance, elevation of left ventricular filling pressure, the extent of dyskinesis estimated angiocardiographically, electrophysiological manifestations of infarction including ST- segment elevation, impairment of exercise tolerance in the convalescent phase, morbidity, and mortality within the first 6 months. For example, among one series of consecutively admitted patients with myocardial infarction without shock and with plasma CK values increasing at the time of admission, mean infarct size in survivors averaged 22 ± 2.5 CK-g-equiv. m^{-2} body surface area (mean $\pm$ SEM) compared to 66 ± 6.9 in non-survivors ($p < 0.001$).[41]

(2) Improved enzymatic estimates can be obtained with CK-MB, a marker more specific to myocardium than total plasma CK activity (Figure 5).

(3) The evolution of myocardial infarction can be slowed or inhibited in experimental animals and in at least some patients such as those with marked systemic arterial hypertension by *early* implementation of therapeutic interventions designed to protect ischaemic myocardium. Enzymatic estimates of infarct size have been used to evaluate several potentially therapeutic interventions on evolving myocardial damage. One approach entails predicting infarct size from projected plasma CK values obtained with curve-fitting techniques.[40,42] After coronary occlusion in conscious dogs, augmentation of myocardial oxygen consumption by electrical pacing or by administration of atropine or isoproterenol is associated with augmentation of infarct size estimated enzymatically. In contrast, administration of propranolol, a β-adrenergic antagonist, leads to a reduction of observed compared to projected enzyme values and apparent protection of ischaemic myocardium. When trimethaphan is administered to hypertensive patients with acute infarction to reduce ventricular afterload and thereby diminish myocardial oxygen consumption, observed infarct size is significantly less than that predicted, suggesting salvage of myocardium that would have otherwise undergone irreversible injury.[42] In addition, early mortality is

reduced. Thus modification of infarction during its evolution appears to be possible. On the other hand, results of several studies suggest that in order for interventions improving the balance between myocardial oxygen requirements and myocardial oxygen supply to be effective in protecting the ischaemic heart, they must be implemented very early after the onset of ischaemia.[10,11] Accordingly, despite favorable electrophysiological and haemodynamic effects with intraaortic balloon counterpulsation in experimental animals,[8] the evolution of necrosis is delayed but not prevented. Similarly, external pressure circulatory assist in patients,[11] administration of dobutamine,[10] and administration of nitroglycerin (W. D. Bussman and colleagues, personal communication) fail to reduce ultimate enzymatically estimated infarct size despite apparent beneficial electrophysiological and haemodynamic effects including decreased ventricular dysrrhythmia (external pressure circulatory assist), reduction of ST- segment elevation on surface electrocardiogram (nitroglycerin), and improved cardiac output (dobutamine). One factor probably contributing to the apparent ineffectiveness of such interventions in limiting the extent of infarction is the delay entailed prior to their implementation because of methodological constraints when projected enzyme curves are employed and because of the well-recognized patient and physician delay slowing admission of patients with chest pain to the hospital after the onset of myocardial ischaemia.

FUTURE DIRECTIONS

Despite the utility of enzymatic estimation of infarct size, the technique entails several limitations suggesting directions for future research. Although it is obvious that total plasma CK activity cannot be used to estimate infarct size when contributions from non-cardiac sources are substantial, pejorative evaluation of the approach has been sometimes based on results obtained in studies disregarding this constraint. Increasing availability of quantitative assays for the CK-MB isoenzyme including chromatographic, elution, adsorption, and radioimmunoassay techniques is facilitating enzymatic estimation of infarct size even among patients with non-cardiac CK release in whom CK-MB curves continue to reflect release of enzyme from the heart apparently exclusively.

Analysis of serial changes in plasma CK-MB has been particularly useful in assessment of infarct size among patients given intramuscular injections and should help to broaden the applicability of enzymatic estimation of infarct size to patients recovering from non-cardiac surgery, trauma, electrical cardioversion, and those with shock as well. As can be seen in Figure 5, even when release of CK from skeletal muscle distorts the total CK time–activity curve, plasma CK-MB curves continue to conform to the pattern characteristic of myocardial infarction.

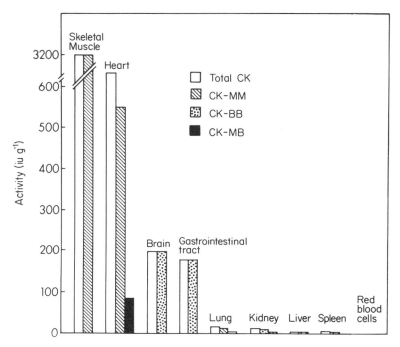

Figure 5. The relative contribution of activity associated with each isoenzyme of CK (MM, BB, and MB) to total CK activity in extracts of human tissue obtained at the time of surgery. As can be seen, the proportion of MB in myocardial extracts is substantially greater than in the corresponding canine preparation, and in contrast to the case in dogs, MB activity does not contribute substantially to total CK activity in extracts from the gastrointestinal tract. [Reproduced by permission of Dun-Donnelley Publishing Co. from *American Journal of Cardiology*, **36**, 433–7 (1975).]

With the increasing information becoming available allowing characterization of physiological processes influencing release of CK from injured myocardium, transport to the circulation, distribution, and removal, development of physiologically based rather than empirical models should be facilitated with consequent refinement of enzymatic estimation of infarct size. The somewhat surprising lack of influence of haemodynamic perturbations on the apparent disappearance of CK suggests that a mechanism relatively independent of any single, specific organ may be responsible for decline of enzyme activity from the circulation. However, improved delineation of physiological processes accounting for disappearance and masking of disappearance by continued release of enzyme into the circulation is needed. When radioactively labelled CK is injected intravenously in conscious dogs, both enzyme and radioactivity decline in parallel for several hours implying that enzyme activity declines because of removal of intact enzyme molecules from the circulation. The absence of radioactivity in acid-precipitable protein in urine and the lack of increased CK enzyme activity in urine under these conditions confirms the expectation that renal excretion of intact CK molecules does not occur, in keeping with the molecular weight of the enzyme of 82,000 daltons.

Results from several laboratories indicate that overall activity of the reticuloendothelial system may influence removal of enzymes from the circulation. When zymosan is administered to conscious dogs previously given CK intravenously, substantial decreases in the observed elimination rate of CK from blood are apparent. It should be recognized, however, that zymosan exerts many other actions besides inhibition of the reticuloendothelial system, including activation of complement and alteration of permeability of barriers between several fluid compartments. Nevertheless, results obtained after injection of zymosan, coupled with those after injection of large doses of pharmacologically active agents such as morphine or 'Valium' that led to diminution of k_d by as much as 50% (although they were virtually without effect in usual therapeutic doses), suggest that activity of the reticuloendothelial system may be involved in removal of CK from the circulation, and that refined enzymatic estimates of infarct size may be possible by appropriate consideration of effects of pharmacological agents on CK disappearance in individual instances.[31]

Among patients exhibiting cardiogenic shock within 48 hours after infarction, total CK activity in plasma declines more slowly than it does in patients without this complication. The blunted decline could reflect continuing release of CK from the heart, release of CK from non-cardiac tissue, or an alteration in the kinetics of CK disappearance due to impaired function of specific organ systems. A substantial contribution from extracardiac CK release seems to be excluded by the fact that patients with cardiogenic shock exhibit sustained elevations of MB as well as total CK and that the decline of CK-MB activity generally parallels that of total CK. It seems probable that slow decline of plasma CK activity associated with cardiogenic shock reflects continuing release of enzyme from the heart rather than a change in k_d, in keeping with effects of profound haemodynamic perturbations on disappearance of enzyme in experimental animals.[12]

Despite their limitations, enzymatic estimates of infarct size have proven useful in delineating relationships between the extent of infarction and clinical sequelae, morbidity, and mortality, and in demonstrating that at least under some conditions the evolution of infarction can be modified favourably. Enzymatic estimates of the ultimate extent of infarction coupled with promising techniques for early quantitative detection of myocardium in jeopardy, such as positron emission transaxial tomography, should facilitate objective evaluation of early implementation of interventions designed to protect ischaemic myocardium. The investigative efforts required to continue to improve and refine enzymatic estimates of infarct size appear to be justified on the basis of the potential value of effective therapeutic interventions in reducing functional impairment and mortality associated with acute myocardial infarction.

REFERENCES

1. Ahmed, S. A., Williamson, J. R., Roberts, R., Clark, R. E., and Sobel, B. E. The association of increased plasma MB CPK activity and irreversible ischemic myocardial injury in the dog. *Circulation*, **54**, 187–93 (1976).
2. Ahumada, G., Roberts, R., and Sobel, B. E. Evaluation of myocardial infarction with enzymatic indices. *Progress in Cardiovascular Diseases*, **18**, 405–20 (1976).
3. Bleifeld, W., Mathey, D., Hanrath, P., Buss, H., and Effert, S. Infarct size estimated from serial serum creatine phosphokinase in relation to left ventricular hemodynamics. *Circulation*, **55**, 303–11 (1977).
4. Carter, C. L. and Amundsen, L. R. Infarct size and exercise capacity after myocardial infarction. *Journal of Applied Physiology: Respiratory, Environmental and Exercise Physiology*, **42**, 782–5 (1977).
5. Clark, G. L., Robison, A. K., Gnepp, D. R., Roberts, R., and Sobel, B. E. Effects of lymphatic transport of enzyme on plasma CK time–activity curves after myocardial infarction. *Circulation Research*, **43**, 162–69 (1978).
6. Clark, G. L., Siegel, B. A., and Sobel, B. E. Qualitative external evaluation of regional cardiac lymph flow in intact dogs. *Physiologist*, **20**(4), 17 (1977).
7. Cox, J. R., Jr., Roberts, R., Ambos, H. D., Oliver, G. C., and Sobel, B. E. Relations between enzymatically estimated myocardial infarct size and early ventricular dysrrhythmia. *Circulation Supplement I*, **53**, I-150–I-155 (1976).
8. DeLaria, G. A., Johansen, K. H., Sobel, B. E., Sybers, H. D., and Bernstein, E. F. Delayed evolution of myocardial ischemic injury after intra-aortic balloon counterpulsation. *Circulation Supplement II*, **50**, II-242–II-248 (1974).
9. Ehsani, A., Ewy, G. A., and Sobel, B. E. Effects of electrical countershock on serum creatine phosphokinase (CPK) isoenzyme activity. *American Journal of Cardiology*, **37**, 12–8 (1976).
10. Gillespie, T. A., Ambos, H. D., Sobel, B. E., and Roberts, R. Effects of dobutamine in patients with actute myocardial infarction. *American Journal of Cardiology*, **39**, 588–94 (1977).
11. Gowda, K. S., Gillespie, T. A., Byrne, J. D., Ambos, H. D., Sobel, B. E., and Roberts, R. Effects of external counterpulsation on enzymatically estimated infarct size and ventricular arrhythmia. *British Heart Journal*, **40**, 308–14 (1978).
12. Gutovitz, A. L., Sobel, B. E., and Roberts, R. Cardiogenic shock: A syndrome frequently due to slowly evolving myocardial injury. *American Journal of Cardiology*, **39**, 322 (1977).
13. Hanrath, P., Bleifeld, W., and Mathey, D. Assessment of left ventricular function and hemodynamic reserve by volume loading in acute myocardial infarction. *European Journal of Cardiology*, **3**, 99–106 (1975).
14. Heng, M. K., Singh, B. N., Norris, R. M., John, M. B., and Elliot, R. Relationship between epicardial ST-segment elevation and myocardial ischemic damage after experimental coronary artery occlusion in dogs. *Journal of Clinical Investigation*, **58**, 1317–26 (1976).
15. Henry, P. D., Roberts, R., and Sobel, B. E. Rapid separation of plasma creatine kinase isoenzymes by batch adsorption on glass beads. *Clinical Chemistry*, **21**, 844–9 (1975).
16. Kjekshus, J. K. and Sobel, B. E. Depressed myocardial creatine phosphokinase activity following experimental myocardial infarction in rabbit. *Circulation Research*, **27**, 403–14 (1970).
17. Kostuk, W. J., Ehsani, A. A., Karliner, J. S., Ashburn, W. L., Peterson, K. L., Ross, J., Jr., and Sobel, B. E. Left ventricular performance after myocardial infarction assessed by radioisotope angiocardiography. *Circulation*, **47**, 242–9 (1973).

18. Lukes, J., Schrader, P., Anastasia, L., Mueller, H., Rao, P., Dutta, D., Yang, J., and Bain, R. Histologic quantitation of infarct size to assess accuracy of estimative creatine phosphokinase formulas. *Clinical Research*, **25**, 235A (1977).
19. Manders, T., Vatner, S., Millard, R., Heyndrickx, G., and Maroko, P. R. Altered relationship between creatine phosphokinase release and infarct size with reperfusion in conscious dogs. *Circulation Supplement II*, **52**, II-5 (1975).
20. Markham, J., Karlsberg, R. P., Roberts, R., and Sobel, B. E. Mathematical characterization of kinetics of native and purified creatine kinase in plasma. In *Computers in Cardiology*, IEEE Computer Society, Long Beach, California, 1976, pp. 3–7.
21. Maroko, P. R., Kjekshus, J. K., Sobel, B. E., Watanabe, T., Covell, J. W., Ross, J., Jr., and Braunwald, E. Factors influencing infarct size following experimental coronary artery occlusions. *Circulation*, **43**, 67–82 (1971).
22. Mathey, D., Bleifeld, W., Hanrath, P., and Effert, S. Attempt to quantitate relation between cardiac function and infarct size in acute myocardial infarction. *British Heart Journal*, **36**, 271–9 (1974).
23. Mimbs, J. W., Yuhas, D. E., Miller, J. G., Weiss, A. N., and Sobel, B. E. Detection of myocardial infarction *in vitro* based on altered attenuation of ultrasound. *Circulation Research*, **41**, 192–8 (1977).
24. Muller, J. E., Maroko, P. R., and Braunwald, E. Evaluation of precordial electrocardiographic mapping as a means of assessing changes in myocardial ischemic injury. *Circulation*, **52**, 16–27 (1975).
25. Norris, R. M., Whitlock, R. M. L., Barratt-Boyes, C., and Small, C. W. Clinical measurement of myocardial infarct size. Modification of a method for the estimation of total creatine phosphokinase release after myocardial infarction. *Circulation*, **51**, 614–20 (1975).
26. Rapaport, E. The fractional disappearance rate of the separate isoenzymes of creatine phosphokinase in the dog. *Cardiovascular Research*, **9**, 473–7 (1975).
27. Roberts, R., Gowda, K. S., Ludbrook, P. A., and Sobel, B. E. Specificity of elevated serum MB creatine phosphokinase activity in the diagnosis of acute myocardial infarction. *American Journal of Cardiology*, **36**, 433–7 (1975).
28. Roberts, R., Henry, P. D., and Sobel, B. E. An improved basis for enzymatic estimation of infarct size. *Circulation*, **52**, 743–54 (1975).
29. Roberts, R., Husain, A., Ambos, H. D., Oliver, G. C., Cox, J., Jr., and Sobel, B. E. Relation between infarct size and ventricular arrhythmia. *British Heart Journal*, **37**, 1169–75 (1975).
30. Roberts, R., Parker, C. W., and Sobel, B. E. Detection of acute myocardial infarction by radioimmunoassay for creatine kinase MB. *Lancet*, **2**, 319–22 (1977).
31. Roberts, R. and Sobel, B. E. Effect of selected drugs and myocardial infarction on the disappearance of creatine kinase from the circulation in conscious dogs. *Cardiovascular Research*, **11**, 103–12 (1977).
32. Roberts, R. and Sobel, B. E. Elevated plasma MB creatine phosphokinase activity. A specific marker for myocardial infarction in perioperative patients. *Archives of Internal Medicine*, **136**, 421–4 (1976).
33. Roberts, R. and Sobel, B. E. Isoenzymes of creatine phosphokinase and diagnosis of myocardial infarction. *Annals of Internal Medicine*, **79**, 741–3 (1973).
34. Roberts, R., Sobel, B. E., and Ludbrook, P. A. Determination of the origin of elevated plasma CPK after cardiac catheterization. *Catheterization and Cardiovascular Diagnosis*, **2**, 329–36 (1976).
35. Roberts, R., Sobel, B. E., and Parker, C. W. Radioimmunoassay for creatine kinase isoenzymes. *Science*, **194**, 855–7 (1976).
36. Roe, C. R., Cobb, F. R., and Starmer, C. F. The relationship between enzymatic and histologic estimates of the extent of myocardial infarction in conscious dogs with permanent coronary occlusion. *Circulation*, **55**, 438–49 (1977).
37. Roe, C.. and Starmer, C. F. A sensitivity analysis of enzymatic estimation of infarct size. *Circulation*, **52**, 1–5 (1975).
38. Rogers, W. J., McDaniel, H. G., Smith, L. R., Mantle, J. A., Russell, R. O., Jr., and Rackley, C. E. Correlation of CPK-MB and angiographic estimates of infarct size in man. *Circulation Supplement II*, **54**, II-28 (1976).

39. Shell, W. E., Kjekshus, J. K., and Sobel, B. E. Quantitative assessment of the extent of myocardial infarction in the conscious dog by means of analysis of serial changes in serum creatine phosphokinase activity. *Journal of Clinical Investigation,* **50,** 2614–25 (1971).
40. Shell, W. E., Lavelle, J. F., Covell, J. W., and Sobel, B. E. Early estimation of myocardial damage in conscious dogs and patients with evolving acute myocardial infarction. *Journal of Clinical Investigation,* **52,** 2579–90 (1973).
41. Shell, W. E. and Sobel, B. E. Biochemical markers of ischemic injury. *Circulation Supplement I,* **53,** I-98–I-106 (1976).
42. Shell, W. E. and Sobel, B. E. Protection of jeopardized ischemic myocardium by reduction of ventricular afterload. *New England Journal of Medicine,* **291,** 481–6 (1974).
43. Sobel, B. E., Bresnahan, G. F., Shell, W. E., and Yoder, R. D. Estimation of infarct size in man and its relation to prognosis. *Circulation,* **46,** 640–8 (1972).
44. Sobel, B. E., Larson, K. B., Markham, J., and Cox, J. R., Jr. Empirical and physiological models of enzyme release from ischemic myocardium. In *Computers in Cardiology,* IEEE Computer Society, Long Beach, California, 1974, pp. 189–95.
45. Sobel, B. E., Markham, J., Karlsberg, R. P., and Roberts, R. The nature of disappearance of creatine kinase from the circulation and its influence on enzymatic estimation of infarct size. *Circulation Research,* **41,** 836–44 (1977).
46. Sobel, B. E., Markham, J., and Roberts, R. Factors influencing enzymatic estimates of infarct size. *American Journal of Cardiology,* **39,** 130–2 (1977).
47. Sobel, B. E., Roberts, R., and Larson, K. B. Considerations in the use of biochemical markers of ischemic injury. *Circulation Research Supplement I,* **38,** I-99–I-106 (1976).
48. Sobel, B. E., Roberts, R., and Larson, K. B. Estimation of infarct size from serum MB creatine phosphokinase activity: Applications and limitations. *American Journal of Cardiology,* **37,** 474–85 (1976).
49. Sobel, B. E. and Shell, W. E. Diagnostic and prognostic value of serum enzyme changes in patients with acute myocardial infarction. In *Progress in Cardiology* (Eds. Yu, Paul N. and Goodwin, J. F.), Lea and Febiger, Philadelphia, 1975, pp. 165–98.
50. Sobel, B. E. and Shell, W. E. Serum enzyme determinations in the diagnosis and assessment of myocardial infarction. *Circulation,* **45,** 471–82 (1972).
51. Sobel, B. E., Weiss, E. S., Welch, M. J., Siegel, B. A., and Ter-Pogossian, M. M. Detection of remote myocardial infarction in patients with positron emission transaxial tomography and intravenous ^{11}C-palmitate. *Circulation,* **55,** 853–7 (1977).
52. Ter-Pogossian, M. M. Limitations of present radionuclide methods in the evaluation of myocardial ischemia and infarction. *Circulation Supplement I,* **53,** I-119–I-121 (1976).
53. Varat, M. A. and Mercer, D. W. Cardiac specific creatine phosphokinase isoenzyme in the diagnosis of acute myocardial infarction. *Circulation,* **51,** 855–9 (1975).
54. Weiss, E. S., Ahmed, S. A., Welch, M. J., Williamson, J. R., Ter-Pogossian, M. M., and Sobel, B. E. Quantification of infarction in cross sections of canine myocardium *in vivo* with positron emission transaxial tomography and ^{11}C-palmitate. *Circulation,* **55,** 66–73 (1977).
55. Yasmineh, W. G., Pyle, R. B., Cohn, J. N., Nicoloff, D. M., Hanson, N. Q., and Steele, B. W. Serial serum creatine phosphokinase MB isoenzyme activity after myocardial infarction. Studies in the baboon and man. *Circulation,* **55,** 733–8 (1977).

APPENDIX

Calculation of Infarct Size from Serial Changes in MB or Total CK (Expressed as CK-g-equiv.)

1. $E(t)$ —activity of CK in blood (iu ml^{-1})
   $f(t)$ —rate of change of CK activity due to enzyme being released by heart (iu min^{-1} ml^{-1})

k_d —fractional rate of disappearance of CK from blood (min^{-1})

$$\frac{dE}{dt} = f(t) - k_d E$$

2. CK_r —cumulative activity of CK released by heart up to time T (iu ml^{-1})

$$CK_r = \int_0^T f(t)\,dt = E(T) + k_d \int_0^T E(t)\,dt$$

Note that CK_r is a function of T.

3. K —proportionality constant (ml CK-g-equiv. kg^{-1} iu^{-1})
   DV —distribution volume/unit body weight (ml kg^{-1})
   P_{CK} —proportion of CK released into blood compared to CK depleted from the heart (iu CK-g-equiv.$^{-1}$)/(iu g^{-1})
   CK_N—CK activity in a homogeneous section of normal myocardium (iu g^{-1})
   CK_I —CK activity in a homogeneous section of infarcted myocardium (iu g^{-1})

$$K = \frac{DV}{P_{CK}(CK_N - CK_I)}$$

4. IS —infarct size (CK-g-equiv.)
   BW —body weight (kg)

$$IS = (K)(BW)(CK_r)$$

5. Given N observed values of CK activity, $E(t_i)$, $i = 1, 2, \ldots, N$, CK_r can be estimated from (see 2 above)

$$CK_r \approx \sum_{i=1}^{N-1} \bar{f}_i \Delta t_i = \sum_{i=1}^{N-1}\left(\frac{\Delta E_i}{\Delta t_i} + k_d \bar{E}_i\right)\Delta t_i = E(t_N) + k_d \sum_{i=1}^{N-1} \bar{E}_i \Delta t_i,$$

where

$$\Delta t_i = t_{i+1} - t_i,$$
$$\Delta E_i = E(t_{i+1}) - E(t_i),$$
$$\bar{E}_i = \frac{E(t_{i+1}) + E(t_i)}{2},$$

and

$$\bar{f}_i = \frac{f(t_{i+1}) + f(t_i)}{2}.$$

Note that $E(t_1) = 0$ is assumed.

Values Currently Used for Constants (Body Weight Expressed in kg)

Infarct size based on CK-MB	Infarct size based on total CK
$k_d(\text{CK-MB}) = 0.0015$	$k_d(\text{CK}) = 0.001$
$DV^a = 44$ ml kg^{-1}	$DV^a = 44$ ml kg^{-1}
$P_{\text{CK-MB}}{}^d = 0.15$	$P_{\text{CK}}{}^d = 0.15$
$\text{CK}_N\text{-MB} = 224$ iu g$^{-1\,b}$	$\text{CK}_N = 1500$ iu g$^{-1\,b}$
$\text{CK}_I\text{-MB} = 44$ iu g$^{-1\,c}$	$\text{CK}_I = 315$ iu g$^{-1\,c}$
$K = 1.6$	$K = 2.5 \times 10^{-1}$

[a] Estimated as plasma volume for man (Nachman, H. M., James, G. W., III, Moore, J. W., and Evans, E. I. Comparative study of red cell volumes in human subjects with radioactive phosphorus tagged red cells and T-1824 dye. *Journal of Clinical Investigation*, **29**, 258 (1950)).

[b] Measured directly.

[c] Calculated from percentage of myocardial CK depleted measured directly in conscious dogs 48 hours after coronary occlusion.

[d] Calculated by analogy based on the empirical relation between CK released calculated from serum changes, and myocardial CK depletion measured directly in conscious dogs with coronary occlusion.

CHAPTER 12

The prediction of infarct size
W. E. Shell and M. F. Groseth-Robertson

INTRODUCTION	291
MODELS OF ACUTE MYOCARDIAL INFARCTION	293
ASSUMPTIONS ASSOCIATED WITH PREDICTION OF INFARCT SIZE	295
USE OF THE LOG-NORMAL FUNCTION TO PROJECT CK CURVES	296
The Log-normal Function	296
Non-linear Solutions	297
Current Methods	298
Initial Estimates	298
Parameter Limitation	303
Definition of Zero Time	303
Computer Algorithms	304
VALIDATION OF INFARCT SIZE PREDICTIONS	305
PROBLEMS ASSOCIATED WITH UTILIZATION OF LOG-NORMAL FUNCTION	310
Choice of Initial Parameter Estimates and Limitation of Parameter Range	310
Defining the Initial Enzyme Elevation	310
Delay in Prediction Time	310
Imprecision of Early Data	311
Correlation Between Observed and Predicted Infarct Size	311
PHYSIOLOGIC MODELS	312
REDUCTION OF PREDICTION TIME	313
DEVIATIONS OF PREDICTED AND OBSERVED CK CURVES	314
CONCLUDING COMMENTS	315
REFERENCES	316

INTRODUCTION

Myocardial infarction is a dynamic process[47] during which a critical reduction in blood flow initiates mechanical dysfunction,[34] electrolyte imbalance,[14] electrical disequilibrium,[24] and dissolution of cellular membrane systems.[13] If the imbalance between oxygen supply and demand is temporary, the injury to the myocardial cells is reversible and the cells will return to their normal function. However, if the imbalance is prolonged, the myocardial injury becomes irreversible and the cells will die. Recent experimental and clinical attention has been directed toward aborting myocardial injury during the reversible stage in order to prevent necrosis.[13,24]

To assess the effects of interventions designed to reduce infarct size, a number of methods have evolved which provide an estimate of the volume of both the zone of initial ischaemic injury and the zone of subsequent necrosis.[11,24,40,41] The methods depend on the quantification of a measurable property of either ischaemic or necrotic myocardium. For example, an early sign of myocardial injury is epicardial or praecordial ST segment elevation.[24,25] The acute electrical injury observed on the epicardial surface is a reversible sign; if myocardial blood flow is rapidly returned to normal, the ST segment elevation returns to normal. Maroko, Braunwald and colleagues[24,25] have constructed an estimate of the area of acute electrical disequilibrium by mapping the number of sites with ST segment elevation on either the epicardial or praecordial surface; this estimate is intended to be a reflection of the extent of acute myocardial injury. Other markers of reversible alteration of the myocardium include altered regional perfusion as reflected by distribution of microspheres,[10] distribution of radiolabelled free fatty acid,[50] and distribution of thallium-201.[52] Markers of the zone of subsequent necrosis have also been utilized (the term 'infarct size' is generally reserved for estimates of the volume of the zone of necrosis while 'protection of ischaemic myocardium' refers to modifications of signs of reversible injury). One of the first methods to assess the mass of the necrotic zone was a determination of the distribution and extent of myocardial creatine kinase (CK) depletion;[16] loss of CK from myocardial tissue reflects cell necrosis[44] and the amount of CK depletion quantitatively defines the amount of necrosis.[16,40] Other markers of the necrotic zone include praecordial Q wave distribution,[29] distribution of technetium 99-m pyrophosphate,[4] histologic mapping,[9] anatomic reconstruction,[17] and distribution of radiolabelled white blood cells.[53] Therapeutic interventions have been assessed by comparing a measure of acute injury to a measure of subsequent necrosis;[24,41] for example, early epicardial ST segment elevation has been extensively compared to myocardial CK depletion 24 hours later.[23,24] The essential concept of these experimental protocols is that an early sign of injury is used to predict subsequent necrosis; the models which have been used to assess potential therapies have intrinsically involved prediction of future events.

The mass of the necrotic zone can be estimated by analysis of serial serum CK activity;[3,40,48] the theory and application of these concepts have been extensively discussed in preceding chapters and will not be repeated. However, serum CK estimates of infarct size are measures of the necrotic zone and thus take hours or days to evolve. Assessment of the efficacy of therapeutic interventions is thereby hindered by the inability to compare signs of acute injury to subsequent signs of completed necrosis. To circumvent this disadvantage, methods for the predicted CK infarct size have been developed.[41] Early serum CK activities are utilized to project, by a digital computer generated curve fit, expected CK activities. The projected CK activities can be used to form an estimate of 'CK infarct size' which represents a prediction of expected events. The predicted CK infarct size is an early estimate of the zone of injury while the observed CK infarct size is a measure of the zone of necrosis. By comparison of predicted CK activities to those subsequently observed, the effects of therapeutic interventions on the rate and amount of CK released into serum can be evaluated. This discussion will focus on the assumptions associated with infarct size prediction, the current techniques utilized to construct CK curve projections, the problems associated with CK predictions, potential models of CK curves, validation of CK projections, utilization of CK prediction to assess potential therapies, and finally, interpretation of CK activities following therapy.

MODELS OF ACUTE MYOCARDIAL INFARCTION

To evaluate any predictive technique, whether it involves CK curve projections, ST segment changes, or distribution of radionuclides, potential models representing the process of acute myocardial infarction must be considered. The simplest model of acute myocardial infarction is acute coronary artery occlusion in a well-trained conscious dog; the onset of necrosis may be precisely timed, the progression of ischaemia to necrosis uniform and, if the dog is well trained to lie quietly, the haemodynamic milieu is stable. In this model, the events are predictable and early predictions are likely to reflect the extent of necrosis which is measured many hours later. Any perturbation of the system can be timed and its effects elucidated (Figure 1).

In man, the process of acute infarction is not as simple especially since the events which precipitate the acute event are unknown. Several theories have been advanced to explain the genesis of myocardial infarction, these include acute thrombotic coronary occlusion,[6] intermittent spasm of a narrowed coronary artery,[18] and a prolonged imbalance between myocardial oxygen supply–demand in the presence of a fixed stenosis.[30] A sudden complete thrombotic occlusion of a major coronary artery represents the closest approximation to simple mechanical occlusion of a coronary artery. A myocardial infarction initiated by a sudden coronary occlusion in which the metabolic, haemodynamic and hormonal milieu either remains stable or progresses through predictable sequences is likely to result in early electrical, enzymatic, or radionuclide-estimated signs which could be used to predict the ultimate volume of necrotic cells. However, if the metabolic, haemodynamic or hormonal environment is non-uniform and unpredictable, early signs of injury are unlikely to be predictive of future degrees of necrosis.

Initiation of myocardial necrosis by intermittent spasm of coronary arteries or reversible imbalances between oxygen supply and demand will introduce further complexities into the relation between early predictions of

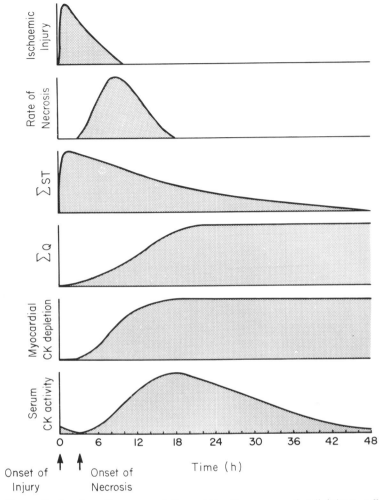

Figure 1. A schematic representation of the time curve of cell injury, cell necrosis and selected markers of the ischaemic process. The presumed model is a single complete coronary occulusion

injury and subsequent estimates of necrosis. If spasm of coronary arteries initiates necrosis and subsequently the spasm is relieved, the injured myocardium will be exposed to ischaemia reflow cycles at unpredictable times and of unpredictable durations. Similarly, if imbalances between oxygen supply and demand initiate myocardial necrosis and if the imbalances are reversible with unpredictable durations, then early events are unlikely to predict reliably conditions hours or days later.

Insufficient attention has been given to the relationship between the events responsible for initiating myocardial necrosis and the reliability of all predictive techniques. These considerations are not unique to utilization of early enzyme changes to predict subsequent necrosis. If early distribution of thallium-201 is used to predict subsequent deposition of technetium 99-m pyrophosphate or if an early praecordial map of ST segment elevation is used to predict the subsequent praecordial Q wave map, then the intervening events must be continuous and predictable. It will be of extreme importance to gather information on the relationship between early estimates of injury and subsequent measures of necrosis in patients exposed to a minimum of therapeutic interventions; assessment of the natural history of these relationships will result in a better appreciation of the mechanisms initiating, propagating, and limiting myocardial necrosis.

ASSUMPTIONS ASSOCIATED WITH PREDICTION OF INFARCT SIZE

Before discussing the specific methods utilized to project CK curves, consideration of assumptions which are implicit in attempts at infarct size prediction would be useful (Table 1). First, the initiation of the process of injury occurs as a discrete event; the event could be a thrombotic occlusion, arterial spasm, or critical imbalance of oxygen supply–demand. The mechanism of initiation of myocardial infarction is unimportant; however, a discrete point in time must exist to serve as a reference point in order to evaluate the precision of any prediction. Second, the time of onset of either injury or necrosis can be defined. The process of injury must have a discrete onset and the technique of prediction must be capable of defining when the process began. If the time of onset cannot be defined, the prediction will be quantitatively incomplete, will only partially reflect total necrosis and may give a misleading impression of the potential zone of necrosis. In this connection, a major practical problem in the interpretation of praecordial ST segment elevation data in man is the inability to relate changes of ST segment elevation to the time–activity profile of necrosis; the investigator never knows whether the observed changes occurred early in the process of necrosis or late in the evolution of injury to necrosis. Similarly, there is virtually no information relating to the dynamic nature of perfusion images to the process of myocardial infarction. Thus as predictive tools these techniques will have limited application until the images can discern

Table 1. Assumptions Associated with CK Infarct Size Predictions

1. Initiation of myocardial necrosis is a discrete event
2. Time of onset of injury–necrosis can be defined
3. The progression of myocardial injury to necrosis is continuous
4. A model exists which describes the continuous process

dynamic changes on a minute–minute basis rather than their current day–day time base.

The third and most important assumption is that the events associated with myocardial necrosis are continuous and not subject to unpredictable perturbations. The reduction in flow is either complete or the return of flow is predictable without wide fluctuations. The haemodynamic milieu is either stable or predictably changing; if there is wide unpredictable fluctuation in blood pressure, wall stress, heart rate, or oxygenation, any predictive technique will fail because the process of acute myocardial injury depends on these phenomena.

The fourth set of assumptions is that a model can be defined which describes the expected continuous processes. The model can be either an empiric model or physiologic model. In empiric models, equations are used which describe the data with reasonable precision; these models provide little direct information concerning the physiologic system. Physiologic models depend on description of a physical system and derivation of equations which describe the system. The observed CK infarct size is an example of a physiologic model based on a one-compartment system. Physiologic models (see also Chapter 11) provide potential insight into the biologic system. However, these systems become mathematically complex because of the complexity of the physiologic systems. Physiologic models have not been used thus far to predict infarct size but may be used in the future. Currently, projections of CK curves utilize an empiric model based on the log-normal function. Although a physiologic model would be desirable, the available models are insufficiently complete to be of predictive value. Accordingly, a discussion of the log-normal model is necessary as a standard.

USE OF THE LOG-NORMAL FUNCTION TO PROJECT CK CURVES

The Log-normal Function

Projection of CK curves[41] has utilized the empiric log-normal model (Figure 2). The log-normal equation is non-linear and has three parameters, i.e. coefficients (in this chapter, the term coefficient and parameter are used interchangeably). The function is geometrically related to the familiar normal distribution except that it is skewed logarithmically to the right. To project a CK curve, enzyme activity observed during the first 7 hours is used to generate a computer curve fit which results in computer-defined values for the coefficients (b, c, and d), and an equation relating time and enzyme activity is derived. The derived equation can then be used to find projected CK values beyond 7 hours: A predicted infarct size is determined from the projected values. In the example in Figure 2 the data from the first 7 hours

THE PREDICTION OF INFARCT SIZE

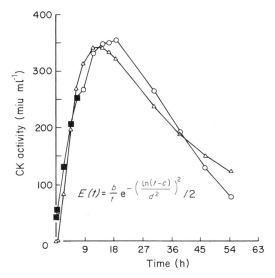

Figure 2. Projection of CK curve from the log-normal function. Seven hours of observed data (■) were utilized to generate the predicted curve (△); the CK activities subsequent to the predicted point are also plotted (○). In the equation $E(t)$ is CK activity at any time t, e the base of the natural logarithm (ln), while b, c, and d represent non-linear coefficients

(solid squares) was used to generate the predicted curve (open triangles) which is compared to the real values (open circles). An observed infarct size was determined from the observed values (solid squares and open circles) which was 37.5 CK-g-equiv. Finally, a predicted CK infarct size was estimated from the computer-generated values (open triangles) which was 37.5 CK-g-equiv. The observed CK infarct size is an estimate of completed necrosis since all observed activities are used (24–72 hours) while the predicted infarct size is an estimate of the jeopardized zone since only early values (7 hours) are used.

Non-linear Solutions

The log-normal function is non-linear so that finding appropriate computer-generated values for the parameters is more complex than when dealing with linear equations.[1] Unambiguous solutions to either a least-squares fit or a chi-square minimization procedure do not exist and a family of potential solutions can be generated. For example, when a straight line is found by linear regression to fit a set of data, only one line will represent the best-fit least-squares solution.[7] However, when an investigator attempts to find a non-linear equation to describe a set of data, a series of solutions can be found depending on the mathematical techniques selected to find the solution. This can account for discrepancies of results between investigators. The important determinants of the non-linear solutions include: (i) choice of initial parameter estimates; (ii) limitation on the range of coefficients; (iii) definition of the first data point, and (iv) nature of the convergence algorithm.

Current Methods

Before discussing the factors influencing precision of CK curve projections, an outline of the current methods utilized by the authors would be useful as a reference for other techniques. The illustration described in Figure 2 was generated from: (i) initial parameter estimates defined by a linear approximation to the log-normal function; (ii) parameter limits set as empiric constants; (iii) initial CK activities taken as the first observed value which showed either a 10% rise in activity or which was less than 200 miu ml^{-1}; and (iv) a computer algorithm based on a gradient search method as described in the BMDX 85 computer program.[40] These series of algorithms have evolved over several years and represent the current approach to generation of CK curve projections. A number of other approaches have been utilized to generate initial estimates, to set parameter limits, and to choose a zero value.[8,27] These different methods will be discussed subsequently and their deviations from this description also considered. The validation of infarct size projection is quite sensitive to the methods used to generate those projections.[36] Any discussion of validation of prediction, therefore, must be evaluated in terms of the methods utilized to generate the predictions. Therefore, a more detailed description of initial estimates, parameter limitation, zero time definition, and computer algorithms is provided.

Initial Estimates

To begin a non-linear least-squares approximation, the computer must be given initial starting values for the parameters of the equation. Initially, a 'biologic algorithm' was utilized for choosing the initial estimates.[41] The parameters of the log-normal function were related to geometric characteristics of the curves (Figure 3). The b parameter was related to the peak activity, the c parameter to the time to peak, and the d parameter to the width of the curve. Empirically, it was observed that the b parameter could be estimated from the CK activity 7 hours after initial CK elevation; thus b was set equal to 3 times the CK value at 7 hours multiplied by an estimate

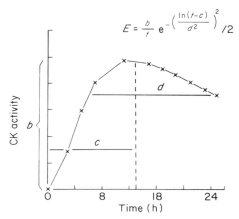

Figure 3. The non-linear coefficients of the log-normal function are related to the geometry of the curve. The b coefficient is related to peak activity, the c coefficient to the time to peak, and the d coefficient to the width of the curve

of time to peak in minutes. The c parameter was related to time to peak: It was empirically observed that the time to peak could be divided between two classes of infarcts—large and small infarcts; a large infarct was one in which the 7 hour CK activity was greater than 300 miu ml^{-1}, whereas a small infarct was one in which the 7 hour value was less than or equal to 300 miu ml^{-1}. When the infarct was large, the time to peak could be approximated as 1800 minutes, and for small infarcts time to peak could be estimated as 1300 minutes. Accordingly, the c parameter could be approximated by taking the natural logarithm of the time to peak. The d parameter was arbitrarily begun at 0.9. This approach was termed the constant biologic algorithm and is an approach utilized by several investigators.[8,27,41,43]

The linear approximation to the log-normal function has recently been evolved[36,51] with which a unique linear solution can be obtained. Although the parameters generated by this approximation are not precisely those generated by the non-linear iterative solution, they are sufficiently close to utilize as initial approximations. Large and small infarcts are again categorized in a manner analogous to the biologic approach; and once an estimate of c parameter is obtained, estimates for b and d can be obtained directly from the linear approximation. The advantage of this approach is that a unique set of initial estimates is obtained which is quite close to the parameters derived from the non-linear least-squares fit.

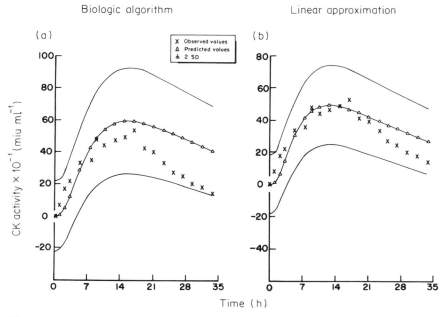

Figure 4. (a) A CK curve projection utilizing the biologic algorithm (see text for explanation). (b) A CK curve from the same data utilizing the linear approximation to the log-normal function to set initial parameter estimates

The linear approximation assures convergence of the solution to a biologically meaningful result. A biologically meaningful result will not always result from the biologic algorithm. For example, when identical data are used to find a solution by either the biologic algorithm or the linear approximation, deviations between the solutions can be frequently obtained (Figure 4). The linear approximation to the log-normal function consistently results in a close correlation between predicted and observed CK curves compared to the biologic algorithm.

When initial estimates for the coefficients are chosen by techniques which differ from either the biologic algorithm or the linear approximation, grossly inconsistent infarct size projections will be obtained (Figure 5). Although this represents an extreme deviation of the choice of initial estimates, substantial error can be introduced into the infarct size projections when relatively minor deviations in the choice of initial estimates are used, and subconscious bias can enter into the construction of an infarct size projection. The effects of therapeutic interventions can be inaccurately defined if systematic unique methods for parameter initialization are not utilized. As

THE PREDICTION OF INFARCT SIZE

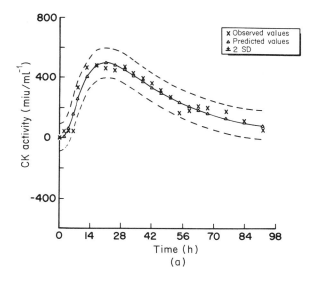

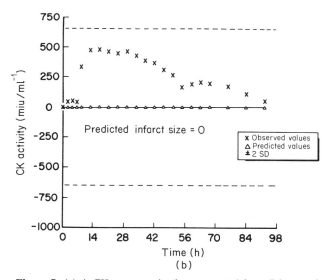

Figure 5. (a) A CK curve projection generated from 7 hours of observed CK data. The initial estimates of the coefficients were generated from a linear approximation to the log-normal function. The initial estimate for the b coefficient was 5×10^6, c was 7.2, and d was 0.9. (b) The data from the same patients were used to generate a totally inconsistent prediction. The initial estimates for the coefficients are taken as $b = 1000$, $c = 4.2$, $d = 0.6$ which are substantially different than the final solution

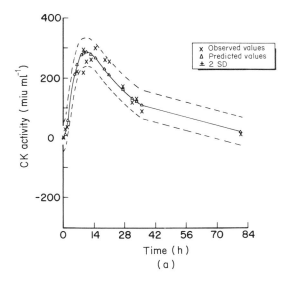

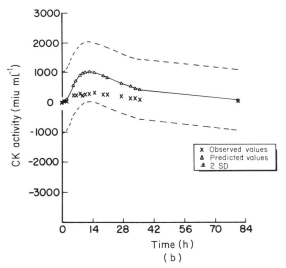

Figure 6. (a) A CK curve projection generated from 7 hours of observed CK data. The limitations on the c parameter were 7.1 to 7.3. (b) The data from the same patient was used to generate a projection. There were no limitations on the range of the c coefficient

methods evolve for projection of CK curves, it will be important to stress consistency of methods and uniqueness of solutions so that subjective biases do not enter into the construction of the predicted curve.

Parameter Limitation

Consistent CK curve projections depend on limitations of the parameter range. The most important coefficient which must be limited is the c parameter. The range of the c parameter must be kept within a narrow limit so that convergence of the solution is obtained. Most investigators have allowed the c parameter to range between 7.1 to 7.3 for small infarcts, and between 7.4 to 7.6 for large infarcts.[41] When the parameter range is not limited, inconsistent infarct size estimates occur, particularly when the biologic algorithm is utilized to select the initial parameter estimate. For example, an appropriate infarct size projection is obtained when limitations of the parameters are placed; however, the projection is completely inconsistent when the limitation of the parameter range has been eliminated (Figure 6).

Definition of Zero Time

An important variable in CK curve projections is the choice of the initial CK value. When total CK activity is examined in patients after acute myocardial infarction, definition of the point of initial CK rise is not a simple problem. There are two phenomena which occur when a patient is put to bed rest after the onset of a myocardial infarction (Figure 7). First, there is an initial fall in serum CK activity as the patient is put to bed rest; the decrease is due to a decline of basal serum CK activity which occurs in normal patients. The falling baseline is counterbalanced by the rising CK activity resulting from enzyme release from the heart. The point at which enzyme activity appears to increase is taken as zero. However, the time of the zero value will depend on the relative relation between the initial falling baseline and the initial rate of increase. With total CK activity, the time from chest pain to the time of the zero value will depend on the physical activity of the patient prior to the infarct in addition to the initial rate of rise of the CK activity induced by the infarction. The problem is further complicated in those patients who enter the hospital after the initial enzyme rise has occurred (Figure 8). In those patients definition of zero will become the first observed data point. However, if the observed first data point is substantially removed from true zero, the curve will have been shifted in time and the estimate of the time to peak from setting the c parameter will be substantially erroneous which will result in a fallacious infarct size projection. For example, when a delayed zero is utilized as the first point,

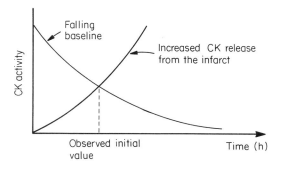

Figure 7. Dependence of the observed initial CK elevation on the falling baseline for CK. Patients placed at bed rest will demonstrate a decline in serum CK activity. Patients with acute infarction will experience a rising serum CK activity. The time of the initial rise will be the result of the rate of fall and the rate of rise. Slight shifts in either rate will change the time of the observed initial value

the infarct size projection is an overestimate of the true predicted CK curve. The incorrect prediction can result in the judgment that a therapeutic intervention is beneficial when in fact there is merely an overestimate of the CK projection by an inappropriate choice of the zero value. In evaluation of all therapeutic interventions utilizing projected CK curves, there must be careful attention to how the zero point is chosen. Currently, we do not feel that infarct size projections are valid when the initial total CK activity is greater than 200. When CK-MB is utilized for infarct size projection, this problem is minimized since there is a stable baseline for CK-MB with an upper limit of normal of 5 miu ml^{-1} (mean ± SD).

Computer Algorithms

The importance of the computer algorithms used to define the non-linear solution has not been investigated. The current infarct size projection programs have utilized the BMDX 85 computer program for arriving at the non-linear solution.[15] This program involves a gradient search method and a non-linear least-squares approach. There are other potential algorithms including chi-square minimization.[46] The influence of these techniques for convergence of the non-linear solutions thus far remains unexamined. It is quite likely that techniques of computer convergence will have substantial influence on the precision of the CK projections.

A summary of current techniques by major centres is provided in Table 2.

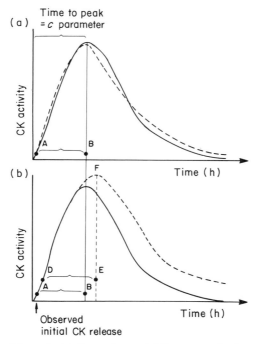

Figure 8. (a) Dependence of CK curve projection on correct identification of point of initial rise. Point A represents the point defined as initial elevation. Point B represents the time of peak CK activity. The segment A–B represents the time to peak activity which is set by limiting the range of the c parameter. The solid line represents an observed curve and the broken line a predicted curve. (b) Delayed identification of initial elevation. If the initial CK rise is observed at point D instead of point A, the time to peak will be estimated as D–E rather than A–B. The predicted curve will then peak at a higher activity with a delayed time to peak. The observed curve will fall below the predicted curve and an inappropriate assertion of decreased CK release made

VALIDATION OF INFARCT SIZE PREDICTIONS

The validation of infarct size predictions depends on a number of observations including: (i) the type of patient or animal model utilized as 'control', (ii) the time of observation, and (iii) the type of mathematical model. Evaluation of the infarct size predictions with early serial CK activities have been examined in the conscious dog,[41] patients with smooth CK curves,[37,41]

Table 2. Current Prediction Algorithms

Investigator	Initial estimate	Parameter limits	Initial CK value	Computer program	CK-MB activity
Cedars-Sinai[35]	Linear approximation	Small infarcts: 7.1–7.3 Large infarcts: 7.4–7.6	10% increase in CK ≤ 200 iu l.$^{-1}$	BMDX 85	No
Washington University[8]	Biologic algorithm	Small infarcts: 7.1–7.3 Large infarcts: 7.4–7.6	-20 iu l.$^{-1}$ increase No exclusion for high values	BMDX 85	No
Cornell University[30]	Biologic algorithm	Small infarcts: 7.1–7.3 Large infarcts: 7.4–7.6	Undefined Exclude if CK > 50 iu l.$^{-1}$	BMDX 85	No

patients with predicted CK curves matched to treated patients,[8,11,33] patients randomly allocated to a non-treated control group,[18,27,51] patients treated with routine but randomly applied therapies,[32,37] and patients randomly allocated to two treatment groups.[20,51] Each attempt at validation has examined a different model of the process of myocardial infarction and provides a different aspect of the process of validation.

The use of the single occlusion in the conscious dog was the first attempt at validation[41] of CK predictions. Data obtained during the first 5 hours after coronary artery occlusion were fitted to the log-normal function and the CK curve projected to 24 hours. In the conscious dogs with uncomplicated myocardial infarction, the CK curves were smooth, with a single peak having uniform decay rates. The correspondence between the observed CK curves and those predicted was quite close (Figure 9). In 11 animals the correlation between the observed CK infarct size and myocardial CK depletion was 0.98 while the correlation between the infarct size predicted 5 hours after coronary artery occlusion and the anatomic infarct size was 0.96.[41] The mean predicted infarct size was 21.4±4.5 (mean±SEM) while the mean observed infarct size was 23.2±5.0; the mean difference was −4.5%. Thus in this simple model with smooth CK curves and uncomplicated myocardial infarction, a close correlation between predicted and observed CK infarct size can be achieved.

The second attempt[37,41] to validate CK curve projections was in patients with smooth CK curves in which there were no obvious second peaks in the CK time–activity curve. These patients were selected to approximate a group of patients with uncomplicated patterns of necrosis—at least as viewed from the enzyme data. The simple question asked of this data

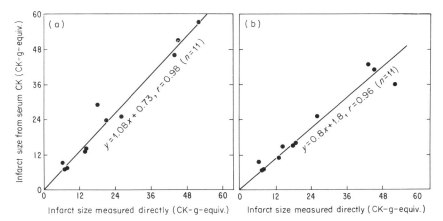

Figure 9. (a) The relations between infarct size calculated from observed serum CK measurements (ordinate) and infarct size estimated from myocardial CK depletion at autopsy (abcissa) in conscious unmedicated dogs. (b) The relation between infarct size predicted 5 hours after coronary artery occlusion (ordinate) and infarct size estimated from myocardial CK depletion 19 hours later (abcissa) in the same dogs

was—if the CK curves are smooth, can they be predicted by the log-normal function? In this group of patients,[41] there was a close correspondence between predicted and observed CK infarct size (Figure 10, $r = 0.93$). The mean observed infarct size was 33 ± 5 while the mean predicted infarct size was 35 ± 6. These data indicated that in patients with apparently uncomplicated profiles of necrosis, smooth CK curves could be projected 7 hours after initial CK rise with the log-normal function.

The third attempt at validation of CK prediction[8,11,33] involved the so-called 'matched control group'. These patients were utilized as a control group to evaluate potentially theraputic interventions.[8,33,43] Therefore, this experimental design was directed towards understanding the effects of therapies on CK release rather than specifically understanding CK curve projections; the 'matched controls' are included in this discussion because of the widespread application of the experimental design. With matched controls, patients with smooth curves were retrospectively selected as control patients; the range of predicted CK infarct sizes was selected to match the predicted values of the treated group. The rationalization for this approach is that the control group verifies the mathematical validity of the predictions within the range of treated subjects; thus, if the treated patients all had small predicted infarcts, the matched controls attempted to illustrate the range of deviations between predicted and observed CK curves which would be expected for small predictions. The choice of matched controls is not based on any presumed model of necrosis but seeks only to control potential

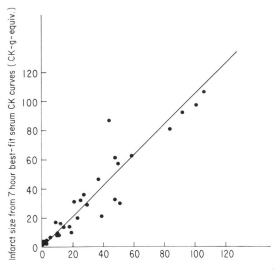

Figure 10. Relation between infarct size predicted 7 hours after initial CK rise (ordinate) and infarct size determined from the subsequently observed CK activities (abcissa) in patients with smooth CK curves

variance introduced by the non-linear mathematical techniques. It is, therefore, not surprising that close correlations between predicted and observed infarct size are achieved by this technique. These correlations merely indicate that if the curves are smooth and are within the range selected, the mathematical variance between predicted and observed curves will be within the demonstrated range.

The fourth attempt at validation[18,27,51] is generation of a non-treated randomly allocated control group of patients. This control group is designed to answer the questions: (i) how often are smooth single-peak curves observed in patients with acute myocardial infarction; (ii) how well does the log-normal function fit randomly generated untreated patients; and (iii) how close are the predicted and observed infarct sizes when a non-treated randomly generated patient group is obtained? There have been two attempts to generate an *untreated* control group; the study of Varonkov et al.[51] indicated a correlation of 0.96 between predicted and observed CK infarct size with a mean 5.3% difference, while the recent study by the authors indicated a correlation of 0.94 with a similar per cent deviation.[37] One of several potential pitfalls in the interpretation of this data is that for patients to remain untreated, their clinical course must remain stable without major complications. This group of patients represents the natural history of necrosis in uncomplicated patients.

The fifth attempt[32,37] to generate patients suitable for validation of CK predictions represents the group of randomly treated control patients. With this approach patients are exposed to the variety of commonly used therapeutic agents including oxygen, analgesics, diuretics, digitalis, and vasodilators without regard to a specific protocol. The assumption is that either the therapies do not influence CK curves or that the effects will be normally distributed between good and deleterious effects. This assumption is untested and unlikely to be true; close correlations between observed and predicted infarct size should not be expected in this population. There have been two series of patients exposed to random therapies. The observations of Mueller and associates[32] indicate that there can be wide deviations between observed and predicted curves; the experience at the authors' centre[37] indicates that the correspondence between predicted and observed curves in this group is poor ($r = 0.74$). These patients represent complicated patients with multiple waves of necrosis who are exposed to multiple vasoactive therapies.

The sixth attempt[20,51] at validation of CK prediction involves selection of two groups of patients who are assigned to two treatment groups—one of the treatment groups receiving only placebo or minimally effective therapy. The deviation of the observed from the predicted CK infarct size in one treatment group can then be compared to the deviations in the second group. In this approach no assumption of the mechanism of occlusion is made but that the multiple potential mechanisms will be randomly allocated between treatment groups. Two trials with this approach at validation have occurred. The studies of Morrison[27] have indicated a close correlation between observed and predicted infarct size in his minimally treated group ($r = 0.96$) and the studies of Magnusson et al.[20] using furosemide as the near placebo also showed close correlation ($r = 0.96$).

The six sets of control studies have several features in common. When smooth curves are selected either in the dog or man, the log-normal function can be used to generate a reliable projection of the CK curve. In uncomplicated randomly generated control patients, the CK curves are usually sufficiently smooth to generate a reliable projection of the CK curve. Patients demonstrating major complications, i.e. new waves of necrosis or those exposed to vasoactive agents, will demonstrate deviations of observed and predicted curves—these deviations representing either spontaneous changes in the rate of necrosis or the effects of the therapeutic agents. The utilization of CK predictions in complicated patients requires either a simultaneous placebo-treated group or comparison of the effects of two therapies. Finally, the projection of a CK curve, 7 hours after initial CK rise assumes that the curves will follow a smooth continuous pattern; deviations of the curves represent either spontaneous deviations in the rate of CK release, i.e. the rate of necrosis, or the influence of therapy.

Any predictive technique will experience similar problems of validation and the greatest utility of such predictions may lie in increasing our understanding of how often deviations in the expected rate of necrosis occur and what precipitates the deviations.

PROBLEMS ASSOCIATED WITH UTILIZATION OF LOG-NORMAL FUNCTION

Choice of Initial Parameter Estimates and Limitation of Parameter Range

Since non-linear curve-fitting techniques are sensitive to both choice of initial parameters and limitations of their range, these methods must be rigidly and consistently defined to avoid variation and subconscious bias. The routine application of the prediction method will require an appreciation and correction of the potential sources of error.

Defining the Initial Enzyme Elevation

A generally unappreciated problem in finding reproducible, precise CK infarct size predictions is the difficulty in defining the initial CK elevation. CK curve projection, utilizing total CK activity, requires 7 hours of observation from the 'initial CK elevation'. When the time of initial CK elevation is poorly defined or if a patient is first observed after CK activities are already increasing, substantial error is introduced into the prediction. The definition of the zero point is difficult utilizing total CK activity and represents a major limitation of this technique. Strict criteria should be used to define the zero point to avoid bias. Future utilization of CK-MB activity to project CK curves may decrease the influence of selection of the zero point since a zero point of 5 miu ml^{-1} can be arbitrarily defined and computer generated.[54]

Delay in Prediction Time

Projection of CK curves has been evolved as a method to aid in determining if potentially therapeutic interventions change the rate or amount of CK released into serum. Successful protection of ischaemic myocardium and reduction of infarct size requires therapy before completion of necrosis. Recent experimental and clinical observations suggest that the critical delay period between coronary artery occlusion and successful therapeutic intervention is less than 8 hours in the dog[2,26] and between 4 and 12 hours in man.[31] Thus intervention 10–15 hours after initial chest pain is likely to have only minimal effects. Infarct size prediction using projected CK curves requires 9–14 hours of observation since there is a 2–6 hour delay between chest pain and initial rise of total CK coupled with a 7 hour observation period required to project the curves. The disappointing results of early therapeutic trials may have resulted from the delayed therapy. Accordingly, utilization of CK predictions will be of marginal utility until the observation period required for projection is reduced.

Imprecision of Early Data

Projection of CK curves depends on the precision of the early data points. The non-linear curve-fitting procedure requires 4–8 data points obtained during the first 7 hours to project data approximately 70 hours into the future. Any imprecision introduced into the data because of imprecision in the enzyme assay, handling of samples, or mislabelling of time will result in difficulty with the curve-fitting algorithms and discrepancies between observed and predicted infarct size. The influence of noise on the precision of predictions will become increasingly important as techniques to reduce prediction time are evaluated.

Correlation Between Observed and Predicted Infarct Size

Although the correlation between observed and predicted CK curves for a group of patients is close and the mean deviation of observed and predicted infarct size is near zero, substantial discrepancy between observed and projected curves can occur in individual patients. The potential reasons for discrepancy between the observed and predicted curves include: (i) CK curves using total CK activity do not conform to a log-normal equation; (ii) definition of the zero point is nearly impossible using total CK activity; (iii) the biology of acute myocardial infarction is not continuous, thereby producing non-smooth, unpredictable CK curves; (iv) the techniques for producing the non-linear curve fit result in a poorly defined log-normal function. Comparison of discrepancies between observed and predicted curves in either animals or patients under a variety of circumstances is likely to lead to important insight into the natural history of acute infarction. Initial interest has been toward developing physiologic models to replace the log-normal function (see next section entitled 'Physiologic Models'). Regardless of the reasons for the discrepancy between individual curves, CK curve predictions can currently be used only with caution in individual patients. A major goal of future investigations will be to improve the correspondence between observed and predicted CK curves in individuals.

PHYSIOLOGIC MODELS

In order to improve the correspondence between observed and predicted CK curves, as well as to better understand the physiologic systems, equations based on physiologic models have been developed. The modelled equations have three goals: (i) to describe the physiologic system; (ii) to increase the correspondence between the amount of CK observed in serum and that lost from myocardium; and (iii) to provide a more precise predictive equation. Two types of physiologic models have been described. In one model,[49] the process of cell death is represented as a series of spheres releasing enzyme into the pericellular space and passive diffusion from the pericellular space followed by clearance of CK from a one-compartment vascular space. The rate-limiting step is considered to be passive diffusion into lymph. Solution of equations representing this model results in a diffusion path of 4 mm. The length of the diffusion path appears inconsistent with anatomic considerations. The distances between individual cells and lymph channels are of the order of microns rather than millimetres. Moreover, comparison of observed CK activities in serum have not been made with either tissue estimates of infarct size or with the results of the simple one-compartment model. The equations based on this model have been insufficiently defined to allow prediction of CK curves from early data and require 24–48 hours of data to solve the equations.

A second model which has been evaluated is the two-compartment model describing CK clearance from serum.[45] In this model, cell death is assumed to be a quantum instantaneous event with immediate non-rate-limiting movement into the vascular compartments. Clearance from the vascular spaces is described as a two-compartment system and the process of clearance is considered rate limiting. Solutions to the two-compartment system have been utilized to compare observed CK clearance rates to those obtained experimentally and to compare total CK appearance in serum to results obtained by the single-compartment model. In the work of Smith and colleagues[45] there is close correspondence between observed CK values and those estimated from the two-compartment model; moreover, the amount of CK observed in serum is 100% greater than that accounted for by the one-compartment model. In the model of Sobel, Markham, and colleagues,[22] the clearance rates defined by the two-compartment model displayed less variance than the clearance rates defined by the one-compartment model; these workers have not compared the estimated CK in serum to myocardial CK depletion. The two-compartment model appears to offer substantial promise as a means of describing the physiologic system; however, the two-compartment model is computationally more complex and it has not been established that these models offer substantial improvement over the simple one-compartment model.

The analyses of physiologic models have been utilized to understand the physiologic systems and to improve the precision of observed CK infarct size measurements. The equations derived from physiologic models have been insufficiently defined to provide predictions of CK curves from early data. Moreover, the equations resulting from any physiologic model which realistically reflects the biologic system are complex with multiple parameters. The initialization of these parameters is a formidable mathematical task. Accordingly, empiric models similar to the log-normal equation are likely to remain the approach to CK curve projection for the foreseeable feature.

REDUCTION OF PREDICTION TIME

An important problem that requires solution before infarct size prediction becomes a useful clinical and research tool is reduction of prediction time. There have been two attempts at reducing the prediction time.[35,38,39] The first attempt[35] utilized the first and second derivatives of CK changes during the first 2 hours after the initial enzyme rise to initialize and limit the parameters of the log-normal function. Total CK activity rather than CK-MB activity was used to generate both the infarct size predictions and the observed infarct sizes. It was recognized that the first derivative of CK activity was related to the d parameter and the second derivative to the b parameter. A correlation coefficient of 0.80 between observed and predicted infarct size was obtained utilizing selected smooth curves by this technique. This technique suffered from several important disadvantages: (i) a true zero point could not be defined and the derivatives were quite sensitive to a zero point; (ii) a full 2 hours of data was required after the zero point which delayed treatment time to a minimum of 4 hours and an average of 6 hours; (iii) the shape of the predicted curves was only approximately related to the data, and (iv) the correlation cofficient of 0.8 was insufficiently precise to be of true predictive value. Thus this technique, although giving major insights into the methods of curve fitting the log-normal function from early data, would be poorly applicable to a general unselected group of patients with acute myocardial infarction.

The second attempt[39] to reduce the prediction time involves modelling the upstroke of the initial portion of the CK time–activity curve to regenerate the first 7 hours of CK data; thereafter, a 7 hour prediction using the log-normal function is computed. This approach depends exclusively on CK-MB activities—a situation in which the baseline is stable and a true zero point can be found. Initial and midphases of the curve behave like a simple exponential, the 7 hour point can be fitted to a linear approximation. This approach requires only a few data points in the first 2 hours. Correlation coefficients between observed and predicted infarct size of 0.98 can be obtained with smooth curves and 0.91 with randomly generated curves from

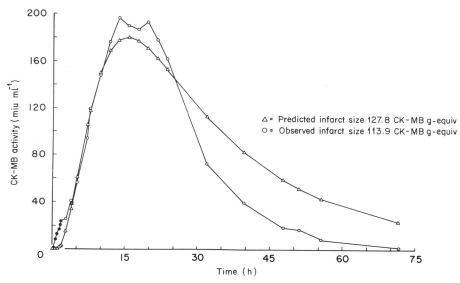

Figure 11. An early CK prediction. In this example, 2 hours of CK data rather than 7 hours of observed data were used to project the CK curve. The data used to find the prediction (●), the predicted curve (△), and the data observed after the prediction (○) are plotted

untreated patients (Figure 11). Accordingly, it appears that the early prediction utilizing CK-MB activity is possible. The development of a precise radioimmunoassay should allow CK curve projection with less than 1 hour of data obtained shortly after chest pain.

DEVIATIONS OF PREDICTED AND OBSERVED CK CURVES

Potential therapeutic agents have been utilized to dissociate observed and predicted CK curves. These experiments have been done in animals and in man. In conscious dogs, propranolol[41] and reperfusion[5] given 5 hours after coronary artery occlusion have been shown to decrease observed CK activity in relationship to predicted activity. Increased heart rate,[42] isoproterenol,[42] trimethaphan,[43] induced hypotension, and reperfusion[5] associated with myocardial haemorrhage increase CK activity as compared to predicted values. Early CK predictions have not been evaluated in dogs, and the effects of interventions when given early following coronary artery occlusion remain unexamined. The paucity of data in experimental animals suggests that further studies are required to define the range of therapies which may be effective.

In man, a number of potential therapies have been evaluated. Agents which increase CK release when applied 11–14 hours after the onset of chest pain include digitalis in non-failing hearts,[51] blood pressure reduction in normotensive patients,[20] and corticosteroids in pharmacologic doses.[33] Agents which do not change CK release when applied at this time include furosemide,[20] corticosteroids in small doses,[27] and dobutamine.[8] Therapies which appear to reduce CK release include propranolol,[31] blood pressure reduction in hypertensive patients,[43] and digitalis in failing hearts.[28]

Blood pressure reduction provides an example of therapy which can either be beneficial[43] or deleterious.[20] When hypertensive patients are treated with either nitroprusside or trimethaphan[43] and blood pressure is reduced to systolic pressure 150–160 mmHg, CK release is reduced below that predicted, implying that the resultant necrosis is less than the original zone of injury. However, when systolic blood pressure is reduced to 120 mmHg so that perfusion decreases,[20] a reduction of coronary blood flow occurs; in this circumstance, an increase in CK release results suggesting an increase in the area of myocardial necrosis.

Thus preliminary studies in experimental animals and in man suggest that the observed CK curves can be dissociated from predicted CK curves, implying that the zone of necrosis can be either extended or reduced with appropriate therapy. Although this assertion is theoretically attractive, a number of observations must be made. Observed CK infarct size is a valid reflection of necrotic zone in untreated animals[12,19,21,40,54] and randomly treated patients.[3] However, accurate determination of observed CK infarct size following therapy requires that the CK disappearance rate[40] and the proportionality constant[40] relating serum to myocardial CK depletion (CK_R/CK_D) remains unchanged by either the therapy or natural history of the infarct. These problems have been considered in detail in this monograph.

CONCLUDING COMMENTS

Projection of CK curves from early CK data has been utilized to predict a CK infarct size. The current techniques are based on an empiric equation and utilize non-linear curve fitting. The mathematical algorithms are complex and require standardization. Physiologic models appear promising but are currently used only to increase the precision of observed CK measurements and not to predict infarct size. Earlier prediction and improved precision will be necessary to make infarct size prediction a practical tool; promising advances have recently occurred which suggest this may be possible. Several potential therapies cause deviation of the predicted and observed curve; however, interpretation of the implication of either an increase or decrease in CK release must be established by further investigations. Infarct size prediction will result in substantial reduction in the number of patients required to ascertain the effectiveness of potential therapies on the rate and amount of CK release. Close correlation of changes in CK release to other indices of infarct size and ultimately patient survival are necessary.

REFERENCES

1. Aird, T. J. Computational solution of global nonlinear least squares problems. *Thesis*, Purdue University, May 1973.
2. Askenazi, J., Hillis, L. D., Diaz, P. E., Davis, M. A., Braunwald, E., and Maroko, P. R. The effects of hyaluronidase on coronary blood flow following coronary artery occlusion in the dog. *Circulation Research*, **40,** 566–71 (1977).
3. Bleifeld, W., Mathey, D., Hanrath, P., Buss, H., and Effert, S. Infarct size estimated from serial serum creatine phosphokinase in relation to left ventricular hemodynamics. *Circulation*, **55,** 303–11 (1977).
4. Botvinick, E. H., Shames, D., Lappin, H., Tyberg, J. V., Townsend, R., and Parmley, W. W. Noninvasive quantitation of myocardial infarction with technetium 99m pyrophosphate. *Circulation*, **52,** 909–17 (1975).
5. Bresnahan, G. F., Roberts, R., Shell, W. E., Ross, J., Jr., and Sobel, B. E. Deleterious effects due to hemorrhage after myocardial reperfusion. *American Journal of Cardiology*, **33,** 82–95 (1974).
6. Chandler, A. B., Chapman, I., Erhardt, L. R., Roberts, W. C., Schwartz, C. J., Sinapius, D., Spain, D. M., Sherry, S., Ness, P. M., and Simon, T. L. Coronary thrombosis in myocardial infarction. Report of a workshop on the role of coronary thrombosis in the pathogenesis of acute myocardial infarction. *American Journal of Cardiology*, **34,** 823–33 (1974).
7. Daniel, C. and Wood, F. S. *Fitting Equations to Data. Computer Analysis of Multifactor Data for Scientists and Engineers*, Wiley-Interscience, New York, 1971.
8. Gillespie, T. A., Ambos, H. D., Sobel, B. E., and Roberts, R. Effects of dobutamine in patients with acute myocardial infarction. *American Journal of Cardiology*, **39,** 588–94 (1977).
9. Ginks, W. R., Sybers, H. D., Maroko, P. R., Covell, J. W., Sobel, B. E., and Ross, J., Jr. Coronary artery reperfusion. II. Reduction of myocardial infarct size at 1 week after the coronary occlusion. *Journal of Clinical Investigation*, **51,** 2717–25 (1972).
10. Heng, M. K., Singh, B. N., Norris, R. M., John, M. B., and Elliot, R. Relationship between ST elevation and experimental ischemic damage. *Journal of Clinical Investigation*, **58,** 1317–26 (1976).
11. Holman, B. L., Lesch, M., Zweiman, F. G., Temte, J., Lown, B., and Gorlin, R. Detection and sizing of acute myocardial infarcts with ^{99m}-Tc (Sn) tetracycline. *New England Journal of Medicine*, **291,** 159–64 (1974).
12. Jarmakani, J. M., Limbird, L., Graham, T. C., and Marks, R. A. Effect of reperfusion on myocardial infarct, and the accuracy of estimating infarct size from serum creatine phosphokinase in the dog. *Cardiovascular Research*, **10,** 245–50 (1976).
13. Jennings, R. B., Herdson, P. B., and Sommers, H. M. Structural and functional abnormalities in mitochondria isolated from ischemic dog myocardium. *Laboratory Investigation*, **20,** 548–54 (1969).
14. Jennings, R. B., Sommers, H. M., Kaltenbach, J. P., and West, J. J. Electrolyte alterations in acute myocardial ischemic injury. *Circulation Research*, **14,** 260–8 (1964).
15. Jennrich, R. I. and Sampson, P. F. Application of stepwise regression to nonlinear estimation. *Technometrics*, **10,** 63–70 (1968).
16. Kjekshus, J. K. and Sobel, B. E. Depressed myocardial creatine phosphokinase activity following experimental myocardial infarction in rabbit. *Circulation Research*, **27,** 403–11 (1970).
17. Kloner, R. A., Reimer, K. A., and Jennings, R. B. Distribution of coronary collateral flow in acute myocardial ischaemic injury: Effect of propranolol. *Cardiovascular Research*, **10,** 81–90 (1976).
18. Lange, R. L., Reid, M. S., and Tresch, D. D. Nonatheromatous ischemic heart disease following withdrawal from chronic industrial nitroglycerin exposure. *Circulation*, **46,** 666–78 (1972).

19. Lukes, J., Schrader, P., Anastasia, L., Mueller, H., Rao, P., Dutta, D., Yang, J., and Bain, R. Histologic quantitation of infarct size to assess accuracy of estimative creatine phosphokinase formulas. *Clinical Research*, **25**, 235A (1977).
20. Magnusson, P. T., Shell, W. E., Forrester, J. S., Charuzi, Y., Singh, B. N., and Swan, H. J. C. Increased creatine phosphokinase release following blood pressure reduction in patients with acute infarction. *Circulation Supplement II*, **54**, II-28 (1976).
21. Manders, T., Vatner, S., Millard, R., Heyndrickx, G., and Maroko, P. R. Altered relationship between creatine phosphokinase release and infarct size with reperfusion in conscious dogs. *Circulation Supplement II*, **51** and **52**, II-5 (1975).
22. Markham, J., Karlsberg, R. P., Roberts, R., and Sobel, B. E. Mathematical characterization of kinetics of native and purified creatine kinase in plasma. *Computers in Cardiology*, Long Beach, IEEE, Computers Society, 1976.
23. Maroko, P. R. and Braunwald, E. Modifications of myocardial infarction size after coronary occlusion. *Annals of Internal Medicine*, **79**, 720–33 (1973).
24. Maroko, P. R., Kjekshus, J. K., Sobel, B. E., Watanabe, T., Covell, J. W., Ross, J., Jr., and Braunwald, E. Factors influencing infarct size following experimental coronary artery occlusions. *Circulation*, **43**, 67–82 (1971).
25. Maroko, P. R., Libby, P., Covell, J. W., Sobel, B. E., Ross, J., Jr., and Braunwald, E. Precordial ST segment elevation mapping: An atraumatic method for assessing alterations in the extent of myocardial ischemic injury. *American Journal of Cardiology*, **29**, 223–30 (1972).
26. Miura, M., Ganz, W., Thomas, R., Singh, B. N., Sokol, T., and Shell, W. E. Reduction of infarct size by propranolol in closed-chest anesthetized dogs. *Circulation Supplement II*, **53** and **54**, II-159 (1976).
27. Morrison, J., Maley, T., Reduto, L., Victa, C., Pyros, I., Brandon, J., and Gulotta, S. Effect of methylprednisolone on predicted myocardial infarction size in man. *Critical Care Medicine*, **3**, 94–102 (1975).
28. Morrison, J., Pizzarello, R., Reduto, L., and Gulotta, S. Effect of digitalis on predicted myocardial infarct size. *Circulation Supplement II*, **53** and **54**, II-28 (1976).
29. Muller, J. E., Maroko, P. R., and Braunwald, E. Evaluation of precordial electrocardiographic mapping as a means of assessing changes in myocardial ischemic injury. *Circulation*, **52**, 16–26 (1975).
30. O'Reilly, R. J. and Spellberg, R. D. Rapid resolution of coronary arterial emboli. *Annals of Internal Medicine*, **81**, 348–50 (1974).
31. Peter, C. T., Norris, R. M., Clarke, E. D., Heng, M. K., Singh, B. N., Williams, B., Howell, D. R., and Ambler, P. K. Reduction of enzyme release after acute myocardial infarction by propranolol. *Circulation*, **57**, 1091 (1978).
32. Rao, P. S. and Mueller, H. Creatine phosphokinase MB profile—Reflection of evolution of myocardial infarct. *Circulation Supplement II*, **54**, II-27 (1976).
33. Roberts, R., DeMello, V., and Sobel, B. E. Deleterious effects of methylprednisolone in patients with myocardial infarction. *Circulation, Supplement I*, **53**, I-204 (1976).
34. Ross, J., Jr. and Franklin, D. Analysis of regional myocardial function, dimensions, and wall thickness in the characterization of myocardial ischemia and infarction. *Circulation Supplement I*, **53**, I-88–92 (1976).
35. Shell, W. E., Groseth-Robertson, M. F., and Vas, R. Infarct size prediction 2 hours after initial CPK rise. *Circulation Supplement II*, **53** and **54**, II-28 (1976).
36. Shell, W. E. and Groseth-Robertson, M. F. Factors influencing the reproducibility of creatine kinase projections by nonlinear curve fitting. *Computers in Biomedical Research*, **11**, 257 (1978).
37. Shell, W. E., Groseth-Robertson, M. F., and Rorke, P. Validation of infarct size prediction in patients with acute myocardial infarction. *American Journal of Cardiology*, in press (1978).
38. Shell, W. E. and Groseth-Robertson, M. F. Early infarct size prediction by MB-CK analysis. *American Federation for Clinical Research*, submitted (1978).
39. Shell, W. E. and Groseth-Robertson, M. F. Early prediction of CK-infarct size. In preparation.
40. Shell, W. E., Kjekshus, J. K., and Sobel, B. E. Quantitative assessment of the extent of

myocardial infarction in the conscious dog by means of analysis of serial changes in serum creatine phosphokinase activity. *Journal of Clinical Investigation*, **50,** 2614–25 (1971).
41. Shell, W. E., Lavelle, J. F., Covell, J. W., and Sobel, B. E. Early estimation of myocardial damage in conscious dogs and patients with evolving acute myocardial infarction. *Journal of Clinical Investigation*, **52,** 2579–90 (1973).
42. Shell, W. E. and Sobel, B. E. Deleterious effects of increased heart rate on infarct size in the conscious dog. *American Journal of Cardiology*, **31,** 474–9 (1973).
43. Shell, W. E. and Sobel, B. E. Protection of jeopardized ischemic myocardium by reduction of ventricular afterload. *New England Journal of Medicine*, **291,** 481–6 (1974).
44. Shell, W. E. and Sobel, B. E. Biochemical markers of ischemic injury. *Circulation Supplement I*, **53,** I-98–106 (1976).
45. Smith, L. R., Turner, M. E., Rogers, W. J., McDaniel, H. G., Blackstone, E. H., and Rackley, C. E. A two-compartment model for the estimation of myocardial damage using creatine kinase-MB time–activity curves. *Clinical Research*, **25,** 255A (1977).
46. Snedecor, G. W. and Cochran, W. E. *Statistical Methods*, Iowa State University Press, Ames 1967.
47. Sobel, B. E. Biochemical and morphologic changes in infarcting myocardium. In *The Myocardium: Failure and Infarction* (Ed. E. Braunwald), H. P. Publishing Co., Inc. New York, 1974, pp. 247–60.
48. Sobel, B. E., Bresnahan, G. F., Shell, W. E., and Yoder, R. D. Estimation of infarct size in man and its relation to prognosis. *Circulation*, **46,** 640–8 (1972).
49. Sobel, B. E., Larson, K. B., Markham, J., and Cox, J. R., Jr. Empirical and physiological models of enzyme release from ischemic myocardium. *Computers in Cardiology*, Long Beach, IEEE, Computers Society, 1974.
50. Sobel, B. E., Weiss, E. S., Welch, M. J., Siegel, B. A., and Ter-Pogossian, M. M. Detection of remote myocardial infarction in patients with positive emission transaxial tomography and intravenous ^{11}C-palmitate. *Circulation*, **55,** 853–7 (1977).
51. Varonkov, Y., Shell, W. E., Smirnov, V., Gukovsky, D., and Chazov, E. I. Augmentation of serum CPK activity by digitalis in patients with acute myocardial infarction. *Circulation*, **55,** 719–27 (1977).
52. Wackers, F. J. T., Becker, A. E., Samson, G., Sokole, E. B., van der Schoot, J. B., Vet, A. J. T. M., Lie, K. I., Durrer, D., and Wellens, H. J. J. Location and size of acute transmural myocardial infarction estimated from thallium-201 scintiscans. A clinicopathological study. *Circulation*, **56,** 72–8 (1977).
53. Weiss, E. S., Ahmed, S. A., Thakus, M. L., Welch, M. J., Coleman, R. E., and Sobel, B. E. Imaging of the inflammatory response in ischemic canine myocardium with 111 indium-labelled luekocytes. *American Journal of Cardiology*, **40,** 195–9 (1977).
54. Yasmineh, W. G., Pyle, R. B., Cohn, J. N., Nicoloff, D. M., Hanson, N. Q., and Steele, B. W. Serial serum creatine phosphokinase MB isoenzyme activity after myocardial infarction. Studies in the baboon and man. *Circulation*, **55,** 733–8 (1977).

CHAPTER 13

Clinical experience with infarct sizing and its value in the prognosis of myocardial infarction

J. C. Kahn, P. Gueret, and J. P. Bourdarias

INTRODUCTION	319
DETAILS OF THE STUDY	320
SERIAL CK DETERMINATIONS	320
INFARCT SIZE AND E.C.G. LOCATION OF INFARCT	321
LEFT VENTRICULAR FAILURE AND INFARCT SIZE	324
CORRELATION BETWEEN INFARCT SIZE AND TIME COURSE OF SERUM ENZYME ACTIVITY	329
IN-HOSPITAL DEATHS	333
FOLLOW-UP STUDY	334
CONCLUDING COMMENTS	335
REFERENCES	336

INTRODUCTION

Accumulating evidence suggests that infarct size is a major determinant of morbidity and mortality following acute myocardial infarction.[2,11,18,23,24,40,41] Accordingly assessment of infarct size may provide both a guide to therapy and an indication of prognosis in patients with infarction. Recently several methods have been developed for indirect assessment of the severity of myocardial injury, including electrocardiographic chest mapping techniques,[16] infarct imaging with radionuclides,[6] and analysis of serial changes in serum enzymes.[37,38] Enzymatic estimation of infarct size by quantitation of total creatine kinase (CK) release is a non-invasive method, based on a mathematical model (see Chapter 11). This model carries several assumptions, some of which are open to criticism.[29,30] Although limited in number, pathological studies both in experimental animals[37] and in man[5] have shown an encouraging correlation between enzymatic estimates and histologic measurements of infarct size ($r = 0.93$; $n = 15$; $p < 0.001$).[5] In addition good agreement between enzymatic and angiographic infarct sizing has been found in patients without prior myocardial infarction.[31]

This chapter described an estimation of infarct size by analysis of serial serum CK changes which was performed routinely in 82 patients with acute myocardial infarction and which was related to the clinical course and to the prognosis.

DETAILS OF THE STUDY

Infarct size was assessed quantitatively by serial changes in serum creatine kinase (CK) in 82 patients with acute myocardial infarction. These patients included 69 males and 13 females with a mean age of 57 ± 10 years (range 33 to 77). Twenty patients had left ventricular hypertrophy, 16 patients previous angina, and 9 had suffered a previous myocardial infarction. Total serum CK activity was measured by Rosalki's method.[32] Since MB isoenzymes were not isolated, patients with cardiogenic shock or those who received intramuscular injection or cardioversion were excluded from the study. Patients admitted more than 12 hours after the onset of symptoms were not included. The diagnosis of acute myocardial infarction was based on a typical history, characteristic e.c.g. criteria of transmural necrosis and elevation of serum enzymes determined daily for 3 consecutive days (creatine kinase, aspartate aminotransferase, and lactate dehydrogenase). Infarct location was classified according to the criteria suggested by the New York Heart Association.[22] Leads V_7, V_8, V_9 were also recorded in patients with an inferior myocardial infarction. Chest X-ray films and systemic blood gases were obtained daily and right heart catheterization using a Swan–Ganz catheter was performed in 26 patients. Left ventricular failure was considered to be present according to the Myocardial Infarction Research Unit criteria, i.e. moist rales, radiological signs of interstitial and/or alveolar pulmonary oedema and Pao_2 less than 60 mmHg. Follow-up information concerning survival and congestive heart failure was obtained from hospital charts or by form letters to referring physicians.

SERIAL CK DETERMINATIONS

A catheter was inserted percutaneously into an antecubital vein and centrally positioned. Blood samples were obtained hourly for the first 12 hours, every 2 hours for the following 12 hours and every 4 hours during the next 48 hours or until CK activity had returned to near-normal levels ($40\ miu\ ml^{-1}$).

Infarct size (IS) expressed as creatine kinase-gram-equivalent (CK-g-equiv.), was calculated using the formula of Sobel et al.:[28]

$$IS = K \cdot BW \cdot CK_r$$

where K is the proportionality constant (ml CK-g-equiv. $kg^{-1}\ iu^{-1}$) = 5.9×10^{-1}

$$BW = \text{body weight (kg)}$$

CK_r (iu ml^{-1}) = cumulative activity of CK released by the heart up to time T

$$CK_r = \int_0^T f(t)\,dt = E(T) + k_d \int_0^T E(t)\,dt$$

k_d = fractional rate of disappearance of CK from blood (min^{-1})
k_d was obtained from a semilogarithmic plot of $E(t)$ versus time and calculated for each patient as suggested by Norris.[23] The mean number of data points used to calculate k_d was 9 (range 5 to 21) and the 95% confidence limits were less than ±5% in 40 cases (49%), less than ±10% in 65 cases (79%), and less than ±15% in 82 cases (100%). The value of k_d in these 82 patients was 0.00073±0.00020 (mean±1 SD) (range 0.00030–0.00130).

A Hewlett-Packard plotter-computer system (HP 9825) was utilized to plot serial serum CK activity levels and to plot and calculate integrated appearance function as a function of time. The rate of rise of the serum CK activity curve was calculated as the ratio of peak serum CK value to peak time and expressed as iu h^{-1}. The CK 'release time' was defined as the time interval (hours) between the first abnormal value of CK and the change in the slope of the disappearance function of CK (or the flattening of the integrated release curve), i.e. the time when the release of myocardial CK had stopped.[17] Data were compared and analysed using Student's test or the chi-square test where appropriate.

INFARCT SIZE AND E.C.G. LOCATION OF INFARCT

As shown in Figure 1, only 20% of the patients had an infarct size greater than 50 CK-g-equiv. since patients with cardiogenic shock were not included in the study. The mean infarct size was 36.2±19.1 CK-g-equiv. (range 4–107 CK-g-equiv.).

When patients were analysed according to infarct location, the average infarct size in 38 patients with anterior myocardial infarction (34.5±22.1 CK-g-equiv.) was not statistically different from the average infarct size in 42 patients with inferior myocardial infarction (37.0±16 CK-g-equiv.) (two patients with both anterior and inferior myocardial infarction were excluded). These findings are consonent with the results of Mathey et al.[17] and Bleifeld et al.[5] In contrast Sobel et al.[41] found a greater infarct size in anterior myocardial infarctions than in inferior myocardial infarctions (Table 1). A more accurate analysis of myocardial infarction location is shown in Figure 2.[13] In the anterior myocardial infarction group, the average infarct size in 12 anteroseptal myocardial infarctions (leads V_1, V_2, V_3) (16.0±6.4 CK-g-equiv.) was significantly smaller than infarct size in 15 anteroapical myocardial infarctions (leads V_1, V_2, V_3, V_4, ±V_5) (35.9±15.9 CK-g-equiv.) ($p<0.001$). In 10 extensive anterior myocardial

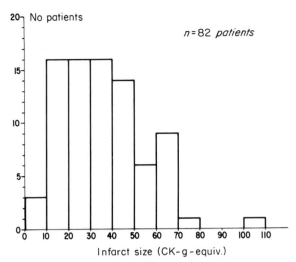

Figure 1. Frequency distribution of infarct size (82 patients). Patients with cardiogenic shock are not included. [Reproduced by permission of Elsevier North Holland from *Journal of Molecular Medicine*, **2**, 223–31 (1977).]

infarctions (leads I, aVL, V_1–V_5, ±V_6) infarct size was greater (57.8 ± 20.1 CK-g-equiv.) than that in anteroapical myocardial infarctions ($p < 0.01$). In the inferior myocardial infarction group, the average infarct size in 10 patients with inferior myocardial infarction (leads II, III, aVF) was 24.7 ± 10.0 CK-g-equiv. compared to 32.8 ± 13.8 CK-g-equiv. in 18 patients with inferoposterior myocardial infarction (leads II, III, aVF, V_7–V_9) (NS).

Table 1. Infarct Size and Location of Myocardial Infarction

Authors	Anterior myocardial infarction	Inferior myocardial infarction	p
Sobel[41]	82 ± 37 ($n = 14$)	50 ± 31 ($n = 14$)	0.02
	64 ± 29[b]	39 ± 24	
Mathey[17]	59 ± 31 ($n = 14$)	66 ± 32.5 ($n = 15$)	NS
Bleifeld[5]	47 ± 33 ($n = 28$)	62 ± 35 ($n = 27$)	NS
Kahn[a,13]	34.5 ± 22 ($n = 38$)	37 ± 16 ($n = 42$)	NS

Results of infarct size (CK-g-equiv.) are expressed as mean ± SD; n = number of patients.

[a] Patients with cardiogenic shock were excluded.

[b] Infarct size recalculated with a proportionality constant $K = 5.9 \times 10^{-1}$.

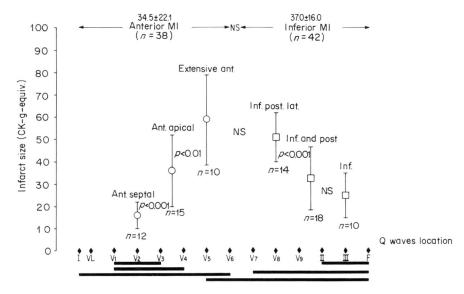

Figure 2. Relation between the number of e.c.g. leads with abnormal Q wave and infarct size (mean ± SD). MI = myocardial infarction; Ant. septal = anteroseptal; Ant. apical = anteroapical; Extensive ant. = extensive anterior; Inf. post. lat. = inferoposterior and lateral; Inf = inferior; Post. = posterior; n = number of patients. [Reproduced by permission of Elsevier North Holland from *Journal of Molecular Medicine*, **2**, 223–31 (1977).]

However, when the lateral wall was also involved (leads II, III, aVF, V_5–V_9, ±I–aVL; $n = 14$) the average infarct size was significantly greater (51.1 ± 11.5 CK-g-equiv.) than that observed in inferoposterior myocardial infarctions ($p < 0.001$). Thus anteroseptal and inferior myocardial infarctions involved a relatively small mass of myocardium (16.0 and 24.7 CK-g-equiv. respectively; $p < 0.05$). On the average, anteroapical and inferoposterior myocardial infarctions showed no difference in infarct size (35.9 and 32.8 CK-g-equiv. respectively) and were of medium size. Extensive anterior and inferoposterolateral myocardial infarctions involved equally large areas of myocardium (57.8 and 51.1 CK-g-equiv. respectively; NS).

This study indicates that in patients without cardiogenic shock the enzymatic estimates of infarct size grossly correlated with the number of e.c.g. leads with abnormal Q waves. This is consistent with previous studies of Selvester *et al.*[36] who, on the basis of e.c.g. criteria, could predict the size and location of an akinetic segment with 85% reliability. Inoue *et al.*[12] also reported that the release of CK calculated from serial specimens corresponded to the amplitude development of abnormal Q waves in the inferior leads, and Awan *et al.*[3] found that praecordial Q wave mapping correlated with the extent of myocardial damage in anterior leads. The finding of an abnormal Q wave in the right praecordial leads (V_1–V_3) or in the inferior leads (II, III, aVF) was associated with the same infarct mass. In addition, the loss of electrical forces in the left praecordial leads (V_4–V_6) was associated with larger infarcts. Previous pathological studies have shown that in patients with anterior myocardial infarction, an abnormal Q wave in leads V_4–V_6 was always associated with circumferential apical infarction,[35] and further extension of the infarct in the lateral free wall is usually indicated by the appearance of Q waves in leads I and aVL.[9] The computer simulation of myocardial activation developed by Selvester et al.[36] has suggested that it would be meaningful to distinguish between 'inferior' (Q waves in leads II, III, aVF) and 'posterior' (increased R waves in leads V_1–V_2, or Q waves in leads V_7–V_9) infarct location. Although this distinction did not prove useful in locating the infarct[35] it provided a more accurate quantitation of infarcts of the inferoposterior wall. Indeed, in this study, inferoposterolateral infarctions involved an equally large amount of myocardium as extensive anterior infarctions.

However because of marked overlap of infarct size values between different e.c.g. locations of myocardial infarction, in the individual patients infarct size could not be predicted accurately from the number of e.c.g. leads with a Q wave. Relatively large scatter of the data could be related to differences in heart position inside the chest, to individual variations in left ventricular wall thickness, and not infrequently, in inferior myocardial infarctions, to the involvement of the inferior wall of the right ventricle unrecognized on the e.c.g. Furthermore the concept that infarction induces alterations in the QRS only if it is transmural has not been validated in several studies. Many patients with definite Q waves in both the anterior and posterior leads may have infarcts limited to the subendocardium[35] whereas on the other hand, relatively large non-transmural infarcts may produce unequivocal Q waves.[8,35,46] In addition, some transmural infarcts cause only ST–T alterations without definite QRS changes.[1]

LEFT VENTRICULAR FAILURE AND INFARCT SIZE

Twenty-two patients (27%) had 2 or 3 criteria of congestive heart failure and were in class II or III of the Killip and Kimball classification.[47] Of the 35 patients with an infarct size of less than 30 CK-g-equiv. only one (3%) had left ventricular failure (this patient had sustained a previous inferior myocardial infarction). Of the 30 patients with an infarct size ranging from 30 to 50 CK-g-equiv., 10 (33%) had left ventricular failure. Eleven of the 17 patients (65%) with an infarct size greater than 50 CK-g-equiv. had left ventricular failure. The left ventricular failure incidence was significantly different among the 3 groups: the greater the infarct size, the higher the

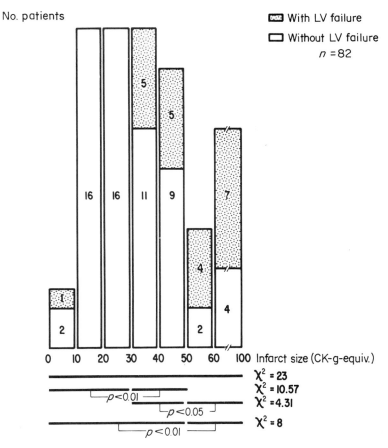

Figure 3. Distribution of left ventricular (LV) failure (Killip and Kimball classification) in relation to infarct size ($n=82$ patients). Patients with cardiogenic shock are not included. [Reproduced by permission of Elsevier North Holland from *Journal of Molecular Medicine*, **2**, 223–31 (1977).]

incidence of left ventricular failure (Figure 3). Mean infarct size was significantly greater ($p<0.001$) in the 22 patients with left ventricular failure (52.9±17.2 CK-g-equiv.), than in the 60 patients without left ventricular failure (30.1±15 CK-g-equiv.). Of the 26 patients with available haemodynamic data, 12 patients with a pulmonary wedge pressure less than 12 mmHg had a mean infarct size of 29.0±11.3 CK-g-equiv., a value significantly smaller than the mean infarct size (57.9±18.3 CK-g-equiv.) in 14 patients with a pulmonary wedge pressure greater than 12 mmHg ($p<0.001$). For the whole series however, pulmonary wedge pressure did not correlate with infarct size.

The above data are in general agreement with earlier studies (Table 2).[12,18,23,41] Although mean infarct size was significantly greater in patients with a cardiothoracic ratio greater than 0.50 and/or a left ventricular filling pressure greater than 12 mmHg, there was considerable scatter in the data. For the whole series, the extent of the infarct was not statistically correlated with the degree of left ventricular dysfunction as assessed by the level of left ventricular filling pressure. Similar findings have been reported by Bleifeld et al.[5] who showed that neither pulmonary artery end-diastolic pressure nor cardiac index correlated with infarct size.

While infarct size is the major determinant of pump dysfunction,[20,27] factors other than the infarct size may contribute to impairment of left ventricular function, e.g. primarily the extent and functional state of residual myocardium. The haemodynamic profile in an individual patient actually depends on the combined effects of old and recent infarcts and possible areas of ischaemic, non-functioning but viable myocardium, whereas serial enzyme changes reflect only the amount of recent necrosis. Indeed, in this study, when 25 patients with previous left ventricular hypertrophy and/or old myocardial infarction were excluded, infarct size averaged 52.7 ± 23.7 CK-g-equiv. in the 11 patients with left ventricular failure compared to 29.3 ± 16.1 CK-g-equiv. in the 46 patients without left ventricular failure ($p < 0.001$), and a significant relationship between infarct size and left ventricular filling pressure was found ($r = 0.75$, $n = 15$, $p < 0.005$), thereby confirming the major relation between infarct size and the degree of depressed left ventricular function. Likewise, Bleifeld et al.[5] found that, out of 10 patients with relatively small infarct size (34 CK-g-equiv. on the average) and high pulmonary artery end-diastolic pressure (mean 22.5 mmHg), 6 had reinfarctions. In a group of 11 patients without prior myocardial infarction, Rogers et al.[31] found that CK-MB infarct size correlated with admission pulmonary artery end-diastolic pressure ($r = 0.78$, $p < 0.005$), with left ventricular ejection fraction ($r = -0.75$, $p < 0.01$) and with angiographic infarct size (percentage of abnormal contracting segment × left ventricular mass) ($r = 0.85$, $p < 0.005$). In contrast CK-MB infarct size did not correlate with these parameters in a second group of 12 patients with prior myocardial infarction.

Several studies have shown that impairment of left ventricular function is greater in patients with anterior than with inferior necrosis.[10,19,24,27] However in our study the incidence of left ventricular failure did not differ in anterior and inferior myocardial infarctions nor in extensive anterior and inferoposterolateral myocardial infarctions (Figure 4). Accordingly, left ventricular impairment correlated with estimated infarct size and not with infarct location. This is in agreement with the observations that in both recent[27] and old[20] myocardial infarction, the left ventricular haemodynamic consequences of anterior or inferior infarcts are similar when the extent of

Table 2. Infarct Size and Congestive Heart Failure (CHF)

Authors	Criteria of CHF	No or mild CHF	Severe CHF	p
Sobel[41]	NYHA	24 ± 13 CK-g-equiv.[a] ($n=19$)	71 ± 12 CK-g-equiv.[a] ($n=4$)	<0.001
Norris[23]	X-Ray pulmonary oedema or congestion	1.7 iu ml^{-1}[b] ($n=17$)	3.4 iu ml^{-1}[b] ($n=14$)	<0.01
Inoue[12]	not defined	664 ± 73 iu ml^{-1}[c] ($n=30$)	998 ± 164 iu ml^{-1}[c] ($n=20$)	<0.01
Kahn[13]	K.K.	30 ± 15 CK-g-equiv. ($n=60$)	53 ± 17 CK-g-equiv. ($n=22$)	<0.001
Mathey[18]	PAEDP	<12 mmHg 13 CK-g-equiv.[a] ($n=19$)	12–20 mmHg 33 CK-g-equiv.[a] ($n=35$) >20 mmHg 77 CK-g-equiv. ($n=28$)	
Kahn[13]	PWP	<12 mmHg 29 ± 11 CK-g-equiv. ($n=12$)	>12 mmHg 58 ± 18 CK-g-equiv. ($n=14$)	<0.001

NYHA = New York Heart Association; K.K. = Killip and Kimball classification;[47] PAEDP = pulmonary artery end-diastolic pressure; PWP = pulmonary wedge pressure; IS = infarct size (CK-g-equiv.); n = number of patients. Results are expressed as mean ± SD.
[a] IS calculated with a proportionality constant $K = 5.9 \times 10^{-1}$.[28]
[b] Mean integrated appearance function (iu ml^{-1}).
[c] Mean total CK released (iu ml^{-1}).

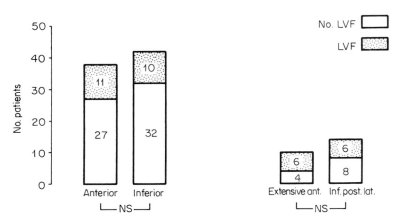

Figure 4. Incidence of left ventricular failure (LVF) in relation to location of myocardial infarction (MI). Extensive ant. = extensive anterior; Inf. post. lat = inferoposterior and lateral. [Reproduced by permission of Elsevier North Holland from *Journal of Molecular Medicine*, **2**, 223–31 (1977).]

necrosis is equivalent. The greater frequency of pump failure in anterior myocardial infarction[26,34] may be explained by the fact that obstructive lesions of the left anterior descending coronary artery result in more extensive necrosis than do similar occlusions of the right coronary or circumflex coronary arteries.[20] Indeed, in a pathological study, Wackers *et al.*[45] demonstrated that in patients dying from cardiogenic shock or refractory congestive heart failure after acute myocardial infarction there is always marked narrowing (greater than 75%) involving a long segment of the left anterior descending artery.

As reported by others[21,44] the finding of a pericardial friction rub in acute myocardial infarction is usually associated with a greater extent of myocardial damage (higher enzyme elevation, higher incidence of arrhythmias and congestive heart failure). In the 15 patients with a pericardial friction rub, infarct size averaged 51.6 ± 23.6 CK-g-equiv. as against 33.0 ± 16.1 CK-g-equiv. in the 66 patients without friction rub ($p < 0.001$). Of these 15 patients with a friction rub, 12 had left ventricular failure and in contrast only 3 out of 66 patients without left ventricular failure had a friction rub.

Sobel *et al.*[42] found that the severity of ventricular dysrrhythmia early after myocardial infarction (premature ventricular complexes and episodes of couplets or ventricular tachycardia) reflected the extent of myocardial injury (Table 3). Furthermore an increase in premature ventricular complexes rate often predicted extension of the infarct.

Table 3. Infarct Size and Ventricular Dysrrhythmia

	Sobel et al.[42]		
	Group I	Group II	Group III
ISI	1–24 ($n=6$)	25–49 ($n=10$)	50 ($n=9$)
PVCs/20 hour	54 ± 21	99 ± 33	635 ± 258
Episodes of couplets or VT	1.5 ± 6	4.3 ± 1.3	9.9 ± 2.2

ISI = Infarct Size Index (CK-g-equiv. m^{-2}); PVCs = premature ventricular complexes; VT = ventricular tachycardia.
Results are expressed as mean ± SEM.
All differences between groups were significant ($p < 0.01$).

CORRELATION BETWEEN INFARCT SIZE AND TIME COURSE OF SERUM ENZYME ACTIVITY

A statistically significant correlation was found between infarct size and the maximal value of aspartate aminotransferase ($r = 0.66$, $p < 0.0001$) and of serum lactate dehydrogenase ($r = 0.76$, $p < 0.0001$). However, blood samples for the determination of aspartate aminotransferase and lactate dehydrogenase were obtained only once a day for 3 or 4 days and thus, peak serum activity was not detected with any great certainty. Furthermore, these enzymes are not specific for myocardial injury. On the other hand, with a greater frequency of serial sampling we were able to detect reliably the peak of serum CK activity. In fact, we observed that the peak CK value correlated more closely with infarct size ($r = 0.85$, $p < 0.001$) than did the other measurements. Three different time courses of serum CK activity were identified. In subgroup I ($n = 61$), CK activity rose steeply, showed an early single peak and a subsequent decline (Figure 5). In subgroup II ($n = 17$) CK activity increased either steeply as in subgroup I patients or more slowly and then levelled off and decreased slowly (Figure 5). In subgroup III ($n = 4$), patients with reinfarction showed a second peak on the descending limb of the CK activity curve. The mean 'CK release time' was significantly shorter in subgroup I (20 ± 5 hours) than in subgroup II (39 ± 7.2 hours) ($p < 0.001$) (Figure 6). The rate of rise was not statistically different between the two subgroups (0.0152 ± 0.0100 iu h^{-1} in subgroup I and 0.0135 ± 0.0065 iu h^{-1} in subgroup II; NS), nor was the fractional disappearance rate k_d (0.00074 ± 0.00021 in subgroup I, and 0.00071 ± 0.00020 in subgroup II; NS). However, infarct size was significantly greater in subgroup II (49.9 ± 13.3 CK-g-equiv.) than in subgroup I (31.7 ± 18.6 CK-g-equiv.; $p < 0.001$). The correlation between peak serum CK and infarct size was further analysed (Figure 7): in subgroup I, peak serum CK value correlated closely with

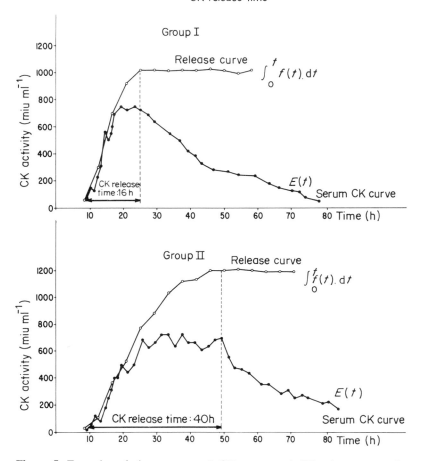

Figure 5. Examples of time courses of CK serum and CK release curves in subgroups I and II. [Reproduced by permission of Elsevier North Holland from *Journal of Molecular Medicine*, **2**, 223–31 (1977).]

infarct size ($r = 0.93$; $p < 0.001$). In contrast, in subgroup II peak serum CK value correlated less closely with infarct size ($r = 0.59$, $p < 0.02$).

Comparison with clinical, electrocardiographic and pathological data suggest that these three time courses reflect different patterns in infarct development.[17,25] Subgroup I, with a 'CK release time' of short duration (20 ± 5 hours), and a CK release curve resembling that seen after experimental coronary artery occlusion, is thought to reflect a single necrosis. With 3 patients in this group that died, necropsy showed homogenous and sharply demarcated necrosis.[17] In our study, when serum CK activity started to

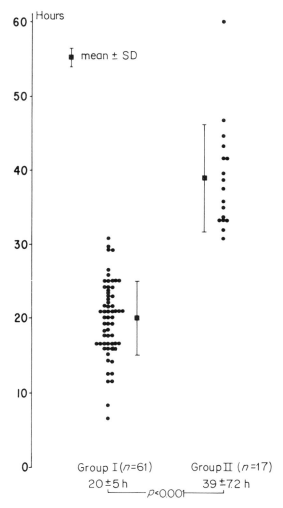

Figure 6. Individual values of 'CK release time' in subgroups I and II. [Reproduced by permission of Elsevier North Holland from *Journal of Molecular Medicine*, **2**, 223–31 (1977).]

decrease within 30 hours (mean ± 2 SD), an individual patient could be classified in this subgroup with 95% reliability. Our calculated 30 hour limit between subgroup I and II is consonant with the experimental findings that the minimum residual CK activity in infarcted myocardium occurs 24 hours after coronary occlusion.[14] Since CK appears in the circulation after a mean period of 4 hours, CK release from a homogeneous necrosis should be

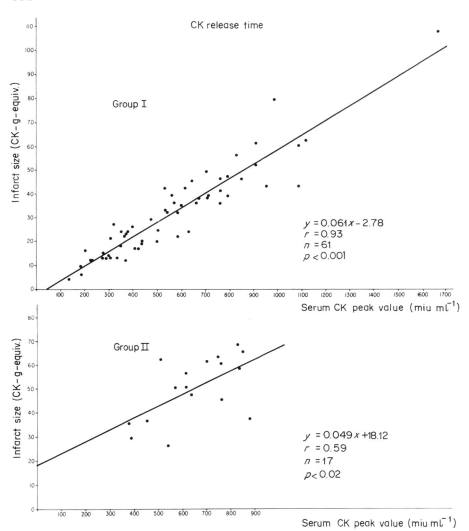

Figure 7. Correlation between peak serum CK value and infarct size in subgroup I (upper panel) and in subgroup II (lower panel). [Reproduced by permission of Elsevier North Holland from *Journal of Molecular Medicine*, **2**, 223–31 (1977).]

completed within 30 hours. In addition, the observed close correlation between infarct size and CK peak value suggests that, in patients with a 'CK release time' of less than 30 hours, infarct size may be estimated with reasonable accuracy from the CK peak value alone. In subgroup II, 'CK release time' was markedly prolonged and the CK activity profile showed a flat single peak, or a plateau or multiple peaks which may suggest the more gradual development of infarction. This is supported by the pathological studies of Mathey et al.[17] who showed scattered foci of necrosis and contiguous areas of ischaemic myocardium. Furthermore infarct size was actually larger in subgroup II than in subgroup I. Several factors may influence 'CK release time' including: (i) delayed CK release; (ii) decreased CK disappearance rate; (iii) different patterns in the development and the extent of the infarction. The rate of rise of CK activity and the CK disappearance rate were not significantly different between the two subgroups. Although no correlation was found between 'CK release time' and infarct size, it might be speculated that larger infarcts require more time to develop and this might account, in part at least, for the prolonged 'CK release time'. This latter suggestion would be in agreement with the report of Inoue.[12] In contrast to subgroup I, only in subgroup II is an attempt to limit the final size of the infarct by pharmacological and/or mechanical procedures likely to be effective.[15,39] In these two subsets, however, although clinical,[12,17] electrocardiographic,[12] pathological,[17] and experimental data[14] suggest different patterns of infarct development, it must be appreciated that no sharp line can be drawn. In this connection, it is still speculative whether or not the time course of acute myocardial infarction represents a continuous spectrum or a bimodal distribution of events.

In subgroup III, the true extension of the infarct is characterized by recurrent chest pain associated with ST segment changes and a secondary rise in serum CK activity. This phenomenon was seen only in 5% of our patients as against 18–22% in other studies.[17,33] It is our current policy to manage patients with impending reinfarction with circulatory assistance by intraaortic balloon pumping and myocardial revascularization whenever possible.[4] Extension of the infarct may account for as much as 24% of the final infarct size[17] and actually be responsible for the 33–40% mortality rate in these patients.[33,43]

IN-HOSPITAL DEATHS

Since patients with cardiogenic shock were not included in this study, the in-hospital mortality rate was relatively low ($n = 6$, 7.5%) and could not be statistically related to infarct size. Three patients died from complications (intractable arrhythmias in 2 patients, and refractory hypoxaemia secondary to a right-to-left shunt through a patent foramen ovale complicating a biventricular myocardial infarction in the third patient). In the 3 remaining patients death may be related to infarct size: in the first patient infarct size was 65 CK-g-equiv.; in the second patient the final infarct size was 35 CK-g-equiv. (recent myocardial infarction) +25 CK-g-equiv. (estimated infarct size of an old inferior myocardial infarction); in the third patient infarct size was 56 CK-g-equiv. (recent myocardial infarction) +20 CK-g-equiv. (estimated infarct size of a subsequent reinfarction of the lateral wall). In all cases infarct size exceeded 60 CK-g-equiv.

In the series of Bleifeld et al.[5] the 65 CK-g-equiv. threshold separated survivors from non-survivors. In Sobel's study,[41] 8 of the 12 patients with infarct size exceeding 51 CK-g-equiv. (calculated with $K = 0.59$) died as opposed to 1 of the 21 patients with infarct size less than 51 CK-g-equiv. ($p < 0.001$) and further confirmation of this distribution has been obtained in more recent studies by this group.[40] In addition, haemodynamic studies have shown that loss of myocardium exceeding 60 CK-g-equiv. is associated with severe impairment of left ventricular function and a markedly restricted cardiac reserve:[18] patients with a pulmonary artery end-diastolic pressure greater than 20 mmHg had an average infarct size of 77 CK-g-equiv. (calculated with $K = 0.59$), substantially decreased cardiac and stroke work indices, and a 60% mortality.

FOLLOW-UP STUDY

Follow-up information was available in 73 (96%) of the 76 patients discharged from hospital for periods ranging from 6 to 45 months (18.1 ± 10.8 months) (Table 4). Of 20 patients with initial left ventricular failure (27% of the 73 survivors) 6 still had left ventricular failure when last examined (and one patient died) and 14 had no signs of left ventricular failure. Of 53 patients without initial left ventricular failure (73%), 49 remained asymptomatic while left ventricular failure developed in 4 (two of these died). Thus at the end of the follow-up period, the percentage of

Table 4. Follow-up of Left Ventricular Failure in 73 Survivors

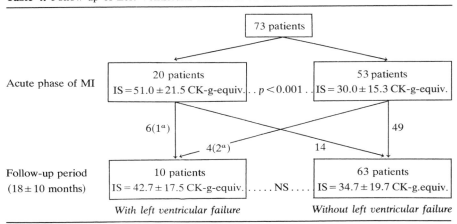

[a] Patients that died.
MI = Myocardial infarction; IS = infarct size.

patients without left ventricular failure was 86% (63/73) and with left ventricular failure was 14% (10/73). Infarct size was not significantly different between these two groups (34.7 ± 19.7 CK-g-equiv. compared to 42.7 ± 17.5 CK-g-equiv.; NS) even when patients with previous left ventricular hypertrophy and/or old myocardial infarction were excluded. Three patients died of intractable left ventricular failure and severe angina pectoris 6, 19, and 22 months after the infarct and in these cases infarct size was 33, 38, and 25 CK-g-equiv. respectively.

The above results would suggest that enzymatic infarct size estimations do not prove useful in predicting either long-term mortality rate or heart failure. However, Carter and Amundsen[7] found a significant relationship between maximal exercise tolerance level and infarct size in patients 3 to 4.5 months after acute myocardial infarction ($r = -0.68$, $n = 15$; $p < 0.05$) and the correlation was even higher after 3 months of cardiac rehabilitation ($r = 0.84$, $n = 11$; $p < 0.01$).

Correlations with clinical, electrocardiographic, and haemodynamic parameters have demonstrated that the estimation of infarct size by serial serum CK analysis is a useful method for quantitation of infarct mass in acute myocardial infarction. However this method takes into account neither the mass nor the functional state of the residual myocardium and cannot be used as a single parameter to predict occurrence of left ventricular failure and/or death. Therefore, it would seem advisable to use calculated infarct size in conjunction with several other parameters in order to derive a more accurate prognostic index in the individual patient. Using a discriminant function analysis with four parameters including infarct size index, clinical class, stroke index, and left heart dimension, Shell and Sobel[40] were able to separate survivors from non-survivors with 99% reliability in a group of 18 patients where all data were available. By combining haemodynamic parameters and infarct size, Bleifeld et al.[5] identified 4 different groups of patients and obtained a better understanding of the influence of a recent infarct on the functional state of the heart and on its prognosis (Table 5). The individualization of different courses of CK release, which may reflect different patterns of myocardial infarction development, might provide a rational basis for early initiation of therapeutic procedures designed to limit the final infarct size.

CONCLUDING COMMENTS

In conclusion, this study, which excluded patients with cardiogenic shock, indicated that in patients with acute myocardial infarction, the enzymatic estimation of infarct size grossly correlated with the number of e.c.g. leads with abnormal Q waves. Our data, and those from other studies would suggest that the incidence of left ventricular failure increased in proportion with infarct size. However, in the individual patient, the degree of left

Table 5. Infarct Size and Left Ventricular Haemodynamics

	IS (CK-g-equiv.)	Bleifeld et al.[5] PAEDP (mmHg)	CI (l. min^{-1} m^{-2})	Mortality rate (%)	No. CS	Previous MI (%)
Class A ($n = 22$)	29	12.8	3.03	9%	0	14
Class B ($n = 13$)	96	26	2.24	54%	2	8
Class C ($n = 5$)	84	17	3.33	20%	0	20
Class D ($n = 10$)	34	22.5	2.9	20%	2	6

IS = Infarct size; PAEDP = Pulmonary artery end-diastolic pressure; CI = cardiac index; No. CS = number of cardiogenic shock; MI = myocardial infarction.

ventricular dysfunction could not be predicted accurately from the estimated infarct size. Factors other than infarct size may of course also contribute to left ventricular impairment, for example, the contractile state and amount of residual myocardium. In patients with a CK release time shorter than 30 hours, a close correlation was found between peak serum CK values and estimated infarct size. Thus in this subset of patients, using a mathematical model based on several assumptions, did not provide a better estimation of infarct size than did the peak serum CK value alone.

REFERENCES

1. Abbott, J. A. and Scheinman, M. M. Nondiagnostic electrocardiogram in patients with acute myocardial infarction. Clinical and anatomic correlations. *American Journal of Medicine*, **55**, 608–13 (1973).
2. Alonso, D. R., Scheidt, S., Post, M., and Killip T. Pathophysiology of cardiogenic shock. Quantification of myocardial necrosis. Clinical, pathologic and electrocardiographic correlations. *Circulation*, **48**, 588–96 (1973).
3. Awan, N. A., Miller, R. R., Vera, Z., Janzen, D. A., Amsterdam, E. A., and Mason, D. T. Non invasive assessment of cardiac function and ventricular dyssynergy by precordial Q wave mapping in anterior myocardial infarction. *Circulation*, **55**, 833–8 (1977).
4. Bardet, J., Rigaud, M., Kahn, J. C., Huret, J. F., Gandjbakhch, I., and Bourdarias, J. P. Treatment of post-myocardial infarction angina by intra-aortic balloon pumping and emergency revascularisation. *Journal of Thoracic and Cardiovascular Surgery*, **74**, 299–306 (1977).
5. Bleifeld, W., Mathey, D., Hanrath, P., Buss, H., and Effert, S. Infarct size estimated from serial serum creatine phosphokinase in relation to left ventricular hemodynamics. *Circulation*, **55**, 303–11 (1977).
6. Bonte, F. J., Parkey, R. W., Graham, K. D., Moore, J., and Stokely E. M. A new method for radionuclide imaging of myocardial infarcts. *Radiology*, **110**, 473–4 (1974).
7. Carter, C. L. and Amundsen, L. R. Infarct size and exercise capacity after myocardial infarction. *Journal of Applied Physiology*, **42**, 782–5 (1977).
8. Cook, R. W., Edwards, J. E., and Pruitt, R. D. Electrocardiographic changes in acute subendocardial infarction. I. Large subendocardial and large non transmural infarcts. *Circulation*, **18**, 603–12 (1958).

9. Dunn, W. J., Edwards, J. E., and Pruitt, R. D. The electrocardiogram in infarction of the lateral wall of the left ventricle: a clinico-pathologic study. *Circulation*, **14,** 540–55 (1956).
10. Hamby, R. I., Hoffman, I. Hilsenrath, J., Aintablian, A., Shanies, S., and Padmanabhan, W. Clinical, hemodynamic and angiographic aspects of inferior and anterior myocardial infarctions in patients with angina pectoris. *American Journal of Cardiology*, **34,** 513–9 (1974).
11. Harnarayan, C., Bennett, M. A., Pentecost, B. L., and Brewer, D. E. Quantitative study of infarcted myocardium in cardiogenic shock. *British Heart Journal*, **32,** 728–32 (1970).
12. Inoue, M., Hori, M., Fukui, S., Abe, H., Minamino, T., Kodama, K., and Ohgitani, N. Evaluation of evolution of myocardial infarction by serial determinations of serum creatine kinase activity. *British Heart Journal*, **39,** 485–92 (1977).
13. Kahn, J. C., Gueret, P., Menier, R., Giraudet, P., Ben Farhat, M., and Bourdarias, J. P. Prognostic value of enzymatic (CPK) estimation of infarct size. *Journal of Molecular Medicine*, **2,** 223–31 (1977).
14. Kjekshus, J. K. and Sobel, B. E. Depressed myocardial creatine phosphokinase activity following experimental myocardial infarction in rabbit. *Circulation Research*, **27,** 403–14 (1970).
15. Maroko, P. R., Kjekshus, J. K., Sobel, B. E., Watanabe, T., Covell, J. W., Ross, J., Jr., and Braunwald, E. Factors influencing infarct size following experimental coronary artery occlusions. *Circulation*, **43,** 67–82 (1971).
16. Maroko, P. R., Libby, P., Covell, J. W., Sobel, B. E., Ross, J., Jr., and Braunwald, E. Precordial ST segment mapping: an atraumatic method for assessing alterations in the extent of myocardial injury. The effects of pharmacologic and hemodynamic interventions. *American Journal of Cardiology*, **29,** 223–30 (1972).
17. Mathey, D., Bleifeld, W., Buss, H., and Hanrath, P. Creatine kinase release in acute myocardial infarction: correlation with clinical, electrocardiographic and pathological findings. *British Heart Journal*, **37,** 1161–8 (1975).
18. Mathey, D., Bleifeld, W., Hanrath, P., and Effert, S. Attempt to quantitate relation between cardiac function and infarct size in acute myocardial infarction. *British Heart Journal*, **36,** 271–9 (1974).
19. Miller, R. R., Hughes, J. L., Salel, A. F., Massumi, R. A., Zelis, R., Mason, D. T., and Amsterdam, E. A. Relation of the electrocardiogram to ventricular function and clinical status in ischemic heart disease (abstr.). *Annals of Internal Medicine*, **76,** 866 (1972).
20. Miller, R. R., Olson H. G., Vismara, L. A., Bogren, H. G., Amsterdam, E. A., and Mason, D. T. Pump dysfunction after myocardial infarction: importance of location, extent and pattern of abnormal left ventricular segmental contraction. *American Journal of Cardiology*, **37,** 340–4 (1976).
21. Niarchos, A. P. and McKendrick, C. S. Prognosis of pericarditis after acute myocardial infarction. *British Heart Journal*, **35,** 49–54 (1973).
22. *Nomenclature and Criteria for Diagnosis of Disease of the Heart and Great Vessels*. The criteria committee of the New York Heart Association. 7th ed., Little, Brown and Company, Boston, 1973, p. 95.
23. Norris, R. M., Whitlock, R. M. L., Barratt-Boyes, C., and Small, Ch. W. Clinical measurement of myocardial infarct size. Modification of a method for the estimation of total creatine phosphokinase release after myocardial infarction. *Circulation*, **51,** 614–20 (1974).
24. Page, D. L., Caufield, J. B., Kastor, J. A., DeSanctis, R. W., and Sanders, C. A. Myocardial changes associated with cardiogenic shock. *New England Journal of Medicine*, **285,** 133–7 (1971).
25. Rao, P. S. and Mueller, H. Creatine phosphokinase MB profile. Reflection of evolution of myocardial infarct (abstr.). *Circulation Supplement II*, **53** and **54,** II-27 (1976).
26. Ratshin, R. A. Massing, G. K., and James, T. N. The clinical significance of the location of acute myocardial infarction. In *Myocardial Infarction* (Eds. Corday, E. and Swan, H. J. C.) Williams and Wilkins, Baltimore, 1973, pp. 77–85.
27. Rigaud, M., Bardet, J., Ben Farhat, M., and Bourdarias, J. P. Cineangiographic assessment of left ventricular performance in acute myocardial infarction. *Symposium in Bordeaux, 22th October (1976): Effets de l'Anoxie et de l'Ischémie Myocardique sur la Fonction Ventriculaire Gauche.*

28. Roberts, R., Henry, P. D., and Sobel, B. E. An improved basis for enzymatic estimation of infarct size. *Circulation*, **52**, 743–54 (1975).
29. Roe, Ch. R., Cobb, F. R., and Starmer, C. F. The relationship between enzymatic and histologic estimates of the extent of myocardial infarction in conscious dogs with permanent coronary occlusion. *Circulation*, **55**, 438–48 (1977).
30. Roe, Ch. R. and Starmer, C. F. A sensitivity analysis of enzymatic estimation of infarct size. *Circulation*, **52**, 1–5 (1975).
31. Rogers, W. J., McDaniel, H. G., Smith, L. R., Mantle, J. A., Russell, R. O., Jr., and Rackley, C. E. Correlation of angiographic estimates of myocardial infarct size and accumulated release of creatine kinase MB isoenzyme in man. *Circulation*, **56**, 199–205 (1977).
32. Rosalki, S. B. Improved procedure for serum creatine phosphokinase determination. *Journal of Laboratory and Clinical Medicine*, **69**, 696–705 (1967).
33. Rosati, R. In discussion of: Pitt, B. Natural history of myocardial infarction and its prodromal syndromes. *Circulation Supplement I*, **53**, (1976).
34. Russel, R. O., Jr., Hunt, D., and Rackley, C. E. Left ventricular hemodynamics in anterior and inferior myocardial infarction. *American Journal of Cardiology*, **32**, 8–16 (1973).
35. Savage, R. M., Wagner, G. S., Ideker, R. E., Podolsky, S. A., and Hackel, D. B. Correlation of post-mortem anatomic findings with electrocardiographic changes in patients with myocardial infarction. Retrospective study of patients with typical anterior and posterior infarcts. *Circulation*, **55**, 279–85 (1977).
36. Selvester, R. H., Wagner, J. O., and Rubin, H. B. Quantitation of myocardial infarct size and location by electrocardiogram and vectocardiogram. In *Boerhave Course in Quantitation in Cardiology* (Ed. Snellen, H. A.), Leyden University Press, The Netherlands, 1972, pp. 31–44.
37. Shell, W. E., Kjekshus, J. K., and Sobel, B. E. Quantitative assessment of the extent of myocardial infarction in the conscious dog by means of analysis of serial changes in serum creatine phosphokinase activity. *Journal of Clinical Investigation*, **50**, 2614–25 (1971).
38. Shell, W. E., Lavelle, J. F., Covell, J. W., and Sobel B. E. Early estimation of myocardial damage in conscious dogs and patients with evolving acute myocardial infarction. *Journal of Clinical Investigation*, **52**, 2579–90 (1973).
39. Shell, W. E. and Sobel, B. E. Protection of jeopardized ischemic myocardium by reduction of ventricular afterload. *New England Journal of Medicine*, **291**, 481–6 (1974).
40. Shell, W. E. and Sobel, B. E. Biochemical markers of ischemic injury. *Circulation Supplement I*, **53**, 98–106 (1976).
41. Sobel, B. E., Bresnahan, G. F., Shell, W. E., and Yoder, R. D. Estimation of infarct size in man and its relation to prognosis. *Circulation*, **46**, 640–8 (1972).
42. Sobel B. E., Roberts, R., Ambos, D., Oliver, G. C. H., and Cox, J. R. The influence of infarct size on ventricular dysrrhythmia (abstr.). *Circulation Supplement III*, **49** and **50**, III-110 (1974).
43. Stenson, R. E., Flamm, M. D., Jr., Zaret, B. L., and McGowan, R. L. Transient ST segment elevation with post-myocardial infarction angina: prognostic significance. *American Heart Journal*, **89**, 449–54 (1975).
44. Thadani, U., Chopra, M. P., Aber, C. P., and Portal, R. W. Pericarditis after acute myocardial infarction. *British Medical Journal*, **2**, 135–7 (1971).
45. Wackers, F. J., Lie, K. I., Becker, A. E., Durrer, D., and Wellens, H. J. J. Coronary artery disease in patients dying from cardiogenic shock or congestive heart failure in the setting of acute myocardial infarction. *British Heart Journal*, **38**, 906–10 (1976).
46. Wilkinson, R. S., Schaefer, J. A., and Abildskov, J. A. Electrocardiographic and pathologic features of myocardial infarction in man. *American Journal of Cardiology*, **11**, 24–35 (1963).
47. Wolk, M. J., Scheidt, S., and Killip, T. Heart failure complicating acute myocardial infarction. *Circulation*, **45**, 1125–38 (1972).

CHAPTER 14

Enzymatic infarct sizing: factors influencing the choice of the marker enzyme

W. Th. Hermens, A. Van der Laarse, and S. A. G. J. Witteveen

INTRODUCTION 339
RELATIONSHIP BETWEEN ENZYME RELEASE AND INFARCT SIZE: POSSIBLE
    INTERFERING FACTORS 340
    Individual Variations in the Enzyme Content of Normal Myocardium 341
    Individual Variations in Factors That Prevent Enzymes from Reaching
        the Blood Stream 342
    Individual Variations in Circulatory Parameters 347
SYSTEMATIC AND NON-SYSTEMATIC ERRORS IN CALCULATIONS OF
    ENZYMATIC INFARCT SIZE 348
CONCLUDING COMMENTS 351
REFERENCES 351

INTRODUCTION

Following the appreciation of the clinical importance of serum enzyme determinations in the diagnosis of acute myocardial infarction, several authors investigated the relation between enzyme release and infarct size. In these studies[11,17,31] aspartate aminotransferase (AST) and lactate dehydrogenase (LD) were measured in patients during the acute phase of myocardial infarction and the results were related to infarct size as assessed morphologically at post-mortem. In these studies a significant, though not impressive, correlation was found. These studies were stimulated by earlier studies in the dog[1,20,27] which also used AST and LD and which demonstrated a correlation. With the exception of one investigation[17] where clearance rates from the circulation were already taken into account, these human studies only used maximal serum activities to estimate enzyme release from the infarcted tissue and the fact that correlations were insufficient to allow accurate determination of infarct size in individual cases, was attributed to such factors as insufficient sampling, differences in clearance rates, and non-cardiac sources of enzyme release. However, in some recent studies in the dog, where some of these problems were avoided, poor[10] or zero[26] correlations between the morphological assessment of infarct size and the calculated release of creatine kinase (CK) were still observed. The only exception to this rather discouraging state of affairs was a recent study[3] in a group of 15 patients, which involved frequent sampling for CK measurements and which reported an excellent correlation between infarct size and serum CK. While these conflicting results may question the validity of the proposed relationship between enzyme release and infarct size, a number of studies in dogs[30] and baboons[40] have shown a good correlation between time-integrated serum CK activities and myocardial CK depletion. In addition, some recent studies[36] in the dog have shown a good correlation between morphometric infarct size and myocardial CK depletion. These combined studies would suggest that the model based on the measurement of enzyme release may well be sound under certain conditions but that a number of factors may interfere in the relationship between cell death and enzyme release thus accounting for the occasionally observed poor correlations.

THE RELATIONSHIP BETWEEN ENZYME RELEASE AND INFARCT SIZE: POSSIBLE INTERFERING FACTORS

There are a number of factors which may interfere with the quantitative relationship between cell death and the appearance of enzymes in the serum. One of these may be the leakage of enzymes from cells which are not dying. The occurrence of such a situation may be supported by studies of reperfusion[16,28] and by histochemical studies[2] of isoproterenol-induced myocardial damage. Thus if reperfusion is able to induce an *additional* release of enzyme the absence or presence of any element of reperfusion during infarction may affect the enzyme release profile. Such a situation may well explain the relatively small enzyme release observed with large infarcts[12,26] where collateral flow may be limited.

Another factor which may distort the relationship between infarct size and the plasma levels of specific enzymes or isoenzymes may be the occurrence of changes in the distribution or nature of the enzyme, an example here may be the the changes in isoenzyme distribution observed in hypoxia[6] and hypertrophy.[21]

Other important factors may be the occurrence of large variations in the normal content of the enzyme between hearts, the occurrence of various effects which may prevent enzymes from reaching the blood stream or the variation of various circulatory parameters such as the magnitude of the extravascular distribution space.

In the following section we will discuss a number of these factors which may influence the serum enzyme level and its relationship to infarct size.

The discussion will be based mainly upon evidence obtained from comparison of three different enzymes measured simultaneously in the same patients and it will lead to the conclusion that interindividual comparison of enzymatic infarct size is justified *if* the appropriate enzymes are chosen. In subsequent sections we will go on to discuss systematic and non-systematic errors in the calculation of enzymatic infarct size. These errors may well be responsible for the conflicting results described in the introduction and they may well depend upon the choice of marker enzyme.

Individual Variations in the Enzyme Content of Normal Myocardium

The enzyme content of tissue can be expressed in many ways and Table 1 shows the variability in the AST, CK, and α-hydroxybutyrate dehydrogenase (HBD) content of freshly obtained porcine heart when measured in 4 different ways. In this study tissue samples of about 0.5 gram wet weight were weighed and homogenized in physiological saline and the enzyme activity of the homogenate was related to wet weight, dry weight, total protein, and to DNA content. Analysis of the simplest experimental procedure namely the expression of enzyme activity per unit wet weight, shows that errors of at least 5% are to be expected. The results in the first column of Table 1 would indicate that differences in enzyme content per unit wet weight of different regions of the same heart are minimal while differences between hearts are less than a few per cent. The somewhat larger variations found in the remaining columns of Table 1 can be adequately explained by the larger experimental errors involved.

Data on other species show the same order of variation. In 32 samples of dog left ventricle a variation of 5% was observed[30] for CK activity. Fifteen samples of rat heart showed standard deviations for AST, CK, and HBD of ±14%, ±14%, and ±13% respectively.[38] Enzyme data on freshly obtained

Table 1. Variability of Enzyme Content of Porcine Myocardium Expressed per Unit of Wet Weight, Dry Weight, DNA, and Total Protein

	Wet weight			Dry weight (freeze drying)			DNA (according to Burton, *Biochem. J.*, **62,** 315–22 (1956))			Total Protein (according to Lowry)		
	AST	CK	HBD	AST	CK	HBD	AST	CK	HBD	AST	CK	HBD
A	7.4	12.5	10.2	14.7	10.5	14.3	13.8	15.6	11.1	9.6	13.5	12.7
B	7.4	14.5	12.9	11.8	13.0	17.1	14.7	15.5	16.9	13.2	14.4	18.0
C	4.6	5.8	7.8	5.1	8.1	7.5	15.2	15.8	16.7	8.4	9.7	10.7

Figures indicate coefficients of variation, i.e. standard deviations expressed as a percentage of mean values.
A: left ventricle samples ($n = 11$) from different hearts.
B: septal samples ($n = 11$) from different hearts.
C: left ventricle samples ($n = 7$) from a single heart.

human myocardium are scarce and may be influenced by the history of the patient, however, available data would suggest that a similar order of variability exists as for animal hearts. In samples obtained from 15 patients at thoracic surgery, standard deviations of ±19%, ±10%, and ±11% of mean values were found for AST, CK, and HBD respectively.[39] After subtraction of the experimental error the results indicate that the interindividual variability of CK and HBD probably does not exceed a few per cent. Furthermore, from the data presented below (see Table 2) on human myocardium it would appear that for HBD at least, the differences in enzyme content between different regions of the left ventricle are minimal.

Individual Variations in Factors that Prevent Enzymes from Reaching the Blood Stream

Incomplete recovery of enzymes from necrotic tissue in blood can be caused by incomplete cullular release or by local denaturation, i.e. inactivation before the blood stream is reached. Localization in organelles or adsorption to local structures can retard release of enzyme and enhance the possibility of local denaturation. For instance, it has been demonstrated in patients that the appearance of mitochondrial enzymes after acute myocardial infarction has a 12–24 hour delay compared to the appearance of cytoplasmic enzymes.[33] It has been shown in rat myocardium that mitochondrial CK was not released into the circulation after hypoxic damage,[2] and the same phenomenon was demonstrated for mitochondrial AST after experimental infarction in dogs.[15] Even cytoplasmic enzymes undergo a time delay in their release into the blood, caused by impairment of local circulation and this increases the possibility of local denaturation. Studies in the dog have demonstrated that a large proportion of the enzyme in necrotic tissue is transported by cardiac lymph[14] and also that CK, in contrast to HBD, undergoes denaturation by exposure to dog lymph.[24]

To investigate the stability of enzymes in human myocardium after circulatory arrest, left ventricular tissue samples were analysed from patients who had died without a previous history of heart disease. Table 2 shows that HBD remains stable for at least 36 hours after death, while CK and AST show an unpredictable but mutually correlated denaturation to levels of about 30% of normal during this period. These data are supported by a study on *in vitro* enzyme release from human myocardium,[8] showing that AST, CK, and HBD were released in quantities representing 35%, 43%, and 82% respectively of the amounts originally present. Only for HBD could the remaining 18% be recovered in the residual tissue. For AST and CK only 33% and 16% could be recovered in the tissue, implying that 32% and 41% respectively were inactivated locally. This shows that HBD is remarkably stable, compared to AST and CK. Recently this latter conclusion was confirmed in studies with cultures of beating rat heart cells.[34]

Table 2. Stability of Enzymes in Human Left Ventricles After Death

Samples	Hours	AST	CK	HBD
1	4.5	77	602	113
2	6.0	87	616	117
3	11.0	79	377	120
4	11.0	59	397	122
5	11.5	41	197	104
6	13.0	79	454	129
7	14.0	76	458	108
8	20.0	58	338	115
9	24.0	58	202	128
10	32.5	70	454	131
11	34.0	52	489	123
12	36.0	54	438	110

Samples were obtained at the indicated time delay after death. Activities were determined using standard methods and expressed in international units (iu) per gram wet weight of tissue.

Enzyme release from myoblasts in these cultures started about 3 hours after onset of anoxia and lasted for about 16–20 hours, after which time the cells had become totally depleted of enzyme. The time course for release of CK was identical to that of HBD, however, CK was inactivated extracellularly much more quickly than was HBD (Figure 1).

From *in vivo* experiments in dogs[22] it was calculated that 24 hours after myocardial infarction only 11% of the CK activity originally present in the infarcted tissue had reached the blood stream while 26% remained in the myocardium. This implies that 63% of the original CK activity was not recovered. Using a different circulatory model to study enzyme release in patients with acute myocardial infarction[38] it was calculated that AST, CK, and HBD are released in quantities representing 25%, 45%, and 80% respectively of the amounts originally present in the damage area. Considering the discrepancies between *in vitro* and *in vivo* recovery of released CK, one has to keep in mind that *in vivo* results depend strongly upon the parameter values of the circulatory model used, such as the magnitude of the extravascular pool and the rate of exchange of enzyme between vascular and extravascular space.[7] As there is considerable uncertainty about these parameters at the present time,[39] the quoted *in vivo* studies do not represent conclusive evidence for local denaturation of AST and CK. Several factors that could cause an underestimation of the amount of enzyme that actually reaches the blood stream will be considered in the next section.

If, due to local denaturation, AST and CK are only partially recovered in the stream, it is of great importance to know whether this process is dependent upon circulatory conditions because, if this is so, significant

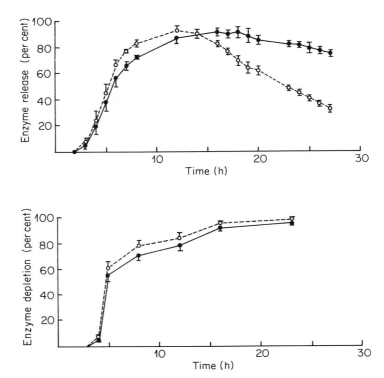

Figure 1. Release of HBD (●) and CK (○) from cultured rat heart cells during anoxic incubation. Duration of anoxia is indicated in hours. Enzyme release into the surrounding medium and enzyme depletion in the cultures is given as a percentage of the enzyme activity in a control culture. Each point represents the mean of 3–6 observations. Vertical bars represent SEM

differences in the recovery between individuals could invalidate any interindividual comparison of enzymatic infarct size measurement. There is evidence however that under clinical conditions such as acute myocardial infarction and cardiac surgery individual variations in the degree of local denaturation do not occur for AST and CK. In support of this, and as can be seen from Table 2, there is a relatively close correlation between the degradation of AST and CK. This implies that if variations in local denaturation cause significant variations in the amount of enzyme reaching the blood stream, then enzymatic infarct size as calculated from AST and CK should show a better correlation than if calculated from a stable enzyme such as HBD. This hypothesis was tested in 14 patients with acute myocardial infarction, total enzyme release was calculated[39] during the first 48 hours after the onset of symptoms and the results were as follows:

correlation HBD/AST: $r = 0.88$
correlation HBD/CK: $r = 0.63$
correlation AST/CK: $r = 0.64$

Thus there was no evidence for a better correlation between infarct sizes calculated from AST and CK.

A second argument against the influence of circulatory conditions on the recovery of CK follows from a comparison of patients with acute myocardial infarction with patients undergoing heart surgery. Figure 2 shows that in myocardium damaged during surgery a much more accelerated washout of enzymes occurred compared with that observed with myocardial infarction. This rapid washout in the surgery group, which is characteristic for reperfusion,[28] leaves no time for local denaturation and should therefore cause a marked increase in the total amount of CK released. Such an increase should become apparent by comparing the total release of CK to the total release of an enzyme such as HBD which is not subject to local denaturation. As skeletal muscle damage is present in the surgery group, the heart-specific isoenzyme CK-MB was compared with HBD. Determination of the linear regression equations between calculated amounts of released CK-MB and HBD gave the following ratios:

Infarction group ($n = 17$):

$$\frac{\text{amount of released CK-MB activity}}{\text{amount of released HBD activity}} = 0.27 \pm 0.08$$

Surgery group ($n = 32$):

$$\frac{\text{amount of released CK-MB activity}}{\text{amount of released HBD activity}} = 0.23 \pm 0.03$$

where the indicated errors represent 95% confidence limits.

It can be concluded that these two groups show no significant difference in the relative amounts of released CK-MB. It would appear that for AST and CK a fixed percentage of the amount of enzyme originally present in the tissue eventually reaches the blood stream. This percentage does not seem to be greatly influenced by circulatory factors. It should be noted here that reperfusion experiments in dogs did show an altered relationship between morphologically assessed infarct size and calculated release of CK.[16] However these experimental interventions probably cause circulatory changes not found under clinical conditions.

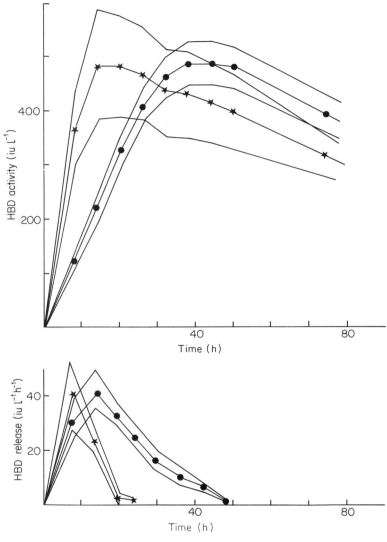

Figure 2. Mean plasma activities in international units per litre and calculated release in international units per litre per hour of HBD in patients suffering from myocardial infarction (●, $n = 24$) and patients undergoing heart surgery (★, $n = 25$). Continuous curves indicate SEM

Individual Variations in Circulatory Parameters

It has been known for a long time that enzymes injected into the blood stream are distributed between an intravascular and an extravascular compartment.[29] The presence of an extravascular enzyme pool causes an underestimation of the clearance rate of that enzyme as calculated from the apparent disappearance rate from the plasma. The true clearance rate is underestimated because during the clearance phase one observes the combined effect of clearance from the plasma and resupply from the extravascular space.[7] If extravascular denaturation of the enzyme occurs then this underestimation is even more pronounced. Extravascular breakdown of CK has also been suggested from experiments in dogs in which CK injected intravascularly showed a monophasic disappearance.[22,23] This implies that no extravascular saturation occurs thus indicating extravascular removal of enzyme. In contrast a biphasic disappearance is observed when AST or LD is injected in dogs[35] or sheep.[4] It should be realized however that if the establishment of an equilibrium between the vascular and extravascular pools is fast compared to metabolic clearance (which is certainly true for HBD) then the effect of extravascular breakdown cannot be distinguished from the presence of a larger extravascular pool. In this case, the essential question is how large is the extravascular pool?

It has been reported[37] that the clearance from the plasma of a number of enzymes (including AST, CK, and HBD) after acute myocardial infarction is remarkably regular. When the clearance was plotted on a semilogarithmic scale, straight lines were observed with correlation coefficients close to unity in most cases. For CK and HBD this observation was substantiated by studies[18,19] which showed that the slope of the semilogarithmic plot (k_d) could be determined within 95% confidence limits of less than 5% in 80% of the cases. This implies that the apparent disappearance rate is strictly proportional to the plasma concentration of the enzyme. This regular clearance is not disturbed in patients suffering from cardiac failure, in which oedema and diuretic therapy are responsible for considerable shifts in fluid balance.[32] Enlargement of extravascular space due to oedema is expected to decrease the value of k_d, but this effect was not observed for CK and HBD.[19,32] In experimental infarction in dogs k_d was shown to be unaffected by severe haemodynamic interventions[22,5] and by pharmacological interventions.[23] The insensitivity of k_d to shifts in fluid balance can only be explained by assuming a small extravascular enzyme pool. For AST, CK, and HBD it has been estimated that only 10–30% of the total enzyme activity is located extravascularly.[39] With reference to this extravascular pool it should be noted that calculations of enzymatic infarct size are not influenced by assuming either a large extravascular fluid volume with low enzyme concentrations or a smaller extravascular volume containing correspondingly higher concentrations of enzyme.[9]

Although k_d can be accurately determined in the majority of cases, several studies have demonstrated large individual variations in this parameter.[18,39] This implies that no fixed mean value of k_d can be used without introducing considerable errors.

In conclusion, we propose that the rate of enzyme clearance from the plasma is not influenced markedly by changes in the patients' conditions and is strictly proportional to plasma enzyme concentrations. This means that calculation of enzyme release into the blood stream from plasma values is probably not disturbed by other unpredictable individual variations than variations in k_d.

SYSTEMATIC AND NON-SYSTEMATIC ERRORS IN CALCULATIONS OF ENZYMATIC INFARCT SIZE

If the conclusions in the preceding section are correct, there are no basic objections to interindividual comparison of quantitated enzyme release after myocardial damage. However, the question remains with what precision can enzymatic infarct size be calculated?

The large potential errors highlighted in several studies[18,25] concern mainly the conversion of calculated enzyme release to grams of destroyed myocardium (since it has been suggested[18] that enzyme release expressed in units per litre of plasma is a better clinical index than when expressed as an amount of tissue).

We will now discuss which errors can occur in the calculation of enzymatic infarct size as defined in the preceding section. The basic equation used for this calculation is

$$A(t) = V_i(t)C_i(t) + V_e(t)C_e(t) + k \int_0^t V_i(\tau)C_i(\tau)\,d\tau \quad (1)$$

where $A(t)$ is the total amount of enzyme released from the infarcted tissue at time t; $V_i(t)$ and $V_e(t)$ are the intra- and extravascular volumes at time t; $C_i(t)$ and $C_e(t)$ are the intra- and extravascular enzyme concentrations at time t; and k is the true clearance constant. Normal steady-state enzyme concentrations are subtracted from the measured values to obtain the net enzyme release associated with myocardial damage. The first two terms in equation (1) represent the amount of enzyme that is still present in the circulation at time t; the last term gives the amount of enzyme that has been cleared prior to time t. Clearance is assumed to be proportional to the plasma enzyme concentration.

In the literature equation (1) is generally simplified to:

$$A(t) = V_i C_i(t) + k_d V_i \int_0^t C_i(\tau)\,d\tau \quad \text{or} \quad Q(t) = C_i(t) + k_d \int_0^t C_i(\tau)\,d\tau \quad (2)$$

where $Q(t)$ is the amount of enzyme released per unit of plasma volume.

As a result the existence of an extravascular pool is neglected in two different ways. In the first place, by simply equating the true clearance constant k to the apparent disappearance constant k_d, the former is underestimated as discussed in the preceding section. In the second place the amount of enzyme present in the extravascular compartment is neglected. Neglect of this extravascular enzyme activity is only justified when t is large enough to allow enzyme concentrations to return to normal values. However calculation of enzyme release over such long periods of time has other drawbacks as will be discussed below.

Simplification of equation (1) to equation (2) introduces still another error because the changes in time of the vascular volume V_i are neglected. During the acute phase of myocardial infarction, plasma volume is generally reduced.[13,32] This reduction however shows large individual variations. As the clearance of enzyme is strictly proportional to its plasma concentration, a reduced plasma volume implies an overestimation of the total amount of enzyme cleared. In a study in which the changes in plasma volume were monitored by measuring the corresponding changes in haematocrit, it was demonstrated[32] that if these changes were not taken into account errors of up to 30% were introduced in enzymatic infarct size calculated for HBD.

Apart from errors introduced by its simplified form, equation (2) contains three sources of error in its last term: k_d, the choice of the time interval of integration, and the numerical approximation of the integration. Although errors in k_d are small in the majority of cases, errors up to 15% are observed.[18] This error will increase considerably if a fixed mean value for k_d is used.

Most authors chose the period of time over which the last term of equation (2) is integrated to be sufficiently long to allow enzyme activities to return to normal values. Although this procedure eliminates the problem of estimation of the extravascular enzyme pool, it should be realized that enzyme release in myocardial infarction ends within approximately 48 hours after first symptoms.[19,39] By extending this period of time to 150 hours (for CK) or to 400 hours (for HBD) a considerable error may be introduced, as small over- or underestimations of the steady-state plasma enzyme activity lead to small over- or underestimations of steady-state enzyme release. Integrated over such long periods (150–400 hours) these errors accumulate resulting in considerable amounts of enzyme being erroneously included in the calculated enzymatic infarct size. In addition to this, there is an increasing risk of including enzyme activity of non-cardiac origin which may be released during the post-infarction period. Table 3 shows that integration of HBD over long time intervals distorts the correlation with results obtained for AST and CK. This is mainly caused by the relatively high steady-state plasma activity of HBD.

The magnitude of the error in the numerical approximation of the

Table 3. Effect of Period of Integration on Correlation of Enzymatic Infarct Sizes ($n = 15$) Calculated for Different Enzymes

Enzymes	Integration over first 48 hours	Integration over total period of observation (150–400 h)
HBD/AST	0.88	0.75
HBD/CK	0.63	0.52
AST/CK	0.64	0.68

Figures indicate correlation coefficients (see text).

integration was estimated by fitting the function $C(t) = Kt^3 \exp(-kt) + C_0$ to curves of CK and HBD obtained from 15 patients and subsequently comparing the exact integral of this function over 48 hours with approximations obtained by the trapezoidal rule, i.e. by simply connecting data points by straight lines and calculating the area enclosed by these lines and the time axis. For time intervals between successive samples up to 6 hours, errors are less than 1 per cent and can be neglected. This error remains small because there is a succession of regions in which the integral is systematically over- and underestimated such that errors partly compensate.

From the data presented above it is evident that the last term of equation (2), the clearance term, may contain individual errors of at least 20%. In contrast, the term $C_i(t)$ only contains the experimental error in enzyme determinations which is 1–3% for AST, CK, and HBD.[38] So it is crucial to obtain conditions in which the relative contribution of the clearance term is small and this is the most important reason for restricting the period of integration to 48 hours. If the integral is calculated over a time interval in which normal steady-state values are reached then the full 20% error may be present in enzymatic infarct. Furthermore, the advantage of using a slowly metabolized enzyme becomes apparent when it is considered that at $t = 48$ hours the contribution of the clearance term will be small for such an enzyme. It has been calculated that for HBD a 15% error in the clearance term introduces only a 2% error in enzymatic infarct size calculated at $t = 48$ hours, while the corresponding error for CK, which is metabolized much faster than HBD can be 11%.[9]

The problems mentioned here for CK are even worse for the heart-specific isoenzyme CK-MB. The clearance of this enzyme is so fast that after $t = 48$ hours plasma levels have often become too low to allow any determination of k_d. This implies that quantitative evaluation of CK-MB release is subject to very large errors.

The correlation between HBD and AST presented in the first column of Table 3 correspond to a mean deviation from a perfect linear relation ($r = 1$) of 17% per data point. These deviations are probably mainly due to errors in the estimation of the amount of released AST, since this enzyme is quickly metabolized compared to HBD. The second column of Table 3 shows that integration over long time intervals (150–400 hours) causes even larger errors in the correlation between AST and HBD. Here a mean deviation of 26% from a perfect linear relation can be calculated. Together with a possible 30% error associated with plasma volume variations (this error was not present in the data from Table 3, as enzymes were measured simultaneously in the same patients) an overall error of 60% may be present in enzymatic infarct size calculated for quickly metabolized enzymes such as CK and AST. This analysis may provide, in part at least, an explanation for the discouraging results mentioned in the literature.

CONCLUDING COMMENTS

In conclusion, errors in the calculation of individual enzymatic infarct sizes can be reduced to less than 10% if (i) corrections are made for changes in plasma volume; (ii) the apparent disappearance constant from the plasma (k_d) is determined for individual patients; (iii) the time delay between successive samples does not exceed 6 hours; (iv) calculation is restricted to the period of actual enzyme release, i.e. about 48 hours for actue myocardial infarction; (v) enzymes are used which are slowly metabolized, such as HBD.

REFERENCES

1. Agress, C. M., Jacobs, H. I., Glassner, H. F., Lederer, M. A., Clark, W. G., Wroblewski, F., Karmen, A., and La Due, J. S. Serum transaminase levels in experimental myocardial infarction. *Circulation*, **XI**, 711–3 (1955).
2. Baba, N., Kim, S., and Farrell, E. C. Histochemistry of creatine phosphokinase. *Journal of Molecular and Cellular Cardiology*, **8**, 599–617 (1976).
3. Bleifeld, W., Mathey, D., Hanrath, P., Buss, H., and Effert, S. Infarct size estimated from serial serum creatine phosphokinase in relation to left ventricular hemodynamics. *Circulation*, **55**, 303–11 (1977).
4. Boyd, J. W. The rates of disappearance of L-lactate dehydrogenase isoenzymes from plasma. *Biochimica et Biophysica Acta*, **132**, 221–31 (1967).
5. Fleisher, G. A. and Wakim, K. G. The fate of enzymes in body fluids—experimental study. I. Disappearance rates of glutamic-pyruvic transaminase under various conditions. *Journal of Laboratory and Clinical Medicine*, **61**, 76–85 (1962).
6. Hammond, G. L., Nadal-Ginard, B., Talner, N. S., and Markert, C. L. Myocardial LDH isoenzyme distribution in the ischemic and hypoxic heart. *Circulation*, **53**, 637–43 (1976).
7. Hermens, W. Th. Dose calculation of human factor VIII and factor IX concentrates for infusion therapy. In *Handbook of Hemophilia* (Eds. Brinkhous, K. M. and Hemker, H. C.), American Elsevier Publishing Company, New York, 1975, pp. 569–89.
8. Hermens, W. Th., Witteveen, S. A. G. J., Hollaar, L., and Hemker, H. C. Effect of a thrombolytic agent (urokinase) on necrosis after acute myocardial infarction. In *Recent Advances in Studies on Cardiac Structure and Metabolism* (Eds. Roy, P-E. and Rona, G.) Vol. 10, University Park Press, Baltimore, 1975, pp. 319–29.
9. Hermens, W. Th. and Witteveen, S. A. G. J. Problems in estimation of enzymatic infarct size. *Journal of Molecular Medicine*, **2**, 233–39 (1977).
10. Jarmakani, J. M., Limbird, L., Graham, T. C., and Marks, R. A. Effect of reperfusion on myocardial infarct, and the accuracy of estimating infarct size from serum creatine phosphokinase in the dog. *Cardiovascular Research*, **10**, 245–53 (1976).

11. Kibe, O. and Nilsson, N. J. Observations on the diagnostic and prognostic value of some enzyme tests in myocardial infarction. *Acta Medica Scandinavica*, **182,** 597–610 (1967).
12. Killen, D. A. and Tinsley, E. A. Serum enzymes in experimental infarcts. *Archives of Surgery*, **92,** 418–22 (1966).
13. Kung-Ming Jan, Shu Chien, and Bigger, J. T. Observations on blood viscosity changes after acute myocardial infarction. *Circulation*, **51,** 1079–84 (1975).
14. Malmberg, P. Aspartase aminotransferase activity in dog heart lymph after myocardial infarction. *Scandinavian Journal of Clinical and Laboratory Investigation*, **30,** 153–8 (1972).
15. Malmberg, P. Enzyme composition of dog heart lymph after myocardial infarction. *Upsala Journal of Medical Science*, **78,** 73–7 (1973).
16. Maroko, P. and Vatner, S. F. Altered relationship between phosphokinase and infarct size with reperfusion in conscious dogs. *Journal of Molecular Medicine*, **2,** 309–15 (1977).
17. Meurman, L. and Ordell, R. Pathophysiological interpretations of serum enzyme activity vs. time curves in myocardial infarction. *Scandinavian Journal of Clinical and Laboratory Investigation*, Supplement, **76,** 70 (1963).
18. Norris, R. M., Whitlock, R. M. L., Barratt-Boyes, C., and Small, C. W. Clinical measurement of myocardial infarct size. *Circulation*, **51,** 614–20 (1975).
19. Norris, R. M., Howell, D., Whitlock, R. M. L., Heng, M. K., and Peter, T. Enzyme release after myocardial infarction: comparison of serial serum α-hydroxy-butyrate dehydrogenase with creatine phosphokinase levels. *European Journal of Cardiology*, **4**/4, 461–8 (1976).
20. Nydick, I., Wroblewski, F., and La Due, J. S. Evidence for increased serum glutamic oxolacetic transaminase (SGOT) activity following graded myocardial infarcts in dogs. *Circulation*, **XII,** 161–8 (1955).
21. Revis, N. W., Thomson, R. Y., and Cameron, A. J. V. Lactate dehydrogenase isoenzymes in the human hypertrophic heart. *Cardiovascular Research*, **11,** 172–6 (1977).
22. Roberts, R., Henry, P. D., and Sobel, B. E. An improved basis for enzymatic estimation of infarct size. *Circulation*, **52,** 743–54 (1975).
23. Roberts, R. and Sobel, B. E. Effect of selected drugs and myocardial infarction on the disappearance of creatine kinase from the circulation in conscious dogs. *Cardiovascular Research*, **11,** 103–12 (1977).
24. Robison, A. K., Gnepp, D. R., and Sobel, B. E. Inactivation of CPK in lymph. *Circulation Supplement II*, **51,** 5 (1975).
25. Roe, R. C. and Starmer, C. F. A sensitivity analysis of enzymatic estimation of infarct size. *Circulation*, **52,** 1–5 (1975).
26. Roe, R. C., Cobb, F. R., and Starmer, C. F. The relationship between enzymatic and histologic estimates of the extent of myocardial infarction in conscious dogs with permanent coronary occlusion. *Circulation*, **55,** 438–49 (1977).
27. Ruegsegger, P., Nydick, I., Freiman, A., and La Due, J. S. Serum activity patterns of glutamic oxaloacetic transaminase, glutamic pyruvic transaminase and lactic dehydrogenase following graded myocardial infarction in dogs. *Circulation Research*, **VII,** 4–10 (1959).
28. Sakai, K., Gebhard, M. M., Spieckermann, P. G., and Bretschneider, M. J. Enzyme release resulting from total ischemia and reperfusion in the isolated, perfused guinea pig heart. *Journal of Molecular and Cellular Cardiology*, **7,** 827–40 (1975).
29. Schultze, H. E. and Heremans, J. F. *Molecular Biology of Human Proteins*, Chap. 2. Elsevier Publishing Company, Amsterdam, 1966.
30. Shell, W. E., Kjekshus, J. K., and Sobel, B. E. Quantitative assessment of the extent of myocardial infarction in the conscious dog by means of analysis of serial changes in serum creatine phosphokinase activity. *Journal of Clinical Investigations*, **50,** 2614–25 (1971).
31. Sjögren, A. Left heart failure in acute myocardial infarction. *Acta Medica Scandinavica Supplement*, **501,** 7–82 (1970).
32. Smith, S. J., Bos, G., Hagemeyer, F., Hermens, W. Th., and Witteveen, S. A. G. J. Influences of changes in plasma volume on quantitation of infarct size in man by means of plasma enzyme levels. *Journal of Molecular Medicine*, **1,** 199–210 (1976).
33. Smith, A. F., Wong, P. C.-P., and Oliver, M. F. Release of mitochondrial enzymes in actue myocardial infarction. *Journal of Molecular Medicine*, **2,** 265–9 (1977).
34. Van der Laarse, Hollaar, L., van der Valk, J. M., and Witteveen, S. A. G. J. Enzyme release from rat heart cultures during anoxia. *Journal of Molecular Medicine*, **3,** 123–31 (1978).

35. Wakim, K. G. and Fleisher, G. A. The fate of enzymes in body fluids—an experimental study. II. Disappearance rates of glutamic-oxalacetic transaminase I under various conditions. *Journal of Laboratory and Clinical Medicine*, **61,** 86–97 (1962).
36. Weiss, E. S., Ahmed, S. A., Welch, M. J., Williamson, J. R., Ter-Pogossian, M. M., and Sobel, B. E. Quantification of infarction in cross-sections of canine myocardium *in vivo* with positron emission transaxial tomography and ^{11}C-palmitate. *Circulation*, **55,** 66–73 (1977).
37. Witteveen, S. A. G. J., Hermens, W. Th., Hemker, H. C., and Hollaar, L. Quantitation of enzyme release from infarcted heart muscle. In *Ischemic Heart Disease* (Eds. De Haas, J. H., Hemker, H. C., and Snellen, H. E.), Leiden University Press, Leiden, 1970, pp. 36–42.
38. Witteveen, S. A. G. J. Assessment of the extent of a myocardial infarction on the basis of plasma enzyme levels. *Thesis*, Leiden, 1972.
39. Witteveen, S. A. G. J., Hemker, H. C., Hollaar, L., and Hermens, W. Th. Quantitation of infarct size in man by means of plasma enzyme levels. *British Heart Journal*, **37,** 795–803 (1975).
40. Yasmineh, W. G., Pyle, R. B., Cohn, J. N., Nicoloff, D. M., Hanson, N. Q., and Steels, B. W. Serial serum creatine phosphokinase MB isoenzyme activity after myocardial infarction. *Circulation*, **55,** 733–8 (1977).

CHAPTER 15

Infarct size quantification, present and future

R. M. Norris

INTRODUCTION	355
PRACTICAL SIGNIFICANCE OF INFARCT SIZE MEASUREMENT	356
BEDSIDE ASSESSMENT OF INFARCT SIZE	356
ELECTROCARDIOGRAPHIC ASSESSMENT OF INFARCT SIZE	357
METABOLIC APPROACHES TO INFARCT SIZING	358
USE OF THE ONE-COMPARTMENT MODEL FOR CALCULATION OF TOTAL CK APPEARANCE	358
PRACTICAL CONSIDERATIONS OF THE ONE-COMPARTMENT MODEL	359
THE USE OF CK—MB: DOES THIS IMPROVE THE ACCURACY OF THE METHOD?	363
CORRELATION OF THE CK METHOD WITH OTHER INDICES OF INFARCT SIZE: RELATION BETWEEN CK APPEARANCE IN BLOOD AND CK DEPLETION FROM MYOCARDIUM	364
CORRELATION OF CK WITH OTHER METABOLIC MARKERS	364
CORRELATION OF CK APPEARANCE WITH CLINICAL INDICES OF SEVERITY OF INFARCTION	365
QUANTIFICATION OF INFARCT SIZE BY MEASUREMENT OF UPTAKE OF METABOLIC SUBSTANCES	367
INCREASE OR REDUCTION OF INFARCT SIZE BY TREATMENT	369
CONCLUDING COMMENTS	371
REFERENCES	373

INTRODUCTION

Ischaemic heart disease is the greatest public health problem of Westernized societies, accounting for about one-third of all deaths and affecting mainly middle-aged people. Community studies from various parts of the world have shown that two-thirds of all deaths from ischaemic heart disease occur outside hospitals and one-third within hospitals, usually in Coronary Care Units. Most of the deaths outside hospital occur from ventricular fibrillation, an unpredictable yet treatable abnormality of heart rhythm which occurs most commonly in patients with severe coronary artery disease but not necessarily in those who have severe myocardial damage. Definitive treatment by electrical defibrillation ensures that few patients now die from this electrical abnormality once they have been admitted to hospital, but in-hospital mortality is still of the order of 15–20% despite effective treatment of arrhythmias. Moreover many hospital survivors have permanent myocardial damage which may limit their activities for the future or make survival from a subsequent infarct less likely.

PRACTICAL SIGNIFICANCE OF INFARCT SIZE MEASUREMENT

The continuing high hospital mortality rate from myocardial infarction occurs because of the mechanical effects of infarction which are manifest as the clinical syndromes of cardiac failure and cardiogenic shock. Autopsy studies[23,69] have shown that these complications are directly caused by loss of a critical amount of functional myocardium, i.e. a large myocardial infarct. Thus the major cause of in-hospital mortality from ischaemic heart disease is massive myocardial damage which is not amenable to known forms of treatment. If we are to learn how to control this damage we must first find ways for its accurate measurement.

BEDSIDE ASSESSMENT OF INFARCT SIZE

Every experienced cardiologist can make an accurate qualitative assessment of the severity of myocardial damage, even during the first few hours after onset of the infarct at a time when the affected myocardium is ischaemic rather than necrotic. Severe and long-continued pain, pallor and weakness, breathlessness and tachycardia indicating left ventricular failure, or a low systolic blood pressure indicating cardiogenic shock, are progressively more ominous indices of severity of cardiac damage. Studies which the author has made of factors affecting prognosis after infarction[62–64] showed that the most important factors which were measurable on first admission to hospital were a low systolic blood pressure, pulmonary venous congestion and oedema, or an enlarged heart—all indices of the cumulative effect of the severity of present and previous infarcts. Thus for routine clinical purposes the physician with his stethoscope and with the aid of a plain chest X-ray can guess intelligently at the size of the infarct and consequently the prognosis and need for therapeutic intervention in a particular case.

Attempts to improve on this qualitative assessment were first made using the traditional tools of the clinical cardiologist, the cardiac catheter and the electrocardiogram. The first of these, the cardiac catheter, has been adapted successfully so that it can be used at the bedside of acutely ill patients in intensive care units without the need for radiological screening. The Swan-Ganz thermodilution catheter[20] has an inflatable balloon at its tip which can be filled with 1 ml of air or CO_2 making it lighter than blood. After passage from an arm vein into one of the great veins, the balloon is inflated and the catheter tip is carried to the lungs allowing measurement of pulmonary wedge pressure (a reliable indirect indicator of left ventricular function) and the cardiac output in litres per minute. Knowing these measurements and the heart rate, left ventricular stroke work can be calculated, and this is related to the oxygen needs of the left ventricle. Measurements using the Swan–Ganz catheter have extended clinical observations enabling trends towards improvement or deterioration to be detected early on. However, in a sense they tell the physician what he knows already; they measure the functional state of the heart which is determined not only by the size of the infarct but also by the extent of previous cardiac damage and the mechanical efficiency of the remaining viable myocardium. Thus haemodynamic measurements allow only an indirect assessment of infarct size.

ELECTROCARDIOGRAPHIC ASSESSMENT OF INFARCT SIZE

The most striking early electrocardiographic manifestation of ischaemia is elevation of the ST segment, this in turn being caused by changes in depolarization and resting potential of myocardial cells associated with loss of intracellular K^+.[89] It was suggested[49,51] that the magnitude of ST segment elevation measured at 15 minutes after the onset of experimental myocardial ischaemia bore a direct relationship to the severity of the ischaemic injury, and that these observations could be extended to patients with anterior transmural infarction,[52] in whom the ST segment deflections could be measured by praecordial mapping over the anterior chest wall. This concept has been challenged on various grounds. First, geometrical considerations suggest that the magnitude of ST elevation will change in opposite directions with increasing ischaemic injury according to whether the recording electrode is situated over the surface of the heart or over the chest wall, and these predictions are borne out by experiment.[10,32] Second, later experiments have failed to show a simple correlation either between epicardial ST segment elevation and subjacent myocardial blood flow and myocardial creatine kinase activity,[26] or with the severity of subjacent histological changes.[34] Finally, the praecordial mapping technique has been applied to groups of patients, and these studies have failed to show any correlation between the extent of ST segment elevation and clinical indices of the severity of infarction.[61,110]

The praecordial electrocardiographic mapping method has been recommended in particular for the assessment of trends towards improvement or deterioration which might occur in individual patients with anterior wall infarction in whom therapeutic intervention is made.[47] It may be that there is a place for this somewhat restricted use of the method and that this brief analysis takes somewhat too pessimistic a view of the whole topic. However, it must be said that theoretical considerations[10,32] argue strongly against any direct relationship between the degree of ST segment elevation and the intensity of the underlying ischaemia or totality of consequent necrosis. Moreover it has been the author's experience that these theoretical objections are borne out by a lack of reliability which has been found in practical experiments. Thus the ST segment method cannot be recommended for the assessment of infarct size or for the intensity of ischaemia which should correlate with this.

METABOLIC APPROACHES TO INFARCT SIZING

Given a mass of tissue which is undergoing rapid and irreversible necrosis, a logical approach is to measure either the appearance in the blood or the uptake of metabolic substances from the infarcting tissue. Both the quantity and the rate of appearance or uptake can then be used as indices of the magnitude or rate of necrosis. Karmen, Wroblewski, and La Due in 1954[36] introduced the use of serum enzyme activity levels in particular aspartate aminotransferase (AST) and lactate dehydrogenase (LD) as aids to the clinical diagnosis of infarction. To these was added the use of creatine kinase (CK) in 1960.[15] It is standard practice in most hospital clinical laboratories for each of these enzymes to be measured once daily for the first 2–3 days after the patient's admission to hospital, a rise in all three enzyme activities being taken as specific for myocardial necrosis.

USE OF THE ONE-COMPARTMENT MODEL FOR CALCULATION OF TOTAL CK APPEARANCE

The historical development, current status, possibilities, and limitations of this model have been discussed in previous chapters. In the present chapter an attempt will be made to summarize the theoretical and practical advantages and disadvantages of this method which are relevant to its use in a Coronary Care Unit. For the purpose of the practical discussion it will be assumed that the one-compartment theory is correct; that is that CK diffuses out of necrotic myocardial cells, travels probably by way of the cardiac lymphatic vessels[42,43] to the blood stream, and is removed by the reticuloendothelial system, removal occurring exponentially as a first-order reaction. It is true that CK metabolism may not be as simple as is suggested by this model and it has been pointed out by Witteveen and colleagues[113] and again by Sobel's group[46,103] that there may be a second or extravascular distribution space for CK as has been postulated for other enzymes and proteins.[16,91] The existence of such an extravascular compartment in addition to the plasma distribution space would imply a rate constant for diffusion of enzyme activity between these two spaces, and in turn the probability that the measured decay rate (k_d) of CK activity in the plasma is under estimated and is not representative of the true k_d of total activity in the body.[98]

The picture is further complicated by the fact that it has recently been found[7,46,103] that k_d for exogenous injected CK in dogs is higher than is k_d for endogenous CK from the animals' own myocardial infarct. Moreover exogenous CK behaves differently after intravenous injection according to whether it is harvested from the myocardium or from the plasma of donor dogs.[103] Present evidence is insufficient to determine whether variation of k_d between endogenous enzyme, exogenous injected enzyme from myocardium, and exogenous enzyme from plasma are 'true' in the sense that they are caused by subtle biochemical differences between the three types of enzyme, or whether the measured plasma k_d is in some of these instances not representative of whole-body degradation. This could happen either because of movement of enzyme activity between vascular and extravascular compartments, or because of continuing appearance of enzyme activity from the infarct during the period when serum activity appears to be falling exponentially.

Although it must be accepted that large and unpredictable movements of enzymes between two or more body compartments would invalidate the one-compartment calculations, the experimental evidence one way or the other is insufficient at present for any judgment to be made. From the viewpoint of the clinician it is perhaps of more practical importance that the validity or otherwise of one-compartment calculations should be correlated in man with other established indices of infarct size or severity. If such comparisons show good correlations, the method can be recommended for practical use.

PRACTICAL CONSIDERATIONS OF THE ONE-COMPARTMENT MODEL

If we accept the one-compartment model as a basis for infarct sizing by the CK appearance method there are a number of factors which affect its validity and which must be clearly understood and appreciated. These may be listed as follows:

(i) It must be reasonably certain that CK appearing in the serum has all come from necrotic myocardial cells. In particular, skeletal muscle contains 5 times as much CK activity as does cardiac muscle,[104] and this may be released after repeated intramuscular injections,[56,115] surgery involving division of muscle fibres, systemic embolism from myocardial infarction, and possibly when tissues are hypoperfused during cardiogenic shock. In the author's hospital total CK release is still used as an index of infarct size if the patient has had only one intramuscular injection before coming into the hospital, but after admission to the Coronary Care Unit all injections are given either intravenously or subcutaneously. CK can also be released after cardioversion,[18] and whether this comes from the skeletal muscles or the heart it almost certainly does not come from irreversibly damaged cells. Infarct sizing by the CK method is accordingly abandoned in patients who have had cardiac arrest or elective cardioversion. If CK-MB is used as the index of infarct size (discussed later) intramuscular injections or cardiogenic shock would not be contraindications although cardioversion might still result in the leakage of some of the MB isoenzyme.[18]

(ii) Potential sources of inaccuracy in the calculation of total CK appearance need to be carefully considered. Total CK appearance is calculated from the expression $E(t) + k_d \sum E \, dt$[94,102] so that 3 variables must be considered.

First, the levels of enzyme activity (E) can be determined quite accurately if a standard method is used and careful attention is paid to both the reaction temperature and the degree of dilution at which activity is measured.[87,105] Second, the frequency of blood sampling for measurement of CK activity (dt) is of great practical importance because very frequent sampling can be distressing for patients, inconvenient for nursing staff, and expensive for the hospital laboratory. In studies with hourly sampling it was found that the accuracy of the calculations was impaired very little if only the 4 hourly levels were included compared with inclusion of all the hourly levels.[66] Blood is now routinely taken every 4 hours by sampling from the patients' intravenous drip with the use of a three-way tap. Third, accurate individualized determination of k_d is essential because calculations of total cumulative enzyme release show that the value that is obtained is directly related to the value for k_d which is entered into the calculation. Thus k_d must be determined for each individual patient, and the confidence limits for its determination must be sufficiently narrow for its value to be meaningful. By far the greatest source of error in the calculation of $E(t) + k_d \sum E \, dt$ lies in the value which is used for k_d.

The author's approach to the measurement of k_d[66] has been to take 5 or more points on the downslope of log CK activity at a time when it is declining most rapidly towards near-normal levels. The points that are included in the calculation are selected by eye, and care is taken to avoid the first part of the downslope; this in practice is nearly always less rapid than the terminal part, presumably because it is distorted by continued CK appearance. Having selected at least 5 (but usually more) 4-hourly points on the descending part of the log curve, the values are entered in to a desk-top computer (Hewlett-Packard 9100B) which is programmed to measure the slope of the line (k_d) and the 95% confidence limits of this slope. Ninety-five per cent confidence limits up to ±15% are accepted, although in practice it is nearly always possible to determine k_d within 95% confidence limits of

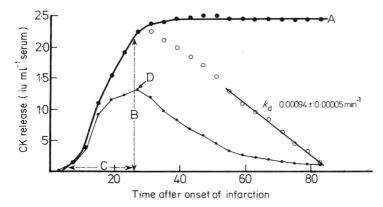

Figure 1. Method for calculation of total enzyme release and the rate of release using individualized k_d. Small closed circles represent 4-hourly measured levels of CK activity in serum from a patient who had a clinically uncomplicated anterior transmural infarct. Open circles represent a semilogarithmic replot of the downslope of measured activity from which the disappearance rate (k_d) is calculated; points on the later part of the downslope, during which time activity is declining rapidly, are used for this calculation. Using the measured activity levels and the individually determined k_d, total cumulative CK release (large closed circles) is plotted. The height of the plateau at A represents total cumulative CK release, and the slope of the upstroke (B/C) is the rate of release. The latter can best be calculated as 90% of total release divided by the time to 90% release, and is an index of the rapidity with which myocardial necrosis is developing. The peak measured enzyme activity (D) can be used as an independent index of infarct severity, which is not dependent on the assumption of a one-compartment distribution volume for CK (see text)

less than ±10%. This modified method using 4-hourly blood sampling and individual k_d is shown in Figure 1.

Some evidence has recently been interpreted to suggest that the use of a mean k_d (calculated from disappearance curves of endogenously released CK in a group of patients or experimental animals) rather than individualized k_d allows equal or greater accuracy in relating CK appearance to CK loss or histological damage to the myocardium.[83,84,114] Whether it is correct to use the individualized k_d depends on the true kinetics of CK disappearance; if the one-compartment theory is correct, and if CK appearance has ceased during the monoexponential decline phase, k_d should be individualized. If CK diffuses into more than one body compartment, or if the enzyme continues to be released from the infarct until serum activity is close to normal, neither an individualized nor a mean k_d will calculate the true CK appearance by this method. Until more detailed information is available about the duration of CK release from the infarct and its movement between body spaces, it is the author's opinion that k_d should be individually and accurately determined for each clinical or experimental study in which total CK appearance is calculated.

(iii) It is necessary to assume that the quantity of CK activity appearing in the circulation is a constant proportion of CK which is lost from the infarcted myocardium. Calculations made from the original data of Kjekshus and Sobel,[94] and assuming a distribution volume for CK of 5% of body weight (equivalent to plasma volume),[79] show that the ratio of CK appearing in the blood stream to CK depleted from the heart muscle is of the order of 15%. If this proportion is constant it can be allowed for in the calculations, but if the ratio varies unpredictably, infarct size could be overestimated if the ratio of CK appearance to CK destruction were high, and conversely it would be underestimated if this ratio were low. A recent experimental study by Cairns and colleagues[7] has shown that the serum entry ratio (proportion of CK lost from the myocardial infarct which appears in the blood) varies inversely ($r = -0.91$) with infarct size. This suggests that CK is degraded locally to a greater extent in large than in small infarcts, possibly because of slower diffusion out of the infarct or more prolonged exposure to cardiac lymph. The effect of this would be that measured enzyme appearance would progressively underestimate enzyme loss from the myocardium the larger the infarct; indeed the authors calculated from their data that for infarcts involving more than 23% of the left ventricle, enzyme appearance would actually *decrease* progressively with increasing infarct size. It must be emphasized that this phenomenon was described after experimental coronary occlusion in dogs and is not necessarily relevant to the clinical situation. Indeed a number of clinical studies have confirmed that enzymatic infarct size is very high in patients who have large infarcts causing circulatory failure, cardiogenic shock, and a poor prognosis. It has to be said that uncertainty about the relationship of appearance to destruction in man could be an important drawback to the use of the CK method; alternatively it is possible that this drawback could be of theoretical rather than practical importance.

(iv) Calculation of $E(t) + k_d \sum E \, dt$ involves the assumption that the distribution space for enzyme activity (assumed to be plasma volume[79]) remains constant. This is not true, particularly for patients with cardiac failure receiving diuretic therapy in whom the venous haematocrit can fall by a mean of 17% after treatment.[101] Failure to correct for changes in haematocrit led in one study to overestimation of infarct size by 15.8% on average in patients with pulmonary oedema and by 8.5% in patients without oedema.[101]

(v) Total CK appearance (iu ml^{-1} serum) can be used as an index of infarct size, or alternatively results can be expressed in terms of 'gram-equivalents of myocardium' which have become infarcted. This final calculation involves multiplication of the figure for total CK appearance by a factor which includes the assumed activity of CK in healthy myocardium, the depletion which occurs at the centre of an infarct, and the distribution space for CK (assumed to be 5% of body weight). Although strictly speaking no further sources of error are included in this further calculation, there is an important conceptual aspect which in the author's opinion argues for the use of total CK appearance rather than 'gram-equivalents' as an index of infarct size. This concerns the well-known fact that heart size is related to body weight.[21] Thus a large infarct (expressed in gram-equivalents) in a large patient is not necessarily more serious than a slightly smaller infarct in a smaller patient who will presumably have a mass of myocardium for which a smaller absolute amount of damage could be equally serious. When infarct size is expressed in gram-equivalents, body weight appears in the numerator, whereas if 'infarct size index' is expressed as iu of CK release ml^{-1} of serum, it is not necessary to correct for body weight. For this reason results are better expressed in units of enzyme activity ml^{-1} serum rather than as gram-equivalents of infarct size.

THE USE OF CK-MB: DOES THIS IMPROVE THE ACCURACY OF THE METHOD?

As has been stated previously, one potential source of inaccuracy of the total CK method relates to the possible release of CK from tissues other than the myocardium, particularly skeletal muscle. Thus the method is unreliable for patients who have had repeated intramuscular injections or tissue hypoperfusion, and it is invalidated for patients who have had recent surgery and in particular coronary artery surgery. Recent investigations in which a large number of tissues were analysed[104] showed that CK-MB was completely specific for the myocardium; moreover it is well established from clinical reports that elevation of CK-MB has both high sensitivity and specificity for the diagnosis of myocardial infarction.[3,99] Quantification of MB release has been reported by Roberts, Sobel, and colleagues and a high positive correlation was found ($r = 0.97$, $n = 12$) between infarct size estimated from CK-MB compared with total CK in patients with uncomplicated infarction.[79] For patients with infarction given intramuscular injections, total CK curves were distorted but CK-MB curves were not apparently affected.

Although determination of CK-MB is superior to the measurement of total CK, for the reasons given, it does have the disadvantage that the proportion of CK-MB to total CK which is released during infarction is small because the proportion of CK-MB in the myocardium is only 14%.[104] Using a batch absorption method[28] for the separation of CK-MB and the standard spectrophotometric method[87] for its kinetic assay it can be difficult to gain sufficient accuracy for determination of k_d when the MB activity is returning to its lower level. However, accuracy can no doubt be increased by use of a fluorometer,[28] and may be improved in the future by radio-immunoassay of the isoenzyme[81] (see Chapters 8, 9, and 10).

CORRELATION OF THE CK METHOD WITH OTHER INDICES OF INFARCT SIZE: RELATION BETWEEN CK APPEARANCE IN BLOOD AND CK DEPLETION FROM MYOCARDIUM

The basic evidence for the use of CK release as an index of infarct size lies in the correlation which was found in experiments using conscious dogs between CK depletion from the infarct and total CK appearance in the blood stream.[94] Does CK depletion from the myocardium in turn correlate with the intensity of ischaemia? In early experiments, Kjekshus and Sobel[38] found a close positive correlation in two dogs between CK activity in pieces of myocardium taken 24 hours after the onset of infarction and local myocardial blood flow measured using labelled microspheres of 25 ± 5 μm diameter. Heng and colleagues later confirmed this finding in a larger number of animals[26] and found a close correlation ($r = 0.93$) in eight dogs between myocardial blood flow and CK depletion when flow was measured 15 minutes after the onset of infarction, and $r = 0.88$ when flow was measured at 24 hours after the onset. Thus if reduction in blood flow is accepted as the definitive lesion of ischaemia, there is an excellent correlation between reduction in flow and loss of CK; this holds even when flow is reduced to very low levels.

The relationship between CK depletion from the myocardium and CK release in blood is altered markedly when ischaemic myocardium is reperfused[35,44,82] or reoxygenated.[24] The evidence for this comes from animal experiments in which a massive release of CK is found presumably from 'washout' occurring at the time of reperfusion. The practical relevance of this phenomenon to myocardial infarction in man is uncertain but unlikely to be important because a secondary rise in the CK release curve in patients is seldom seen in the absence of clear clinical evidence of reinfarction.

CORRELATION OF CK WITH OTHER METABOLIC MARKERS

Surprisingly little information is available on quantification of metabolic substances other than enzymes which are released into the blood stream after myocardial infarction. In particular the detection of myoglobin in urine[90] or serum[108] has been found to be a sensitive test for the detection of infarction in patients; however there appears to be little information on quantification of myoglobin appearance nor on its time relationship with the appearance of CK. Correlation of CK with other enzymes has the problem of possible lack of specificity of other enzymes for myocardium and in particular the possibility that enzymes such as lactate dehydrogenase might be released from inflammatory cells migrating into the necrotic tissue rather than from the myocardium itself.[38]

The author has found a good relationship, however ($r = 0.85$), between total CK and total hydroxybutyrate dehydrogenase (HBD) appearance in eight patients with acute infarction who were free from cardiac failure.[65] Prolongation of HBD activity in the serum of these patients was entirely explicable by the slow degradation rate for this enzyme, so that it was not necessary to postulate any prolongation of the release phase over that for CK. Patients who had cardiac failure on the other hand did not show any correlation between total CK and total HBD, there being a tendency towards greater release of both enzymes and prolongation of the duration of release of HBD. Prolonged HBD release in patients with failure might have been due to hepatic venous congestion.

Similarly, Witteveen and colleagues have shown a close quantitative correlation among five enzymes namely lactate dehydrogenase, HBD, aspartate aminotransferase, phosphohexose isomerase, and CK in 15 patients.[113] These workers have argued (see Chapter 14) that quantification of cumulative HBD appearance might be of greater value than cumulative CK because of the slower clearance rate of the former enzyme.[29] Thus the lower k_d for HBD than for CK makes the peak measured HBD activity closer to the total cumulative appearance than does the peak activity of the more rapidly cleared CK. According to this argument, CK-MB which has a faster clearance than CK-MM[79] would be even less satisfactory for quantification.

CORRELATION OF CK APPEARANCE WITH CLINICAL INDICES OF SEVERITY OF INFARCTION

The original clinical studies of Sobel and colleagues[102] showed that CK appearance could be used as an index of prognosis. Of twelve patients with infarct size exceeding 65 CK-g-equiv., eight died, while of 21 patients with infarct size less than 65 CK-g-equiv., only one died, after an average follow-up of 5.3 months. Moreover, survivors with symptoms of cardiac decompensation (New York Heart Association Class 3 and 4) had a significantly larger infarct size (91 ± 8 CK-g-equiv.) than survivors without cardiac decompensation (Class 1 and 2) who had an infarct size of 31 ± 4 CK-g-equiv.

Infarct size calculated from total CK appearance also parallels the severity of infarction judged by clinical criteria in surviving patients. CK appearance is significantly greater in patients with left ventricular failure than in those without failure[65,66] and is greater in transmural than in subendocardial infarction.[66,102] Serious ventricular arrhythmias are associated with large infarct size calculated by the CK method,[80] and patients who develop right bundle branch block in association with anterior infarction—a known marker of massive septal damage—also have high enzyme appearance.[66] Total CK cannot be used as an index of infarct size when the skeletal muscles are hypoperfused in cardiogenic shock, but CK-MB is of particular value in this situation, and massive release of this isoenzyme has been shown.[22] Further evidence that a large infarct size is associated with arrhythmias comes from experiments using anaesthetized dogs[4] in which ventricular fibrillation threshold during acute myocardial infarction produced by occlusion of the left anterior descending artery was correlated with infarct size measured both by the CK method and at necropsy. Significant correlations were found between infarct size measured by either method and the reduction in ventricular fibrillation threshold after coronary ligation from the baseline before ischaemia.

Good correlations have also been reported between CK infarct size and indirect indices of left ventricular function assessed either during the acute stage of infarction or subsequently. Thus Rogers and colleagues[85] found a positive correlation ($r = 0.78$) between pulmonary artery end-diastolic pressure during the acute phase and total CK-MB appearance in 11 patients with a first myocardial infarct. Not surprisingly, this correlation did not hold for patients suffering from a recurrent infarction. Correlation between pulmonary artery end-diastolic pressure and infarct size recorded by Bleifeld and colleagues[2,55] was less close; however their data show that of 16 patients with a first infarct and infarct size greater than 65 CK-g-equiv., 14 had an end-diastolic pressure of 18 mmHg or greater, while of 23 patients with first infarct of less than 65 CK-g-equiv., end-diastolic pressure was greater than 18 mmHg in only 4. Similarly, 7 out of 8 patients with an infarct size of 80 CK-g-equiv. or greater had a cardiac index less than $2.5 \, l \, min^{-1} \, m^{-2}$ while only 2 of 23 patients with an infarct size less than 80 CK-g-equiv. had a cardiac index below this figure (Table 1, ref. 2).

Definitive measurement of infarct size can be made only at autopsy; yet even at autopsy infarct size will vary with the time after onset of infarction, and precise separation of necrotic, ischaemic, and normal myocardium can be difficult using purely histological criteria.[17] A promising approach to measurement of non-functioning myocardium during life is the measurement of non-contractile segments of the left ventricle studied at biplane cineangiography.[19] This invasive method can be used with safety only in convalescent patients, so that it may measure the size of the scar rather than the size of the original infarct; a more promising approach would be by the non-invasive method of gated blood pool scanning[92] which can be used in acutely ill patients. Reported correlations between these definitive measurements and CK infarct size have in general been good,[2,4,35,85] although one report[83] denies this correlation.

Two experimental studies in dogs[4,35] and one clinical study[2] have shown good correlations between anatomical infarct size and total CK release although these were quite large systematic differences for absolute values of infarct size determined by these two methods. In one experimental study[4] a correlation coefficient of $r = 0.90$ was obtained in 11 dogs between measurements made by the two methods, but infarct size in CK-g-equiv. by the enzyme method was nearly twice as great as the anatomical infarct size. In the other study[35] the relationship in 9 dogs was $r = 0.87$, but infarct size was again overestimated more than twofold by the enzyme method. Moreover the relationship between anatomical and estimated infarct size was changed markedly after reperfusion of the infarct. In the only reported autopsy study in man,[2] 15 patients in whom total CK release had been expressed in CK-g-equiv. infarct size during life, subsequently died. Autopsy with myocardial staining by nitro blue tetrazolium showed a high correlation ($r = 0.93$) with little systematic difference between the two methods. Complementary to this autopsy study is the angiographic study of Rogers and colleagues[85] in which the size of the scar on biplane cineangiograms of the left ventricle performed 34 days after a first myocardial infarct in 11 patients showed a good correlation ($r = 0.85$) with enzymatic infarct size measured using CK-MB.

In contrast to these generally favourable reports, one experimental study of conscious dogs with occlusion of the circumflex coronary artery has notably failed to show a relationship between enzymatic and later histologic estimates of infarct size.[83] In these experiments histologic infarct size was assessed by a point counting method done in 8 transverse rings of myocardium obtained 5-6 days after infarction. Enzymatic infarct size was calculated by various techniques from serial measurements of CK and the MB isoenzyme. No correlation could be found between the two methods, and analysis of the data showed that lack of correlation was mainly due to a relatively small enzyme release in four out of five dogs which had a large anatomical infarct size (greater than 20 grams). Exclusion of the 5 animals with large infarcts and subtraction of background non-cardiac CK activity improved the correlation for the remaining 9 animals to $r = 0.84$ ($r^2 = 0.71$). This observation, that total myocardial CK release correlated reasonably well with anatomical measurements for small infarcts but not for large infarcts, is at variance with evidence from the experimental and clinical studies which have been described in the earlier part of this section. It could be explained, however, by increased local degradation of CK in large infarcts as described by Cairns and colleagues[7] in dogs, and discussed earlier in this review.

QUANTIFICATION OF INFARCT SIZE BY MEASUREMENT OF UPTAKE OF METABOLIC SUBSTANCES

That this review has been concerned mainly with enzymatic determination of infarct size by the CK method is a reflection of the work that has been done on this subject and the aims of this volume. Although the direction of movement of organic substances from dying cells is outward rather than inward, it has been known that the ischaemic myocardium can take up substances, quantification of which could provide an index of infarct size. For instance calcium is taken up by the ischaemic myocardium[11] and myocardial uptake of radioactive organic mercury compounds,[8] tetracycline,[33] and the bone-scanning agents technetium 99-m pyrophosphate[70] and diphosphonate[116] have all been used experimentally and clinically in the diagnosis and localization of infarctions. Many studies have shown that uptake of these and other substances provides a reliable method for detection of infarctions occurring spontaneously or after cardiac surgery. A number of factors, namely uncertainty about geometrical projection of the scintigraphic image, uptake of the imaging agent by surrounding normal organs, and uncertainty about mechanisms of uptake of the imaging agent by ischaemic cells, together with lack of knowledge about the relationship of quantity of uptake to intensity of ischaemia, have all precluded use of these methods for clinical measurement of infarct size up until the present. The first two of these problems should be overcome by advances in technology such as multiple projections and computer subtraction techniques. The third difficulty, however, may be more serious, and a great deal more basic work is needed before uptake of any imaging agent can be equated with intensity of ischaemia or size of infarct, particularly as the duration of ischaemia is almost certainly also a factor of critical importance for the degree of uptake of the substance in question.

Several experimental studies of the bone-scanning agent technetium pyrophosphate[5,6,45,73] have investigated the relationship between the intensity of uptake of technetium 99-m and other indices of severity of tissue ischaemia, namely gross infarct size, histology, myocardial CK content, and myocardial blood flow. Although a good correlation has been reported between the area of anterior infarcts measured at post-mortem in dogs and the area of the scintigraphic image,[5,107] there remains a problem in sizing inferior infarcts which cannot be imaged in profile. Moreover correlation of more sensitive indices of local ischaemia, namely regional histology, regional myocardial CK content, and regional myocardial blood flow, have in general not shown a good relationship between intensity of ischaemia assessed by these indices and the regional intensity of technetium 99-m uptake.[6,45,73] In clinical practice it is well known that the image is often 'doughnut shaped' because of reduced uptake of the imaging agent at the centre of the infarct, and this has been attributed to poor delivery of the imaging agent because of the very low blood flow at the centre. If this were always the case, the central area of no uptake could be included as part of the infarct for the purpose of clinical infarct sizing; however the lack of a quantitative relationship between intensity of ischaemia and intensity of uptake of the imaging agent must remain a serious drawback to the use of this method for infarct sizing.

INCREASE OR REDUCTION OF INFARCT SIZE BY TREATMENT

The whole thrust of the current interest on infarct sizing must ultimately be directed to the idea, first put forward by Maroko, Braunwald, and colleagues in 1971,[51] that infarct size can be reduced with resulting benefit to patients. During the 7 years which have passed since 1971 many conflicting views have been put forward, and there is still no agreement as to whether the size of infarcts can be modified either in the experimental laboratory or in the Coronary Care Unit.

Broadly speaking the many agents which have been suggested for reduction in infarct size could be effective in one of two ways—either they could improve the supply of oxygen or other metabolites to the ischaemic myocardium, or they could reduce myocardial demand for these substances. A third group of agents could act in some ill-understood way either by reducing tissue oedema or cellular swelling or by a direct 'stabilizing' effect on the myocardial cell membrane. Some publications describing the use of these agents are summarized in Table 1. Whether or not they have been found to be effective depends in large measure on what end point has been taken as the index of infarct size or the intensity of ischaemia. Thus many agents have been shown to reduce ST segment elevation, but the relevance of this to infarct size is questionable in the context of evidence from a number of studies which show that ST elevation does not correlate with infarct size or the intensity of ischaemia.[10,26,32,34,61,110] Moreover, it has been shown in the author's laboratory that three therapeutic agents, namely verapamil,[97] propranolol,[71] and glucose–insulin–potassium infusions[27] all reduce experimental ST elevation without changing either local myocardial blood flow or tissue CK activity in any favourable way.

The most reliable evidence at present available that infarct size can be altered comes from morbid anatomical or enzyme studies. From the viewpoint of the former of these methods, the work of Jennings and colleagues[77,78] is outstanding in consistently showing reduction of tissue damage assessed by light microscopy in the posterior papillary muscle after circumflex artery ligation in the dog when the ischaemic myocardium is protected by previous treatment with propranolol. A recent report[76] suggests that verapamil may also be valuable in this regard.

In the author's and colleagues' experimental studies of myocardial protection in dogs,[27,71,97] CK activity of pieces of myocardium taken from the border and centre of the infarct at 24 hours after ligation of the left anterior descending coronary artery has been compared with the blood flow of contiguous pieces of myocardium measured either at 15 minutes or at 24

Table 1. Some Therapeutic Agents which Might Reduce Infarct Size or Myocardial Ischaemic Injury

Therapeutic agent	Postulated mechanism of action	Favourable effect claimed (reference no.)[a]	No. or unfavourable effect (reference no.)[a]
Oxygen inhalation	Metabolic support	40, 54	
Glucose–insulin–potassium	Metabolic support	13, 53, 68, 86, 109	25, 27, 75, 88
β-Pyridyl carbinol	Metabolic support (inhibition of lipolysis)	37	
β-adrenergic blockade	Reduction of myocardial oxygen consumption	51, 58, 59, 67, 72 77, 78	71
Sodium nitroprusside, glyceryl trinitrate, trimethaphan	Reduction of pre-load and afterload to myocardial contraction	1, 12, 31	9, 41
Intraaortic balloon counterpulsation	Reduction of afterload; improvement of diastolic myocardial perfusion	48, 57	
Verapamil	Vasodilation; reduction of transmembrane Ca^{2+} flux	60, 76, 100	97
Methylprednisolone	Membrane stabilizing effect	39, 106	14, 111
Hyaluronidase	Reduction of myocardial oedema	30, 50	
Mannitol	Reduction of cellular swelling	74, 112	

[a] Numbers in bold type refer to animal experiments and numbers in italics to studies in patients with myocardial infarction.

hours after coronary ligation.[26] Therapeutic interventions have been tested by comparison of untreated animals with those in which either pre-treatment was given immediately before coronary ligation, or intervention was made 15 minutes after ligation. For interventions made after coronary ligation a shift in the relationship, established in control animals, has been sought between myocardial blood flow (measured immediately before intervention) and regional CK activity measured 24 hours later; a shift towards greater CK activity for a given level of blood flow in treated animals could be taken as evidence for myocardial protection. This comparison is not possible in pre-treatment experiments in which no baseline measurement of infarct blood flow is possible; in these cases, CK activities of myocardium from the centre and border of the infarct are compared between treated and control animals. Using this model the effects of verapamil given both before and after coronary ligation,[97] glucose–insulin–potassium infusions commencing after coronary ligation,[27] and a high dose of propranolol (5 mg kg^{-1})

given before ligation[76] have all been studied. Disappointingly, no therapeutic benefit could be demonstrated from these agents using the above variables as indices of the local intensity of ischaemia (blood flow at 15 minutes) or the consequent totality of necrosis (CK depletion at 24 hours).

The author's approach to the investigation of reduction of infarct size in patients has been that of the controlled clinical trial in which patients are selected at random for treatment by the therapeutic agent or as control cases, and total CK release is calculated from 4-hourly enzyme activity levels and individualized k_d.[66] Patients judged to have early uncomplicated full thickness infarcts have been selected in whom reduction of infarct size would be of practical benefit but in whom the necessity for other methods of treatment would be unlikely to blur the distinction between treated and control cases. Using this method, no tendency has been shown for total CK release to be reduced by glucose–insulin–potassium infusions.[25] Recently, however, it has been encouraging to find that propranolol, given in a dose of $0.1\,\mathrm{mg\,kg^{-1}}$ intravenously followed by 320 mg orally over 24 hours and started within 4 hours of the onset of the most severe chest pain, reduced total CK appearance, the rate of enzyme appearance, and the peak measured activity levels in treated compared with control patients.[72] When the propranolol was started more than 4 hours after the onset of infarction, no good effect was seen. Enzyme curves from this recent trial of propranolol are shown in Figure 2. These results suggest that β-adrenergic blockade may reduce infarct size in man if treatment is started within 4 hours of the onset.

An alternative method, namely prediction of total enzyme release from the first part of the release curve (discussed in Chapter 12) has the advantage that each patient can be used as his own control.[96] However this advantage is outweighed by the disadvantage that treatment cannot begin until the baseline observations have been made; according to the originally described method it is necessary to take hourly blood samples for 7 hours before intervention,[95] although it was later suggested that this interval might be reduced to 2 hours.[93] Basic considerations argue strongly that if infarct size is to be reduced, intervention must be given as early as possible, and immediately after clinical examination has been completed.

CONCLUDING COMMENTS

This section of the review attempts to assess the present place of enzyme measurements for infarct sizing in relation to other methods which are available. Although some evidence is conflicting, there is support for the concept that both depletion of CK from the ischaemic myocardium and its appearance in the blood stream correlate with clinical and pathological measurements of the severity of infarction. The CK method is the best documented of enzymatic methods, but it is possible that quantification of

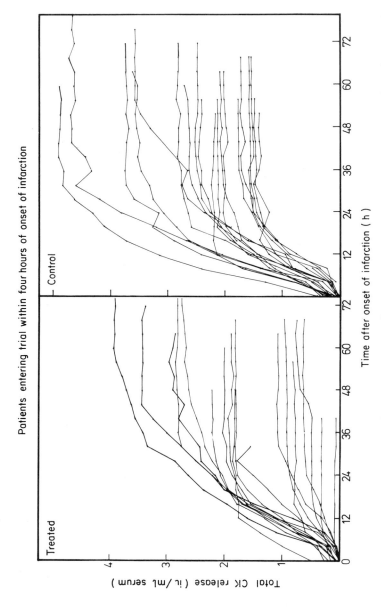

Figure 2. Total CK release curves (derived as in Figure 1) for 18 patients with uncomplicated transmural infarction treated within 4 hours of onset in a controlled clinical trial with intravenous and oral propranolol, compared with 19 untreated control patients. Both total calculated CK release ($p<0.025$), and the rate of enzyme release ($p<0.005$) were reduced in patients treated with propranolol, as were the peak measured CK activity levels ($p<0.01$, data not shown).

other enzymes or other metabolic substances might be even more valuable. Methods involving measurement of uptake of substances by ischaemic myocardium are not yet sufficiently reliable for quantification, but the position may change. Although present reports are inconclusive, there is hope that some method for reduction of infarct size will become established for therapeutic use with resultant benefit for patients and reduction in mortality from acute myocardial infarction.

REFERENCES

1. Banka, V. S., Bodenheimer, M. M., and Helfant, R. H. Nitroglycerin in experimental myocardial infarction. Effects on regional left ventricular length and tension. *American Journal of Cardiology*, **36,** 453–8 (1975).
2. Bleifeld, W., Mathey, D., Hanrath, P., Buss, H., and Effert, S. Infarct size estimated from serial serum creatine phosphokinase in relation to left ventricular hemodynamics. *Circulation*, **55,** 303–11 (1977).
3. Blomberg, D. J., Kimler, W. D., and Burke, M. D. Creatine kinase isoenzymes: predictive value in the early diagnosis of acute myocardial infarction. *American Journal of Medicine* **59,** 464–9 (1975).
4. Bloor, C. M., Ehsani, A., White, F. L., and Sobel, B. E. Ventricular fibrillation threshold in acute myocardial infarction and its relation to myocardial infarct size. *Cardiovascular Research*, **9,** 468–72 (1975).
5. Botvinick, E. H., Shames, D., Lappin, H., Tyberg, J. V., Townsend, R., and Parmley, W. W. Noninvasive quantitation of myocardial infarction with technetium 99m pyrophosphate. *Circulation*, **52,** 909–15 (1975).
6. Bruno, F. P., Cobb, F. R., Rivas, F., and Goodrich, J. K. Evaluation of 99m technetium stannous pyrophosphate as an imaging agent in acute myocardial infarction. *Circulation*, **54,** 71–8 (1976).
7. Cairns, J. A., Missirlis, E., and Fallen, E. L. Myocardial infarction size from serial CPK: variability of CPK serum entry ratio with size and model of infarction. *Circulation Supplement III*, **56,** 70 (1977).
8. Carr, E. A., Beierwaltes, W. H., Patno, M. E., Bartlett, J. D., and Wegst, A. V. The detection of experimental myocardial infarcts by photoscanning. A preliminary report. *American Heart Journal*, **64,** 650–60 (1962).
9. Chatterjee, K., Swan, H. J. C., Forrester, J., Kaushik, V. S., and Gray, R. Vasodilator therapy in patients with acute myocardial infarction complicated by severe pump failure—immediate and long-term prognosis. *Circulation Supplement II*, **51** and **52,** 108 (1975).
10. Cohen, M. V. and Kirk, E. S. Reduction of epicardial ST-segment elevation following increased myocardial ischemia: experimental and theoretical demonstration. *Clinical Research*, **22,** 269A (1974).
11. D'Agostino, A. N. and Chiga, M. Mitochondrial mineralization in human myocardium. *American Journal of Clinical Pathology*, **53,** 820–4 (1970).
12. da Luz, P. L., Forrester, J. S., Wyatt, H. L., Tyberg, J. V., Chagrasulis, R., Parmley, W. W., and Swan, H. J. C. Hemodynamic and metabolic effects of sodium nitroprusside on the performance and metabolism of regional ischemic myocardium. *Circulation*, **52,** 400–7 (1975).
13. De Leiris, J., Opie, L. H., and Lubbe, W. F. Effects of free fatty acid and glucose on enzyme release in experimental myocardial infarction. *Nature*, **253,** 746–7 (1975).
14. de Mello, V. R., Roberts, R., and Sobel B. E. Deleterious effects of multiple dose methyl prednisolone on evolving myocardial infarction. *Circulation*, **52,** 106 (1975).
15. Dreyfus, J. C., Schapira, G., Resnais, J., and Scebat, L. La créatine-kinase sérique dans le diagnostic de l'infarctus myocardique. *Revue Francais d'Etudes Cliniques et Biologiques*, **5,** 386 (1960).

16. Dunn, M., Martin, J., and Reissmann, K. R. The disappearance rate of glutamic oxalo-acetic transaminase from the circulation and its distribution in the body's fluid compartments and secretions. *Journal of Laboratory and Clinical Medicine*, **51,** 259–65 (1958).
17. Edwards, J. E. What is myocardial infarction? *Circulation Supplement* 4, **39** and **40,** 5–12 (1969).
18. Ehsani, A., Eiwy, G. A., and Sobel, B. E. Effects of electrical countershock on serum creatine phosphokinase enzyme activity. *American Journal of Cardiology*, **37,** 12–8 (1976).
19. Feild, B. J., Russell, R. O., Dowling, J. T., and Rackley, C. E. Regional left ventricular performance in the year following myocardial infarction. *Circulation*, **46,** 679–89 (1972).
20. Forrester, J. S., Ganz, W., Diamond, G., McHugh, T., Chonette, D. W., and Swan, H. J. C. Thermodilution cardiac output determination with a single flow-directed catheter. *American Heart Journal*, **83,** 306–11 (1972).
21. Gorlin, R. In *The Heart Arteries and Veins* (Ed. Willis Hurst, J.) McGraw-Hill, 1974. p. 109.
22. Gutowitz, A. L., Sobel, B. E., and Roberts, R. Cardiogenic shock: a syndrome frequently due to slowly evolving myocardial injury. *American Journal of Cardiology*, **39,** 322 (1977).
23. Harnaryan, C., Bennett, M. A., Pentecost, B. L., and Brewer, D. B. Quantitated study of infarcted myocardium in cardiogenic shock. *British Heart Journal*, **32,** 728–32 (1970).
24. Hearse, D. J., Humphrey, S. M., Nayler, W. G., Slade, A., and Border, D. Ultrastructural damage associated with reoxygenation of the anoxic myocardium. *Journal of Molecular and Cellular Cardiology*, **7,** 315–24 (1975).
25. Heng, M. K., Norris, R. M., Singh, B. N., and Barratt-Boyes, C. The effects of glucose and glucose–insulin–potassium on haemodynamics and enzyme release after acute myocardial infarction. *British Heart Journal*, **39,** 748–57 (1977).
26. Heng, M. K., Singh, B. N., Norris, R. M., John, M. B., and Hurley, P. J. Relationship between epicardial ST-segment elevation and myocardial ischemic damage after experimental coronary occlusion in dogs. *Journal of Clinical Investigation*, **58,** 1317–26 (1976).
27. Heng, M. K., Singh, B. N., Peter, T., Nisbet, H. D., and Norris, R. M. The effects of glucose–insulin–potassium on experimental myocardial infarction in the dog. *Cardiovascular Research*, **12,** 429–35 (1978).
28. Henry, P. D., Roberts, R., and Sobel, B. E., Rapid separation of plasma creatine kinase isoenzymes by batch absorption on glass beads. *Clinical Chemistry*, **21,** 844–9 (1975).
29. Hermens, W. T. and Witteveen, S. A. G. J. Problems in estimation of enzymatic infarct size. *Journal of Molecular Medicine*, **2,** 233–9 (1977).
30. Hillis, L. D., Askenazi, J., Braunwald, E., Radvany, P., Muller, J. E., Fishbein, M. C., and Maroko, P. R. Use of changes in the epicardial QRS complex to assess intervention which modify the extent of myocardial necrosis following coronary artery occlusion. *Circulation*, **54,** 591–8 (1976).
31. Hirshfield, J. W., Borer, J. S., Goldstein, R. E., Barrett, M. J., and Epstein, S. E. Reduction in severity and extent of myocardial infarction when nitroglycerine and methoxomine are administered during coronary occlusion. *Circulation*, **49,** 291–7 (1974).
32. Holland, R. P. and Brooks, H. Precordial and epicardial surface potentials during myocardial ischemia in the pig. A theoretical and experimental analysis of the TQ and ST segments. *Circulation Research*, **37,** 471–80 (1975).
33. Holman, B. L., Lesch, M., Zweiman, F. G., Temte, J., Lown, B., and Gorlin R. Detection and sizing of acute myocardial infarcts with ^{99m}Tc (Sn) tetracycline. *New England Journal of Medicine*, **291,** 159–63 (1974).
34. Irvin, R. G. and Cobb, F. R. Relationship between epicardial ST segment elevation, regional myocardial blood flow and extent of myocardial infarction in awake dogs. *Circulation*, **55,** 825–32 (1977).
35. Jarmakani, J., Limbird, L., Graham, T. C., and Marks, R. A. Effect of reperfusion on myocardial infarct, and the accuracy of estimating infarct size from serum creatine phosphokinase in the dog. *Cardiovascular Research*, **10,** 245–53 (1976).
36. Karmen, A., Wroblewski, F., and La Due, J. S. Transaminase activity in human blood. *Journal of Clinical Investigation*, **34,** 126–33 (1955).

37. Kjekshus, J. K. and Mjos, O. D. Effect of inhibition of lipolysis on infarct size after experimental coronary artery occlusion. *Journal of Clinical Investigation*, **52**, 1770–8 (1973).
38. Kjekshus, J. K. and Sobel, B. E. Depressed myocardial creatine phosphokinase activity following experimental myocardial infarction in rabbit. *Circulation Research*, **27**, 403–14 (1970).
39. Libby, P., Maroko, P. R., Bloor, C. M., Sobel, B. E., and Braunwald, E. Reduction of experimental myocardial infarct size by corticosteroid administration. *Journal of Clinical Investigation*, **52**, 599–607 (1973).
40. Madias, J. E., Madias, N. E., and Hood, W. B. Precordial ST-segment mapping. 2. Effects of oxygen inhalation on ischemic injury in patients with acute myocardial infarction. *Circulation*, **53**, 411–7 (1976).
41. Magnusson, P., Shell, W. E., Forrester, J. S., Charuzi, V., Singh, B. N., and Swan, H. J. C. Increased creatine phosphokinase release following blood pressure reduction in patients with acute infarction. *Circulation Supplement II*, **53** and **54**, 28 (1976).
42. Malmberg, P. Aspartate aminotransferase activity in dog heart lymph after myocardial infarction. *Scandinavian Journal of Clinical and Laboratory Investigation*, **30**, 153–8 (1972).
43. Malmberg, P. Time course of enzyme escape via heart lymph following myocardial infarction in the dog. *Scandinavian Journal of Clinical and Laboratory Investigation*, **30**, 405–9 (1972).
44. Manders, T., Vatner, S., Millard, R., Heyndrickx, G., and Maroko, P. R. Altered relationship between creatine phosphokinase release and infarct size with reperfusion in conscious dogs. *Circulation Supplement II*, **51** and **52**, 5 (1975).
45. Marcus, M. L., Tomanek, R. J., Erhardt, J. C., Kerber, R. E., Brown, D. D., and Abboud, F. M. Relationships between myocardial perfusion, myocardial necrosis, and technetium-99m pyrophosphate uptake in dogs subjected to sudden coronary occlusion. *Circulation*, **54**, 647–53 (1976).
46. Markham, J., Karlsberg, R. P., Roberts, R., and Sobel, B. E. Mathematical characterization of kinetics of native and purified creatine kinase in plasma. In *Computers in Cardiology*, Long Beach, IEEE Computer Society, 1976.
47. Maroko, P. R. Assessing myocardial damage in acute infarcts. *New England Journal of Medicine*, **290**, 158–9 (1974).
48. Maroko, P. R., Bernstein, E. F., Libby, P., De Laria, G. A., Covell, J. W., Ross, J., and Braunwald, E. Effects of intra-aortic balloon counterpulsation on the severity of myocardial ischemic injury following acute coronary occlusion. *Circulation*, **45**, 1150–9 (1972).
49. Maroko, P. R. and Braunwald, E. Modification of myocardial infarction size after coronary occlusion. *Annals of Internal Medicine*, **79**, 720–33 (1973).
50. Maroko, P. R., Davidson, D. M., Libby, P., Hagan, A. D., and Braunwald, E. Effects of hyaluronidase administration on myocardial ischemic injury in acute infarction. A preliminary study in 24 patients. *Annals of Internal Medicine*, **82**, 516–20 (1975).
51. Maroko, P. R., Kjekshus, J. K., Sobel, B. E., Watanabe, T., Covell, J. W., Ross, J., and Braunwald, E. Factors influencing infarct size following experimental coronary artery occlusions. *Circulation*, **43**, 67–82 (1971).
52. Maroko, P. R., Libby, P., Covell, J. W., Sobel, B. E., Ross, J., and Braunwald, E. Precordial ST-segment elevation mapping: An atraumatic method for assessing alterations in the extent of myocardial ischemic injury. The effects of pharmacologic and hemodynamic interventions. *American Journal of Cardiology*, **29**, 223–30 (1972).
53. Maroko, P. R., Libby, P., Sobel, B. E., Bloor, C. M., Sybers, H. D., Shell, W. E., Covell, J. W., and Braunwald, E. Effect of glucose–insulin–potassium infusion on myocardial infarction following experimental coronary artery occlusion. *Circulation*, **45**, 1160–75 (1972).
54. Maroko, P. R., Radvany, P., Braunwald, E., and Hale, S. L. Reduction of infarct size by oxygen inhalation following acute coronary occlusion. *Circulation*, **52**, 360–8 (1975).
55. Mathey, D., Bleifeld, W., Hanrath, P., and Effert, S. Attempt to quantitate relation between cardiac function and infarct size in acute myocardial infarction. *British Heart Journal*, **36**, 271–9 (1974).

56. Meltzer, H. Y. Mrozak, S., and Beyer, M. Effects of intramuscular injections on serum creatine phosphokinase activity. *American Journal of Medical Science*, **259**, 42–8 (1970).
57. Mueller, H., Ayres, S. M., Conklin, E. F., Gianneli, S., Mazzara, J. R., Grace, W. T., and Nealon, T. F. The effects of intra-aortic counterpulsation on cardiac performance and metabolism in shock associated with acute myocardial infarction. *Journal of Clinical Investigation*, **50**, 1885–900 (1971).
58. Mueller, H. S., Ayres, S. M., Religa, A., and Evans, R. G. Propranolol in the treatment of aucte myocardial infarction. Effect on myocardial oxygenation and hemodynamics. *Circulation*, **49**, 1078–87 (1974).
59. Muira, M., Ganz, W., Thomas, R., Singh, B. N., Sokol, T., and Shell, W. E. Reduction of infarct size by propranolol in closed-chest anesthetized dogs. *Circulation Supplement II*, **54**, 159 (1976).
60. Nayler, W. G., Grace, A., and Slade A. A proctective effect of verapamil on hypoxic heart muscle. *Cardiovascular Research*, **10**, 650–62 (1976).
61. Norris, R. M., Barratt-Boyes, C., Heng, M. K., and Singh, B. N. Failure of ST segment elevation to predict severity of acute myocardial infarction. *British Heart Journal*, **38**, 85–92 (1976).
62. Norris, R. M., Brandt, P. W. T., Caughey, D. E., Lee, A. J., and Scott, P. J. A new coronary prognostic index. *Lancet*, **I**, 274–8 (1969).
63. Norris, R. M., Caughey, D. E., Deeming, L. W., Mercer, C. J., and Scott, P. J. Coronary prognostic index for predicting survival after recovery from acute myocardial infarction. *Lancet*, **II**, 485–8 (1970).
64. Norris, R. M., Caughey, D. E., Mercer, C. J., and Scott, P. J. Prognosis after myocardial infarction: 6 year follow-up. *British Heart Journal*, **36**, 786–90 (1974).
65. Norris, R. M., Howell, D., Whitlock, R. M. L., Heng, M. K., and Peter, T. Enzyme release after myocardial infarction: Comparison of serial serum alpha hydroxybutyrate dehydrogenase with creatine phosphokinase levels. *European Journal of Cardiology*, **4**, 461–8 (1976).
66. Norris, R. M., Whitlock, R. M. L., Barratt-Boyes, C., and Small, C. W. Clinical measurement of myocardial infarct size: modification of a method for the estimation of total creatine phosphokinase release after myocardial infarction. *Circulation*, **51**, 614–20 (1975).
67. Obeid, A., Spear, R., Mookherjee, S., Warner, R., and Eich, R. The effects of propranolol on myocardial energy stores during myocardial ischemia in dogs. *Circulation Supplement II*, **54**, 159 (1976).
68. Opie, L. H., Bruynell, K., and Owen, P. Effects of glucose insulin and potassium infusion on tissue metabolic changes within first hour of myocardial infarction in the baboon. *Circulation*, **52**, 49–57 (1975).
69. Page, D. L., Caulfield, J. B., Kaster, J. A., De Sanctis, R. W., and Sanders, C. A. Myocardial changes associated with cardiogenic shock. *New England Journal of Medicine*, **285**, 133–7 (1971).
70. Parkey, R. W., Bonte, F. J., Meyer, S. L., Atkins, J. M., Curry, G. L., Stokely, E. M., and Willerson, J. T. A new method for radionuclide imaging of acute myocardial infarction in humans. *Circulation*, **50**, 540–6 (1974).
71. Peter, T., Heng, M. K., Singh, B. N., Ambler, P., Nisbet, H., Elliot, R., and Norris, R. M. Failure of high doses of propranolol to reduce experimental myocardial ischemic damage. *Circulation*, **57**, 534–40 (1978).
72. Peter, T., Norris, R. M., Clarke, E. D., Heng, M. K., Singh, B. N. Williams, B., Howell, D. R., and Ambler, P. Reduction of enzyme levels after acute myocardial infarction by propranolol. *Circulation*, **57**, 1091–1095 (1978).
73. Peter, T., Norris, R. M., John, M. B., Heng, M. K., Williams, D., and Hurley, P. J. Regional technetium 99m diphosphonate uptake in experimental dog heart infarcts: relation to duration and severity of ischaemia. Submitted for publication.
74. Powell, W. J., Di Bona, D., Flores, J., and Leaf, A. The protective effect of hyperosmotic mannitol in myocardial ischemia and necrosis. *Circulation*, **54**, 603–15 (1976).
75. Reduto, L., Galitta, S., and Morrison, M. Ineffectiveness of glucose–insulin–potassium infusion on ischemic myocardium. *Clinical Research*, **22**, 685A (1974).

76. Reimer, K. A., Lowe, J. E., and Jennings, R. B. Effect of the calcium antagonist verpamil on necrosis following temporary coronary artery occlusion in dogs. *Circulation*, **55**, 581–7 (1977).
77. Reimer, K. A., Rasmussen, M. M., and Jennings, R. B. Reduction by propranolol of myocardial necrosis following temporary coronary artery occlusion in dogs. *Circulation Research*, **33**, 353–63 (1973).
78. Reimer, K. A., Rasmussen, M. M., and Jennings, R. B. On the nature of protection by propranolol against myocardial necrosis after temporary coronary occlusion in dogs. *American Journal of Cardiology*, **37**, 520–7 (1976).
79. Roberts, R., Henry, D. D., and Sobel, B. E. An improved basis for enzymatic estimation of infarct size. *Circulation*, **52**, 743–54 (1975).
80. Roberts, R., Husain, A. H., Amlos, H. D., Oliver, G. C., Cox, J. R., and Sobel, B. E. The relationship between infarct size and ventricular arrhythmia. *British Heart Journal*, **37**, 1169–75 (1975).
81. Roberts, R., Sobel, B. E., and Parker, C. W. Radioimmunoassay for creatine kinase isoenzymes. *Science*, **194**, 855–7 (1976).
82. Roe, R., Cobb, F. R., and Starmer, C. F. Peripheral enzyme appearance rates following acute myocardial infarction: an index of perfusion of infarcted regions. *Circulation Supplement II*, **51** and **52**, 5 (1975).
83. Roe, C. R., Cobb, F. R., and Starmer, C. F. The relationship between enzymatic and histologic estimates of the extent of myocardial infarction in conscious dogs with permanent coronary occlusion. *Circulation*, **55**, 438–49 (1977).
84. Roe, C. R., Starmer, C. F., and Cobb, F. R. Mathematical modifications fail to improve CPK estimates of extent of infarct. *Circulation*, **55**, 678–9 (1977).
85. Rogers, W. J., McDaniel, H. G., Smith, L. R., Mantle, J. A., Russell, R. O., and Rackley, C. E. Correlation of angiographic estimates of myocardial infarct size and accumulated release of creatine kinase MB isoenzyme in man. *Circulation*, **56**, 199–205 (1977).
86. Rogers, W. J., Stanley, A. W., Breinig, J. B., Pratter, J. W., McDaniel, H. G., Moraski, R. E., Mantle, J. A., Russell, R. O., and Rackley, C. E. Reduction of hospital mortality rate of acute myocardial infarction with glucose–insulin–potassium infusion. *American Heart Journal*, **92**, 441–54 (1976).
87. Rosalki, S. B. An improved procedure for serum creatine phosphokinase determination. *Journal of Laboratory and Clinical Medicine*, **69**, 696–705 (1967).
88. Rovetto, M. J., Whitmer, J. T., and Neely, J. R. Comparison of the effects of anoxia and whole heart ischemia on carbohydrate utilization in isolated working rat hearts. *Circulation Research*, **32**, 699–711 (1973).
89. Samson, W. E. and Scher, A. M. Mechanism of ST-segment alteration during acute myocardial injury. *Circulation Research*, **8**, 780–7 (1960).
90. Saranchak, H. J. and Bernstein, S. H. A new diagnostic test for acute myocardial infarction.The detection of myoglobinuria by radial immunodiffusion assay. *Journal of the American Medical Association*, **228**, 1251–5 (1974).
91. Schultze, H. E. and Heremans, J. F. *Molecular Biology of Human Proteins*, Vol. 1, *Nature and Metabolism of Extracellular Proteins*, Elsevier, Amsterdam, 1966.
92. Schulze, R. A., Rouleau, J., Rigo, P., Bowers, S., Strauss, H. W., and Pitt, B. Ventricular arrhythmias in the late hospital phase of acute myocardial infarction. Relation to left ventricular function detected by gated cardiac blood pool scanning. *Circulation*, **52**, 1006–11 (1975).
93. Shell, W. E., Groseth-Robertson, R., and Vas, R. Infarct size prediction 2 hours after initial CPK rise. *Circulation, Supplement 2*, **54**, 28 (1976).
94. Shell, W. E., Kjekshus, J. K., and Sobel, B. E. Quantitative assessment of the extent of myocardial infarction in the conscious dog by means of analysis of serial changes in serum creatine phosphokinase activity. *Journal of Clinical Investigation*, **50**, 2614–25 (1971).
95. Shell, W. E., Larelle, J. F., Covell, J. W., and Sobel, B. E. Early estimation of myocardial damage in conscious dogs and patients with evolving acute myocardial infarction. *Journal of Clinical Investigation*, **52**, 2579–90 (1973).
96. Shell, W. E. and Sobel, B. E. Protection of jeopardized ischemic myocardium by reduction of ventricular afterload. *New England Journal of Medicine*, **291**, 481–6 (1974).

97. Singh, B. N., Heng, M. K., Peter, T., and Norris, R. M. Epicardial ST-segment reduction by verapamil without change in experimental infarct size. *American Journal of Cardiology*, **37**, 173 (1976).
98. Slutzky, A. S. Individualized values for the disappearance rate parameter (k_d) in the enzymatic estimation of infarct size. *Circulation*, **56**, 545–7 (1977).
99. Smith, A. F., Radford, D., Wong, C. P., and Oliver, M. F. Creatine kinase MB isoenzyme studies in diagnosis of myocardial infarction. *British Heart Journal*, **38**, 225–32 (1976).
100. Smith, H. J., Singh, B. N., Nisbet, H. D., and Norris, R. M. Effects of verapamil in infarct size following experimental coronary occlusion. *Cardiovascular Research*, **9**, 569–78 (1975).
101. Smith, S. J., Bos, G., Hagemeyer, F., Hermens, W. J., and Witteveen, S. A. G. J. Influences of changes in plasma volume on quantitation of infarct size in man by means of plasma enzyme levels. *Journal of Molecular Medicine*, **1**, 199–210 (1976).
102. Sobel, B. E., Bresnahan, G. F., Shell, W. E., and Yoder, R. D. Estimation of infarct size in man and its relation to prognosis. *Circulation*, **46**, 640–8 (1972).
103. Sobel, B. E., Markham, J., Karlsberg, R. P., and Roberts, R. The nature of disappearance of creatine kinase from the circulation and its influence on enzymatic estimation of infarct size. *Circulation Research*, **41**, 836–44 (1977).
104. Sobel, B. E., Roberts, R., and Larson, K. B. Estimation of infarct size from serum MB creatine phosphokinase activity: applications and limitations. *American Journal of Cardiology*, **37**, 474–85 (1976).
105. Sobel, B. E. and Shell, W. E. Serum enzyme determinations in the diagnosis and assessment of myocardial infarction. *Circulation*, **45**, 471–82 (1972).
106. Spath, J. A., Lane, D. L., and Lefer, A. M. Protective action of methylprednisolone on the myocardium during experimental myocardial ischemia in the cat. *Circulation Research*, **35**, 44–51 (1974).
107. Stokely, E. M., Buja, L. M., Lewis, S. E., Parkey, R. W., Bonte, F. J., Harris, R. A., and Willerson, J. T. Measurement of acute myocardial infarcts in dogs with 99m Tc-stannous pyrophosphate scintigrams. *Journal of Nuclear Medicine*, **17**, 1–5 (1976).
108. Stone, M. J., Willerson, J. T., Gomez-Sanchez, C. E., and Waterman, M. R. Radioimmunoassay of myoglobin in human serum. Results in patients with acute myocardial infarction. *Journal of Clinical Investigation*, **56**, 1334–9 (1975).
109. Sybers, H. D., Maroko, P. R., Ashraf, M., Libby, P., and Braunwald, E. The effect of glucose–insulin–potassium on cardiac ultrastructure following acute experimental coronary occlusion. *American Journal of Pathology*, **70**, 401–20 (1973).
110. Thompson, P. L. and Katavatis, V. Acute myocardial infarction. Evaluation of praecordial ST segment mapping. *British Heart Journal*, **38**, 1020–4 (1976).
111. Vogel, W. M., Zannoni, V. G., Abrams, G. D., and Luchessi, B. R. Inability of methylprednisolone sodium succinate to decrease infarct size or preserve enzyme activity measured 24 hours after coronary occlusion in the dog. *Circulation*, **55**, 588–95 (1977).
112. Willerson, J. T., Powell, W. J. Guiney, T. E., Stark, J. J., Sanders, C. A., and Leaf, A. Improvement in myocardial function and coronary blood flow in ischemic myocardium after mannitol. *Journal of Clinical Investigation*, **51**, 2989–98 (1972).
113. Witteveen, S. A. G. J., Hemker, H. C., Hollaar, L., and Hermens, W. T. Quantitation of infarct size in man by means of plasma enzyme levels. *British Heart Journal*, **37**, 795–803 (1975).
114. Yasmineh, W. G., Pyle, R. B., Cohn, J. N., Nicoloff, D. M., Hanson, N. Q., and Steele, B. W. Serial creatine phosphokinase MB isoenzyme activity after myocardial infarction: studies in the baboon and man. *Circulation*, **55**, 733–8 (1977).
115. Zener, J. C. and Harrison, D. C. Serum enzyme values following intramuscular administration of lidocaine. *Archives of Internal Medicine*, **134**, 48–9 (1974).
116. Zweiman, F. G., Holman, B. L., O'Keefe, A., and Idoine, J. Selective uptake of 99mTc complexes and ^{67}Ga in acutely infarcted myocardium. *Journal of Nuclear Medicine*, **16**, 975–9 (1975).

CHAPTER 16

Experimental models for the study of myocardial tissue damage and enzyme release

A. Waldenström and Å. Hjalmarson

INTRODUCTION	379
METHODS FOR INDUCING ISCHAEMIA	380
Open Chest Techniques for Regional Ischaemia	380
Ligation of the Left Anterior Descending Artery (LAD)	380
Ligation of the Circumflex Artery	381
Ligation of a Papillary Muscle	382
Ligation of a Coronary Artery in Small Animals	382
Progressive Constriction of a Coronary Artery	383
Closed Chest Techniques for Regional Ischaemia	383
Induction of Ischaemia in the Dog	383
Induction of Ischaemia in the Pig	385
Induction of Ischaemia in the Rat	385
Whole Heart Ischaemia *in situ*	385
Whole Heart or Regional Ischaemia *in vitro*	387
METHODS FOR INDUCING ANOXIA	389
Isolated Perfused Hearts	389
Isolated Incubated Muscles	389
SPECIAL CONSIDERATIONS	390
Choice of Animal	390
Ischaemia *vs* Anoxia	391
Experimental Technique	391
Measurement of Enzyme Release	392
CONCLUDING COMMENTS	392
REFERENCES	395

INTRODUCTION

This chapter will deal with experimental models for studying acute myocardial infarction. There is no perfect animal model that covers all aspects of acute myocardial infarction in man and thus it is necessary to devise different models for the elucidation of different facets of the myocardial infarction process. The interpretation of experimental data has to be related to the model used, which means that the findings are often not directly applicable to the situation in man.

There are many variables that correlate well with the degree of ischaemic damage. Among the most frequently used ones are praecordial e.c.g. mapping,[41] loss of myocardial tissue enzymes,[20] and appearance of myocardial enzymes into the blood.[32,57] In this chapter it was decided to review various models of myocardial infarction, not all of these have been used for measurement of enzyme release but in general they should be amenable to such studies. Since a large proportion of cardiac research is carried out with anoxic or hypoxic models, these will be discussed in addition to ischaemic models.

METHODS FOR INDUCING ISCHAEMIA

Open Chest Techniques for Regional Ischaemia

Ligation of the left anterior descending artery (LAD)

Ligation of the LAD is the most commonly used method for the production of myocardial infarction in the dog. Experimental ligation of a coronary artery was described as early as 1894.[52] It was found, however, that abrupt ligation of the proximal LAD often led to ventricular fibrillation.[11,35,52] Harris[12] therefore introduced the two-stage occlusion model in which early ventricular fibrillation was almost totally prevented. A midline sternotomy is first performed and then the anterior descending artery is dissected free near the distal edge of the left auricular appendage. Double ligature is then passed under the artery and cut to form two ligatures. The first ligature is gently tied around the artery and a 20-gauge hypodermic syringe needle. The needle is then immediately withdrawn, leaving the artery constricted but still permitting some blood to flow. The second ligature can then be tied 30 minutes later, thus completely and permanently occluding the artery.

Despite the development of this two-stage procedure, the one-stage method (performed by placing a single ligature at the same site as above) is still used probably as a result of the introduction of d.c. conversion for ventricular fibrillation and the development of anti-arrhythmic agents such as lidocaine. As long ago as in 1935, Tennant and Wiggers[60] described the use of d.c. countershock for converting ventricular fibrillation to sinus rhythm after clamping the LAD. This method has since been used by many investigators, for example, by Hood et al.[16] for testing pharmacological agents during ischaemia, by Braunwald and colleagues in numerous studies[40–43] (Figure 1), and by Shell et al.[57] in their extensive investigations on enzyme release from infarcting myocardium. It has been reported[58] that if the left coronary artery is tied within 2 cm of its origin and just beyond the first medium-sized branches then the ventricular fibrillation which so often occurs with more proximal ligations can be prevented.

In addition to the dog, other animals have been used for myocardial infarction studies. Brooks et al.,[4] for example, used the open chest pig.

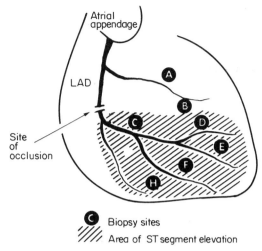

Figure 1. Dog heart showing site of ligature of LAD (left anterior descending coronary artery). Some authors ligate several side branches of LAD to obtain a sufficient reduction of flow to the ischaemic area. [Modified text to figure reproduced by permission of American Heart Association, Inc. from *Circulation*, **46**, 430–7 (1972).]

During anaesthesia and artificial respiration the heart was exposed by a median sternotomy and the left anterior descending coronary artery was ligated once with a silk snare so as to give an anteroseptal myocardial infarction. Minipigs have also been used, for example, Winbury et al.,[66] anaesthetized minipigs and after left thoracotomy they ligated the LAD. Bruyneel and Opie[5] ligated the distal third of the LAD in baboons and obtained an infarct representing about 10% of the total heart weight. In this preparation, the distal part of the ligated coronary artery remained empty with no signs of the retrograde refilling which is often seen in the dog. Massion et al.[44] have described experimental myocardial infarction in cats following ligation of branches of the LAD and similar techniques have been used by Krug et al.[29] and by Mathes et al.[45]

Ligation of the circumflex artery

When quantitative biochemical and morphological studies are required it is desirable to obtain a homogeneous infarct. Simply ligating any major coronary artery in the dog leads to a heterogeneous mixture of necrotic and living cells and this varies from animal to animal. The reason for this heterogeneity is the extensive collateral coronary circulation which is

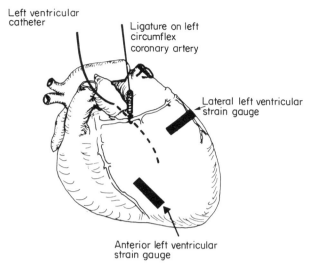

Figure 2. Dog heart showing the site of ligature of the left circumflex coronary artery. [Modified text to figure reproduced, with permission, from *Acta Medica Scandinavica Supplement*, **587**, 71–82 (1975).]

characteristic of the dog. However, as pointed out by Burchell[6] ligation of the circumflex branch of the left coronary artery leads to a homogeneous necrosis of the posterior papillary muscle (Figure 2). The method has been further developed and standardized by Jennings and Wartman.[19] After anaesthesia, the dogs are given procainamide as anti-arrhythmic prophylaxis, thoracotomy is performed and the circumflex artery is ligated with silk and the chest is then closed. In dogs allowed to survive for 24 hours or longer after ligation a uniform infarct of the posterior papillary muscle was observed (see also refs. 20, 21, 59).

Ligation of a papillary muscle

An alternative way of producing homogenous infarction of a papillary muscle has been described by Miller *et al.*[46] Under anaesthesia and after left thoracotomy, the left atrial appendage is exposed and opened to allow the index finger to pass into the atrium and through the mitral valve to detect the papillary muscle. Ligatures may then be placed around the four margins of the anterior papillary muscle and infarction be induced.

Ligation of a coronary artery in small animals

A good technique for ligation of coronary arteries in rats as well as mice, hamsters, and guinea-pigs has been described by Johns and Olson.[23] Positive pressure anaesthesia is delivered through a face mask connected to an ether oxygen gas supply. A left intercostal thoracotomy is performed and the left coronary artery is ligated. For an experienced operator, the whole procedure can be accomplished in 5–7 minutes. Changes in heart rate or rhythm seldom occur and ventricular fibrillation is extremely rare in contrast to dogs where serious arrhythmias frequently occur after ligation of a coronary artery. In these small animal preparations, the margins of the infarct are sharply defined and are remarkably constant from one animal to another, especially in mice and rats. A similar technique has been described by Bajusz and Jasmin.[2]

Progressive constriction of a coronary artery

As the occlusion of a coronary artery is thought to develop over a long period of time, different devices have been designed to permit the slow and progressive constriction of an artery. Litvak et al.[37] produced a gradual occlusion of the LAD by means of a hygroscopic material called 'Ameroid'. This is composed of compressed casein which has been cured in formalin. The 'Ameroid' is encased in a stainless steel sleeve in order to contain the expansive force of the material and direct it towards the lumen. A hydraulic occlusive device has been described by Khouri et al.[26] A cuff is placed around a coronary artery and connected to a tube that is exteriorized to a stainless steel micrometric piston. The whole system is filled with mercury. After recovery from surgery, the dogs may be subjected to graded ischaemia. A similar system has been described by Jageneau[18] but in this preparation the local vein is catheterized to facilitate measurement of local myocardial metabolism during ischaemia.

Slow or delayed occlusion can also be achieved by placing a silastic balloon around a coronary artery and attaching it to a Dacron-reinforced belt. The balloon is bonded to a silastic tube and is exteriorized on the dogs' back. After closure of the thorax and a suitable recovery period saline may be injected into the balloon to occlude the artery. The advantage of this type of preparation is that the onset of infarction is well controlled and the dogs are conscious and are well recovered from the surgical trauma.[7,24,30] A similar technique for occlusion and reperfusion studies has been described by Sham et al.[56] Bresnahan et al.[3] have modified this technique including a lidocaine infusion at the onset of occlusion and during reperfusion in an attempt to prevent occlusion or reperfusion-induced arrhythmias.

Closed Chest Techniques for Regional Ischaemia

Induction of ischaemia in the dog

Anaesthesia and surgical trauma are known to influence the results of experimental coronary occlusion. In order to eliminate these factors, a number of techniques for inducing experimental myocardial infarction without thoracotomy have been introduced. In most of these techniques it is not necessary to anaesthetize the animals.

Some methods involve the injection of sclerosing solutions or other necrotizing agents into a coronary artery in order to produce tissue necrosis. However, the distribution of the necrosis may be patchy and unpredictable. An alternative technique described by Salazar[54] is the electrical induction of thrombus formation via an intracoronary catheter. In this preparation a modified West catheter is passed from the carotid artery into the left coronary artery and is further advanced to either the LAD or the left circumflex artery. Under fluoroscopic guidance, a stainless steel catheter is then advanced through the West catheter to the desired location in the coronary tree. A direct current from a 3 volt battery is then introduced into the circuit. The intensity of the current is gradually increased to between 100 and 900 microamperes. The current is continued until definite electrocardiographic evidence of myocardial ischaemia is observed. The presence of complete occlusion may then be confirmed by coronary arteriography. A similar technique has been described by Kordenat et al.,[28] here the inner catheter is made of copper or magnesium alloy and it is helical in shape and it is the exposed length of the inner catheter which determines the clotting time. In another method Nakhjavan et al.[49] used a stainless steel cylinder positioned in the LAD or in a circumflex branch to induce ischaemia in mongrel dogs. A catheter is first placed in the desired position and a 'Teflon'-coated guide wire is inserted into the catheter. The catheter is then removed, a stainless steel cylinder inserted over the wire and the catheter is then reinserted to push the cylinder into the desired position in the coronary artery. The catheter and guide wire are then withdrawn and electrocardiographic evidence of myocardial ischaemia appears 6–8 hours later.

A more acute induction of ischaemia can be achieved by occluding a coronary artery with a radiopaque sphere, to achieve this the sphere is placed on a Sones catheter and placed in a suitable position under fluoroscopic guidance.[10] Instead of using a sphere, the coronary artery may also be occluded by the injection of a bolus of mercury (0.2 ml) through the catheter.[38] A more diffuse form of ischaemic damage may be produced by injecting graded microspheres into the aortic root. Here, a balloon catheter is placed with its tip at the level of the coronary ostia. The balloon catheter is then transiently inflated in order to completely obstruct the aorta. The microspheres are then rapidly injected through the catheter to enter the coronary circulation. Fifteen seconds later the balloon is deflated and the catheter withdrawn.[1]

Occlusion of a coronary artery with a catheter *per se* has also been used by many investigators. Kurien et al.,[31] for example, induced myocardial ischaemia by placing a catheter in the wedge position in the left circumflex coronary artery. If the catheter does not wedge at the desired position then a catheter-tip balloon can be inflated at that point. In their studies Corday et al.[8,9] have often used an inflatable catheter-tip balloon. In such preparations, the tip protrudes 1 cm beyond the balloon thus enabling distal arterial pressure measurements to be made and blood samples to be taken. The balloon may be deflated at various times after the onset of ischaemia, thus facilitating studies of graded ischaemia and reperfusion. In addition to the dog, the pig has also been used for closed chest occlusion of a coronary artery using an inflatable balloon catheter positioned in the LAD under fluoroscopy.[47]

Induction of ischaemia in the pig

Another, entirely different model for myocardial infarction has been described by Johansson et al.[22] Here, a short-acting peripheral muscle relaxant is given 15–20 minutes before electrical stimulation via an electric 'animal pusher' (an electric rod which is used in slaughter houses). Electrical stimulation is repeated 5 or 6 times over a 20 minute period. Electrocardiographic signs of ischaemia, such as T wave inversion and ST segment elevation, are detectable during the stimulation period and small necrotic foci or confluent necrosis can be detected. These are disseminated throughout the heart but are most prominent in the inner third of the muscle wall.

Induction of ischaemia in the rat

A less specific model for inducing myocardial ischaemic damage was introduced by Rona et al.[53] It was found that the administration of isoproterenol to rats resulted in the development of myocardial necrosis. This necrosis may be caused by one or more of several mechanisms such as the myocardial metabolic demand being caused to exceed the supply of oxygen or substrate fluid; microembolism caused by catecholamine-induced increases in platelet aggregation; alternatively, increased lipolysis may play a role in the development of damage. This isoproterenol model is probably of considerable relevance since the release of myocardial norepinephrine may well be related to the development of myocardial infarction.

Whole Heart Ischaemia *in situ*

Whole heart ischaemia of the swine heart *in situ* has been suggested as a good model for studying ischaemic myocardial metabolism. In this preparation the pig is anaesthetized, intubated, and artificially ventilated. After transthoracotomy, the heart is exposed and the azygos and hemiazygos veins ligated. Both the venae cavae and the pulmonary artery are cannulated and connected with silicone rubber tubing to form a right heart bypass. Antegrade flow is maintained by a Sarns pump which is adjusted to match the systemic venous return. A right ventriculotomy is then performed and a drainage pump is inserted to collect coronary effluent blood. This blood is then oxygenated via a Bentley oxygenator and is returned for perfusion of both the cannulated coronary arteries. The coronary circuit thus established is driven by another Sarns pump and total coronary perfusion can be controlled by the pump and can be adjusted to any desired rate[36] (Figure 3). Coronary blood samples are easily collected for the measurement of enzymes or metabolites.

A graded ischaemia in the open chest dog heart has been described by Lekven et al.[34] The dogs are maintained at a constant level of anaesthesia and artificially ventilated and after a left-sided thoracotomy, the descending

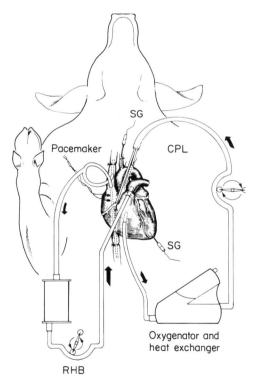

Figure 3. Schematic representation of working swine heart model. RHB = right heart bypass; CPL = coronary perfusion loop; SG = strain gauge. [Reproduced, with permission, from *American Journal of Physiology*, **228**, 655–62 (1975).]

and circumflex left coronary arteries are cannulated and perfused through an extracorporeal shunt from the left carotid artery. The coronary flow may then be varied via an adjustable clamp.

Whole Heart or Regional Ischaemia *in vitro*

Perfusion of isolated hearts was described as early as 1895 by Langendorff[33] and a model for graded ischaemia in the isolated rat heart has been described by Kligfield et al.[27] In this model the isolated heart is perfused in a modified Langendorff retrograde mode with Krebs–Ringer bicarbonate buffer containing glucose (5.5 mmol l.$^{-1}$) noradrenaline (0.1 mmol l.$^{-1}$), lactate, and insulin. Graded ischaemia may be induced by reducing the coronary flow to a predetermined value by perfusion through a variable pump. The fall in perfusion pressure which accompanies the decrease in flow approximates to the clinical state in which vascular pressure, distal to a partially occluded artery, is decreased in response to an obstruction. This model permits steady-state regulation of both the severity and duration of ischaemia.

A graded ischaemia can also be produced by lowering the perfusion pressure of Langendorff perfused rat hearts[64] and in this system the perfusate need not be recirculated. A more physiologic isolated perfused rat heart preparation was introduced by Neely and coworkers in 1967,[50] here, both the proximal aorta and the left atrium are cannulated. A peristaltic pump carries perfusion fluid to a bubble trap (which acts as an atrial filling reservoir) and the overflow is returned to an oxygenating chamber. The atrial cannula is connected to the bubble trap by tygon tubing and the vertical distance from the overflow to the left atrium determines the left atrial filling pressure. The left ventricle ejects the perfusate to the top of the oxygenating chamber (vertical distance 80 cm) and this distance determines the cardiac afterload. Regional ischaemia may be induced in this working heart preparation by ligating coronary arteries as described by Kannengiesser et al.[25] using the coronary artery ligation technique described by Bajusz and Jasmin.[2]

In 1973, Neely et al.,[51] introduced a new technique for the induction of whole heart ischaemia in the isolated working rat heart. The same perfusion basis system as described in the preceding paragraph was used but there were certain modifications. In particular, a one-way ball valve is introduced into the aortic cannula just distal to the coronary orifices. This valve is designed to prevent diastolic backflow (normally, myocardial perfusion takes place during diastole and prevention of this leads to the development of ischaemia). It should be pointed out that hearts perfused with erythrocyte-free buffer are dependent upon the oxygen which is physically dissolved in the buffer. This requires a relatively high coronary flow for the delivery of

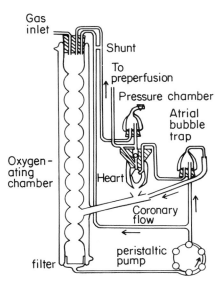

Figure 4. Perfusion apparatus for the isolated working rat heart under ischaemic and non-ischaemic conditions. The aorta and the left atrium are cannulated, a one-way ball valve is placed just distal to the coronaries but this ball valve can be bypassed. A Windkessel effect is provided by a partly air-filled chamber connected to the aortic tube. The heart pumps perfusate to the top of the oxygenating chamber and a peristaltic pump provides the atrial bubble trap with perfusate. The height of this bubble trap is adjustable and the height chosen determines the atrial filling pressure. By clamping the bypass of the one-way valve, diastolic perfusion of the coronaries will be inhibited, thus inducing ischaemia. Retrograde perfusion with constant perfusion pressure is provided by opening a connection between the pump overflow tube and the aortic tube [Modified text to figure reproduced, with permission, from *Recent Advances in Studies of Cardiac Structure and Metabolism*, **10**, 307–13 (1975).]

sufficient oxygen. Hence, under both normal and ischaemic conditions, coronary flow is rather high. However, on some occasions in this preparation, coronary flow will be reduced to such an extent that sampling of coronary effluent for biochemical studies is not possible. To circumvent this problem Waldenström and Hjalmarson[61] proposed a slight modification to the perfusion system. Adequate coronary reperfusion was made possible in the retrogradely perfused heart by opening the ball valve bypass tube and introducing a shunt from the pump overflow to the aortic tube (see Figure 4). This perfusion technique permits whole heart ischaemia which is characterized by an even distribution of the reduced flow in the myocardium. In this modified preparation the aortic output, intracavity pressures and dP/dt are easily measured and are determined by the function of the ischaemic myocardium and not by the combination of ischaemic and normal myocardium (as is the case when a coronary artery is ligated).

METHODS FOR INDUCING ANOXIA

Isolated Perfused Hearts

There has been much confusion about the difference between anoxia and ischaemia (see definitions in Chapter 1 of this book) and it is well known that metabolism during these two states is not the same. Many experimental myocardial infarction studies have been performed using Langendorff perfused hearts during anoxia or hypoxia. The main advantage of this technique is its simplicity. It merely entails switching the gas mixture from 95% O_2 + 5% CO_2 to 95% N_2 + 5% CO_2 for anoxia, or for hypoxia an intermediate gas mixture is used. Anoxic perfusion of rat hearts may be exemplified by the work of Hearse and collaborators.[13] Henry et al.[15] used the isolated perfused guinea-pig heart for studying metabolism during anoxia. The isolated heart is perfused according to the Langendorff technique with Krebs–Henseleit buffer using glucose as substrate. Anoxia is instituted by gassing the perfusate with 95% N_2 and 5% CO_2. The hearts can be electrically paced at any desired rate.

Isolated Incubated Muscles (Perifusion)

Isolated papillary muscle from cats and other species has been used to study the effect of anoxia and hypoxia. These isolated muscle strips or papillary muscles are maintained in an organ bath and they have been widely used particularly for the study of myocardial function. The work of Willerson et al.[65] serves as a good example of the use of these techniques. After removal of the papillary muscle from anaesthetized cats it was held at both ends by spring-loaded clips and was then placed in a bath containing

Krebs–Ringer bicarbonate buffer with glucose (18 mmol l.$^{-1}$) as a substrate. The perifusion fluid in the bath was maintained by suitable reservoirs and the fluid and the tissue were oxygenated with a gas mixture of 95% O_2 and 5% CO_2. Anoxia can be induced by changing the gas mixture to 95% $N_2 + 5\% CO_2$, whereupon the Po_2 of the perifusate falls to 30 mmHg or less. During experiments the muscle strips can be stimulated with d.c. impulses at a rate of 12 contractions per minute.

A new model for whole heart anoxia and ischaemia has been described by Ingwall et al.[17] Fifteen- to 22-day-old foetal mouse hearts are placed in an organ bath while they are beating. These preparations are not perfused but the composition of the perifusate can be altered so as to create various combinations of ischaemia and anoxia. This technique has the advantage of simplicity but it must be emphasized that it is restricted to small foetal hearts where the myocardial metabolism may differ from that of the adult heart. A similar method has been used by Mustafa et al.[48] who used 16-day-old chick embryo cells obtained from foetal hearts which had been cut into 4 pieces. Isolated single cells can also be maintained in perifusion medium and the effects of anoxia studied on the leakage of enzymes to the medium.

SPECIAL CONSIDERATIONS

Choice of Animal

The dog was probably the first, and still is the most widely used animal for experimental studies of myocardial infarction. The dog is large and is easily operated upon and can provide large amounts of material for biochemical studies. It is also very easy to obtain sufficient and repeated blood samples from the dog for analysis. However, the dog, like other large animals, is expensive and cumbersome. The porcine heart is suggested to have close similarity to the human heart in relation to coronary architecture and collateral circulation and the pig is therefore the preferred experimental animal for some authors. The inconvenience of this rather large animal can, however, be overcome by the use of minipigs. It is probably the baboon heart which provides the closest model for the coronary anatomy of man[5] and this species is obviously preferred in areas where it is relatively easy to obtain, but for most investigators the baboon is not a practical choice.

Smaller animals hold many attractions, they are cheap and easy to handle and many experiments can be performed simultaneously so that groups for statistical analysis can be obtained in a very short time. However, it must be remembered that hearts from small animals may differ from the human heart in many ways. Well-known examples of this problem are the absence of the staircase phenomenon in the rat, the resistance to arrhythmias in the rat, the sensitivity to arrhythmias in the guinea-pig, and the absence of the release of endogenous catecholamines during ischaemia of the guinea-pig heart contrary to other species.[67] In the final analysis therefore, the choice of animal must represent a balance between cost, practicality, and suitability, and clearly, no one animal is ideal.

Ischaemia vs Anoxia

Ischaemia is defined as decreased perfusion of a tissue, resulting in an inadequate supply of fluid, oxygen, and substrates as well as an inadequate washout of metabolites. Anoxia is defined as normal perfusion of a tissue with normal supply of substrates but without oxygen. In anoxic perfusion local accumulation of potassium, hydrogen ions, and lactate is prevented. This, together with major metabolic differences, represents the fundamental difference between the anoxic and the ischaemic state and these differences must be taken into account when interpreting experimental data.

Experimental Technique

As the trigger mechanism for initiation of the myocardial infarction process is unknown, the choice of experimental technique may be of crucial importance. At the present time it is very controversial whether coronary thrombosis is the primary or a secondary event. For example, Waldenström et al.[62] have suggested that the infarct may start with a rapid release of myocardial noradrenaline. If this were the case then one should ask whether ligation of a coronary artery is an acceptable model for infarction? The answer may be that ligation may be suited to the study of certain stages of the myocardial infarction process—which is certainly not an instantaneous event but something that develops over a certain period of time.

Experimental infarction in whole animals may more closely resemble infarction in man. But the effects of nervous reflexes and circulating hormones are difficult to control in the experimental situation. In this respect the isolated heart, muscle strip, or myocardial cell culture are to be preferred. Open chest techniques are influenced by the surgical trauma and anaesthesia but these problems can be overcome by closed chest techniques. In general, bigger animals more closely resemble man but are difficult to keep under well-controlled conditions, whereas small animals and isolated hearts or parts of the hearts lend themselves to well-controlled experiments at the expense of closeness to the human heart.

The effect of reperfusion on tissue damage and enzyme release has been under debate recently.[14] This very interesting phenomenon may be studied in models where a ligature is placed around a coronary artery or a coronary artery is occluded by an inflatable balloon catheter or just occluded by the catheter tip. The ischaemic perfusion model as described by Neely is also well suited for such studies Anoxic heart perfusions may of course be subjected to reoxygenation but it must be remembered that the consequences of reoxygenation may not be the same as those of reperfusion (see Chapter 18).

Measurement of Enzyme Release

Determination of enzyme leakage as an indicator of myocardial damage is discussed in detail elsewhere in this book. In large animal preparations, enzyme leakage can be measured by blood samples from the coronary sinus, from the systemic circulation or from the heart lymph or from the reduction of tissue content. Myocardial enzymes appear early in heart lymph and their concentration is considerably higher than that in blood. Measurement of enzymes in cardiac lymph requires cannulation of lymphatic vessels and this may be difficult or laborious and usually coronary sinus or systemic measurements are preferred. Measuring arteriovenous differences of enzymes in coronary sinus blood or cardiac lymph may overcome measurement problems arising from enzyme release due to surgical trauma. This problem can also be overcome by measuring a cardiospecific enzyme such as CK-MB. In isolated heart preparations or incubated muscle preparations, the choice of enzyme is less. In all instances, however, factors influencing diffusion, distribution, and degradation of the enzymes must be taken into account (see Chapter 5). The enzyme leakage profile will also differ depending upon the model used and where the sample for assay was taken. The time of peak enzyme release may differ between different enzymes when samples are taken from peripheral blood in whole animals or man but when different enzymes are measured in dog heart lymph[39] or from perfused rat hearts[63] all enzymes appear at the same time. Different patterns of distribution and different rates of degradation probably explain some of the differences. Some profiles for enzyme release obtained with some fundamentally different infarction models are shown in Figures 5–7.

CONCLUDING COMMENTS

The mechanism for the development of myocardial infarction is not known and thus when studying this process experimentally the choice of an appropriate model is difficult but important. Various experimental models may only reflect in part the clinical situation during acute myocardial infarction. For this reason different models, using different species should be used to elucidate the various aspects of myocardial infarction in man. A number of models, considered to be of special interest, are reviewed in this chapter, these include ligation of a coronary arteries *in vivo* in large and

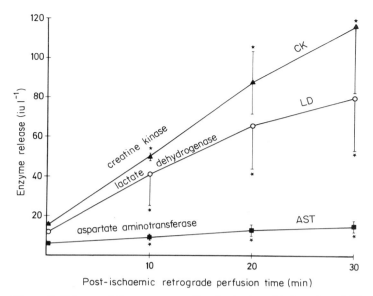

Figure 5. Release of three enzymes after 5 minutes of ischaemia and 30 minutes of reperfusion. Ischaemia was induced in isolated perfused rat hearts according to the method of Neely. Each value represents the mean ±SEM for 5–8 hearts. * = significantly different from initial value ($p \leq 0.05$). [Reproduced, with permission, from *Acta Medica Scandinavica*, **201,** 533–8 (1977).]

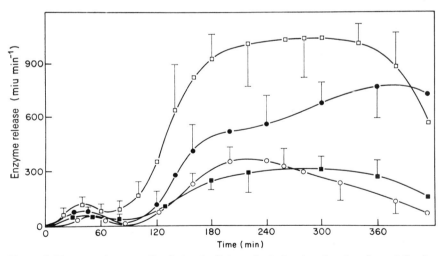

Figure 6. Enzyme release profiles (miu min^{-1}) from the isolated perfused rat heart following the onset of anoxia ($t = 0$). (●) = α-Hydroxybutyrate dehydrogenase; (□) = creatine kinase; (○) = adenylate kinase; (■) = aspartate aminotransferase. Each point represents the mean for 4 hearts and the bars represent the SEM. [Reproduced, with permission, from *Journal of Molecular and Cellular Cardiology*, **7,** 463–82 (1975).]

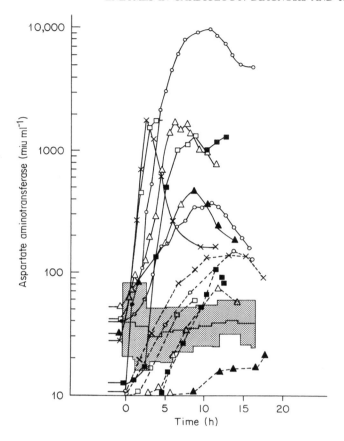

Figure 7. Aspartate aminotransferase in heart lymph (unbroken lines) and in serum (dashed lines) in individual dogs with myocardial infarction. In addition, mean values and range of the enzyme activity in the heart lymph of 5 control dogs is indicated (boxes). The time zero in the infarction group indicates the time of ligation of a coronary artery. In the control group the time zero indicates the start of the collection of heart lymph. [Reproduced, with permission, from *Scandinavian Journal of Clinical and Laboratory Investigation*, **30**, 153–8 (1972).]

small animals, *in vitro* perfusion techniques with hearts from small animals as well as perifusion techniques for isolated papillary muscles and isolated myocytes. While each of these models have their own inherent advantages and disadvantages, they are all amenable to studies of tissue damage by measurement of enzyme release.

REFERENCES

1. Agress, C. M., Rosenberg, M. J., Jacobs, H. I., Binder, M. J., Schneiderman, A., and Clark, W. G. Protracted shock in the closed-chest dog following coronary embolization with graded microspheres. *American Journal of Physiology*, **170,** 536–49 (1952).
2. Bajusz, E. and Jasmin, G. Histochemical studies on the myocardium following experimental interference with coronary circulation in the rat. I. Occlusion of coronary artery. *Acta Histochemica*, **18,** 222–37 (1964).
3. Bresnahan, G. F., Roberts, R., Shell, W. E., Ross, J. Jr., and Sobel, B. E. Deleterious effects due to hemorrhage after myocardial reperfusion. *American Journal of Cardiology*, **33,** 82–6 (1974).
4. Brooks, H., Al-Sadir, J., Schwartz, J., Rich, B., Harper, P., Resnekov, L., and Dellenbaugh, J. Biventricular dynamics during quantitated anteroseptal infarction in the porcine heart. *American Journal of Cardiology*, **36,** 765–75 (1975).
5. Bruyneel, K. and Opie, L. H. Use of baboons in studies of acute myocardial infarction and effects of infusions of glucose, insulin and potassium (GIK). *Acta Medica Scandinavica Supplement*, No. 587, 65–8 (1975).
6. Burchell, H. B. Adjustments in the coronary circulation after experimental coronary occlusion. *Archives of Internal Medicine*, **65,** 240–62 (1940).
7. Chimoskey, J. E., Szentivanyi, M., Zakheim, R., and Barger, A. C. Temporary coronary occlusion in conscious dogs: collateral flow and electrocardiogram. *American Journal of Physiology*, **212,** 1025–32 (1967).
8. Corday, E., Lang, T. W., Crexells, C., and Meerbaum, S. Intracoronary balloon—a new model for induction of myocardial ischemia. *American Journal of Cardiology*, **29,** 301 (1972).
9. Corday, E., Lang, T. W., Meerbaum, S., Gold, H., Hirose, S., Rubins, S., and Dalmastro, M. Closed chest model of intracoronary occlusion for study of regional cardiac function. *American Journal of Cardiology*, **33,** 49–59 (1974).
10. Gensini, G. G., Palacio, A., Buonenno, C., Kelly, A. E., and Muller, W. F. Superselective coronary occlusion under cinefluoroscopic control in experimental animals: Technique and results. *Circulation Supplement III*, **33** and **34,** III-108–9 (1966).
11. Harris, A. S. Terminal electrocardiographic patterns in experimental anoxia, coronary occlusion, and hemorrhagic shock. *American Heart Journal*, **35,** 895–909 (1948).
12. Harris, A. S. Delayed development of ventricular ectopic rhythms following experimental coronary occlusion. *Circulation*, **1,** 1318–28 (1950).
13. Hearse, D. J., Humphrey, S. M., and Chain, E. B. Abrupt reoxygenation of the anoxic potassium-arrested perfused rat heart: A study of myocardial enzyme release. *Journal of Molecular and Cellular Cardiology*, **5,** 395–407 (1973).
14. Hearse, D. J. Reperfusion of the ischaemic myocardium. *Journal of Molecular and Cellular Cardiology*, **9,** 605–13 (1977).
15. Henry, P. D., Sobel, B. E., and Braunwald, E. Protection of hypoxic guinea pig hearts with glucose and insulin. *American Journal of Physiology*, **226,** 309–13 (1974).
16. Hood, W. B. Jr., McCarthy, B., and Lown, B. Myocardial infarction following coronary ligation in dogs. *Circulation Research*, **21,** 191–9 (1967).
17. Ingwall, J. S., DeLuca, M., Sybers, H. D., and Wildenthal, K. Fetal mouse hearts: A model for studying ischemia. *Proceedings of the National Academy of Sciences (USA)*, **72/7,** 2809–13 (1975).
18. Jageneau, A. H. M., van Gerven, W., Kruger, R., van Belle, H., and Reneman, R. S. An improved animal model for studying the effect of drugs on myocardial metabolism during ischemia. In *Recent Advances in Studies on Cardiac Structure and Metabolism*, Vol. 10, *The Metabolism of Contraction* (Eds. Roy, P. E. and Rona, G. University Park Press, Baltimore 1975, pp. 331–41.
19. Jennings, R. B. and Wartman, W. B. Production of an area of homogenous myocardial infarction in the dog. *American Medical Association Archives of Pathology*, **63,** 580–5 (1975).
20. Jennings, R. B., Kaltenbach, J. P., and Smetters, G. W. Enzymatic changes in acute myocardial ischemic injury. *American Medical Association Archives of Pathology*, **64,** 10–6 (1957).

21. Jennings, R. B. and Ganote, C. E. Structural changes in myocardium during acute ischemia. *Circulation Research Supplement III*, **34** and **35**, III-156-68 (1974).
22. Johansson, G., Jönsson, L., Lannek, N., Blomgren, L., Lindberg, P., and Poupa, O. Severe stress-cardiopathy in pigs. *American Heart Journal*, **87**, 451-7 (1974).
23. Johns, T. N. P. and Olson, B. J. Experimental myocardial infarction. I. A method of coronary occlusion in small animals. *Annals of Surgery*, **140**, 675-82 (1954).
24. Joison, J., Kumar, R., Hood, W. B., Jr., and Norman, J. C. An implantable system for producing left ventricular failure for circulatory-assist device evaluation. *Transactions American Society for Artificial Internal Organs*, **15**, 417-23 (1969).
25. Kannengiesser, G. J., Lubbe, W. F., and Opie, L. H. Experimental myocardial infarction with left ventricular failure in the isolated perfused rat heart. Effects of isoproterenol and pacing. *Journal of Molecular and Cellular Cardiology*, **7**, 135-51 (1975).
26. Khouri, E. M., Gregg, D. E., and Lowensohn, H. S. Flow in the major branches of the left coronary artery during experimental coronary insufficiency in the unanesthetized dog. *Circulation Research*, **23**, 99-109 (1968).
27. Kligfield, P., Horner, H., and Brachfeld, N. A model of graded ischemia in the isolated perfused rat heart. *Journal of Applied Physiology*, **40/6**, 1004-8 (1976).
28. Kordenat, R. K., Kezdi, P., and Stanley, E. L. A new catheter technique for producing experimental coronary thrombosis and selective coronary visualization. *American Heart Journal*, **83**, 360-4 (1972).
29. Krug, A., du Mesnil de Rochemont, W., and Korb, G. Temporary coronary occlusion. *Circulation Research*, **19**, 57-62 (1966).
30. Kumar, R., Hood, W. B., Jr., Joison, J., Norman, J. C., and Abelmann, W. H. Experimental myocardial infarction. II. Acute depression and subsequent recovery of left ventricular function: serial measurements in intact conscious dogs. *Journal of Clinical Investigation*, **49**, 55-62 (1970).
31. Kurien, V. A., Yates, P. A., and Oliver, M. F. The role of free fatty acids in the production of ventricular arrhythmias after acute coronary artery occlusion. *European Journal of Clinical Investigation*, **1**, 225-41 (1971).
32. LaDue, J. S. Wroblewski, F., and Karmen, A. Serum glutamic oxaloacetic trnsaminase activity in human acute transmural myocardial infarction. *Science*, **120**, 497-9 (1954).
33. Langendorff, O. Untersuchungen am überlebenden Säugethierherzen. *Pflügers Archiv für die Gesamte Physiologie*, **61**, 291-332 (1895).
34. Lekven, J., Mjøs, O. D., and Kjekshus, J. K. Compensatory mechanisms during graded myocardial ischemia. *American Journal of Cardiology*, **31**, 467-73 (1973).
35. Lewis, T. Experimental production of paroxysmal tachycardia and the effects of ligation of the coronary arteries. *Heart*, **1**, 98-137 (1909).
36. Liedtke, A. J., Hughes, H. C., and Neely, J. R. Metabolic responses to varying restrictions of coronary blood flow in swine. *American Journal of Physiology*, **228**, 655-62 (1975).
37. Litvak, J., Siderides, L. E., and Vineberg, A. M. The experimental production of coronary artery insufficiency and occlusion. *American Heart Journal*, **53**, 505-18 (1957).
38. Lluch, S., Moguilevsky, H. C., Pietra, G., Shaffer, A. B., Hirsch, L. J., and Fishman, A. P. A reproducible model of cardiogenic shock in the dog. *Circulation*, **39**, 205-18 (1969).
39. Malmberg, P. Enzyme composition of dog heart lymph after myocardial infarction. Enzymes in heart lymph after infarction. *Upsala Journal of Medical Science*, **78**, 73-7 (1973).
40. Maroko, P. R., Kjekshus, J. K., Sobel, B. E., Watanabe, T., Covell, J. W., Ross, J., Jr., and Braunwald, E. Factors influencing infarct size following experimental coronary artery occlusions. *Circulation*, **43**, 67-82 (1971).
41. Maroko, P. R., Libby, P., Covell, J. W., Sobel, B. E., Ross, J., Jr., and Braunwald, E. Precordial S-T segment elevation mapping: An atraumatic method for assessing alterations in the extent of myocardial ischemic injury. The effects of pharmacologic and hemodynamic interventions. *American Journal of Cardiology*, **29**, 223-30 (1972).
42. Maroko, P. R., Libby, P., Sobel, B. E., Bloor, C. M., Sybers, H. D., Shell, W. E., Covell, J. W., and Braunwald, E. Effect of glucose–insulin–potassium infusion on myocardial infarction following experimental coronary artery occlusion. *Circulation*, **45**, 1160-75 (1972).
43. Maroko, P. R., Libby, P., and Braunwald, E. Effect of pharmacologic agents on the function of the ischemic heart. *American Journal of Cardiology*, **32**, 930-6 (1973).

44. Massion, W., Blümel, G., Erhardt, W., Krueger, P., Petrowicz, O., Wendt, P., and Heinkelmann, W. Die Wirkung von Proteinase-inhibitoren in der Frühphase nach experimentellem Myokardinfarkt. *Münchener medizinische Wochenschrift*, **118**, 197–202 (1976).
45. Mathes, P., Romig, D., Sack, D., and Erhardt, W. Experimental myocardial infarction in the cat. I. Reversible decline in contractility of noninfarcted muscle. *Circulation Research*, **38**, 540–6 (1976).
46. Miller, G. E., Jr, Cohn, K. E., Kerth, W. J., Selzer, A., and Gerbode, F. Experimental papillary muscle infarction. *Journal of Thoracic and Cardiovascular Surgery*, **56**, 611–6 (1968).
47. Most, A. S., Capone, R. J., and Mastrofrancesco, P. A. Failure of hyaluronidase to alter the early course of acute myocardial infarction in pigs. *American Journal of Cardiology*, **38**, 28–33 (1976).
48. Mustafa, S. J., Rubio, R., and Berne, R. M. Uptake of adenosine by dispersed chick embryonic cardiac cells. *American Journal of Physiology*, **228**, 62–7 (1975).
49. Nakhjavan, F. K., Shedrovilzky, H., and Goldberg, H. Experimental myocardial infarction in dogs. Description of a closed chest technique. *Circulation*, **38**, 777–82 (1968).
50. Neely, J. R., Liebermeister, H., Battersby, E. J., and Morgan, H. E. Effect of pressure development on oxygen consumption by isolated rat heart. *American Journal of Physiology*, **212**, 804–14 (1967).
51. Neely, J. R., Rovetto, M. J., Whitmer, J. T., and Morgan, H. E. Effects of ischemia on function and metabolism of the isolated working rat heart. *American Journal of Physiology*, **225**, 651–8 (1973).
52. Porter, W. T. On the results of ligation of the coronary arteries. *Journal of Physiology*, **15**, 121–38 (1894).
53. Rona, G., Chappel, C. I., Balazs, T., and Gaudry, R. An infarct-like myocardial lesion and other toxic manifestations produced by isoproterenol in the rat. *American Medical Association Archives of Pathology*, **67**, 443–55 (1959).
54. Salazar, A. E. Experimental myocardial infarction. Induction of coronary thrombosis in the intact closed-chest dog. *Circulation Research*, **9**, 1351–6 (1961).
55. Shahab, L., Wollenberger, A., Krause, E. G., and Genz, S. The effect of acute ischaemia on catecholamines and cyclic AMP levels in normal and hypertrophied myocardium. In *Effect of Acute Ischaemia on Myocardial Function* (Eds, Oliver, M. F., Julian, D. G., and Donald, K. W.) Churchill Livingstone, Edinburgh and London, 1972, pp. 97–107.
56. Sham, G. B., White, F. C., and Bloor, C. M. A constrictive and occlusive cuff for medium and large blood vessels. *Journal of Applied Physiology*, **28**, 510–2 (1970).
57. Shell, W. E., Kjekshus, J. K., and Sobel, B. E. Quantitative assessment of the extent of myocardial infarction in the conscious dog by means of analysis of serial changes in serum creatine phosphokinase activity. *Journal of Clinical Investigation*, **50**, 2614–25 (1971).
58. Shnitka, T. K. and Nachlas, M. M. Histochemical alterations in ischemic heart muscle and early myocardial infarction. *American Journal of Pathology*, **42**, 507–21 (1963).
59. Sommers, H. M. and Jennings, R. B. Experimental acute myocardial infarction. Histologic and histochemical studies of early myocardial infarcts induced by temporary or permanent occlusion of a coronary artery. *Laboratory Investigation*, **13/12**, 1491–503 (1964).
60. Tennant, R. and Wiggers, C. J. The effect of coronary occlusion on myocardial contraction. *American Journal of Physiology*, **112**, 351–61 (1935).
61. Waldenström, A. P. and Hjalmarson, Å. C. Myocardial enzyme release from ischemic isolated perfused working rat heart. In *Recent Advances in Studies on Cardiac Structure and Metabolism*, Vol. 10, *The Metabolism of Contraction* (Eds, Roy, P. E. and Rona, G.), University Park Press, Baltimore, 1975, pp. 307–15.
62. Waldenström, A. P., Hjalmarson, Å. C., and Thornell, L. A possible role of noradrenaline in the development of myocardial infarction. *American Heart Journal*, **95**, 43–52 (1978).
63. Waldenström, A. P., Hjalmarson, Å. C., Jodal, M., and Waldenström, J. Significance of enzyme release from ischemic isolated rat heart. *Acta Medica Scandinavica*, **201**, 533–8 (1977).
64. Weisfeldt, M. L. and Shock, N. W. Effect of perfusion pressure on coronary flow and oxygen usage of nonworking heart. *American Journal of Physiology*, **218**, 95–101 (1970).

65. Willerson, J. T., Weisfeldt, M. L., Sanders, C. A., and Powell, J., Jr. Influence of hyperosmolar agents on hypoxic cat papillary muscle function. *Cardiovascular Research*, **8,** 8–17 (1974).
66. Winbury, M. M., Hausler, L. M., Prioli, N. A., and Zitowitz, L. Effect of epinephrine on cardiac rhythm in the miniature pig before and after coronary occlusion. *Journal of Pharmacology and Experimental Therapeutics*, **138,** 287–91 (1962).
67. Wollenberger, A. and Krause, E.-G. Metabolic control characteristics of the acutely ischemic myocardium. *American Journal of Cardiology*, **22,** 349–59 (1968).

CHAPTER 17

The effect of calcium on myocardial tissue damage and enzyme release

T. J. C. Ruigrok and A. N. E. Zimmerman

INTRODUCTION	399
ISOPROTERENOL-INDUCED MYOCARDIAL NECROSIS	400
CALCIUM-FREE PERFUSION	400
REPERFUSION WITH CALCIUM	402
The Calcium Paradox	402
Factors Influencing the Calcium Paradox	404
Energy Consumption	404
Energy Dependence	406
General Comments	407
REOXYGENATION OF ISCHAEMIC OR ANOXIC MYOCARDIUM	410
POSSIBLE ROLE OF MITOCHONDRIA	411
CLINICAL RELEVANCE	412
CONCLUDING COMMENTS	413
REFERENCES	413

INTRODUCTION

Calcium ions play an essential role in excitation–contraction coupling in cardiac muscle. During depolarization calcium enters the cells and accumulates at specific sites within each cell, which ultimately leads to contraction of the myofibrils. It has been claimed that intracellular calcium overload, as a result of an excessive influx of calcium, causes depletion of high-energy stores. This is due to the activation of calcium-dependent ATPases and the impairment of the phosphorylating capacity of mitochondria, resulting in myocardial necrosis.[13] Several situations may result finally in an increased cardiac calcium content: such as exposure of cardiac muscle to solutions containing very high calcium concentrations;[28] administration of cardiac glycosides[34] and catecholamines;[42] a transient period of ischaemia;[48,49] and successive perfusion with calcium-free and calcium-containing solutions (calcium paradox).[57,58] In this chapter different types of calcium-induced cell damage will be described and current theories on the mechanism of calcium-induced myocardial necrosis will be discussed.

ISOPROTERENOL-INDUCED MYOCARDIAL NECROSIS

The cardiotoxic action of the synthetic β-adrenergic catecholamine isoproterenol, first described by Rona et al.,[42] has been investigated thoroughly. It has been shown that large doses of isoproterenol lead to an increased mitochondrial calcium content in the rat heart[36] and depletion of high-energy phosphate stores.[12,52] Twenty-four hours after subcutaneous injection of 25 mg kg^{-1} isoproterenol in rats, myocardial creatine phosphate and ATP were decreased by 50%.[52] In another study, 30 mg kg^{-1} isoproterenol produced an 85% loss of myocardial creatine phosphate and a 50% loss of ATP within 2 hours.[12] The main causes of the exhaustion of creatine phosphate and ATP were an excessive breakdown of ATP by calcium-dependent ATPases and an impaired oxidative phosphorylation in the mitochondria. It has been postulated[13] that this high-energy phosphate depletion is a crucial point in the aetiology of β-adrenergic catecholamine-induced myocardial necrosis.

The isoproterenol-induced intracellular calcium overload can be prevented by the calcium antagonists verapamil, D600, and prenylamine, and by the calcium-antagonistic cations K^+, Mg^{2+}, and H^+, while Na^+ aggravates the cardiotoxicity of isoproterenol.[13,43,46] The protective action of potassium has been ascribed to the ability of potassium to inhibit the catecholamine-stimulated activation of adenylate cyclase and thus the production of cyclic AMP.[10] In this way potassium would indirectly influence the influx of calcium, since the positive inotropic action of catecholamines is supposed to be mediated by cyclic AMP.[50] Magnesium has been suggested to have a protective effect on the phosphorylating mechanism of mitochondria during active calcium uptake.[51]

CALCIUM-FREE PERFUSION

When an isolated heart is perfused with a calcium-free medium, the heart loses its contractile properties,[41] but initially not its electrical activity[29,31] (Figure 1). The amplitude and the frequency of the e.c.g. of rat hearts decrease[21,32] and after 2–3 minutes the heart shows atrioventricular conduction defects (Figure 2). After 3–30 minutes of calcium-free perfusion electrical activity changes into flutter and fibrillation and after 70–90 minutes all recordable activity disappears.[32]

During the first 3 minutes of calcium-free perfusion no ultrastructural alterations are perceptible.[32,53,56] After 5 minutes separation of the intercalated discs occurs,[56] followed by dilation of the tubular system.[21,58] After 30–40 minutes mitochondria and sarcoplasmic reticulum are swollen and myofibrils are in a contracted state.[53] The basement membrane, which normally appears as a layer of diffuse material adherent to the sarcolemma,

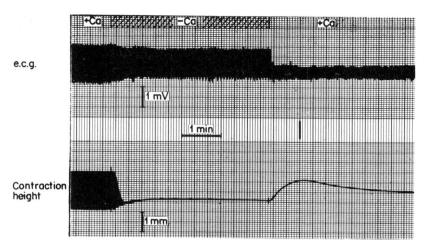

Figure 1. Simultaneous recording of e.c.g. and contraction height of an isolated perfused rat heart during successive perfusion with a calcium-containing, calcium-free, and again calcium-containing medium. Calcium-free perfusion causes electromechanical dissociation. Reperfusion with calcium results in sudden cessation of the electrical activity of the ventricles, while the atrial activity is unaffected. The heart develops contracture. [Reproduced by permission of Elsevier North Holland from *European Journal of Cardiology*, **7**, 241–56 (1978).

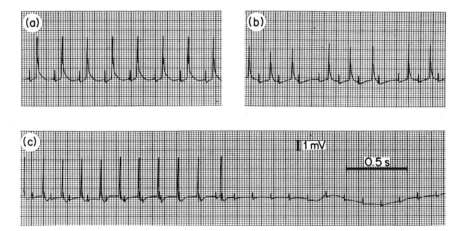

Figure 2. Recording of the e.c.g. of an isolated perfused rat heart during successive perfusion with (a) a calcium-containing, (b) calcium-free, and (c) calcium-containing medium. Electrodes were sutured to the right atrium and left ventricle. The duration of the calcium-free perfusion was 4 min; after 1 min the heart developed failure of atrioventricular conduction (b). Five seconds after reintroduction of calcium the ventricular activity disappeared; the atrial activity was unaffected (c). [Reproduced by permission of Elsevier North Holland from *European Journal of Cardiology*, **7**, 241–56 (1978).

becomes separated from the plasma membrane.[32,53] Further continuation of calcium-free perfusion finally results in complete separation of the cells.[32]

Even after 60 minutes of calcium-free perfusion rat hearts do not release measurable amounts of creatine kinase (CK) into the coronary effluent. Rabbit and guinea-pig hearts, on the contrary, release substantial amounts of CK from the very onset of exposure to a calcium-free medium.[20] In the anoxic rat heart, however, omission of calcium from the medium exacerbates the anoxia-induced release of CK.[35]

It is not feasible to indicate how few calcium ions a calcium-free medium may contain to make a heart susceptible to the deleterious effect of reperfusion with calcium (calcium paradox), as the effectiveness of a calcium-free perfusion is also determined by pH,[1] temperature,[21] and ionic composition[56] of the medium as well as by the duration of the calcium-free perfusion.[1,57]

It can be concluded that perfusion of isolated hearts with a calcium-free medium may introduce severe electrophysiological, contractile, ultrastructural, and biochemical alterations. Omission of calcium from the perfusion medium for only 3 minutes has been reported to be sufficient to induce the calcium paradox in rat heart on reintroduction of calcium.[57,58] The most conspicuous changes, observed during this short period, are electromechanical dissociation and a sharp decrease of that part of the tissue calcium content which is supposed to be the pool related to the maintenance of contractility.[53] It has been suggested that during calcium-free perfusion there is loss of membrane-bound proteins responsible for the 'calcium channel'.[21] It has also been suggested, on the other hand, that the calcium pool responsible for the maintenance of the structural integrity of the membranes, is not depleted until at least 10 minutes perfusion with calcium-free medium has elapsed.[53] Additional studies are needed therefore to get more insight into the biochemical and ultrastructural consequences of calcium-free perfusion, that render a heart so vulnerable to the readmission of calcium.

REPERFUSION WITH CALCIUM

The Calcium Paradox

Readmission of calcium to a rat heart, after an adequate calcium-free perfusion, results in a sudden cessation of the electrical activity of the ventricles, while the electrical activity of the atria is unaffected (Figure 2). After a brief burst of uncoordinated mechanical activity the heart develops contracture[21,56,57] (Figure 1). The heart loses its normal colour and acquires a pale and mottled appearance. This phenomenon, whereby the heart maintains its red colour and electrical activity when the perfusion medium is changed from normal to calcium-free, but loses these properties on return to

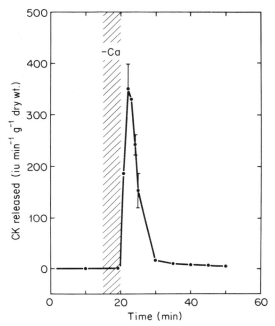

Figure 3. Effect of reperfusion of isolated rat hearts with calcium after a calcium-free period (shaded bar) on the release of CK (iu min^{-1} g^{-1} dry weight). Reperfusion with calcium results in an immediate and massive release of CK. Values are given as mean ±SEM ($n = 4$)

normal, was first described by Zimmerman and Hülsmann and named the calcium paradox.[57] The bleaching of the heart is accompanied by a massive release of myoglobin (which colours the effluent yellowish-brown), creatine kinase (Figure 3), and lactate dehydrogenase.[15,21,44,57] The tissue levels of sodium and calcium increase[8,21] and those of potassium and magnesium decrease.[8] The activities of adenylate cyclase, Ca^{2+}-ATPase, Mg^{2+}-ATPase, and Na^+/K^+-ATPase of the sarcolemma decrease during calcium-free perfusion and show a further and sharp decrease upon reperfusion with calcium.[8] Microsomal calcium binding and uptake activities decrease during reperfusion with calcium, while the mitochondrial calcium binding and uptake activities show a marked increase.[25] It has been suggested that upon reperfusion with calcium the mitochondria are altered in such a way that they serve as an adaptive mechanism to handle the massive influx of calcium, resulting from defects at the level of the sarcolemma, as well as to compensate for the loss of the calcium pump mechanism of the sarcoplasmic reticulum.[25]

Reperfusion with a calcium-containing medium introduces an extensive irreversible ultrastructural damage such as a severe contracture with a considerable disruption of the contractile elements, and swelling of mitochondria which may contain electron-dense particles. Disruption of plasma membranes and intercalated discs, and even complete separation of myocardial cells have been seen.[15,21,33,56,58]

Factors Influencing the Calcium Paradox

The original calcium paradox experiments were performed at pH 7.35 and 37 °C.[57] A lower pH increases, and a higher pH decreases the length of the calcium-free perfusion period, necessary to evoke the calcium paradox. It has been proposed that lowering of the pH might hamper the calcium movements across the plasma membrane and/or at subcellular level.[1] When perfusion with a calcium-free medium is performed at 4 °C, both contractility and electrical activity disappear. Readmission of calcium to these hearts permits rapid recovery of the excitation–contraction coupling and normal contractility upon rewarming to 37 °C.[21] In contrast with calcium-free perfusion at 37 °C, calcium-free perfusion at 4 °C does not lead to ultrastructural damage and uptake of calcium-45 by the myocardium once perfusion with calcium is resumed. It has been suggested that the damage to the sarcolemma, occurring during calcium-free perfusion at 37 °C, consists of loss of membrane-bound proteins responsible for the physiological 'calcium channel'. At 4 °C their escape from the bilayer would be retarded by a more sluggish shift in conformation from the calcium-bound to the calcium-free state, and a more crystalline membrane.[21]

Another factor, influencing the occurrence of the calcium paradox, is the ionic composition of the perfusion medium.[56] Reducing the sodium concentration in the calcium-free medium has been found to augment the recovery of calcium-deprived hearts in addition to preventing the ultrastructural damage upon reperfusion with calcium. The beneficial effects of low sodium in the calcium-free medium have been explained[56] on the basis of a decrease of the alterations in the plasma membrane, and a delay in the depletion of the intracellular calcium due to its actions on the calcium influx and efflux.[30,40] In the same study it was found that omission of magnesium from the calcium-free medium did not alter the recovery of contractility, while reduction of the potassium concentration was deleterious to the ability of the heart to recover contractility on reperfusion with calcium.

Energy Consumption

Sudden and severe decline of myocardial high-energy phosphates is one of the features of the calcium paradox and has been demonstrated not only in

Table 1. Content of Creatine Phosphate, Creatine, and the Adenine Nucleotides in Isolated Rat Heart at the end of a 15 minute Perfusion Period with Standard Medium, after 4 minutes of a Subsequent Perfusion with Calcium-free Medium, and at Different Times of Reperfusion with Calcium-containing Medium. [Reproduced by permission of Academic Press from *Journal of Molecular and Cellular Cardiology*, **8,** 973–9 (1976).]

	Perfusion sequence					
	plus Ca^{2+} 15 min	minus Ca^{2+} 4 min	plus Ca^{2+}			
			0.5 min	1 min	2 min	4 min
CP	29.6 ±2.1	35.6 ±2.9	12.0±5.3	4.6±3.2	4.0±2.3	1.6±1.2
Creatine	31.7 ±5.2	32.1 ±2.7	37.4±7.5	32.9±4.6	23.0±4.7	9.8±5.2
ATP	18.8 ±0.8	20.8 ±1.4	11.3±2.4	6.3±2.7	4.8±1.8	2.9±1.0
ADP	3.2 ±0.4	3.3 ±0.3	6.1±0.6	4.7±0.5	3.2±0.4	2.4±0.8
AMP	0.21±0.04	0.18±0.06	5.2±2.8	5.1±1.9	1.8±0.5	1.1±0.5
$\sum A\sim$	22.2	24.3	22.6	16.1	9.8	6.4

Values are given in $\mu mol\, g^{-1}$ dry weight±SD ($n=4$).

rat heart[2,21,56] but also in mouse, guinea-pig, and rabbit heart.[20] These findings are in agreement with the theory on intracellular calcium overload and depletion of high-energy stores, as developed by Fleckenstein.[13] After 30 seconds of reperfusion with calcium, myocardial CP and ATP levels of isolated rat hearts were decreased by 65% and 45%, respectively. In the same period there was an increase in creatine (15%), ADP (85%), and AMP (2800%) (Table 1). During continued reperfusion with calcium the concentration of all compounds decreased gradually. The effluent fluid, collected during the first 2 minutes of reperfusion with calcium-containing medium, contained large amounts of creatine and AMP, and only small amounts of CP and ATP (Table 2). It has been argued that reperfusion with calcium, after a short calcium-free period, produces an abrupt consumption of myocardial high-energy stores, prior to the release of these compounds.[2]

The available data on the occurrence of the calcium paradox in rabbit hearts are controversial. It has been reported[26] that the calcium paradox as

Table 2. Release of Creatine Phosphate, Creatine, and the Adenine Nucleotides into the Effluent Perfusion Medium, collected during 2 minutes after Reintroduction of Calcium in the Medium. [Reproduced by permission of Academic Press from *Journal of Molecular and Cellular Cardiology*, **8,** 973–9 (1976).]

CP	1.40±0.40
Creatine	32.6 ±3.9
ATP	0.73±0.29
ADP	0.80±0.24
AMP	3.50±0.76

Values represent μmol released g^{-1} dry tissue±SD ($n=4$).

such cannot be evoked in rabbit heart and that reperfusion with calcium after a calcium-free period of even 12 minutes, does not result in a decrease of myocardial CP. Calcium paradox phenomena (irreversible contracture, influx of calcium, loss of CP and ATP) were observed, however, when rabbit hearts were perfused with normal medium after a calcium-free, hypokalaemic period. Ten minutes of reperfusion with calcium resulted in a decrease of myocardial CP (90%), creatine (75%), and ATP (50%), and an increase of ADP (60%) and AMP (100%). Inexplicably large amounts of CP (corresponding with 88% of the decrease of myocardial CP), however, were found in the effluent medium, which points to leakage of cell components without a preceding substantial consumption of high-energy phosphates. In a recent investigation,[20] on the other hand, data have been presented similar to the results as obtained from rat hearts.[2] Two minutes of reperfusion with calcium led to a decrease of myocardial CP (95%) and ATP (85%), and an increase of creatine (30%), ADP (50%), and AMP (2150%). The effluent medium contained only small amounts of CP (corresponding with 2% of the decrease of myocardial CP).

Energy Dependence

In the original calcium paradox experiments[57,58] isolated rat hearts were perfused with oxygenated glucose-containing medium (Figure 4; series a). This procedure enabled the hearts to maintain their high-energy stores until the end of the calcium-free period at a fairly high level by oxidation of glucose. By perfusing isolated rat hearts under anoxic conditions with and without substrate, and by reoxygenating hearts after an anoxic period, the effect of different energy states on the capacity of myocardial tissue to develop the calcium paradox could be studied.[45]

After 35 minutes of anoxic perfusion in the presence of glucose (Figure 4; series b), the hearts still contained 80% of the ATP which was present at the end of the control perfusion period (Table 3). No measurable amount of CK was released from the hearts during this period of anoxic perfusion. Reintroduction of calcium to the anoxic glucose-containing medium resulted in a massive CK release from the hearts (Figure 5).

After 35 minutes of anoxic perfusion in the absence of glucose (Figure 4; series c), the myocardial ATP content was only 2% of the content at the end of the control perfusion period (Table 3). The hearts did not contain creatine phosphate but relatively large amounts of AMP. The absence of glucose in the perfusion medium caused a slow release of CK. After reintroduction of calcium to the anoxic glucose-free medium, enzyme release increased slightly which indicates that the calcium paradox did not occur in hearts subjected to anoxic substrate-free perfusion (Figure 6). This was not due to an inadequate calcium-free perfusion, since the characteristic

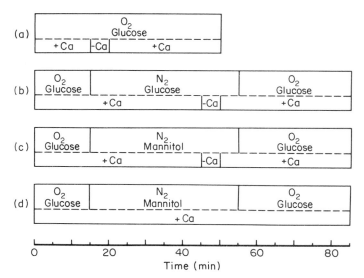

Figure 4. Perfusion sequence in four different series of experiments. Rat hearts were perfused with a modified Tyrode solution at 37 °C, according to Langendorff, and at a pressure of 100 cm of water. During normoxic perfusion the medium was equilibrated with 95% O_2–5% CO_2 and contained glucose. During anoxic perfusion the medium was equilibrated with 95% N_2–5% CO_2 and contained either glucose or mannitol

enzyme peak was observed as soon as the hearts were reoxygenated. This oxygen-induced calcium paradox was completely inhibited, when KCN (5 mmol l.$^{-1}$ [15]) was present in the anoxic media during the last 15 minutes of anoxia and during the reoxygenation period. This indicates that reactivation of the electron transport system could be responsible for the sudden cell damage.

These data show that the calcium paradox fails to appear in the absence of energy, but occurs as soon as the electron transport system is reactivated (Figure 6). Electron transport is not a condition for the calcium paradox; in the absence of electron transport the calcium paradox occurs, provided that ATP is present (Figure 5).

General Comments

After the first reports on the destructive effect of successive perfusion with calcium-free and calcium-containing medium on cardiac muscle,[1,33,57,58] it has been claimed that the calcium paradox as originally defined,[57] does not occur in rabbit heart[26,27] and not even in rat heart.[7] During the last few years, however, the calcium paradox has been reaffirmed.[2,20,21,44,45,56] It has

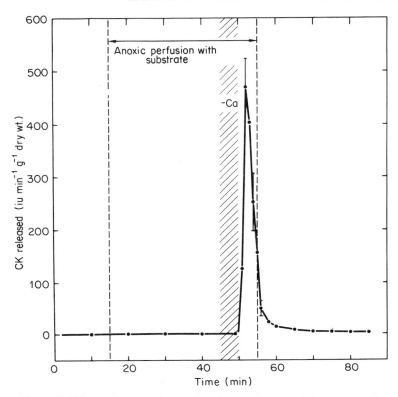

Figure 5. Effect of reperfusion with calcium after a calcium-free period (shaded bar), in the anoxic rat heart in the presence of glucose, on the release of CK (iu min^{-1} g^{-1} dry weight). Reperfusion with calcium results in an immediate and massive release of CK. Values are given as mean ±SEM ($n = 4$). [Reproduced by permission of University Park Press, Baltimore from *Recent Advances in Studies on Cardiac Structure and Metabolism*, **11**, 565–9 (1978).]

Table 3. Effect of Anoxic Perfusion in the Presence and Absence of Glucose on Creatine Phosphate and Adenine Nucleotides Levels in Isolated Rat Heart

Perfusion sequence	CP	ATP	ADP	AMP
15 min control perfusion	29.6±2.1	18.8 ±0.8	3.2±0.4	0.21±0.04
15 min control perfusion 35 min anoxic perfusion with glucose (30 min with Ca^{2+}; 5 min without Ca^{2+})	7.9±2.7	14.9 ±3.3	4.1±0.9	1.1 ±0.9
15 min control perfusion 35 min anoxic perfusion without glucose (30 min with Ca^{2+}; 5 min without Ca^{2+})	a	0.39±0.13	1.2±0.3	6.1 ±1.0

a Did not differ significantly from zero.
Values are given in μmol g^{-1} dry weight ± SD ($n = 4$).

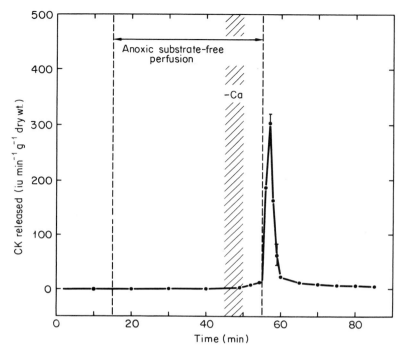

Figure 6. Effect of reperfusion with calcium after a calcium-free period (shaded bar), in the anoxic rat heart in the absence of glucose, on the release of CK (iu min^{-1} g^{-1} dry weight). Massive release of CK does not occur on reperfusion with calcium, but only on reoxygenation of the heart. Values are given as mean ±SEM ($n = 4$). [Reproduced by permission of University Park Press, Baltimore from *Recent Advances in Studies on Cardiac Structure and Metabolism*, **11**, 565-9 (1978).]

further been established that, after a calcium-free perfusion, the calcium paradox will occur as soon as calcium is readmitted in a concentration that exceeds the intracellular systolic calcium concentration ($>10^{-5}$ mol l.$^{-1}$).[9]

It has also been claimed that enzyme release, which is one of the characteristics of the calcium paradox, does not result from a paradoxical effect of calcium but is dependent on the recovery of contractile activity.[7] It is well established, however, that upon reperfusion with calcium a heart does not recover its contractile activity, but develops contracture.[21,56,57] In addition, it was found (T. J. C. Ruigrok/unpublished results) that the calcium antagonist verapamil (1 mg l.$^{-1}$) (which was added to the calcium-free medium and the reperfusion medium in an attempt to protect the heart against the calcium paradox) reduced the calcium paradox contracture, whereas the massive enzyme release upon reperfusion with calcium was unaffected by the presence of verapamil. That mechanical activity does not play a major role in this type of cell damage is further supported by the finding that the calcium paradox can also be evoked in the kidney.[37]

REOXYGENATION OF ISCHAEMIC OR ANOXIC MYOCARDIUM

The consequences of myocardial ischaemia and reperfusion on tissue damage and enzyme release will be reviewed elsewhere. It has been mentioned[2,14,15] that the biochemical and ultrastructural changes, induced by reoxygenation of anoxic myocardium[11,14,17,18] and by reperfusion after a period of ischaemia,[24,55] show a striking similarity to the calcium paradox phenomena. The calcium paradox as well as the reperfusion and reoxygenation damage have been attributed to energy-dependent transmembrane calcium fluxes.[19] Strong evidence has been provided that not the mere readmission of oxygen, but reactivation of the electron transport system and a consequent accumulation of calcium by the mitochondria are responsible for the paradoxical extension of ischaemic damage upon reperfusion of the coronary system.[15,19,48,49] Apart from the similarity between the above-mentioned phenomena, the alterations of the sarcolemma induced by calcium-free perfusion are different from those resulting from energy depletion. During a 60 minute period of calcium-free perfusion rat hearts do not release measurable amounts of creatine kinase, while during the same

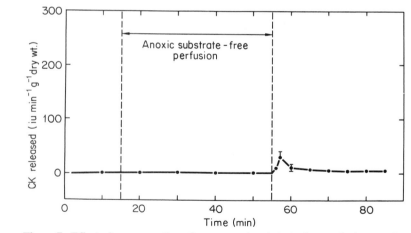

Figure 7. Effect of reoxygenation after an anoxic substrate-free perfusion, on the release of CK ($iu\,min^{-1}\,g^{-1}$ dry weight) from the isolated perfused rat heart. Reoxygenation results in a relatively slight release of CK. Values are given as mean ±SEM ($n = 4$). [Reproduced by permission of University Park Press, Baltimore from *Recent Advances in Studies on Cardiac Structure and Metabolism*, **11**, 565–9 (1978).]

period of anoxia a significant loss of enzymes has been observed.[17] Upon reperfusion with calcium after only 5 minutes of calcium-free perfusion, however, there is nearly 12 times as much release of creatine kinase (Figure 3), as upon reoxygenation after an anoxic substrate-free perfusion of 40 minutes (Figure 7). Superimposition of the same short calcium-free perfusion upon 40 minutes of anoxia, furthermore, exacerbates the reoxygenation-induced enzyme release substantially (Figures 6 and 7).

POSSIBLE ROLE OF MITOCHONDRIA

The energy-linked accumulation of calcium by isolated mitochondria has been studied in great detail during the last two decades.[5] The mitochondrial accumulation of calcium together with phosphate is supported by electron transport and can be inhibited by cyanide. During electron transport supported calcium accumulation mitochondria are not able to phosphorylate ADP. In the absence of electron transport, calcium uptake can be supported by ATP.[4] During this process electron-dense granules of calcium phosphate are deposited in the mitochondrial matrix.

In situ loading of mitochondria with calcium is considered to be an important defence mechanism by which cells try to control an increased influx of calcium from the extracellular spaces.[6] Although the cytoplasmic concentration of ionized calcium increases during ischaemia as a result of progressive membrane damage, the mitochondria do not accumulate calcium due to the absence of energy. Upon reperfusion, however, electron transport is reactivated and mitochondria start to accumulate massive amounts of calcium and phosphate ions[48,49] in preference to phosphorylating ADP.[39]

It is generally accepted that during calcium-free perfusion of isolated hearts the sarcolemma is influenced in such a way that reperfusion with calcium results in a massive influx of calcium into the cells. The mitochondria most likely play an important role in the origin of the calcium paradox, since readmission of calcium leads to an increased tissue level of calcium,[8,21] an abrupt decline of high-energy stores,[2,20] and the formation of intramitochondrial electron-dense particles.[15,21,56,58] Furthermore, it has been demonstrated that the calcium paradox only occurs when at least one of the conditions for mitochondrial calcium accumulation is fulfilled: the presence of electron transport or ATP.[45]

It remains to be elucidated why the readmission of oxygen to ischaemic myocardium and calcium to calcium-deprived hearts is so extremely damaging. An explanation may be found in the overall reaction for 'massive loading' of calcium by mitochondria:[5]

$$3\,Ca^{2+} + 2\,HPO_4^{2-} \rightarrow Ca_3(PO_4)_2\downarrow + 2\,H^+$$

The efflux of protons from the mitochondria and the resulting acidification of the cytoplasm may account for the sudden cell damage.[2,19]

CLINICAL RELEVANCE

It is generally accepted that calcium plays an important role in the production of various forms of cardiac necrosis. Recently, a theory concerning the course of events in angina and myocardial infarction has been developed.[16] It has been suggested that an increased myofibrillar concentration of calcium, in consequence of an impairment of the sarcolemmal integrity and also by the action of catecholamines, may initiate vicious circles which could lead to myocardial infarction.

Metabolic[38,47] and surgical[19] interventions are widely used to reduce or to halt the consequences of myocardial ischaemia. Infusions of glucose, insulin, and potassium, for instance, have been shown to exert a beneficial influence on the effects of coronary artery ligation by reducing the decrease of the high-energy stores and the K^+/Na^+ ratio of ischaemic tissue.[38] In this way, a deleterious high-energy phosphate deficiency, which has been considered as a crucial point in the aetiology of myocardial necrosis,[13] may be retarded. When intracellular ATP is still present and when the sarcolemma is still functionally intact, reperfusion of ischaemic myocardium will result in resumption of contractile activity. When ATP stores are depleted and damage of the sarcolemma has occurred, on the other hand, reperfusion may accelerate the ischaemic damage, in consequence of energy-dependent transmembrane calcium fluxes.[19]

There is ample evidence that the use of calcium-free solutions to arrest and protect the heart during open heart surgery may be hazardous by making the heart susceptible to the calcium paradox.[22,54] Although the calcium paradox is not limited to the rat heart,[20] it is as yet unknown to what extent the human heart is also vulnerable to the calcium paradox phenomena. As to calcium-free cardioplegic infusates, it should be considered that possible deleterious effects will depend on a number of factors, such as temperature,[21] pH,[1] and ionic composition[56] of the solution, and possibly on the presence of membrane-stabilizing compounds. Fortunately, these solutions are characterized by a low sodium content,[3,23] which has been shown to provide some protection against the calcium paradox.[56] Since the calcium paradox is such an extreme example of the damaging effect of an imbalance in calcium homeostasis, calcium-free cardioplegic infusates should not be used without a thorough knowledge of the calcium paradox characteristics.

It can be concluded that calcium-free perfusion and reperfusion with calcium form valuable models for studying the consequences of intracellular calcium deficiency and overload, which have appeared to be of increasing importance in relation to certain clinical situations.

CONCLUDING COMMENTS

Intracellular calcium overload, as a result of an excessive influx of calcium, causes depletion of myocardial high-energy stores due to activation of calcium-dependent ATPases and uncoupling of the oxidative phosphorylation. It has been postulated that this high-energy phosphate depletion is a crucial point in the aetiology of myocardial necrosis, produced by a number of interventions.

Reoxygenation of ischaemic myocardium and particularly readmission of calcium to calcium-deprived hearts can be extremely damaging. The damaging effects of these interventions have been attributed to sudden energy-dependent transmembrane calcium fluxes, in consequence of an increased permeability of the sarcolemma for calcium caused by energy or calcium deprivation.

Inability of the myocardium to maintain adequate calcium homeostasis has appeared to be of increasing importance in relation to certain clinical situations.

REFERENCES

1. Bielecki, K. The influence of changes in pH of the perfusion fluid on the occurrence of the calcium paradox in the isolated rat heart. *Cardiovascular Research*, **3**, 268–71 (1969).
2. Boink, A. B. T. J., Ruigrok, T. J. C., Maas, A. H. J., and Zimmerman, A. N. E. Changes in high-energy phosphate compounds of isolated rat hearts during Ca^{2+}-free perfusion and reperfusion with Ca^{2+}. *Journal of Molecular and Cellular Cardiology*, **8**, 973–9 (1976).
3. Bretschneider, H. J., Hübner, G., Knoll, D., Lohr, B., Nordbeck, H., and Spieckermann, P. G. Myocardial resistance and tolerance to ischemia: physiological and biochemical basis. *Journal of Cardiovascular Surgery*, **16**, 241–60 (1975).
4. Brierley, G. P., Murer, E., and Bachmann, E. Studies on ion transport. III. The accumulation of calcium and inorganic phosphate by heart mitochondria. *Archives of Biochemistry and Biophysics*, **105**, 89–102 (1964).
5. Bygrave, F. L. Mitochondrial calcium transport. In *Current Topics in Bioenergetics* (Ed. Sanadi, D. R.), Vol. 6, Academic Press, New York, 1977, pp. 259–318.
6. Carafoli, E. Mitochondrial uptake of calcium ions and the regulation of cell function. In *Biochemical Society Symposia*, Vol. 39, The Biochemical Society, London, 1974, pp. 89–109.
7. De Leiris, J. and Feuvray, D. Factors affecting the release of lactate dehydrogenase from isolated rat heart after calcium and magnesium free perfusions. *Cardiovascular Research*, **7**, 383–90 (1973).
8. Dhalla, N. S., Tomlinson, C. W., Singh, J. N., Lee, S. L., McNamara, D. B., Harrow, J. A. C., and Yates, J. C. Role of sarcolemmal changes in cardiac pathophysiology. In *Recent Advances in Studies on Cardiac Structure and Metabolism* (Eds. Roy, P. E. and Dhalla, N. S.), Vol. 9, University Park Press, Baltimore; 1976, pp. 377–94.
9. Farmer, B. B., Harris, R. A., Jolly, W. W., Hathaway, D. R., Katzberg, A., Watanabe, A. M., Whitlow, A. L., and Besch, H. R. Isolation and characterization of adult rat heart cells. *Archives of Biochemistry and Biophysics*, **179**, 545–58 (1977).
10. Fedeléšová, M., Ziegelhöffer, A., Luknárová, O., and Kostolanský, S. Prevention by K^+, Mg^{2+}-aspartate of isoproterenol-induced metabolic changes in the myocardium. In *Recent Advances in Studies on Cardiac Structure and Metabolism* (Eds. Fleckenstein, A. and Rona, G.), Vol. 6, University Park Press, Baltimore, 1975, pp. 59–73.
11. Feuvray, D. and De Leiris, J. Ultrastructural modifications induced by reoxygenation in the anoxic isolated rat heart perfused without exogenous substrate. *Journal of Molecular and Cellular Cardiology*, **7**, 307–14 (1975).
12. Fleckenstein, A., Döring, H. J., and Leder, O. The significance of high-energy phosphate

exhaustion in the etiology of isoproterenol-induced cardiac necroses and its prevention by iproveratril, compound D600 or prenylamine. In *International Symposium on Drugs and Metabolism of Myocardium and Striated Muscle* (Eds. Lamarche, M. and Royer, R.), Nancy 1969, pp. 11–22.
13. Fleckenstein, A. Specific inhibitors and promoters of calcium action in the excitation–contraction coupling of heart muscle and their role in the prevention or production of myocardial lesions. In *Calcium and the Heart* (Eds. Harris, P. and Opie, L. H.), Academic Press, London, 1971, pp. 135–88.
14. Ganote, C. E., Seabra-Gomes, R., Nayler, W. G., and Jennings, R. B. Irreversible myocardial injury in anoxic perfused rat hearts. *American Journal of Pathology*, **80,** 419–50 (1975).
15. Ganote, C. E., Worstell, J., and Kaltenbach, J. P. Oxygen-induced enzyme release after irreversible myocardial injury. Effects of cyanide in perfused rat hearts. *American Journal of Pathology*, **84,** 327–50 (1976).
16. Harris, P. A theory concerning the course of events in angina and myocardial infarction. *European Journal of Cardiology*, **3,** 157–63 (1975).
17. Hearse, D. J., Humphrey, S. M., and Chain, E. B. Abrupt reoxygenation of the anoxic potassium-arrested perfused rat heart: a study of myocardial enzyme release. *Journal of Molecular and Cellular Cardiology*, **5,** 395–407 (1973).
18. Hearse, D. J., Humphrey, S. M., Nayler, W. G., Slade, A., and Border, D. Ultrastructural damage associated with reoxygenation of the anoxic myocardium. *Journal of Molecular and Cellular Cardiology*, **7,** 315–24 (1975).
19. Hearse, D. J. Reperfusion of the ischemic myocardium. *Journal of Molecular and Cellular Cardiology*, **9,** 605–16 (1977).
20. Hearse, D. J., Humphrey, S. M., Boink, A. B. T. J., and Ruigrok, T. J. C. The calcium paradox: metabolic, electrophysiological, contractile and ultrastructural characteristics in four species. *European Journal of Cardiology*, **7,** 241–56 (1978).
21. Holland, C. E. and Olson, R. E. Prevention by hypothermia of paradoxical calcium necrosis in cardiac muscle. *Journal of Molecular and Cellular Cardiology*, **7,** 917–28 (1975).
22. Jynge, P., Hearse, D. J., and Braimbridge, M. V. Myocardial protection during ischemic cardiac arrest: a possible hazard with calcium-free cardioplegic infusates. *Journal of Thoracic and Cardiovascular Surgery*, **73,** 848–55 (1977).
23. Kirsch, U., Rodewald, G., and Kalmár, P. Induced ischemic arrest. Clinical experience with cardioplegia in open-heart surgery. *Journal of Thoracic and Cardiovascular Surgery*, **63,** 121–30 (1972).
24. Kloner, R. A., Ganote, C. E., Whalen, D. A., and Jennings, R. B. Effect of a transient period of ischemia on myocardial cells. II. Fine structure during the first few minutes of reflow. *American Journal of Pathology*, **74,** 399–422 (1974).
25. Lee, S. L. and Dhalla, N. S. Subcellular calcium transport in failing hearts due to calcium deficiency and overload. *American Journal of Physiology*, **231,** 1159–65 (1976).
26. Lee, Y. C. P. and Visscher, M. B. Perfusate cations and contracture and Ca, Cr, PCr, and ATP in rabbit myocardium. *American Journal of Physiology*, **219,** 1637–41 (1970).
27. Lee, Y. C. P. and Visscher, M. B. Influx of calcium into rabbit myocardium in relation to its ionic environment. *Proceedings of the National Academy of Sciences (USA)*, **66,** 603–6 (1970).
28. Legato, M. J., Spiro, D., and Langer, G. A. Ultrastructural alterations produced in mammalian myocardium by variation in perfusate ionic composition. *Journal of Cell Biology*, **37,** 1–12 (1968).
29. Locke, F. S. and Rosenheim, O. Contributions to the physiology of the isolated heart. The consumption of dextrose by mammalian cardiac muscle. *Journal of Physiology*, **36,** 205–20 (1907).
30. Lüttgau, H. C. and Niedergerke, R. The antagonism between Ca and Na ions on the frog's heart. *Journal of Physiology*, **143,** 486–505 (1958).
31. Mines, G. R. On functional analysis by the action of electrolytes. *Journal of Physiology*, **46,** 188–235 (1913).
32. Muir, A. R. The effects of divalent cations on the ultrastructure of the perfused rat heart. *Journal of Anatomy*, **101,** 239–61 (1967).

33. Muir, A. R. A calcium-induced contracture of cardiac muscle cells. *Journal of Anatomy*, **102,** 148–9 (1968).
34. Nayler, W. G. Calcium exchange in cardiac muscle: a basic mechanism of drug action. *American Heart Journal*, **73,** 379–94 (1967).
35. Nayler, W. G., Grau, A., and Slade, A. A protective effect of verapamil on hypoxic heart muscle. *Cardiovascular Research*, **10,** 650–62 (1976).
36. Nirdlinger, E. L. and Bramante, P. O. Subcellular myocardial ionic shifts and mitochondrial alterations in the course of isoproterenol-induced cardiopathy of the rat. *Journal of Molecular and Cellular Cardiology*, **6,** 49–60 (1974).
37. Nozick, J. H., Zimmerman, A. N. E., Poll, P., and Mankowitz, B. J. The kidney and the calcium paradox. *Journal of Surgical Research*, **11,** 60–7 (1971).
38. Opie, L. H., Bruyneel, K., and Owen, P. Effects of glucose, insulin and potassium infusion on tissue metabolic changes within first hour of myocardial infarction in the baboon. *Circulation*, **52,** 49–57 (1975).
39. Peng, C. F., Kane, J. J., Murphy, M. L., and Straub, K. D. Abnormal mitochondrial oxidative phosphorylation of ischemic myocardium reversed by Ca^{2+}-chelating agents. *Journal of Molecular and Cellular Cardiology* **9,** 897–908 (1977).
40. Reuter, H. and Seitz, N. The dependence of calcium efflux from cardiac muscle on temperature and external ion composition. *Journal of Physiology*, **195,** 451–70 (1968).
41. Ringer, S. A further contribution regarding the influence of the different constituents of the blood on the contraction of the heart. *Journal of Physiology*, **4,** 29–42 (1883).
42. Rona, G., Chappel, C. I., Balazs, T., and Gaudry, R. An infarct-like myocardial lesion and other toxic manifestations produced by isoproterenol in the rat. *Archives of Pathology*, **67,** 443–55 (1959).
43. Rona, G., Chappel, C. I., and Gaudry, R. Effect of dietary sodium and potassium content on myocardial necrosis elicited by isoproterenol. *Laboratory Investigation*, **10,** 892–7 (1961).
44. Ruigrok, T. J. C., Burgersdijk, F. J. A., and Zimmerman, A. N. E. The calcium paradox: a reaffirmation. *European Journal of Cardiology*, **3,** 59–63 (1975).
45. Ruigrok, T. J. C., Boink, A. B. T. J., Spies, F., Blok, F. J., Maas, A. H. J., and Zimmerman, A. N. E. Energy dependence of the calcium paradox. *Journal of Molecular and Cellular Cardiology*, in press (1978).
46. Selye, H. The Chemical Prevention of Cardiac Necroses, The Ronald Press Co, New York, 1958.
47. Shatney, C. H., Maccarter, D. J., and Lillehei, R. C. Effects of allopurinol, propranolol and methylprednisolone on infarct size in experimental myocardial infarction. *American Journal of Cardiology*, **37,** 572–80 (1976).
48. Shen, A. C. and Jennings, R. B. Myocardial calcium and magnesium in acute ischemic injury. *American Journal of Pathology*, **67,** 417–40 (1972).
49. Shen, A. C. and Jennings, R. B. Kinetics of calcium accumulation in acute myocardial ischemic injury. *American Journal of Pathology*, **67,** 441–52 (1972).
50. Sobel, B. E. and Mayer, S. E. Cyclic adenosine monophosphate and cardiac contractility, *Circulation Research*, **32,** 407–14 (1973).
51. Sordahl, L. A. and Silver, B. B. Pathological accumulation of calcium by mitochondria: modulation by magnesium. In *Recent Advances in Studies on Cardiac Structure and Metabolism* (Eds. Fleckenstein, A. and Rona, G.), Vol. 6, University Park Press, Baltimore, 1975, pp. 85–93.
52. Takenaka, F. and Higuchi, M. High-energy phosphate contents of subepicardium and subendocardium in the rat treated with isoporterenol and some other drugs. *Journal of Molecular and Cellular Cardiology*, **6,** 123–35 (1974).
53. Tomlinson, C. W., Yates, J. C., and Dhalla, N. S. Relationship among changes in intracellular calcium stores, ultrastructure, and contractility of myocardium. In *Recent Advances in Studies on Cardiac Structure and Metabolism* (Ed. Dhalla, N. S.), Vol. 4, University Park Press, Baltimore, 1974, pp. 331–45.
54. Tyers, G. F. O. Metabolic arrest of the ischemic heart. *Annals of Thoracic Surgery*, **20,** 91–4 (1975).
55. Whalen, D. A., Hamilton, D. G., Ganote, C. E., and Jennings, R. B. Effect of a transient

period of ischemia on myocardial cells. I. Effects on cell volume regulation. *American Journal of Pathology*, **74,** 381–98 (1974).
56. Yates, J. C. and Dhalla, N. S. Structural and functional changes associated with failure and recovery of hearts after perfusion with Ca^{2+}-free medium. *Journal of Molecular and Cellular Cardiology*, **7,** 91–103 (1975).
57. Zimmerman, A. N. E. and Hülsmann, W. C. Paradoxical influence of calcium ions on the permeability of the cell membranes of the isolated rat heart. *Nature*, **211,** 646–7 (1966).
58. Zimmerman, A. N. E., Daems, W., Hülsmann, W. C., Snijder, J., Wisse, E., and Durrer, D. Morphological changes of heart muscle caused by successive perfusion with calcium-free and calcium-containing solutions (calcium paradox). *Cardiovascular Research*, **1,** 201–9 (1967).

CHAPTER 18

Reoxygenation, reperfusion and the calcium paradox: studies of cellular damage and enzyme release*

D. J. Hearse, S. M. Humphrey, and G. R. Bullock

INTRODUCTION	417
MATERIALS AND METHODS	418
Perfusion Techniques	418
Perfusion Medium	418
Perfusion Time Sequence in Reoxygenation Studies	419
Perfusion Time Sequence in Reperfusion Studies	419
Perfusion Time Sequence in Calcium Paradox Studies	420
Enzyme Leakage Analysis	420
Tissue Preparation for Light and Electron Microscopy	420
RESULTS	421
Enzyme Leakage	421
Anoxia and Reoxygenation	421
Ischaemia and Reperfusion	423
Calcium Depletion and Calcium Repletion	423
Morphological Changes	423
Fresh Heart Control Tissue	424
Calcium-free Hearts	424
Calcium-repleted Hearts	425
Anoxic Hearts	426
Reoxygenated Hearts	426
Ischaemic Hearts and Reperfused Hearts	427
Factors Influencing the Calcium Paradox and the Reoxygenation Phenomenon	427
Time	427
Stepwise Readmission of Oxygen or Calcium	431
Temperature	433
DISCUSSION	439
REFERENCES	443

INTRODUCTION

Zimmerman and colleagues[28,29] have reported that the perfusion of rat hearts with calcium-free media for very short periods of time, creates a

* Parts of this chapter are reprinted from the *Journal of Molecular and Cellular Cardiology*, **10,** 641–68 (1978), by kind permission of Academic Press.

situation such that upon readmission of calcium there is massive tissue disruption, enzyme release, and the development of contracture.[18,27,29] This phenomenon has been called the 'calcium paradox'[28] and its damaging effects have been attributed to sudden transmembrane fluxes of calcium.

We[5-9] and others[3,4,13,14,24,25] have reported that reperfusion or reoxygenation of the ischaemic or hypoxic myocardium in a number of species is able to induce a massive and sudden extension of ultrastructural damage involving the intensification of contracture bands, disruption of myofibrils and sarcolemma, rapid cell swelling, rupture and loss of some mitochondria, and the appearance of intramitochondrial calcium phosphate particles. In our studies we described this oxygen-dependent exacerbation of tissue damage as the 'reoxygenation phenomenon' but suggested[6,11] that the damage may be caused in some way by oxygen-induced, sudden transmembrane calcium fluxes.

The striking similarity[11] between many of the characteristics and attributed mechanisms of the calcium paradox and the reoxygenation or reperfusion phenomena have led us to investigate whether they may in fact be different facets of the same problem. The following paper therefore, details a comparison of the characteristics of these phenomena in terms of enzyme release and ultrastructure. In addition, in an attempt to understand some of the underlying principles involved, a number of different factors which influence the occurrence or magnitude of the phenomena have been investigated.

MATERIALS AND METHODS

Male rats (280–300 g body weight) of the Wistar strain, maintained on a standard diet, were used throughout the experiments.

Perfusion Techniques

Rats were lightly anaesthetized with a diethyl ether–oxygen mixture. The left femoral vein was exposed and heparin (200 iu) was administered intravenously. One minute after the administration of heparin, the heart was excised[20] and placed in ice-cold perfusion medium until contraction had ceased. The aorta was attached to a stainless steel cannula, the pulmonary artery incised to allow complete coronary drainage and the heart subjected to non-recirculating Langendorff[16] perfusion (85 cmH$_2$O perfusion pressure) at 37 °C.

Perfusion Medium

Krebs–Henseleit[15] bicarbonate buffer (pH 7.4 when gassed at 37 °C with oxygen or nitrogen gas mixtures containing 5% CO_2) was the standard perfusion fluid. During all aerobic perfusion periods the buffer was gassed with 95% O_2 + 5% CO_2 and glucose (11 mmol l.$^{-1}$) was added to the medium. During anoxic perfusion periods the buffer was gassed with 95% N_2 and 5% CO_2. During the calcium-free perfusion periods, precautions were taken to ensure[22] that both the perfusion fluid and the apparatus were free of any detectable calcium.

During the preparation of all calcium-containing buffers, precautions[26] were taken to prevent the precipitation of calcium. Before use, all perfusion fluid was filtered through a cellulose acetate filter of pore size 5.0 μm. During all perfusion periods atmospheric gas contamination of the heart was prevented by completely enclosing the heart in a water-jacketed (37 °C) chamber which was continually gassed (20 ml min^{-1}) with the same gas mixture (water-saturated) as that used for the perfusion fluid.

Perfusion Time Sequence in Reoxygenation Studies

Immediately after mounting, the hearts were perfused (standard perfusion fluid plus glucose) aerobically for a 5 min washout and equilibration period. Hearts were then subjected to various periods of anoxic perfusion (glucose-free) followed by reoxygenation with oxygenated glucose-free perfusion fluid. The presence on the aortic cannulae of side arms linked to secondary perfusion fluid reservoirs facilitated rapid changes of perfusate composition. During the entire experimental period coronary flow was recorded and samples of coronary effluent were taken at regular intervals for immediate analysis of creatine kinase which had leaked from the myocardial cells to the perfusate. At the end of the experimental period the hearts were taken for dry weight determination. In parallel studies of reoxygenation and morphology, hearts were perfusion-fixed at various times in the experimental period and tissue samples were taken for light and electron microscopy.

Perfusion Time Sequence in Reperfusion Studies

Immediately after mounting, the hearts were perfused (standard perfusion fluid plus glucose) aerobically for a 5 min washout and equilibration period. Hearts were then subjected to various periods of ischaemic perfusion. This was achieved by perfusing the hearts (standard buffer, glucose-free, gassed with 95% O_2 + 5% CO_2) via an infusion pump which reduced the coronary flow to 0.06 ml min^{-1} (less than 1% of the flow rate in the preceding period of unrestricted perfusion). After the period of ischaemia, hearts were reperfused with glucose-free standard buffer (gassed with either 95% O_2 + 5% CO_2 or 95% N_2 + 5% CO_2) from a reservoir at a perfusion pressure equivalent to 85 cmH_2O. During the entire experimental period samples of the coronary effluent were taken for enzyme analysis and at the end of the experimental period the hearts were taken for dry weight determination. In a parallel series of studies hearts were perfusion-fixed at various times and tissue samples were taken for light and electron microscopy.

Perfusion Time Sequence in Calcium Paradox Studies

Immediately after mounting, the hearts were perfused (standard perfusion fluid plus glucose) aerobically for a 5 min washout and equilibration period. Hearts were then subjected to various periods of calcium-free perfusion (glucose-containing perfusion fluid) followed by periods of perfusion with buffer to which calcium had been restored to its previous concentration (2.4 mmol l.$^{-1}$). During the entire experimental period coronary flow was recorded and samples of coronary effluent were taken at regular intervals. At the end of the experimental period the hearts were taken for dry weight determination. In parallel studies hearts were perfusion-fixed at various times and tissue samples were taken for light and electron microscopy.

Enzyme Leakage Analysis

Creatine kinase (ATP; creatine phosphotransferase) and α-hydroxybutyrate dehydrogenase (D-2 hydroxybutyrate: NAD oxidoreductase) which leaked into the perfusate were measured as previously described.[5] Enzyme leakage was expressed in iu min^{-1} g^{-1} dry weight.

Tissue Preparation for Light and Electron Microscopy

Hearts were fixed by gentle perfusion (4 ml min^{-1} for 4 min via a side arm of the aortic cannula) with cold (4 °C) fixative. The fixative, which was designed to achieve optimal tissue preservation with minimal tissue shrinkage or loss of ground material, was 0.08 M phosphate buffer containing 2.5% glutaraldehyde plus 4.0% formaldehyde (prepared freshly from paraformaldehyde). The pH of the solution was 7.2 and the osmolarity of the buffer was 167 mosmol kg^{-1} water.

Immediately after perfusion–fixation the heart was removed from the perfusion cannula, immersed in the same fixative solution and a portion of the left ventricular wall, close to an identifiable branch of the left anterior descending coronary artery, was taken. This sample was then cut into small pieces (approximately 2–3 mm^3) and fixed for a further 2 h at 4 °C. The tissue pieces were then transferred to fresh phosphate buffer and were stored at 4 °C overnight. The tissue samples were than cut into smaller pieces and were subjected to post-fixation in phosphate buffer containing 1.0% osmic acid for 1 hour. Following dehydration through a series of alcohols the tissues were embedded in an 'Epon'–'Araldite' resin mixture[17] via propylene oxide. The resin was polymerized at 60 °C for 48 h after which sections were cut using a diamond knife.

For light microscopic examination, large block faces (2–3 mm^2) were cut to give sections of 1.0 μm thickness. The sections were transferred to glass slides, warmed to flatten the sections, and were then stained with 0.1% toluidine blue in 0.1% borax using heat to aid the penetration of the stain. After staining and rinsing, the sections were blotted dry and mounted under a coverslip using fresh resin.

For electron microscopy, small block faces (0.4–1.0 mm^2) were cut and the thin sections (silver-grey) were mounted on copper grids and were then stained with uranyl acetate (5% aqueous solution) and lead citrate.[21] The sections were examined under an AE1-EM6B electron microscope.

For light microscopy and electron microscopy, at least three hearts were used for each test group and at least two blocks were taken from each heart. Each block was then cut at several depths to achieve representative sampling. Both thick and thin sections were cut at each level (for light and electron microscopy respectively). For light microscopy 15 random sections were cut from each block and were examined after staining. For each section four fields were examined and the numbers of normal and abnormal cells within each field estimated and compared with control tissue. For electron microscopy ten grids were cut per block and six fields per grid were photographed at constant magnification (×7500).

RESULTS

Enzyme Leakage

Anoxia and reoxygenation

Hearts ($n = 6$) were subjected to 100 min of uninterrupted substrate-free anoxic perfusion. Figure 1(a) reveals that during this period there was a progressive increase in enzyme leakage reaching 2.0 iu min^{-1} g^{-1} dry wt. after 100 min. During this 100 min period a total of 72 ± 4 iu of creatine kinase was released to the perfusate. In order to assess the effects of reoxygenation, a second series of hearts ($n = 6$) was subjected to 60 min of anoxic substrate-free perfusion which was then followed by 40 min aerobic substrate-free perfusion. The results (Figure 1(a)) reveal that reoxygenation induces a sudden and massive exacerbation of enzyme release such that within 2 min of reoxygenation the amount of creatine kinase activity released to the coronary effluent was increased from 1 to 52 iu min^{-1} g^{-1} dry wt. Calculation of the cumulative enzyme release revealed that in the absence of reoxygenation 62 ± 4 iu g^{-1} dry wt. were released during the period 60–100 min whereas reoxygenation at the onset of this period led to a total enzyme release of 414 ± 30 iu g^{-1} dry wt. This major increase in

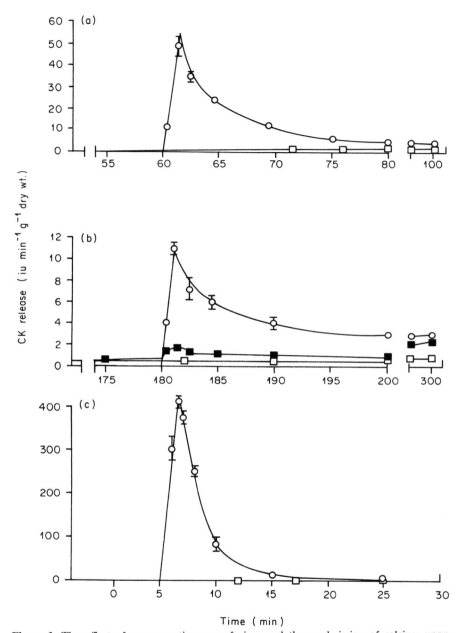

Figure 1. The effect of reoxygenation, reperfusion, and the readmission of calcium upon creatine kinase leakage in the isolated perfused rat heart. (a) Hearts were subjected to: (□) 100 min anoxia; (○) 60 min anoxia plus 40 min reoxygenation. (b) Hearts were subjected to: (□) 300 min ischaemia with coronary flow per heart reduced to 0.06 ml min^{-1}; (○) 180 min ischaemia plus 120 min aerobic reperfusion; or (■) 180 min ischaemia plus 120 min anoxic reperfusion. (c) Hearts were subjected to: (□) 25 min perfusion with calcium-free media; (○) 5 min perfusion with calcium-free media plus 20 min perfusion with calcium-containing media. Six hearts were used for each condition and the bars indicate the standard error of the mean

enzyme leakage would suggest a sudden exacerbation of tissue damage and since there were no changes in coronary flow, perfusion pressure, heart rate (zero) or substrate availability this exacerbation must be primarily attributable to the readmission of molecular oxygen.

Ischaemia and reperfusion

Figure 1(b) illustrates the effect of 300 min of ischaemia and also 180 min of ischaemia + 120 min of reperfusion upon enzyme leakage. As with reoxygenation (although not as great) reperfusion causes a sudden increase in creatine kinase leakage with total leakage during the period 180–300 min increasing from 109 ± 10 to 353 ± 40 iu g^{-1} dry wt. While this effect may be attributed to change in coronary flow, perfusion pressure, or oxygen availability, the similarity of the result to that obtained with reoxygenation study would suggest that the readmission of an unrestricted supply of oxygen may be the causative agent. This was confirmed in studies in which hearts ($n = 6$) were subjected to 180 min of ischaemia followed by anoxic reperfusion for 120 min with buffer which was gassed with 95% N_2 + 5% CO_2. Figure 1(b) reveals that although there was a small increase in enzyme release at the time of anoxic reperfusion, the total enzyme release during the reperfusion period (140 ± 20 iu g^{-1} dry wt.) was only slightly greater than that observed during continued ischaemia (109 ± 10 iu g^{-1} dry wt.) and was considerably less than that observed during aerobic reperfusion (353 ± 40 iu g^{-1} dry wt.).

Calcium depletion and calcium repletion

Figure 1(c) illustrates the consequences of 5 min of perfusion with calcium-free perfusate (containing glucose and gassed with 95% O_2 + 5% CO_2) followed by 20 min of perfusion with calcium-containing perfusate. The results reveal that the readmission of calcium leads to a sudden and major leakage of creatine kinase activity which increases to greater than 400 iu min^{-1} g^{-1} dry wt. in less than 2 min and during the subsequent 20 min accounts for the release of 1200 ± 120 iu g^{-1} dry wt. of creatine kinase. This leakage of enzyme activity, which is several times greater than that observed with reoxygenation, is comparable to the estimate[7] of the entire myocardial content of creatine kinase and as such would be indicative of the occurrence of very extensive tissue damage. Since the calcium content of the perfusion fluid was the only variable in this series of studies, this effect must be attributed directly to the readmission of calcium.

Morphological Changes

The consequences of calcium repletion, reoxygenation, and reperfusion were examined in tissue samples obtained just before and just after calcium repletion, reoxygenation, and reperfusion and the morphological characteristics compared with control tissue.

Fresh heart control tissue

In order to confirm the adequacy of tissue preparation procedures, hearts were subjected to 5 min of aerobic perfusion with glucose-containing buffer followed by immediate perfusion fixation. Plate 1 shows a light micrograph of a toluidine blue stained thick resin section in which the fibres are fully relaxed and clear vascular and intercellular spaces are apparent. Plates 2, 3, and 4 show electron micrographs of thin sections of fresh tissue. Plate 2 shows well-preserved, aligned mitochondria, relaxed well-aligned myofibrils, electron-dense lipid droplets are seen adjacent to some mitochondria and there are numerous glycogen granules, the sarcoplasmic reticulum is clearly defined and there is no evidence of intracellular oedema. Plate 3 additionally shows an intercalated disc which is generally closely adposed with tight junctions but occasional vesicles are apparent. Plate 4 is a high-magnification photomicrograph of the sarcolemma and basement membrane both of which are intact and closely follow the edge of the muscle fibre.

Calcium-free hearts

Hearts in this group were subjected to 5 min calcium-free perfusion followed by immediate perfusion–fixation. Plate 5 shows a light micrograph of a toluidine blue stained section. In contrast to the control tissue in Plate 1, this tissue shows considerable shrinkage of fibres, increased intercellular space, and a marked increase in vesiculation in the region of the intercalated disc. The muscle fibres are relaxed. Plates 6, 7, and 8 show electron micrographs of the same tissue. Plate 6 reveals that the mitochondria are well defined and well aligned as are the myofibrils. However, marked changes have occurred in the region of the intercalated disc with the appearance of large vesicular structures and there is evidence for the separation of opposing faces of the intercalated disc. Elsewhere in the cell there is evidence of the dilation and extension of the T-tubular system. Plate 7 shows the most striking and consistent feature of the calcium-free tissue, that is clear separation in all cells of the basement membrane from the muscle cell and the sarcolemma. Both the sarcolemma and the basement membrane remain intact but are separated by large vacuolar structures. The basement membrane appears to retain its original shape and as illustrated in Plates 7 and 8 the separation appears to result from the formation of large fluid-filled vacuoles between the sarcolemma and the basement membrane.

Calcium-repleted hearts

Hearts in this group were subjected to 5 min calcium-free perfusion followed by the 1 min period of calcium repletion and immediate fixation. Plate 9 shows a light photomicrograph of a number of muscle fibres and reveals the very striking morphological changes (in contrast to Plate 5) which occur in the 1 min of calcium repletion. Severe contracture of the fibres is evident with the creation of large non-staining areas into which many hundreds of mitochondria are discharged. Careful examination of Plate 9 reveals a membranous continuity between and around these areas of intense contracture with either the sarcolemma or the basement membrane retaining its original location and now appearing to stretch between and around areas of intense staining. Electron microscopic examination (Plates 10, 11, and 12) of the various areas observed in the light micrographs confirms the onset of massive tissue damage. Plate 10 shows a photomicrograph of the end region of a muscle cell. The discharge of mitochondria is clearly shown as are concentric contracture bands (with merging of the Z bands) which radiate from the end of the cell into the less damaged centrally located cytoplasmic regions. Plate 10 also shows how the sarcolemma and basement membrane remain to envelop the region of maximal damage. Despite the massive disruption and dissolution of the cytoplasm and myofibrils it is of interest that the mitochondria remain essentially intact and well preserved. Although not reported in detail here the coronary perfusate was examined for mitochondria and mitochondrial enzymes. None were found and this would be consistent with our observation that calcium repletion does not result in mitochondrial disruption and although apparently being discharged from the main cytoplasmic mass the mitochondria are still contained within the sarcolemma and basement membrane.

Plates 9 and 10 would indicate a polarity of damage which appears to commence in the region of the intercalated disc and which then spreads longitudinally along the fibre towards the midpoint of each cell. Thus in these micrographs, which were taken only 1 min after the onset of damage, relatively intact regions of the cell still persist and Plate 11 shows centrally located tissue where little damage has yet occurred.

Plates 10 and 11 show the basement membrane and sarcolemma in close juxtaposition to each other and the body of the fibre. However, in the preceding period of calcium-free perfusion (Plates 7 and 8) the basement membrane was consistently found to be separated from the sarcolemma. The reassociation of these two bodies may possibly result from the (Plate 9) stretching of the cell membranes which apparently occurs during the repletion of calcium. This stretching may result from cell swelling and the onset of contracture. Plate 12 shows a high-magnification electron micrograph of the sarcolemma and basement membrane from the area of maximal damage at the end of a muscle fibre, while the continuity of the membrane complex may be intact there is evidence of considerable damage and possible fragmentation of the sarcolemma.

Thus the readmission of calcium following a short period of calcium depletion causes massive ultrastructural damage characterized by cellular dissolution, the onset of severe contracture, the expulsion of intact mitochondria and the disruption of the plasma membrane. These changes would readily account for the massive leakage of enzymes and proteins described in the preceding section.

Anoxic hearts

Hearts from this group were subjected to 60 min of anoxic perfusion followed by immediate fixation. Plate 13 shows a light micrograph of a number of muscle fibres which, in contrast to control tissue (Plate 1), exhibit some damage with an increase in intercellular space. The pattern of damage exhibits a characteristic heterogeneity with some cells staining normally while other cells appear to be resistant to staining. The heterogeneity of damage also occurred between hearts in this series, one heart exhibited considerably more damage than the others. However, electron microscopic examination (Plate 14) of all tissue showed that after this relatively short period of anoxia, glycogen granules were depleted, the myofibrils were partially contracted and the Z bands were distorted. Mitochondria were distorted but retained their alignment. There was evidence of intracellular oedema and in particular there was extensive vesiculation in the region of the intercalated disc. The sarcolemma and basement membrane were intact.

Reoxygenated hearts

Hearts in this group were subjected to 60 min of anoxic perfusion plus 1 min of reoxygenation followed by immediate fixation. Plate 15 shows a light micrograph of this tissue and reveals that the 1 min period of reoxygenation induces significant morphological changes, including a small increase (less than 15%) in intercellular space. A striking feature of this reoxygenation damage is the occurrence of irregular undulations of the longitudinal edges of some fibres. Examination of this tissue under the electron microscope (Plate 16) reveals that the undulations are a result of the extrusion of mitochondria into the subsarcolemmal region. Reoxygenation also induces an intensification of contraction bands, an increase in cellular oedema and further distortion and misalignment of mitochondria. This damage, and in particular the distortion of the cell membrane induced by the mitochondria may account for the sudden increase in enzyme leakage.

There appears to be a good correlation between the extent of enzyme

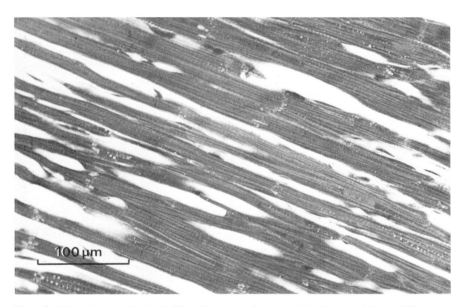

Plate 1. Light micrograph of toluidine blue stained section of fresh control tissue. Fibres are relaxed. Magnification × 240

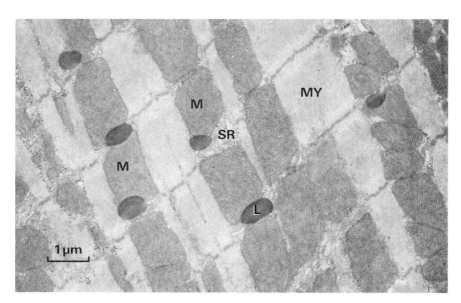

Plate 2. Electron micrograph of fresh control tissue. M = mitochondria; SR = sarcoplasmic reticulum: L = lipid droplet; MY = myofibrils

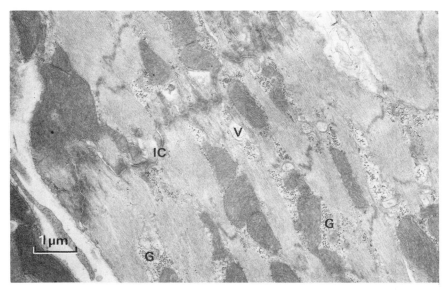

Plate 3. Electron micrograph of fresh control tissues. IC = intercalated disc; V = vesicles; G = glycogen granules

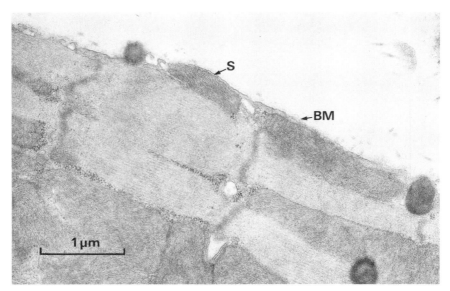

Plate 4. Electron micrograph of fresh control tissue. S = sarcolemma; BM = basement membrane

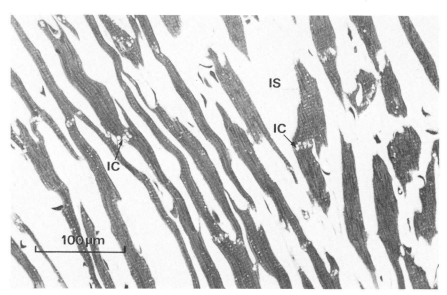

Plate 5. Light micrograph of toluidine blue stained section of tissue obtained after a 5 min period of calcium-free perfusion. IS = intercellular space; IC = intercalated disc with prominent vesiculation. Magnification × 240

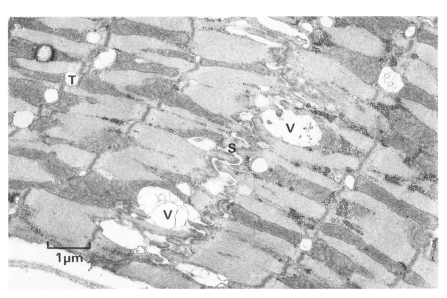

Plate 6. Electron micrograph of tissue obtained after a 5 min period of calcium-free perfusion. V = vesicular structures in the region of the intercalated disc; S = separation of opposing faces of the disc; T = swollen T-tubules

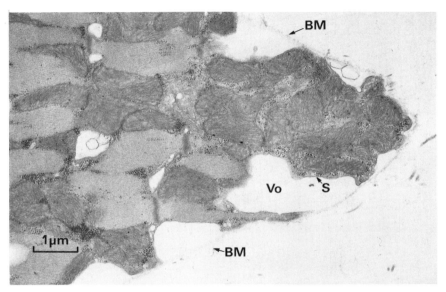

Plate 7. Electron micrograph of tissue obtained after a 5 min period of calcium-free perfusion. BM = separated basement membrane; Vo = vacuole; S = sarcolemma

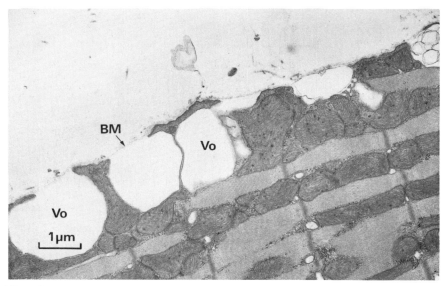

Plate 8. Electron micrograph of tissue obtained after a 5 min period of calcium-free perfusion. BM = separated basement membrane; Vo = vacuole

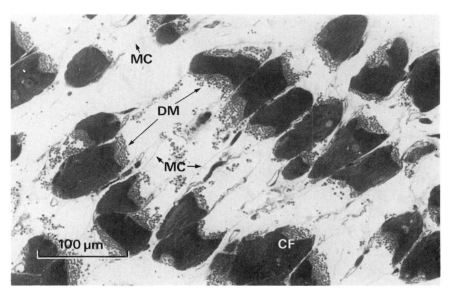

Plate 9. Light micrograph of toluidine blue stained section of tissue obtained following a 5 min period of calcium depletion plus a 1 min period of calcium repletion. CF = contracted fibres; DM = discharging mitochondria; MC = membranous continuity around and between contracted tissue. Magnification ×240

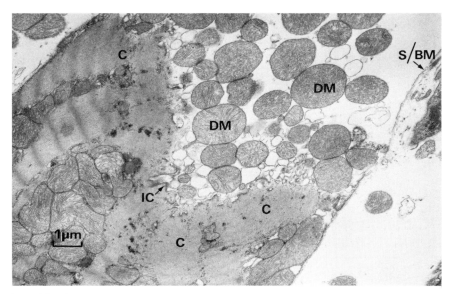

Plate 10. Electron micrograph of tissue at the end of a muscle fibre obtained following a 5 min period of calcium depletion plus a 1 min period of calcium repletion. DM = discharging mitochondria; C = concentric contracture bands; IC = intercalated disc; S/BM = sarcolemma/basement membrane enveloping the damaged tissue

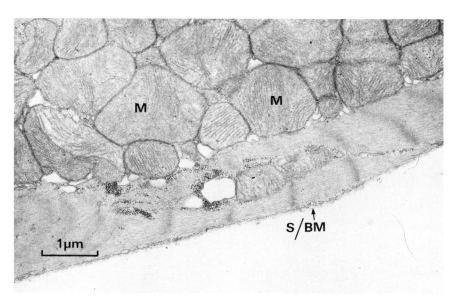

Plate 11. Electron micrograph of tissue at the centre of a muscle fibre obtained following a 5 min period of calcium depletion plus a 1 min period of calcium repletion. S/BM = intact sarcolemma basement membrane; M = intact mitochondria

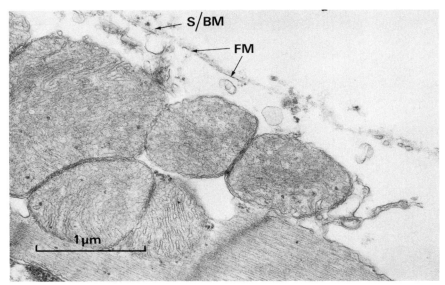

Plate 12. Electron micrograph of tissue at the end of a muscle fibre obtained following a 5 min period of calcium depletion plus a 1 min period of calcium repletion. S/BM = sarcolemma/basement membrane; FS = possible fragmentation of sarcolemma

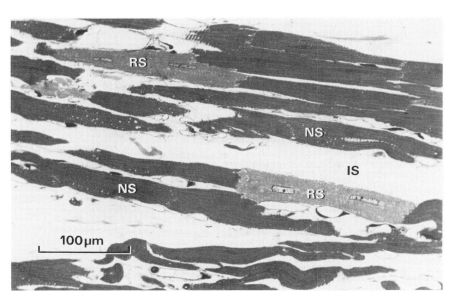

Plate 13. Light micrograph of toluidine blue stained section of tissue obtained following a 60 min period of anoxia. NS = normally heavily staining cells; IS = intercellular space; RS = cells which are resistant to staining. Magnification × 240

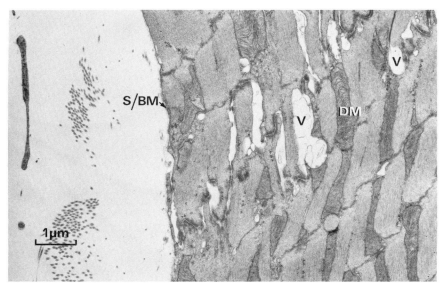

Plate 14. Electron micrograph of tissue obtained after 60 min anoxia. V = vesiculation in the region of the intercalated disc; DM = distorted mitochondria; S/BM = sarcolemma/basement membrane

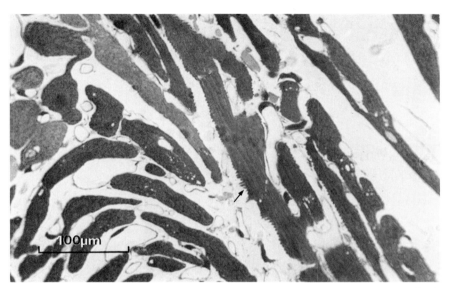

Plate 15. Light micrograph of toluidine blue stained section of tissue obtained after 60 min anoxia plus 1 min reoxygenation. Arrows show undulations of muscle fibre edge. Magnification × 240

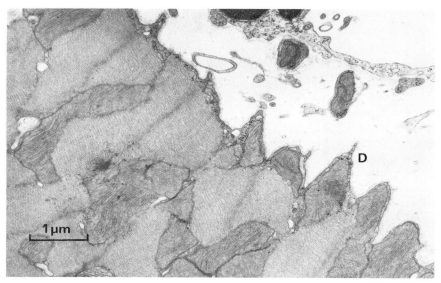

Plate 16. Electron micrograph of tissue obtained after 60 min plus 1 min reoxygenation. D = distension of sarcolemma and basement membrane by mitochondria

leakage and the degree of ultrastructural damage. Thus in the present paper the calcium paradox induces massive ultrastructural damage and an equally massive enzyme leakage while reoxygenation induces substantially less ultrastructural damage and correspondingly less enzyme leakage. The enzyme leakage observed in the reperfusion studies was less than in the reoxygenation studies and, as detailed below, so too were the morphological changes.

Ischaemic hearts and reperfused hearts

Hearts in this group were subjected to 180 min ischaemia followed by immediate fixation. Light and electron micrographs of ischaemic damage were very similar to those obtained for anoxic tissue in the preceding section. Aerobic reperfusion resulted in an extension of damage which like the increase in enzyme release was not so great as that observed in the anoxic and reoxygenated hearts.

Reperfusion of the ischaemic hearts with anoxic fluid resulted in no increase in damage and as described earlier no significant increase in enzyme leakage. Thus as with the anoxia and reoxygenation, any exacerbation of damage induced by reperfusion of ischaemic hearts can most probably be attributed to the readmission of molecular oxygen.

The results presented thus far would indicate that although differing in magnitude, the consequences of calcium depletion and repletion are in many ways similar to the consequences of oxygen depletion and repletion. It was therefore decided to extend the comparison between these two phenomena by investigating the effects of a number of potential moderating factors. Since the magnitude of damage induced by oxygen repletion in reoxygenation is greater than that in reperfusion and thus is closer to that observed with calcium repletion, it was decided to use anoxia and reoxygenation as the model for the following comparative studies.

Factors Influencing the Calcium Paradox and the Reoxygenation Phenomenon

Time

We have previously reported[5] that hearts subjected to reoxygenation after 30 min of anoxia failed to exhibit the reoxygenation phenomenon. A series of studies was therefore carried out to determine whether the occurrence of the reoxygenation damage was an 'all-or-none' phenomenon or whether it progressively increased with the duration of anoxia. Hearts ($n = 6$ for each group) were subjected to 30, 35, 40, 45, 50, 55, 60, 65, 80, 100, 120, and 150 min of anoxia followed by reoxygenation for 20 min. Creatine kinase leakage was measured during anoxia and reoxygenation. The enzyme release profiles are shown in Figure 2 and the total enzyme activity released

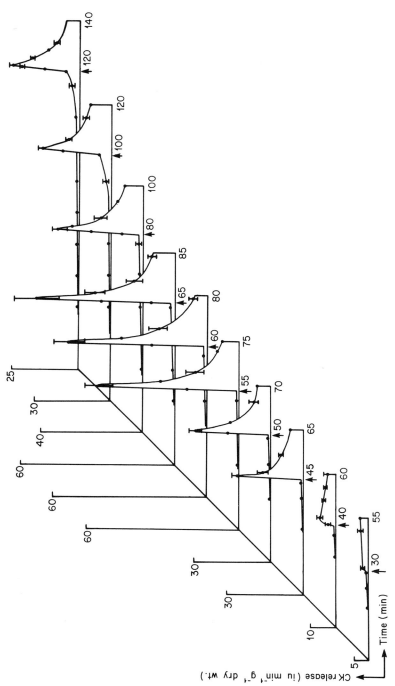

Figure 2. The extent of reoxygenation-induced exacerbation of creatine kinase leakage as a function of the duration of the preceding period of anoxia. Hearts were subjected to 30, 35, 40, 45, 50, 55, 60, 65, 80, 100, 120, and 150 min anoxia followed by reperfusion (indicated by the arrows) for 20 min. Each result is the mean of six hearts and the bars represent the standard error of the mean

Table 1. The Extent of the Reoxygenation-induced Exacerbation of Creatine Kinase Leakage as a Function of the Duration of the Preceding Period of Anoxia

Duration of anoxia (min)	Creatine kinase activity released during subsequent reoxygenation (iu $(20\ min)^{-1}\ g^{-1}$ dry wt.)
30	25 ± 3
35	50 ± 3
40	87 ± 10
45	200 ± 25
50	221 ± 25
55	345 ± 33
60	338 ± 25
65	367 ± 42
80	291 ± 15
100	267 ± 26
120	228 ± 16
150	203 ± 10

Each result is the mean of six hearts and the standard error of the mean is indicated.

during each 20 min reoxygenation phase is shown in Table 1. Reoxygenation after 30 or 35 min of anoxia causes only a small increase in enzyme release (20–50 iu g^{-1} dry wt.). Between 35 and 45 min of anoxia there is a large increase in the amount of enzyme released (rising to 200 iu g^{-1} dry wt.). By 55 min total enzyme release has increased to 345 ± 33 iu g^{-1} dry wt. and thereafter, increasing durations of anoxia do not result in any significant increase in total enzyme release. In fact, after approximately 1 h of anoxia there is a progressive reduction in the amount of enzyme release which can be induced by reoxygenation (possibly a result of enzyme inactivation). Thus these results would indicate that while the onset and magnitude of the reoxygenation phenomenon is dependent upon the duration of anoxia, the time between no effect and a maximal effect is very short (15–20 min).

There have been reports[12] that the calcium paradox is also time dependent over a very short period. Hearts ($n = 6$ for each group) were therefore subjected to 1.0, 2.0, 2.5, 3.0, 3.5, 4.0, 5.0, and 20.0 min of calcium-free perfusion followed by 10 min perfusion with calcium-containing buffer. The resulting creatine kinase release profiles are shown in Figure 3 and the cumulative enzyme release is shown in Table 2. The results indicate that with only 1 min of calcium-free perfusion the calcium paradox did not occur. With 2.0 or 2.5 min calcium-free perfusion, the readmission of calcium caused some enzyme release but this was small compared with a full calcium paradox and the increase in the enzyme release was not as abrupt. With 3.0 min of calcium-free perfusion there was major and sudden enzyme release but it was not until the hearts were subjected to 3.5 min or longer of

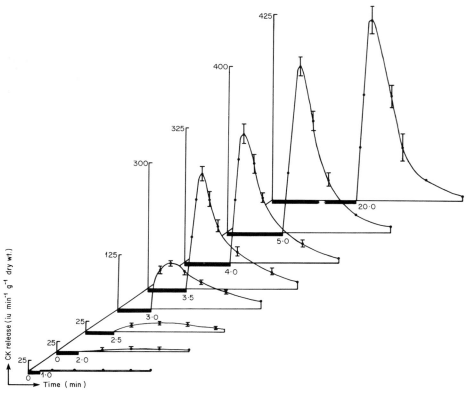

Figure 3. The relationship between the duration of calcium depletion and the extent of creatine kinase leakage induced by calcium repletion. Hearts were subjected to 1.0, 2.0, 2.5, 3.0, 3.5, 4.0, 5.0, and 20.0 min periods of calcium-free perfusion (black bars) followed by 10 min perfusion with calcium-containing medium. Each result is the mean of six hearts and the bars represent the standard error of the mean

Table 2. The Relationship between the Duration of Calcium Depletion and the Extent of Creatine Kinase Leakage induced by Calcium Repletion

Duration of calcium-free perfusion (min)	Creatine kinase activity released during subsequent calcium repletion (iu $(10\ min)^{-1}\ g^{-1}$ dry wt.)
1.0	0
2.0	50 ± 2
2.5	158 ± 5
3.0	650 ± 10
3.5	1062 ± 25
4.0	1150 ± 35
5.0	1350 ± 50
20.0	1512 ± 100

Each result is the mean of six hearts and the standard error of the mean is indicated.

calcium-free perfusion that the maximal effects of the paradox were observed. Under these conditions in excess of 1000 iu of creatine kinase was released over the 10 min period following the readmission of calcium. Thus under the conditions of this study the calcium paradox is time dependent but this dependency occurs over an extremely narrow time band of 1-2 min.

It is again important to bear in mind that the maximum amount of enzyme release induced by the calcium paradox is 3-4 times greater than that induced by reoxygenation, an observation which is consistent with the degree of observed ultrastructural damage.

Stepwise readmission of oxygen or calcium

In previous studies[7] we have shown that the reoxygenation phenomenon could not be totally prevented by the partial or gradual reoxygenation of the anoxic myocardium. Thus hearts ($n = 6$ for each group) were subjected to 150 min anoxia followed by 30 min reoxygenation at a series of different oxygen tensions. The total enzyme release (in this instance α-hydroxybutyrate dehydrogenase but this was comparable to creatine kinase) during the 30 min reoxygenation period is shown in Figure 4 and in Table 3. The results indicate a direct relationship between the magnitude of enzyme release and the PO_2 at the time of reoxygenation. In previous studies[7] of the effects of partial reoxygenation, hearts were subjected to a second reoxygenation step to bring the reoxygenation PO_2 to the maximal level associated with 95% O_2 + 5% CO_2 (>650 mmHg). In these studies there was an additional exacerbation which was inversely related to the PO_2 used during the first reoxygenation period and was directly related to the PO_2 step of the second reoxygenation. Thus the sum of the enzyme release during the two periods of reoxygenation was comparable in each instance and approximately equivalent to that which would have occurred with a single and full reoxygenation step.

In an attempt to determine whether a comparable stepwise situation occurred with the calcium paradox, hearts ($n = 6$ for each group) were subjected to 5 min of calcium-free perfusion followed by 20 min of perfusion with buffer containing 0.012, 0.024, 0.050, 0.100, 0.150, 0.180, 1.00, and 2.40 mmol l.$^{-1}$ calcium. Creatine kinase leakage was measured over the 20 min period of calcium repletion and the results are shown in Table 4 and Figure 5. In contrast with the stepwise reoxygenation studies which showed a progressive increase in enzyme leakage over a wide range of PO_2 values, the calcium repletion studies indicate a relatively narrow calcium concentration range over which enzyme leakage increased from very low to very high levels. Thus between 0 and 0.024 mmol l.$^{-1}$ calcium there is no significant exacerbation of enzyme leakage. Between 0.024 and 0.10 mmol l.$^{-1}$ calcium there is a substantial amount of enzyme activity released but the leakage is

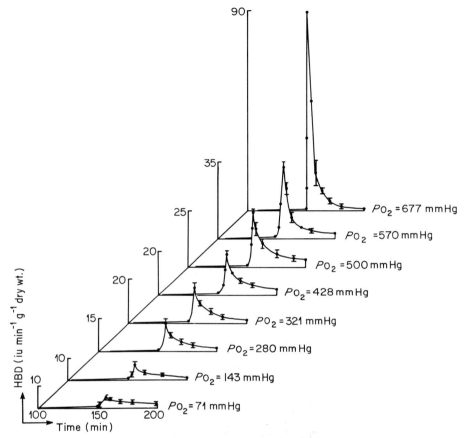

Figure 4. The relationship between the PO_2 at the time of reoxygenation and the extent of enzyme leakage. Hearts were subjected to 150 min of anoxic perfusion followed by a 50 min period of reoxygenation with perfusion fluid having a PO_2 of 71, 143, 280, 321, 428, 500, or 677 mmHg. Each result is the mean of six hearts and the bars represent the standard error of the mean

not abrupt and the total enzyme released is only between a quarter and a third of that seen with a full paradox. Between 0.100 and 0.150 mmol l.$^{-1}$ calcium enzyme leakage becomes abrupt and total release increases to 846 ± 127 iu $(20\ \text{min})^{-1}\ g^{-1}$ dry wt., a value which increases to a maximum paradox value of 1205 ± 130 iu $(20\ \text{min})^{-1}\ g^{-1}$ dry wt. when the calcium concentration is increased to 0.180 mmol l.$^{-1}$. Between this latter concentration and 2.40 mmol l.$^{-1}$ there is no significant increase in the total enzyme leakage but there is an increase (Figure 5) in the initial rate of leakage and peak amount released. Thus there appears to be a relatively sharp threshold (0.10–0.150 mmol l.$^{-1}$ calcium) over which the calcium paradox may be

Table 3. The Effect of the PO_2 during Reoxygenation upon the Extent of Induced Enzyme Leakage

PO_2 during reoxygenation (mmHg)	α-Hydroxybutyrate dehydrogenase released (iu $(30\ min)^{-1}\ g^{-1}$ dry wt.)
<50	13± 2
71	19± 3
143	26± 4
280	36± 8
321	39± 7
428	44± 7
500	57±15
570	66± 9
>650	112±11

Hearts were subjected to 150 min anoxic perfusion followed by a 30 min period of reoxygenation. Each result is the mean of six hearts and the standard error of the mean is indicated.

induced. However, the initial rate of enzyme leakage appears to be more dose-dependent over a wider range of calcium concentrations.

Temperature

It has been reported[12] that hypothermia (4 °C) is able to abolish the calcium paradox. The protective effects of hypothermia are well known[10] and are in general related to the degree of hypothermia although the relationship is not necessarily a linear one.[10] It was therefore decided to

Table 4. The Relationship between the Concentration of Calcium and the Extent of Creatine Kinase leakage during a 20 min period of Calcium Repletion following a 5 min period of Calcium-free Perfusion

Concentration of calcium during repletion period (mmol $l.^{-1}$)	Creatine kinase activity released during repletion period (iu $(20\ min)^{-1}\ g^{-1}$ dry wt.)
0.012	28± 12
0.024	31± 2
0.050	102± 5
0.100	355± 67
0.150	846±127
0.180	1205±130
1.00	1001±102
2.40	1200±120

Each result is the mean of six hearts and the standard error of the mean is indicated.

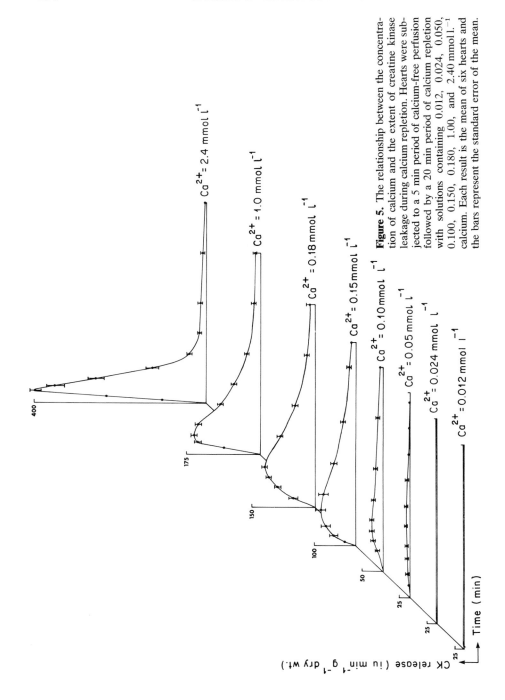

Figure 5. The relationship between the concentration of calcium and the extent of creatine kinase leakage during calcium repletion. Hearts were subjected to a 5 min period of calcium-free perfusion followed by a 20 min period of calcium repletion with solutions containing 0.012, 0.024, 0.050, 0.100, 0.150, 0.180, 1.00, and 2.40 mmol l^{-1} calcium. Each result is the mean of six hearts and the bars represent the standard error of the mean.

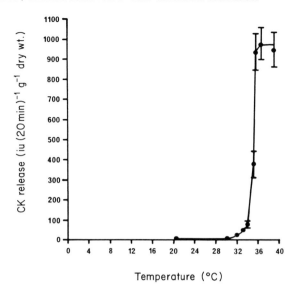

Figure 6. The relationship between the temperature during the period of calcium-free perfusion and the extent of creatine kinase leakage during the period of calcium repletion. Hearts were subjected to a 10 min period of calcium-free perfusion at 4, 20, 30, 32, 33, 34, 35, 36, 37, and 39 °C followed by a 10 min period of calcium repletion at 37 °C. Each result is the mean of six hearts and the bars represent the standard error of the mean

characterize the relationship between the degree of hypothermia and the occurrence of the calcium paradox and investigate whether a comparable relationship exists for the reoxygenation phenomenon.

Although the hearts were maintained at 4 °C during both calcium depletion and repletion periods, Holland and Olsen[12] suggested that the protective effects of hypothermia occurred during the calcium-free period. In our first series of temperature-response studies, therefore, the hearts were maintained at various temperatures (4, 20, 30, 32, 33, 34, 35, 36, 37, and 39 °C) during the 10-min period of calcium-free perfusion but calcium was readmitted at 37 °C. The results (Figure 6, Table 5) for creatine kinase leakage over the 10 min repletion period reveal a striking relationship. Between 4 and 30 °C calcium repletion does not induce any creatine kinase release. Between 30 and 34 °C there is a small enzyme release but insufficient to represent a calcium paradox. However, between 34 and 36 °C there is a sharp threshold with the induction of a full calcium paradox.

In the second series of temperature-response studies the hearts were subjected to 10 min of calcium-free perfusion followed by a 10 min period of calcium repletion at different temperatures (37, 20, and 4 °C). The period

Table 5. The Relationship between the Temperature during the Period of Calcium-free Perfusion and the Extent of Creatine Kinase Leakage during the Period of Calcium Repletion at 37 °C

Temperature during the period of calcium-free perfusion (°C)	Creatine kinase activity released during repletion period (iu $(10 \min)^{-1}$ g^{-1} dry wt.)
4	1.10 ± 0.02
20	1.11 ± 0.09
30	1.70 ± 0.35
32	31.0 ± 3.8
33	45.1 ± 8.9
34	71.8 ± 22.7
35	384.0 ± 68.0
36	940.0 ± 89.0
37	977.0 ± 85.0
39	950.0 ± 83.0

Each result is the mean of six hearts and the standard error of the mean is indicated.

of calcium-free perfusion was divided into two stages, an initial 5 min period of perfusion at 37 °C followed by a second period of calcium-free perfusion at the temperature selected for the subsequent period of calcium repletion. This precaution permitted the hearts to equilibrate to the correct temperature prior to the sudden readmission of calcium and thus allowed a clear distinction to be made between the effects of temperature and calcium.

The results (Table 6) reveal that hypothermia during the period of calcium repletion is unable to abolish the calcium paradox even at 4 °C. While there was some small temperature-related reduction in the peak levels of enzyme released, the total activity released over the 10 min repletion period was essentially unchanged.

Table 6. The Relationship between Temperature and the Extent of Creatine Kinase Leakage during the Period of Calcium Repletion

Temperature during the period of calcium repletion (°C)	Creatine kinase activity released during repletion period (iu $(10 \min)^{-1}$ g^{-1} dry wt.)
4	614 ± 53
20	898 ± 75
30	945 ± 89
37	977 ± 85

Each result is the mean of six hearts and the standard error of the mean is indicated.

Table 7. The Relationship between Temperature and the Extent of Creatine Kinase Leakage during the Period of Reoxygenation

Temperature during the reoxygenation period (°C)	Creatine kinase activity released during reoxygenation period (iu $(40 \text{ min})^{-1}$ g^{-1} dry wt.)
4	277 ± 17
20	427 ± 28
30	320 ± 30
37	400 ± 35

Each result is the mean of six hearts and the standard error of the mean is indicated.

These results would lend strong support to the suggestion[12] that the fine but critical changes which predispose to the calcium paradox, occur during the period of calcium-free perfusion. Furthermore they would tend to indicate that the damaging effects of repletion are perhaps passive and not dependent upon an active or energy-requiring metabolic process.

In two parallel series of experiments hearts were subjected to 60 min of anoxic perfusion followed by 40 min of reoxygenation. In the first series of experiments hearts were subjected to reoxygenation at various degrees of hypothermia (4, 20, 30, and 37 °C). The preceding period of anoxia was divided into two phases: an initial period of 55 min perfusion at 37 °C followed by a period of anoxic perfusion at whatever temperature was to be used for the following period of reoxygenation. As with the calcium paradox studies this precaution ensured that the effects of temperature and reoxygenation could be distinguished. The results (Table 7) revealed that hypothermia during the period of reoxygenation did not abolish the sudden exacerbation of damage. As with hypothermia during calcium repletion there was a temperature-related reduction in peak enzyme release levels but the total activity released over the 40 min reoxygenation period was essentially unchanged.

In the second series of reoxygenation studies hearts were maintained at various degrees of hypothermia (4, 20, 30, 33, 34, 35, 36, 37, and 39 °C) for the entire 60 min of anoxia. Reoxygenation was carried out at 37 °C and the results (Figure 7, Table 8) reveal a situation almost identical with that observed (Figure 6) with hypothermia during calcium-free perfusion. Essentially there was no reoxygenation-induced enzyme leakage between 4 and 30 °C. Between 30 and 33 °C reoxygenation induced a very small release. However, between 33 and 37 °C the reoxygenation phenomenon was suddenly induced with massive and maximal enzyme release. Thus both the reoxygenation phenomenon and the calcium paradox appear to have a sharp threshold between 33 and 36 °C.

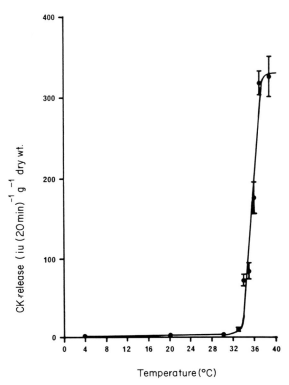

Figure 7. The relationship between the temperature during the anoxic period and the creatine kinase leakage during the reoxygenation period. Hearts were subjected to 60 min of anoxia at 4, 20, 30, 33, 34, 35, 36, 37, and 39 °C followed by a 40 min period of reoxygenation at 37 °C. Each result is the mean of six hearts and the bars represent the standard error of the mean

Table 8. The Relationship between the Temperature during the Anoxic Period and the Creatine Kinase Leakage during the Reoxygenation Period at 37 °C

Temperature during the anoxic period (°C)	Creatine kinase activity released during reoxygenation period (iu (40 min)$^{-1}$ g^{-1} dry wt.)
4	2.5 ± 1.5
20	3.0 ± 1.5
30	3.4 ± 0.7
33	8.5 ± 2.0
34	74.0 ± 7.0
35	83.3 ± 10.8
36	175.0 ± 19.0
37	318.0 ± 15.0
39	325.0 ± 25.0

Each result is the mean of six hearts and the standard error of the mean is indicated.

DISCUSSION

The results of this study reveal a remarkable similarity between the characteristics of both the oxygen and the calcium paradox. Consideration of these characteristics may shed some further light on the mechanisms responsible for the induction of damage in both phenomena.

With reoxygenation, reperfusion, and calcium repletion there is an abrupt increase in enzyme leakage which can be associated with the sudden extension of morphological damage. While the time base and exact nature of these changes may vary, the consequences of the repletion of calcium after a short period of calcium depletion or the repletion of oxygen after a relatively long period of anoxia are in many ways similar. In both instances numerous intracellular vesicles develop, the myofibrils rapidly develop contracture, mitochondria are translocated to the cell border and may be partially or totally extruded and with both phenomena there is extensive damage to the cell membrane. The striking results obtained for the calcium paradox in Plate 9 (which shows the progressive contracture of the myocardial cells into polarized masses with the expulsion of intact mitochondria at the end of the cell), amply illustrate the suddenness and severity of the damage which occurs in the course of the calcium paradox.

In considering the calcium paradox, perhaps the most important and interesting morphological change which we observed was the separation of the basement membrane from the sarcolemma which occurred during the period of calcium-free perfusion (Plates 7 and 8). Associated with this separation was the development of large fluid-filled vacuoles at the cell periphery. The exact relationship of the sarcolemma to the basement membrane and the nature of the basement membrane is complex. The mucopolysaccharide basement membrane is characterized by a vast number of fixed anionic sites which like an ion exchange resin are able to bind and control cations in solution. In this way the basement membrane has been implicated, in part at least, in the control of cellular calcium homeostasis.[19] The abrupt removal of free calcium ions from an anionic matrix of this type would affect the physicochemical characteristics of the basement membrane and may alter such factors as its degree of hydration, its control over the transmembrane movement of water and ions and possibly even its physical structure such that, like an ion exchange resin, swelling and matrix dimension changes may occur. In this way, it is possible to speculate that a period of calcium-free perfusion may result in the separation of the basement membrane from the cell surface, possibly modifying any control it may have had over water and ion movements (this may lead to the observed vacuolation) and permitting the loss of calcium from cytoplasm to the calcium-free extracellular space. Under these conditions, with changes in the basement membrane and abnormally low intracellular calcium, when repletion of calcium occurs, the tight control which is normally exerted over calcium influx and cell volume regulation may be lost and as a result, a massive calcium influx may be permitted which may in turn lead to contracture, cell swelling, and tissue damage.

In considering the possible site of entry of the calcium and a likely reason for the apparent subcellular polarization of damage, the electron micrographs obtained during the calcium-free perfusion period (Plate 6) showed a characteristic separation of opposing faces at the intercalated disc and the vesiculation of the cell in the region of the intercalated disc, a finding previously observed by Yates and Dhalla.[27] Electron micrographs obtained following calcium readmission (Plate 10) would indicate that the area of initial and maximal damage is also in this region and as such might implicate the intercalated disc as the site of the primary lesion. The intercalated disc has very different characteristics to the longitudinal sarcolemma and thus may provide a site for the initial influx of calcium and thus account for the polarization of damage towards the cell centre. Certainly, the sudden and massive structural damage associated with the calcium paradox, and in particular the fragmentation of the sarcolemma, explains adequately the very large leakage of enzyme activity observed during calcium repletion.

In considering the oxygen paradox, it is also possible to speculate on the role of calcium redistribution and sudden calcium fluxes. During ischaemia or anoxia, there is likely to be a slow but progressive increase in cytoplasmic ionized calcium levels due to dechelation from ATP and citrate (the latter compounds declining under conditions of oxygen deprivation), reduced sequestration by the sarcoplasmic reticulum and mitochondria, and possibly also a net cellular influx of calcium as cell membrane integrity progressively deteriorates. As has been discussed previously[11] reoxygenation may lead to a re-energization of the mitochondrial electron transport chain and associated with this may be massive mitochondrial calcium uptake. Accompanying this calcium uptake there may be a major efflux of protons from the mitochondria to the cytoplasm. This latter effect would result in cytoplasmic acidification, a factor which may contribute significantly to the development of cellular damage.

The possibility that disturbances of calcium homeostasis may predispose towards both the oxygen and calcium paradoxes, and that in both cases the damage may be attributed to sudden transmembrane (sarcolemma and/or mitochondrial) calcium fluxes, may be supported by our studies of factors which moderate the two phenomena.

The factor of time is important in both phenomena. In the calcium paradox a minimum period (approximately 3 min) of calcium-free perfusion is required before the paradox can be induced and with the oxygen paradox a minimum period of anoxia (approximately 45 min) is required before reoxygenation can induce major damage and enzyme release. In both instances however the time span between minimum and maximum damage is relatively short (between 3 and 4 min after the onset of calcium-free perfusion and between 40 and 50 min after the onset of anoxic perfusion) and suggests that there may be a threshold which must be reached before the phenomena can be induced. In the case of the calcium paradox it is possible that the threshold is reached when the basement membrane has separated from the plasma membrane and when adequate intracellular calcium loss has occurred. In the case of the oxygen paradox the threshold may be reached when cytoplasmic calcium levels have increased sufficiently and when mitochondrial and/or plasma membrane damage is such that normal control of cell volume and transmembrane ion movements are lost. These changes would take considerably longer than those of the calcium paradox. Possible evidence for the calcium paradox threshold may be derived from our studies of the stepwise readmission of calcium. In these studies there appeared to be a threshold concentration $(0.10-0.15 \text{ mmol l.}^{-1}$ calcium) below which paradox could not be induced but above which a full paradox was observed.

Some information about the nature of the calcium fluxes, together with perhaps the strongest evidence for an association of oxygen and calcium paradoxes may be derived from our study of the effect of temperature upon the initiation of damage. These results revealed two important facts. Firstly, if calcium or oxygen depletion is carried out at 37 °C then hypothermia, even as low as 4 °C, is unable to reduce repletion damage in either phenomenon. This finding was surprising in view of the suggestion by ourselves[11] and others[2,23] that the damaging calcium influxes, or at least their triggering, were energy dependent, consuming as suggested by Boink and colleagues,[2] large amounts of ATP and creatine phosphate. However, the inability of hypothermia, even at 4 °C to abolish the calcium paradox would lead us to question whether energy is required during the *repletion* period since such ATP-dependent processes would almost certainly be temperature sensitive. It would therefore seem more likely that the ATP and creatine phosphate degradation observed during calcium repletion is not supportive of the tissue damage but results from it (possible non-metabolic acid hydrolysis arising as a result of calcium-induced acidification[11] of the cytoplasm).

The second important fact to emerge from the hypothermia studies is that hypothermia *during* the calcium-free or anoxic phase is able to modify repletion damage. Furthermore, the characteristics of this modification (Figures 6 and 7) are very striking and identical for both the oxygen and the calcium paradox. In both instances there is a very sharp threshold in the range 33 °C–36 °C. If hearts are subjected to anoxic or calcium-free perfusion at or above this very narrow threshold then oxygen or calcium repletion induces major damage, whereas below the threshold little or no damage results. This sudden transition is suggestive of a specific temperature-dependent change in cellular or subcellular membrane fluidity. The function of membrane protein carrier systems or receptors can be influenced by the membrane–lipoprotein complex in which these systems are located.[1] The physical state of the lipids in these membrane complexes is known to be very temperature sensitive such that they can exist in either a 'gel' or a 'crystalline' form. The transition between these forms, or 'melting' may occur over a very narrow range which is characteristic for each lipoprotein complex. Holland and Olsen[12] have suggested that the abolition of the calcium paradox at 4 °C may be explained by the sarcolemma existing in a more crystalline form at low temperatures, under which condition the membrane-bound lipoprotein system responsible for the calcium channel is less able to undergo the conformational changes associated with calcium transport. We would extend the hypothesis put forward by Holland and Olsen[12] and propose that the calcium and oxygen paradoxes are dependent upon the redistribution of calcium during the calcium-free or oxygen-free period, and that this process is energy related. Thus energy is required for the loss of cytoplasmic calcium during calcium-free perfusion whereas it is the absence of energy during anoxia which permits cytoplasmic calcium accumulation. In both cases, inhibitors of ATP production or hypothermia affect the phenomena.[11] We would further suggest that these processes are mediated by membrane calcium transporting systems, the activity of which can be greatly influenced by membrane fluidity. The fluidity characteristics of these membranes are very temperature sensitive and change markedly over a very narrow temperature range. Since the profiles for these temperature-dependent transitions for the oxygen and calcium paradoxes are essentially superimposable then very similar calcium transporting systems must be responsible for predisposing the cell to damage from both calcium and oxygen depletion.

While the primary lesion which predisposes the cell to sudden tissue damage in both the calcium and the oxygen paradox is likely to be a relatively small, relatively slow redistribution of calcium, either from or to the cytoplasm, the primary agent responsible for the repletion damage is likely to be a massive and rapid transmembrane calcium flux. While both phenomena depend upon these fluxes, there may be fine differences in their nature. In the calcium paradox the major event is likely to be the passive, non-energy-dependent movement of calcium down a concentration gradient from the extracellular space to the cytoplasm and the sudden ultrastructural damage which occurs is probably mechanical in nature resulting from sudden, calcium-induced contracture coupled with explosive cell swelling. In contrast, the primary event in the oxygen paradox is likely to be the oxygen-induced re-energization of the electron transport chain which could trigger uncontrolled mitochondrial calcium uptake from the cytoplasm. This process is passive, does not require energy and cannot be blocked by hypothermia or inhibitors of ATP production but can be blocked by compounds which prevent reoxidation of the reduced electron transport chain intermediates (D. J. Hearse and S. M. Humphrey, unpublished results. Linked to the calcium uptake there may be major proton efflux and consequent cytoplasmic acidification which may contribute to cytoplasmic damage.

In conclusion, we would propose that the calcium and oxygen paradoxes are closely related and are both dependent upon critical changes which occur in cellular membranes and which are linked to calcium transport and the maintenance of calcium homeostasis. The final elucidation of these phenomena will undoubtedly require detailed investigations of the molecular response of membranes to various injurious factors.

REFERENCES

1. Bashford, C. L., Harrison, S. J., Radda, G. K., and Mehdi, Q. The relation between lipid mobility and the specific hormone binding of thyroid membranes. *Biochemical Journal*, **146,** 473–99 (1975).
2. Boink, A. B. T. J., Ruigrok, T. J. C., Maas, A. H. J., and Zimmerman, A. N. E. Changes in high energy phosphate compounds of isolated rat hearts during calcium-free perfusion and reperfusion with calcium. *Journal of Molecular and Cellular Cardiology*, **9,** 973–9 (1976).
3. Feuvray, D. and De Leiris, J. Ultrastructural modifications induced by reoxygenation in the anoxic isolated rat heart perfused with exogenous substrate. *Journal of Molecular and Cellular Cardiology*, **7,** 307–14 (1975).
4. Ganote, C. E., Seabra-Gomes, R., Nayler, W. G., and Jennings, R. B. Irreversible myocardial injury in anoxic perfused rat hearts. *American Journal of Pathology*, **80,** 419–50 (1975).
5. Hearse, D. J., Humphrey, S. M., and Chain, E. B. Abrupt reoxygenation of the anoxic potassium arrested perfused rat heart: a study of myocardial enzyme release. *Journal of Molecular and Cellular Cardiology*, **5,** 395–407 (1973).
6. Hearse, D. J., Humphrey, S. M., Nayler, W. G., Slade, A. and Border, D. Ultrastructural damage associated with reoxygenation of the anoxic myocardium. *Journal of Molecular and Cellular Cardiology*, **7,** 315–24 (1975).
7. Hearse, D. J. and Humphrey, S. M. Enzyme release during myocardial anoxia: a study of metabolic protection. *Journal of Molecular and Cellular Cardiology*, **7,** 463–82 (1975).
8. Hearse, D. J., Humphrey, S. M., and Garlick, P. B. Species variation in myocardial anoxic enzyme release, glucose protection and reoxygenation damage. *Journal of Molecular and Cellular Cardiology*, **8,** 329–39 (1976).
9. Hearse, D. J., Humphrey, S. M., Feuvray, D., and De Leiris, J. A biochemical and ultrastructural study of the species variation in myocardial cell damage. *Journal of Molecular and Cellular Cardiology*, **8,** 659–778 (1976).
10. Hearse, D. J., Stewart, D. A., and Braimbridge, M. V. Cellular protection during myocardial ischaemia, the development and characterization of a procedure for the induction of reversible ischaemic arrest. *Circulation*, **54,** 193–202 (1976).
11. Hearse, D. J. Reperfusion of the ischaemic myocardium. *Journal of Molecular and Cellular Cardiology*, **9,** 605–16 (1977).
12. Holland, C. E. and Olsen, R. E. Prevention by hypothermia of paradoxical calcium necrosis in cardiac muscle. *Journal of Molecular and Cellular Cardiology*, **7,** 917–28 (1975).
13. Jennings, R. B. and Ganote, C. E. Structural changes in myocardium during acute ischaemia. *Circulation Research Supplement III*, **34** and **35,** 156–72 (1974).

14. Jennings, R. B. and Ganote, C. E. Mitochondrial structure and function in acute myocardial ischaemic injury. *Circulation Research Supplement 1*, **38**, 80–9 (1976).
15. Krebs, H. A. and Henseleit, K. Untersuchungen über die Harnstoffbildung im Tierkorper. *Hoppe-Seyler's Zeitschrift fur Physiologische Chemie*, **210**, 33–66 (1932).
16. Langendorff, O. Untersuchungen am uberlebenden Säugertierherzen. *Pflügers Archiv für die Gesamte Physiologie des Menschen und der Tiere*, **61**, 291–332 (1895).
17. Mollenhaur, H. H. Plastic embedding mixtures for use in electron microscopy. *Stain Technology*, **39**, 11 (1964).
18. Muir, A. R. A calcium induced contracture of cardiac muscle cells. *Journal of Anatomy*, **102**, 148 (1968).
19. Nayler, W. G. The cardiac cell. In *Contraction and Relaxation of the Myocardium* (Ed. Nayler, W. G.) Academic Press, London, 1975, pp. 1–28.
20. Neely, J. R., Liebermeister, H., Battersby, D. J., and Morgan, H. E. Effect of pressure development on oxygen consumption by the isolated rat heart. *American Journal of Physiology*, **212**, 804–14 (1967).
21. Reynolds, E. S. The use of lead citrate at high pH as an electron opaque stain in electron microscopy. *Journal of Cell Biology*, **17**, 208–12 (1963).
22. Ruigrok, T. J. C., Burgersdijk, F. J. A., and Zimmerman, A. N. E. The calcium paradox: a reaffirmation. *European Journal of Cardiology*, **3**, 59–63 (1975).
23. Ruigrok, T. J. C., Boink, A. B. T. J., and Zimmerman, A. N. E. Influence of ATP, or oxygen plus substrate on the occurrence of the calcium paradox. In *Recent Advances in Studies of Cardiac Structure and Metabolism*, Vol. 11, University Park Press, Baltimore; 1978, p. 565.
24. Shen, A. C. and Jennings, R. B. Myocardial calcium and magnesium in acute ischaemic injury. *American Journal of Pathology*, **67**, 417–40 (1972).
25. Shen, A. C. and Jennings, R. B. Kinetics of calcium accumulation in acute myocardial ischaemic injury. *American Journal of Pathology*, **67**, 441–52 (1972).
26. Umbreit, W. W., Burris, R. H., and Stauffer, J. F. In *Manometric Techniques*, Burgess, Minneapolis, 1964, p. 132.
27. Yates, J. C. and Dhalla, N. S. Structural and functional changes associated with failure and recovery of hearts after perfusion with Ca^{2+}-free medium. *Journal of Molecular and Cellular Cardiology*, **7**, 91–103 (1975).
28. Zimmerman, A. N. E. and Hulsmann, W. C. Paradoxical influence of calcium ions on the permeability of the cell membranes in the isolated rat heart. *Nature*, **211**, 646–7 (1966).
29. Zimmerman, A. N. E., Daems, W., Hulsmann, W. C., Snyder, J., Wisse, E. and Durrer, D. Morphological changes of heart muscle caused by successive perfusion with calcium-free and calcium containing solutions (calcium paradox). *Cardiovascular Research*, **1**, 201–9 (1967).

CHAPTER 19

Morphological correlates of myocardial enzyme release

J. de Leiris and D. Feuvray

INTRODUCTION	445
ANOXIA	446
Anoxia without Reoxygenation	446
Anoxia Plus Reoxygenation	447
Glucose Protection in Anoxia	448
Pharmacological Protection against Anoxia	449
Verapamil	449
Reserpine	449
Species Differences in Anoxic Damage	450
ISCHAEMIA	451
CALCIUM DEPRIVATION	454
CONCLUDING COMMENTS	455
REFERENCES	456

INTRODUCTION

The leakage of myocardial enzymes occurs in a number of pathological conditions.[20] Most of these conditions have been shown to be associated with modifications in the morphology of myocardial cells. In this chapter we attempt firstly, to describe the time course and nature of the ultrastructural changes occurring in pathological conditions in which cellular enzymes are released from the heart; and secondly, to show to what extent such enzyme release can be correlated with detectable ultrastructural changes.

Most of the observations presented here were made in anoxia, with or without reoxygenation, in ischaemia, with or without reflow, after metabolic interventions during anoxia or ischaemia, and after calcium deprivation. In all experimental situations, ultrastructural studies lead to the major problem of tissue fixation (see Chapter 20). Using a Langendorff isolated perfused heart, Hatt and Moravec[19] have studied the relationship between the ultrastructure of the mitochondria and the rapidity of fixative penetration into the tissue. They observed that when fixative perfused through the coronary arteries reaches the heart without any delay, the mitochondria show a dense matrix, numerous intracristal bodies and regular cristae. If, however, the fixative reaches the heart after a 1 minute delay, the mitochondria show a less opaque matrix. When the delay exceeds 3 minutes, both the mitochondria and the sarcoplasmic reticulum show an hypoxic-like swelling. Therefore, intracoronary perfusion of the heart with fixative, *in situ* or *in vitro*, is the only reliable technique which should be used to study mitochondrial and cellular ultrastructure.[14]

ANOXIA

Anoxia without Reoxygenation

In isolated hearts, in which the oxygen tension of the perfusion fluid is reduced to cause hypoxia or anoxia, there is a marked effect upon myocardial function[9] and metabolism.[8,10,56,64,67,68,76] As described in Chapter 1 of this book, creatine phosphate (CP) and adenosine triphosphate (ATP) levels decrease, whereas adenosine diphosphate (ADP) and adenosine monophosphate (AMP) content increases.[8,21,42,57,76,77] The availability of adequate supplies of ATP is important for both contractile activity and the maintenance of cellular integrity and when ATP drops below a critical level the release of enzymes occurs[22,73] and this is associated with the onset of ultrastructural modifications. In view of the normal magnitude of enzyme release it has been generally thought that the leakage most probably results from some specific ultrastructural damage and not from some simple change in membrane permeability characteristics.[12]

As shown in Plate 1, isolated rat hearts perfused with oxygenated media containing glucose, maintain their ultrastructural integrity over a long period of time.[12,15] Myofibrils are in parallel array, glycogen is widely distributed throughout the cytosol, and mitochondria appear dense and normal in size and configuration. Nuclear chromatin is distributed evenly throughout the nucleus, no contraction bands are visible, Z lines appear normal, transverse tubules are rounded and the sarcoplasmic reticulum resembles flattened sacs.

In spontaneously beating rat heart, perfused without exogenous substrate, cardiac contraction briefly increases after the onset of the anoxic perfusion and then, gradually decreases in both force and frequency and ceases after approximately 5 minutes.[15] Appreciable creatine kinase (CK) release does not occur until approximately 30 minutes after the onset of anoxia. During this period some ultrastructural changes become apparent, mitochondria appear slightly swollen and are less dense than those of control; in addition, normal matrix granules are less evident. Occasionally, some cytoplasmic swelling may be observed and some vacuoles may be detected in the cytoplasm and the number of pinocytotic vesicles is reduced from that seen in control hearts. Between 30 and 55 minutes of anoxia, the number of detectably damaged cells increases as a linear function of time. After 60

Plate 1. Isolated rat heart perfused for 60 minutes with oxygenated buffer. Myofibrillar pattern appears normal with distinct A, I, and M bands, as well as dense mitochondria, sarcoplasmic reticulum, T tubules, and sarcolemma. Numerous glycogen granules are visible

Plate 2. Isolated rat heart after 150 minutes of anoxic, glucose-free perfusion. Note the intracellular oedema, the disintegration of some myofibrils and tubular system. The mitochondria are slightly swollen

minutes of anoxic perfusion, sustained release of CK occurs and specific ultrastructural changes are evident. Nuclear chromatin is clumped, mitochondria become more swollen, glycogen stores are depleted, and the sarcoplasmic reticulum becomes dilated and forms small vesicular structures. In some cells, Z lines appear thickened and contraction bands are visible. The transverse tubular system is often markedly dilated.[15]

In isolated, potassium-arrested rat hearts, perfused without exogenous substrate, anoxia induces a biphasic enzyme release.[23] The first phase of enzyme release usually occurs during the first 10–60 minutes of anoxia, reaching a maximum after approximately 30 minutes of anoxia. This small phase represents 2–5% of the total enzyme released over an 8 hour period of anoxia.[23] The remaining enzyme release occurs during a secondary phase which starts after about 60–100 minutes of anoxic perfusion and lasts for several hours. At the peak of phase 1 enzyme release ultrastructural integrity appears to be well maintained[13] with the exception of a very slight contraction of some sarcomeres which may be observed. After 100–150 minutes of anoxia, i.e. when the secondary phase of enzyme release has commenced, more marked ultrastructural damage is observed (see Plate 2). Myofibrils are contracted with indistinct I bands and the Z lines may be distorted. The sarcoplasmic reticulum appears enlarged or vesiculated, glycogen and intramitochondrial bodies disappear and mitochondria are slightly swollen but their cristae remain well defined. Some tissue oedema is apparent but there is no disruption of the plasma membrane.[13] After 300 minutes of anoxia (see Plate 3), extensive ultrastructural damage occurs; there is extensive intracellular oedema, no glycogen granules are apparent, contraction bands are evident as are distorted Z lines and disrupted myofibrils with reduced actin density. Mitochondria are swollen and distorted and may be aggregated thus losing their characteristic alignment. Mitochondrial cristae may be vesiculated, there is aggregation of nuclear chromatin, vesiculation of the sarcoplasmic reticulum and possibly also some distortion of the intercalated disc.[24] Associated with this extensive morphological damage, there is a massive release of enzymes.[24]

Thus in the anoxic isolated rat heart perfused without exogenous substrate, both ultrastructural changes and enzyme release increase with the duration of the anoxic perfusion.

Anoxia Plus Reoxygenation

In potassium-arrested rat hearts, perfused without exogenous substrate, Hearse et al.[23] showed that reoxygenation of the heart during the first-phase enzyme release had a marginal effect upon the rate of release, while reoxygenation during the second phase resulted in a sudden and very large (100- to 200-fold) increase in myocardial enzyme release. Morphological studies[13] revealed (Plate 4) that reoxygenation after 30 minutes of anoxia had no effect upon cellular ultrastructure. In contrast in hearts which are reoxygenated during the second phase of enzyme release (for example after 100–150 minutes of anoxia), there was a sudden and major exacerbation of the tissue damage (Plate 5) including extensive intracellular oedema, disruption of the cell membrane, and considerable and characteristic mitochondrial damage (the mitochondria appearing distorted or broken with very irregular outlines, vesiculation and disruption of cristae). In addition myofibrils are severely disrupted.

In spontaneously beating rat hearts (in which anoxia consistently induces earlier ultrastructural alterations than in potassium-arrested hearts), Ganote et al.[15,16] have reported that reoxygenation as soon as 50 minutes after the onset of anoxia greatly accelerates the rate of CK release (up to 100-fold) and also increases the number of cells exhibiting severe ultrastructural damage. The cells are markedly swollen and most have disrupted membranes. After a long period of reoxygenation, these severely damaged cells show additional progressive degenerative changes including the formation of amorphous densities in mitochondria (as well as granular dense bodies of the type associated with calcium accumulation) and intense contraction bands. These changes would be characteristic of irreversible injury.[15]

In general it would therefore appear that the changes observed in enzyme release during anoxia and reoxygenation parallel in their time course and magnitude the ultrastructural changes revealed by electron microscopy.

Glucose Protection in Anoxia

In isolated, potassium-arrested rat hearts, the inclusion of glucose (11 mmol l.$^{-1}$) in the perfusate during anoxia has striking metabolic[21] and ultrastructural protective effects. After 300 minutes of anoxic perfusion, enzyme release is reduced by 95% and myocardial morphology remains well preserved.[24] As shown in Plate 6, mitochondria are electron dense, well aligned, and the cristae appear intact. The muscle fibres are relaxed, the Z lines are not distorted and there is little evidence of intracellular oedema or damage to the cell membrane, sarcoplasmic reticulum, or nucleus.[24] Despite the severe anoxia, intracellular glycogen granules are maintained (an observation which has been confirmed biochemically) but autophagic vacuoles can be detected.

The protective effect of exogenous glucose (see Chapter 21) most likely depends upon the ability of glucose to act as a substrate for glycolytic ATP production. While glucose may also exert an osmotic effect this would appear to be relatively unimportant, since mannitol at the same concentration and under similar experimental conditions, fails to protect the hearts against anoxic damage.[15] However it should be noted that hyperosmolar

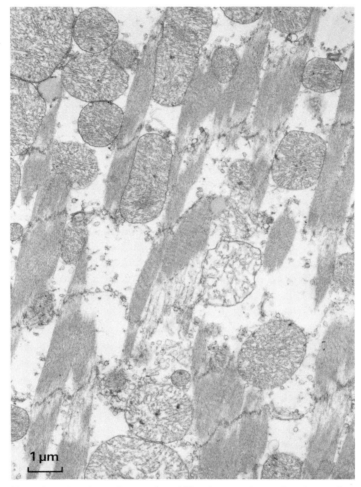

Plate 3. Isolated rat heart after 300 minutes of anoxic, glucose-free perfusion. Cells are markedly swollen, with myofibrillar degeneration and completely disorganized tubular system. The mitochondria are swollen

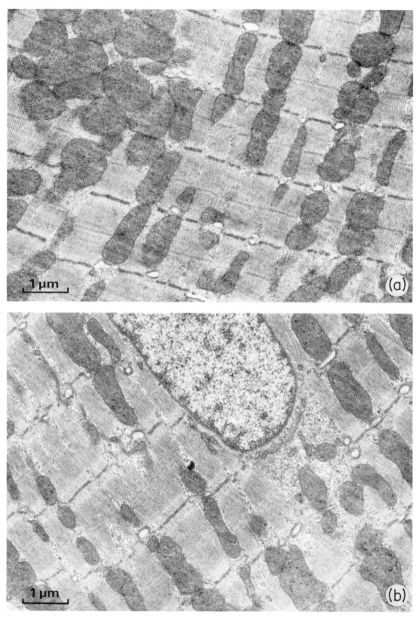

Plate 4. Isolated rat heart (a) after anoxia (30 minutes) and reoxygenation (2 minutes); (b) after anoxia (30 minutes). [Reproduced by permission of Academic Press from *Journal of Molecular and Cellular Cardiology*, **7,** 307–14 (1975).]

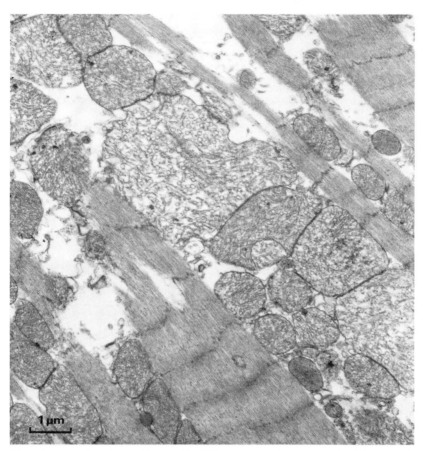

Plate 5. Isolated rat heart after anoxia (150 minutes) and reoxygenation (2 minutes). During peak enzyme release, very large distorted mitochondria with vesiculated cristae, and disrupted membranes are observed in addition to the cellular damage observed after anoxia alone

Plate 6. Isolated rat heart after 300 minutes of anoxic perfusion with buffer containing glucose (11 mmol l.$^{-1}$). This plate shows well-preserved cellular ultrastructure with glycogen granules. [Reproduced by permission of Academic Press from *Journal of Molecular and Cellular Cardiology*, **8,** 759–78 (1976).]

mannitol has been reported[7] to significantly reduce post-anoxic ultrastructural changes in the isolated rat heart.

Once again, changes in enzyme release and ultrastructural damage would appear to parallel each other closely.

Pharmacological Protection against Anoxia

Verapamil

The inclusion of verapamil (0.5–1.0 mg l.$^{-1}$) in anoxic, substrate-free perfusion fluid, can result (see Plate 7) in a marked preservation of the fine morphology of the isolated rat heart and associated with this is a decreased rate (3-fold) of enzyme leakage.[59] Thus after 60 minutes of anoxia in the presence of verapamil, relatively few cells have disrupted membranes, the percentage of damaged mitochondria is significantly reduced, and fewer contraction bands are apparent. It should be noted however that although fewer mitochondria (28% compared with 84%) are swollen and vacuolated, there is a significant increase in the number of intramitochondrial bodies.

Verapamil has been shown to block slow calcium channels,[43] furthermore in the presence of verapamil it has been shown that the calcium-accumulating activity of isolated cardiac mitochondria remains within normal limits even after 60 minutes of hypoxia.[58] Since the deterioration of mitochondrial function can be associated with an increased rate of calcium accumulation, it has been suggested that the protective effect of verapamil may be dependent upon the ability of this drug to prevent cardiac mitochondria from accumulating calcium at an excessively rapid rate.[58] It has also been reported that when verapamil is added to a perfusion fluid at the onset of anoxia, then the rate at which the endogenous stores of ATP and CP are depleted can be significantly reduced.[59] The protective effect of verapamil could therefore also be explained in terms of its sparing effect upon the rate of depletion of endogenous energy reserves (so that sufficient ATP remains available for maintenance of membrane integrity).

Reserpine

It has been suggested that the release of endogenous catecholamines may play an important role in the development of anoxia-induced myocardial damage.[70] In this connection, reserpine pre-treatment (which depletes cardiac catecholamines by up to 90%) has been shown to markedly reduce the enzyme release which can be induced by a number of stresses.[46,70] Thus potassium-arrested perfused hearts from reserpine-pre-treated rats, show a reduction of tissue damage after 100 minutes of anoxia, enzyme release is reduced by more than 50%, CP and ATP levels are better preserved and glycogen reserves remain (this latter observation has been confirmed both biochemically and morphologically). In addition, it has been shown that these hearts maintain active anaerobic glycolysis with a high level of lactate production.[47] As shown in Plate 8, cellular morphology remains well preserved, with mitochondria, myofibrils, intercalated discs, and tubular systems, all appearing normal; however membrane-limited autophagic vacuoles are apparent. Reserpine pre-treatment also reduces the exacerbation of enzyme release and the sudden increase in ultrastructural damage which occurs at the time of the reoxygenation.[47]

Species Differences in Anoxic Damage

Species differences in resistance to anoxia have frequently been reported. For example the turtle heart can produce sufficient ATP by anaerobic glycolysis to support normal contractile activity during several hours of anoxia[5] whereas the anoxic rat heart is unable to contract for more than a few minutes. Species differences (rat, mouse, guinea-pig, and rabbit) in susceptibility to anoxic damage, reoxygenation damage, and glucose protection have been investigated in isolated, substrate-free perfused hearts by Hearse et al.[24] In these studies, myocardial damage was assessed by enzyme release and by ultrastructural changes and these were related to changes in myocardial glycogen, CP, ATP, and lactate. From these and other studies, these authors have suggested that species differences in the speed of onset, the magnitude of early tissue damage, and the extent to which reoxygenation can exacerbate damage, may be related to characteristics of cardiac cell membranes. In contrast to the rat and the mouse, the rabbit and the guinea-pig appear to be more susceptible to the onset of early tissue damage during anoxic arrest. In contrast, reoxygenation (after extended anoxia) induces a great exacerbation of enzyme release and ultrastructural damage in the rat and mouse (see Plates 5 and 9), whereas it has little effect in the guinea-pig and rabbit. These latter two species exhibit an earlier onset of anoxic damage and thus may be less susceptible to damage induced by the later readmission of oxygen.

Major species differences also exist in the extent to which glucose can protect the anoxic myocardium and thereby reduce tissue damage and enzyme release. In the rat and guinea-pig, glucose consistently reduces enzyme release and maintains cardiac morphology after 300 minutes of anoxic perfusion (see Plates 6 and 10). By contrast, glucose affords little protection to the mouse myocardium and after 300 minutes of anoxia there is extensive ultrastructural damage and enzyme release (see Plate 11). In the rabbit, the response to anoxia and glucose protection is complex and varies between individuals (see Plate 12) but within any single individual there is a close concordance between ultrastructural damage, enzyme release, and tissue protection.

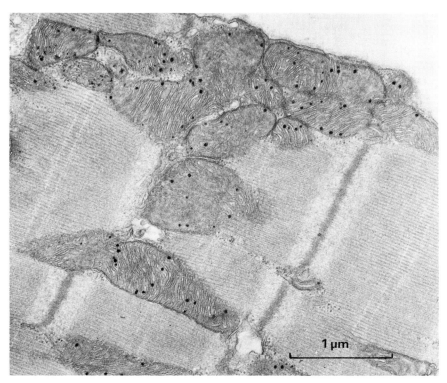

Plate 7. Isolated rat heart after 60 minutes of anoxic perfusion in the presence of verapamil. [Reproduced by permission of Elsevier North Holland from *Journal of Molecular Medicine*, **2**, 299–308 (1977).]

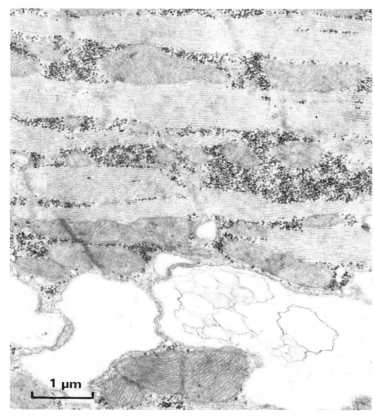

Plate 8. Anoxically perfused (100 minutes) heart muscle from reserpine-pretreated rat. Cells are packed with many glycogen granules. Note the presence of large cytoplasmic vacuoles

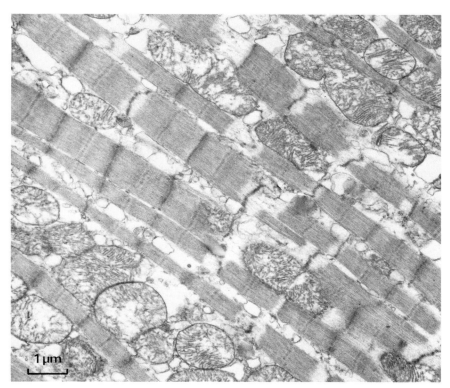

Plate 9. Isolated mouse heart after anoxia (150 minutes) and reoxygenation (2 minutes). Note the swollen mitochondria with clear matrices and severe disruption of mitochondrial membranes and cristae

Plate 10. Isolated guinea-pig heart after 300 minutes of anoxic perfusion with buffer containing glucose (11 mmol l.$^{-1}$). Cellular ultrastructure appears well preserved with the presence of glycogen granules. [Reproduced by permission of Academic Press from *Journal of Molecular and Cellular Cardiology*, **8,** 759–78 (1976).]

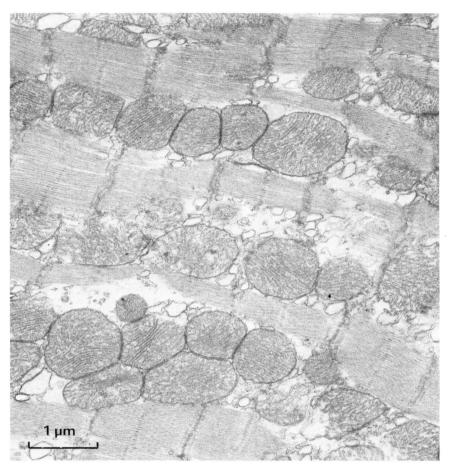

Plate 11. Isolated mouse heart after 300 minutes of anoxic perfusion with buffer containing glucose (11 mmol l.$^{-1}$). Moderate intracellular oedema, contracted myofibrils, vesiculated sarcoplasmic reticulum and mild swelling of mitochondria are observed. No glycogen granules can be seen. [Reproduced by permission of Academic Press from *Journal of Molecular and Cellular Cardiology*, **8,** 759–78 (1976).]

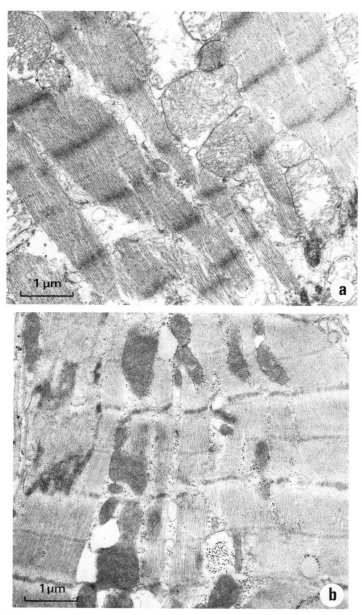

Plate 12. Isolated rabbit heart after 300 minutes of anoxic perfusion with buffer containing glucose (11 mmol l.$^{-1}$): (a) subgroup 1: extensive ultrastructural damage is observed; (b) subgroup 2: well-preserved cellular ultrastructure. Glycogen granules are present. [Reproduced by permission of Academic Press from *Journal of Molecular and Cellular Cardiology*, **8**, 759–78 (1976).]

ISCHAEMIA

In ischaemia, the low myocardial tissue oxygen tension generally results from a reduction in coronary flow. Associated with this reduction in oxygen and substrate supply is a corresponding reduction in the washout of metabolites such as lactate and protons.[15,28] As with anoxia, myocardial ischaemia is characterized by a rapid shift of cellular metabolism from the aerobic to the much less efficient anaerobic form. Cellular energy levels decrease and are insufficient to support normal contractile activity or other specialized cellular functions.[30] Ischaemia and anoxia or hypoxia all lead to similar changes in the energy state of the tissue.[18,27,61,76] The levels of ATP and CP decrease and the levels of ADP and AMP increase. However, as discussed in other chapters some of the metabolic changes associated with ischaemia are quite different to those observed in anoxia or hypoxia.[3,4,38,63,69]

The most widely used experimental model for studying morphological changes during myocardial ischaemia involves the occlusion of one or more branches of the coronary arteries followed by fixation of the tissue and examination with the electron microscope.[25,27,28,29,33] Very many *in vivo* experiments have been carried out by Jennings and coworkers (see Chapter 2) who have extensively studied the fine structural changes produced in the posterior papillary dog muscle by permanent or temporary occlusion of the circumflex coronary artery.[34] Morphological studies have shown that cellular damage induced by ischaemia depends on the duration of the ischaemic period. Once occlusion is maintained for more than 20 minutes, restoration of blood flow does not consistently reverse cell damage. After 60 minutes of permanent ischaemia the myofibrils appear markedly relaxed. There is an absence of glycogen, margination of nuclear chromatin, and sarcolemmal scalopping. Tubular systems are severely disorganized, the mitochondria are swollen and show disrupted cristae, abnormally clear matrices and in addition, contain dense amorphous granules of approximately 80 nm diameter.[1,25]

Restoration of blood flow in severely injured cells leads to the development of more extensive structural damage. These changes would parallel rapidly developing abnormalities of tissue electrolyte distribution[32] and increased enzyme leakage. The increased enzyme leakage may be due in part to the decreased time which is available for local degradation of enzyme in the myocardium because reperfusion increases the rate of transport of enzyme to the systemic circulation.[55] Examination of this reperfused tissue reveals widespread contraction bands, disruption of the sarcolemmal membrane and intracellular organelles, translocation and disorganization of mitochondria and the appearance of large, dense, intramitochondrial calcium phosphate granules. This latter observation indicates that alterations of cell membrane permeability are sufficient to allow calcium and phosphate ions to enter the cell in large quantities. These ions may then become available for active accumulation by the mitochondria of the damaged cells.[25,29,30]

After more than 60–120 minutes of coronary occlusion, reperfusion of the ischaemic tissue either does not occur, or is greatly impeded. This phenomenon has been called 'no-reflow' and is associated with extensive capillary damage and myocardial cell swelling.[40,66] Elevation of osmolality by mannitol results in improved function and increased coronary flow probably by preventing or reducing cell swelling.[41,66] Dihydroxyacetone and allopurinol have also been shown to improve reperfusion of ischaemic rat hearts by reducing myocardial cell swelling and tissue oedema.[11]

In experimental myocardial ischaemia *in situ*, vascular reflexes, hormone changes, and platelet aggregation are some of the factors which may, directly or indirectly, influence myocardial metabolism and ultrastructural damage. These factors are difficult to control in the intact animal.[75] In relation to this latter problem, various isolated perfused heart models offer a means of controlling or eliminating a number of variables which may influence the progression of ischaemic damage. Small animal models of acute myocardial ischaemia (such as isolated atrially perfused rat hearts[60] with left coronary artery ligation[37,45]) have several practical advantages. Firstly, it is relatively easy to study large infarcts because sustained ventricular fibrillation does not develop (as would be the case with a lesion of comparable size in a larger animal such as the dog or the baboon). Secondly, it is easy to fix isolated rat hearts by perfusing the fixative solution through the coronary vessels (a procedure which has been shown[14] to permit the best fixation for cardiac tissue). The possibility that the uneven distribution of coronary flow, associated with regional ischaemia, may lead to inhomogeneous fixation must be considered. However, since there is evidence[37] that there may be substantial (as much as 20%) collateral flow even in the centre of the infarct, fixation problems may be minimal.

In isolated working rat hearts, perfused with glucose-containing media, extensive ultrastructural changes and enzyme release are observed 45 minutes after left coronary artery ligation.[45,48] In the subendocardial part of the ischaemic area (see Plate 13) major oedema develops, clear interfibrillar spaces and lysis of myofilaments are observed and the actin filaments are pulled apart from the *fascia adherens* of the intercalated discs. Glycogen disappears and the longitudinal tubular system appears to be disrupted and vesiculated. Nuclear chromatin is aggregated on the periphery of the nucleus, mitochondria are distorted and swollen and most of them have disrupted membranes. No intramitochondrial granules can be seen and cristae are widely separated or disrupted.[45]

In Neely's whole heart ischaemia preparation,[61] the dissolution of gap junctions has been reported to occur as a function of the time after

Plate 13. Myocardium from ischaemic isolated rat heart perfused with glucose. Major oedema has developed with unduly marked interfibrillar spaces and disorganized tubular system. Mitochondria appear swollen with clear matrices and disrupted cristae. Glycogen is absent. [Reproduced by permission of Academic Press from *Journal of Molecular and Cellular Cardiology*, **9,** 365–73 (1977).

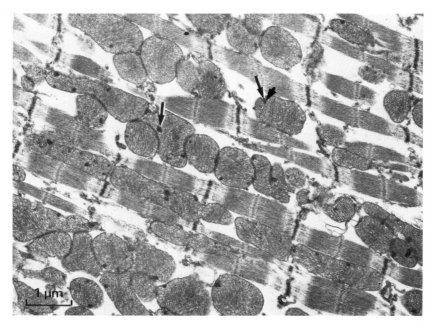

Plate 14. Myocardium from ischaemic isolated rat heart, perfused with palmitate (1.5 mmol l.$^{-1}$) bound to albumin (0.45 mmol l.$^{-1}$). Similar alterations as in Plate 13 can be observed, except that we can note the presence of amorphous densities inside mitochondria (arrows). [Reproduced by permission of Academic Press from *Journal of Molecular and Cellular Cardiology*, **9**, 365–73 (1977).

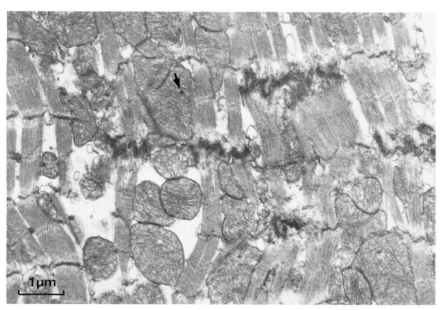

Plate 15. Myocardium from ischaemic isolated rat heart perfused with palmitate–albumin plus glucose (11 mmol l.$^{-1}$) and insulin (2 miu ml.$^{-1}$). The number of mitochondria bearing amorphous densities is markedly reduced by the addition of glucose and insulin (arrow), expressing the beneficial effect of the addition of these substrates in this model of ischaemia. [Reproduced by permission of Academic Press from *Journal of Molecular and Cellular Cardiology*, **9,** 365–73 (1977).

Plate 16. Myocardium from isolated rat heart after 30 minutes of ischaemic cardiac arrest and 5 minutes of reperfusion. Before ischaemic cardiac arrest, coronary bed was infused with St. Thomas' solution (see reference 35) for 2 minutes. Cellular ultrastructure appears well preserved except for the development of a slight intracellular oedema

myocardial failure.[6] This damage has been correlated with the breakdown of primary lysosomes and is thought to be coincident with the onset of irreversible tissue damage.[6] Such structural alterations would add a significant resistance to the nexus and this may cause a decrement in the conduction velocity of the action potential across the myocardial cell. Whether or not tissue damage and membrane dissolution in the ischaemic heart are a result of direct lysosomal action, are a result of high-energy phosphate depletion, or are attributable to some other cause remains open to question.[6,17]

It has been suggested (see Chapter 21) that the outcome of myocardial ischaemia and the extent of tissue damage may be dependent upon the nature of the substrate reaching the ischaemic zone.[62,63] In isolated rat hearts this suggestion has been supported by the observation that the rate of enzyme release occurring after 60–120 minutes of coronary artery ligation is quantitatively dependent upon the nature of the perfusion substrate.[48,50,51] Thus for example, enzyme release is considerably higher when long-chain fatty acids are the substrate instead of glucose.[48] Morphologically it is not possible to distinguish the patterns of damage in glucose or fatty acid perfused hearts during the first 45 minutes of ischaemia. In both instances there is extensive intracellular oedema, alterations to the myofibrils, tubular systems, and nucleus as well as glycogen depletion and mitochondrial swelling.[45] However some differences are apparent, particularly in relation to mitochondrial structure. In fatty acid perfused hearts amorphous electron-dense opacities appear in 50% of the mitochondria while in contrast, such opacities are never observed in glucose or glucose–albumin perfused hearts (see Plate 14). It is of interest that the addition of glucose and/or insulin to fatty acid–albumin perfused hearts (see Plate 15) exerts a marked protective effect which is characterized by a reduction in the amount of enzyme leakage and a considerable reduction in the development of mitochondrial opacities.[45,49] Thus these amorphous intramitochondrial densities (which may correspond with those observed[30,31] in ischaemic dog heart) might act as good markers of the degree of cellular damage and may be correlated further with the extent of enzyme leakage.[48–51]

The beneficial effects of glucose or glucose–insulin–potassium[64,65,74] in myocardial ischaemia has already been discussed here and elsewhere in this book. In dog hearts with experimental myocardial infarction which had been treated with glucose–insulin–potassium infusion, 36% of the samples taken from the centre of the infarcted area were found to have normal ultrastructure. In contrast, in untreated dogs, structural abnormalities were observed in all samples. These findings were complemented by the observation of considerably less tissue depletion of creatine kinase in the treated group.[54]

Studies with hyaluronidase also reinforce the association between ultrastructural damage and enzyme release. Hyaluronidase is thought to increase diffusion through the extracellular space and as such may facilitate the delivery of substrate to ischaemic cells. This drug has been shown to decrease experimental myocardial infarct size, improve collateral flow, and reduce the depression of myocardial creatine kinase activity.[53] This protective effect has been associated with some degree of ultrastructural protection.[39] Moreover hyaluronidase has been shown to protect the microvasculature which may contribute to an improved collateral flow.[39,53]

Currently, a number of cardioplegic infusates are in clinical use for myocardial protection during ischaemic arrest (see reference 35). One of these solutions (St. Thomas' solution) is based on the additive protective effects of moderately raised extracellular concentrations of potassium and magnesium and the inclusion of various protective agents.[35] In isolated rat hearts this solution has been shown to reduce cellular damage during ischaemia and improve functional recovery at the time of reperfusion. In these studies (see Plate 16) improvements in cardiac function and ultrastructural preservation have been closely related to the extent of myocardial enzyme leakage.[36]

CALCIUM DEPRIVATION

Calcium-free perfusion of the isolated rat heart, or of other species, causes a rapid cessation of contractile activity with electromechanical dissociation.[52] If the duration of calcium-free perfusion exceeds 3 minutes, then upon the repletion of calcium there is a sudden onset of irreversible cellular necrosis[2,80] and large amounts of enzymes and other intracellular components leak from the heart.[44,80] This phenomenon (see Chapters 17 and 18) has been called the 'calcium paradox'[80] and has been shown to be associated with the sudden onset of extensive ultrastructural damage including severe contracture and mitochondrial alterations.[79] Thus electron microscopic examination of hearts fixed after 1–3 minutes of calcium-free perfusion does not reveal any major ultrastructural damage.[78] Calcium repletion after this 1–3 minute period results in the marked separation of intercalated discs but there is no major damage.[78] Calcium-free perfusions of 4–10 minutes duration induce various degrees of cell damage including the separation of intercalated discs[78] and some dilatation of the tubular systems.[26] Calcium repletion after these periods results in extensive and irreversible ultrastructural damage including the complete separation of myocardial cells, hypercontraction and disruption of contractile elements, mitochondrial swelling, and disorganization of mitochondrial cristae.[78] These striking changes are matched by equally striking changes in enzyme release.

These ultrastructural and enzymatic changes can be modified by, for example, lowering the temperature during calcium-free perfusion (see Chapter 18). This results in minimal ultrastructural damage and although some swelling of the tubular systems occurs, the myofibrillar structure is generally well preserved, glycogen granules are abundant and intercalated discs appear normal.[26] These hypothermia-induced protective effects include a major reduction in enzyme leakage.[44]

CONCLUDING COMMENTS

In isolated hearts, various experimental conditions such as reoxygenation after extended anoxia, reperfusion after ischaemia, or calcium repletion after a period of calcium depletion, all induce extensive enzyme leakage from myocardial cells. In each instance major ultrastructural changes are also observed. In general, the higher the enzyme release, the greater is the ultrastructural damage. Metabolic or pharmacological interventions which reduce enzyme release also preserve myocardial morphology.

It should be stressed however that the interpretation of myocardial cell damage as revealed by electron microscopy, and the relating of this information to enzyme release characteristics may be limited by a number of considerations.

Firstly, it is not necessarily possible to visualize all membrane changes which may permit cellular enzyme leakage.

Secondly, it is difficult to establish a strict relationship between the development of morphological injury and the time course of enzyme release. Furthermore it is necessary to ask, whether enzyme leakage may represent a major loss of enzyme from a small proportion of irreversibly damaged cells or whether it may represent a moderate loss of enzyme from a large number of cells which may only be reversibly damaged? In this connection it has been suggested that in apparently uniform whole heart anoxia, ultrastructural damage develops earlier in some areas than in others, such that in a mildly damaged area, some cells appear severely damaged while others appear almost normal.[13] Similarly, in ischaemic conditions, it has been suggested that while comparable sequential changes occur in both severely and mildly ischaemic cells, different time scales for the development may exist.[31,72] Thus any attempt to relate enzyme release and tissue damage to ultrastructural change must involve analysis of representative tissue ideally using rigorous morphometric techniques.

A third consideration is that none of the alterations normally encountered in the myocardium can be specifically related to any given pathological condition. Glycogen depletion, tubular system dilatation, mitochondrial changes, alterations of myofibrils, and nuclear changes are a feature of a variety of abnormal conditions and as such must be taken as relatively non-specific signs of myocardial cell injury.[19] Although ultrastructural changes may not identify a specific lesion they may be valuable in the assessment of the progression of that lesion. Cellular damage is generally of a progressive nature and the rate of progression may depend upon the intensity or duration of the abnormal condition. This is well illustrated by the progressive mitochondrial changes which occur with increasing durations of ischaemia or hypoxia.[13,30,54,71]

In conclusion therefore, these limitations of electron microscopy do not allow us to use morphological observations as a single test for the precise assessment of myocardial cell damage and we remain dependent upon a combination of biochemical (such as ATP and CP levels, or serial enzyme determinations), electrophysiological (such as ST segment changes), and morphological assessments.

REFERENCES

1. Armiger, L. C., Gavin, J. B., and Herdson, P. B. Mitochondrial changes in dog myocardium induced by neutral lactate *in vitro*. *Laboratory Investigation*, **31**, 29–33 (1974).
2. Boink, A. B. T. J., Ruigrok, T. J. C., Maas, A. H. J., and Zimmerman, A. N. E. Changes in high-energy phosphate compounds of isolated rat hearts during calcium-free perfusion and reperfusion with calcium. *Journal of Molecular and Cellular Cardiology*, **8**, 973–9 (1976).
3. Braasch, W., Gudbjarnason, S., Puri, P., Ravens, K. G., and Bing, R. J. Early changes in energy metabolism in the myocardium following acute coronary artery occlusion in anesthetized dogs. *Circulation Research*, **23**, 429–38 (1968).
4. Brachfeld, N. and Scheuer, J. Metabolism of glucose by the ischemic dog heart. *American Journal of Physiology*, **212**, 603–6 (1967).
5. Brachfeld, N., Ohtaka, Y., Klein, I., and Kawade, M. Substrate preference and metabolic activity of the aerobic and the hypoxic turtle heart. *Circulation Research*, **31**, 453–67 (1972).
6. McCallister, L. P., Munger, B. L., and Neely, J. R. Electron microscopic observations and acid phosphatase activity in the ischemic rat heart. *Journal of Molecular and Cellular Cardiology*, **9**, 353–64 (1977).
7. Christodoulou, J., Erlandson, R., Smithen, C., Killip, T., and Brachfeld, N. Effects of mannitol on cardiac ultrastructure and microcirculation following anoxia. *American Journal of Physiology*, **229**, 853–60 (1975).
8. Cornblath, M., Randle, P. J., Parmeggiani, A., and Morgan, H. E. Regulation of glycogenolysis in muscle: effects of glucagon and anoxia on lactate production, glycogen content and phosphorylase activity in the perfused isolated rat heart. *Journal of Biological Chemistry*, **238**, 1592–7 (1963).
9. Dhalla, N. S., Yates, J. C., Walz, D. A., McDonald, V. A., and Olson, R. E. Correlation between changes in the endogenous energy stores and myocardial function due to hypoxia in the isolated perfused rat heart. *Canadian Journal of Physiology and Pharmacology*, **50**, 333–45 (1972).
10. Evans, J. R. Cellular transport of long chain fatty acids. *Canadian Journal of Biochemistry*, **42**, 955–68 (1964).
11. Fabiani, J. N. The no-reflow phenomenon following early reperfusion of myocardial infarction and its prevention by various drugs. *Heart Bulletin*, **7**, 134–42 (1976).
12. Feuvray, D. and de Leiris, J. Effect of short dimethylsulfoxide perfusions on ultrastructure of the isolated rat heart. *Journal of Molecular and Cellular Cardiology*, **5**, 63–9 (1973).
13. Feuvray, D. and de Leiris, J. Ultrastructural modifications induced by reoxygenation in the anoxic isolated rat heart perfused without exogenous substrate. *Journal of Molecular and Cellular Cardiology*, **7**, 307–14 (1975).
14. Forssmann, W. G., Siegrist, G., Orci, L., Girardier, L., Picket, R., and Rouiller, C. Fixation par perfusion pour la microscopie électronique. Essai de généralisation. *Journal de Microscopie*, **6**, 279–304 (1967).
15. Ganote, C. E., Seabra-Gomes, R., Nayler, W. G., and Jennings, R. B. Irreversible myocardial injury in anoxic perfused rat hearts. *American Journal of Pathology*, **80**, 419–38 (1975).

16. Ganote, C. E., Worstell, J., and Kaltenbach, J. P. Oxygen-induced enzyme release after irreversible myocardial injury. Effects of cyanide in perfused rat hearts. *American Journal of Pathology* **84**, 327–50 (1976).
17. Gazitt, Y., Ohad, I., and Loyter, A. Changes in phospholipid susceptibility toward phospholipases induced by ATP depletion in avian and amphibian erythrocyte membranes. *Biochimica et Biophysica Acta*, **382**, 65–72 (1975).
18. Gudbjarnason, S., Mathes, P., and Ravens, K. G. Functional compartmentation of ATP and creatine phosphate in heart muscle. *Journal of Molecular and Cellular Cardiology*, **1**, 325–39 (1970).
19. Hatt, P. Y. and Moravec, J. Acute hypoxia of the myocardium. Ultrastructural changes. *Cardiology*, **56**, 73–84 (1972).
20. Hearse, D. J. Enzymes and isoenzymes in the diagnosis and assessment of myocardial injury. *Journal of Molecular Medicine*, **2**, 185–200 (1977).
21. Hearse, D. J. and Chain, E. B. The role of glucose in the survival and recovery of the anoxic isolated perfused rat heart. *Biochemical Journal*, **128**, 1125–33 (1972).
22. Hearse, D. J. and Chain, E. B. The effect of glucose on enzyme release from, and recovery of the anoxic myocardium. In *Myocardial Metabolism: Recent Advances in Studies on Cardiac Structure and Metabolism*, Vol. 3, University Park Press, Baltimore, 1973, pp. 763–72.
23. Hearse, D. J., Humphrey, S. M., and Chain, E. B. Abrupt reoxygenation of the anoxic potassium-arrested perfused rat heart: a study of myocardial enzyme release. *Journal of Molecular and Cellular Cardiology*, **5**, 395–407 (1973).
24. Hearse, D. J., Humphrey, S. M., Feuvray, D., and de Leiris, J. A biochemical and ultrastructural study of the species variation in myocardial cell damage. *Journal of Molecular and Cellular Cardiology*, **8**, 759–78 (1976).
25. Herdson, P. B., Sommers, H. M., and Jennings, R. B. A comparative study of normal and ischemic dog myocardium with special reference to early changes following temporary occlusion of a coronary artery. *American Journal of Pathology*, **46**, 367–86 (1965).
26. Holland, C. E. and Olson, R. E. Prevention by hypothermia of paradoxical calcium necrosis in cardiac muscle. *Journal of Molecular and Cellular Cardiology*, **7**, 917–28 (1975).
27. Jennings, R. B. Symposium on the pre-hospital phase of acute myocardial infarction. Part II: Early phase of myocardial ischemic injury and infarction. *American Journal of Cardiology*, **24**, 753–65 (1969).
28. Jennings, R. B. Myocardial ischemia observations, definitions and speculations. *Journal of Molecular and Cellular Cardiology*, **1**, 345–9 (1970).
29. Jennings, R. B., Baum, J. H., and Herdson, P. B. Fine structural changes in myocardial ischemic injury. *American Journal of Pathology*, **79**, 135–43 (1965).
30. Jennings, R. B. and Ganote, C. E. Ultrastructural changes in acute myocardial ischaemia. In *Effect of Acute Ischaemia on Myocardial Function* (Eds. by Oliver, M. F., Julian, D. G., and Donald, K. W. Churchill Livingstone, Edinburgh and London, 1972, pp. 50–67.
31. Jennings, R. B. and Ganote, C. E. Mitochondrial structure and function in acute myocardial ischemic injury. *Circulation Research Supplement 1*, **38**, 80–9 (1976).
32. Jennings, R. B., Sommers, H. M., Kaltenbach, J. P., and West, J. J. Electrolyte alterations in acute myocardial ischemic injury. *Circulation Research*, **14**, 260–9 (1964).
33. Jennings, R. B., Sommers, H. M., Herdson, P. B., and Kaltenbach, J. P. Ischemic injury of myocardium. *Annals of the New York Academy of Sciences*, **156**, 67–78 (1969).
34. Jennings, R. B., Wartman, W. B., and Zudik, Z. E. Production of an area of homogenous myocardial infarction in the dog. *American Medical Association Archives of Pathology*, **63**, 580–5 (1957).
35. Jynge, P., Hearse, D. J., and Braimbridge, M. V. Myocardial protection during ischemic cardiac arrest. A possible hazard with calcium-free cardioplegic infusates. *Journal of Thoracic and Cardiovascular Surgery*, **73**, 848–55 (1977).
36. Jynge, P., Hearse, D. J., de Leiris, J., Feuvray, D., and Braimbridge, M. V. Protection of the ischemic myocardium: ultrastructural, enzymatic and functional assessment of the efficacy of various cardioplegic infusates. *Journal of Thoracic and Cardiovascular Surgery*, **76**, 1–15 (1978).
37. Kannengiesser, G. J., Lubbe, W. F., and Opie, L. H. Experimental myocardial infarction

with left ventricular failure in the isolated perfused rat heart: effect of isoproterenol and pacing. *Journal of Molecular and Cellular Cardiology*, **7**, 135–51 (1975).
38. Katz, A. M. and Hecht, H. H. The early 'pump' failure of the ischemic heart. *American Journal of Medicine*, **47**, 497–501 (1969).
39. Kloner, R. A., Fishbein, M. C., McLean, D., Braunwald, E., and Maroko, P. R. Effect of hyaluronidase on myocardial ultrastructure following coronary artery occlusion in the rat. *Circulation Supplement 2*, **54**, 88 (1976).
40. Kloner, R. A., Ganote, C. E., and Jennings, R. B. The 'no-reflow' phenomenon after temporary coronary occlusion in the dog. *Journal of Clinical Investigation*, **54**, 1496–508 (1974).
41. Kloner, R. A., Reimer, K. A., Willerson, J. T., and Jennings, R. B. Reduction of experimental infarct size with hyperosmolar mannitol. *Proceedings of the Society for Experimental Biology and Medicine*, **151**, 677–83 (1976).
42. Kübler, W. and Spieckermann, P. G. Regulation of glycolysis in the ischemic and anoxic myocardium. *Journal of Molecular and Cellular Cardiology*, **1**, 351–77 (1970).
43. McLean, M. J., Shigenobu, K., and Sperelakis, N. Two pharmacological types of cardiac slow Na^+ channels as distinguished by verapamil. *European Journal of Pharmacology*, **26**, 379–82 (1974).
44. de Leiris, J. and Feuvray, D. Factors affecting release of lactate dehydrogenase from isolated rat heart after calcium and magnesium free perfusions. *Cardiovascular Research*, **7**, 383–90 (1973).
45. de Leiris, J. and Feuvray, D. Ischaemia-induced damage in the working rat heart preparation: the effect of perfusate substrate composition upon subendocardial ultrastructure of the ischaemic left ventricular wall. *Journal of Molecular and Cellular Cardiology*, **9**, 365–73 (1977).
46. de Leiris, J., Feuvray, D., and Come, C. Acetylchline-induced release of lactate dehydrogenase from isolated perfused rat heart. *Journal of Molecular and Cellular Cardiology*, **4**, 357–65 (1972).
47. de Leiris, J., Gauduel, Y., and Feuvray, D. Role of myocardial catecholamines in the development of anoxic cell damage. *XXVIIth International Congress of Physiological Sciences*, Paris (1978).
48. de Leiris, J., Lubbe, W. F., and Opie, L. H. Effects of free fatty acids and glucose on enzyme release in experimental myocardial infarction. *Nature (London)*, **253**, 746–7 (1975).
49. de Leiris, J. and Opie, L. H. Beneficial effects of glucose, insulin and potassium and detrimental effects of free fatty acid on enzyme release and on mechanical performance of isolated rat heart with coronary artery ligation. *Cardiovascular Research*, in press (1978).
50. de Leiris, J., Opie, L. H., and Feuvray, D. Effect of substrate on enzyme release and electron microscopic appearances after coronary artery ligation in isolated rat heart. *Acta Medica Scandinavica Supplement*, **587**, 137–9 (1975).
51. de Leiris, J., Opie, L. H., Lubbe, W. F., and Bricknell, O. Effect of substrate on enzyme release after coronary artery ligation in isolated rat heart. In *Recent Advances in Studies on Cardiac Structure and Metabolism: The Metabolism of Contraction* (Eds. Roy, P. E. and Rona, G.), vol. 10, University Park Press, 1975, pp. 291–3.
52. Locke, F. S. and Rosenheim, O. Contribution to the physiology of the isolated heart. *Journal of Physiology*, **36**, 205–20 (1907).
53. Maroko, P. R., Libby, P., Bloor, C. M., Sobel, B. E., and Braunwald, E. Reduction by hyaluronidase of myocardial necrosis following coronary artery occlusions. *Circulation*, **46**, 430–7 (1972).
54. Maroko, P. R., Libby, P., Sobel, B. E., Bloor, C. M., Sybers, H. D., Shell, W. E., Covell, J. W., and Braunwald, E. The effect of glucose–insulin–potassium infusion on myocardial infarction following experimental coronary artery occlusion. *Circulation*, **45**, 1160–75 (1972).
55. Maroko, P. R. and Vatner, S. F. Altered relationship between phosphokinase and infarct size with reperfusion in conscious dogs. *Journal of Molecular Medicine*, **2**, 309–16 (1977).
56. Morgan, H. E., Henderson, M. J., Regen, D. M., and Park, C. R. Regulation of glucose uptake in heart muscle from normal and alloxan-diabetic rats: effects of insulin, growth

hormone, cortisone and anoxia. *Annals of the New York Academy of Sciences,* **82,** 387–402 (1959).
57. Morgan, H. E., Randle, P. J., and Regen, D. M. Regulation of glucose uptake by muscle: III. Effects of insulin, anoxia, salicylate and 2:4-dinitrophenol on membrane transport and intracellular phosphorylation of glucose in the isolated rat heart. *Biochemical Journal,* **73,** 573–9 (1959).
58. Nayler, W. G. and Fassold, E. Hypoxic-induced changes in the ultrastructure and metabolism of cardiac muscle. *Journal of Molecular Medicine,* **2,** 299–308 (1977).
59. Nayler, W. G., Grau, A., and Slade, A. A protective effect of verapamil on hypoxic heart muscle. *Cardiovascular Research,* **10,** 650–62 (1976).
60. Neely, J. R., Liebermeister, H., Battersby, E. J., and Morgan, H. E. Effect of pressure development on oxygen consumption by isolated rat heart. *American Journal of Physiology,* **212,** 804–14 (1967).
61. Neely, J. R., Rovetto, M. J., Whithmer, J. T., and Morgan, H. E. Effects of ischemia on ventricular function and metabolism of the isolated working rat heart. *American Journal of Physiology,* **225,** 651–8 (1973).
62. Oliver, M. F. Metabolic response during impending myocardial infarction. I-Clinical implications. *Circulation,* **45,** 491–500 (1972).
63. Opie, L. H. Metabolic response during impending myocardial infarction. II. Relevance of studies of glucose and fatty acid metabolism in animals. *Circulation,* **45,** 483–89 (1972).
64. Opie, L. H. Effects of regional ischemia on metabolism of glucose and free fatty acids. Relative rates of aerobic and anaerobic energy production during myocardial infarction and comparison with effect of anoxia. *Circulation Research Supplement 1,* **38,** 52–68 (1976).
65. Opie, L. H. and Owen, P. Effects of glucose–insulin–potassium infusions on arteriovenous differences of glucose and of free fatty acids and on tissue metabolic changes in dogs with developing myocardial infarction. *American Journal of Cardiology,* **38,** 310–21 (1976).
66. Powell, W. J., Dibona, D. R., Flores, J., and Leaf, A. The protective effect of hyperosmotic mannitol in myocardial ischemia and necrosis. *Circulation,* **54,** 603–15 (1976).
67. Randle, P. J. and Smith, G. H. Regulation of glucose uptake by muscle: I. Effects of insulin, anaerobiosis, and cell poisons on the uptake of glucose and release of potassium by isolated rat diaphragm. *Biochemical Journal,* **70,** 490–500 (1958).
68. Randle, P. J. and Smith, G. H. Regulation of glucose uptake by muscle: II. Effects of insulin, anaerobiosis, and cell poisons on the penetration of isolated rat diaphragm by sugars. *Biochemical Journal,* **70,** 501–8 (1958).
69. Rovetto, M. J., Whithmer, J. T., and Neely, J. R. Comparison of the effects of anoxia and whole heart ischemia on carbohydrate utilization in isolated working rat hearts. *Circulation Research,* **32,** 699–711 (1973).
70. Sakai, K. and Spieckermann, P. G. Effects of reserpine and propranolol on anoxia-induced enzyme release from the isolated perfused guinea-pig heart. *Naunyn-Schmiedeberg's Archiv Pharmacology,* **291,** 123–30 (1975).
71. Schaper, J., Hehrlein, M., Schlepper, M., and Thiedemann, K. U. Ultrastructural alterations during ischemia and reperfusion in human hearts during cardiac surgery. *Journal of Molecular and Cellular Cardiology,* **9,** 175–89 (1977).
72. Schaper, W. and Schaper, J. Pathophysiology of myocardial perfusion. In *Pathobiology Annual* (Ed. Ioachim, H. L., Appleton Century Crofts, 1976, pp. 317–63.
73. Spieckermann, P. G., Gebhard, M. M., Kalbow, K., Knoll, D., Kohl, F., Nordbeck, H., Sakai, K., and Bretschneider, H. J. Freisetzung von Enzymen aus der Herzmuskelzelle wahrend Sauerstoffmangel. *Verhandlungen der Deutschen Gesellschaft für Kreislaufförschung,* **39,** 193–8 (1973).
74. Sybers, H. D., Maroko, P. R., Ashraf, M., Libby, P., and Braunwald, E. The effect of glucose–insulin–potassium on cardiac ultrastructure following acute experimental coronary occlusion. *American Journal of Pathology,* **70,** 401–11 (1973).
75. Waldenström, A. Factors influencing experimental myocardial infarction. *Thesis,* Göteborg, 1976.
76. Williamson, J. R. Glycolytic control mechanisms: II. Kinetics of intermediate changes during the aerobic–anoxic transition in perfused rat heart. *Journal of Biological Chemistry,* **241,** 5026–36 (1966).

77. Wollenberger, A. and Krause, E. G. Metabolic control characteristics of the acutely ischemic myocardium. *American Journal of Cardiology*, **22,** 349–59 (1968).
78. Yates, J. C. and Dhalla, N. S. Structural and functional changes associated with failure and recovery of hearts after perfusion with calcium-free medium. *Journal of Molecular and Cellular Cardiology*, **7,** 91–103 (1975).
79. Zimmerman, A. N. E., Daems, W., Hülsmann, W. C., Snidjer, J., Wisse, E., and Durrer, D. Morphological changes of heart muscle caused by successive perfusion with calcium-free and calcium-containing solutions (calcium paradox). *Cardiovascular Research*, **1,** 201–9 (1967).
80. Zimmerman, A. N. E. and Hülsmann, W. C. Paradoxical influence of calcium ions on the permeability of the cell membranes of the isolated rat heart. *Nature (London)*, **211,** 646–7 (1966).

CHAPTER 20

Preservation of myocardium for ultrastructural and enzymatic studies

G. R. Bullock

INTRODUCTION	461
GENERAL TECHNIQUES FOR FIXING HEART MUSCLE	462
Immersion Fixation	462
Perfusion-fixation of the *in situ* Heart	462
Perfusion-fixation of the Isolated Heart	467
PARAMETERS AFFECTING THE QUALITY OF FIXATION	467
Choice of Fixative	467
Formaldehyde	467
Glutaraldehyde	467
Osmium Tetroxide	468
Osmolality of Buffer	468
Control of pH	472
Control of Temperature	473
ANALYSIS OF RESULTS	474
Distinction between Fixation Artefacts and Pathological Changes	474
Morphometric Analysis	474
RETENTION OF IONS WITHIN THE TISSUE	477
CONCLUDING COMMENTS	477
REFERENCES	477

INTRODUCTION

For the morphological study of human cardiac muscle, the only tissue available is that obtained by biopsy, the processing of which must commence with immersion fixation. In contrast, animal studies provide the possibility of alternative techniques and are therefore important in setting standards and giving an insight into how the handling of biopsy material could be improved. Thus studies on perfusion techniques, different fixatives, temperature of fixation, the pH of the solution used, osmolality of solutions, and retention of enzyme activity all have their relevance to studies on biopsies.

In general many published works contain illustrations of human cardiac muscle which have suffered considerably from mishandling from the time material was removed. Artefacts resulting from inattention to detail, delay in processing after removal from the heart, resulting in post-mortem changes, and lack of reference to the many good published papers are inexcusable.

In this paper, it is the intention to show the effects of various parameters on obtaining good fixation with reference both to other authors and work currently going on in this laboratory. The main emphasis will be on *in vitro* studies but reference will also be made to *in vivo* work. In addition, the effect of various techniques on the enzymatic activity of the tissue as demonstrated histochemically will also be discussed. This will be referred to as the histochemical enzyme activity (HEA).

GENERAL TECHNIQUES FOR FIXING HEART MUSCLE

Immersion Fixation

The two major problems encountered in immersion fixation of cardiac muscle are (i) lack of uniform penetration of the fixative into the tissue and (ii) contracture of the tissue. One cause of uneven fixative penetration is use of large tissue samples which should ideally be not more than 1 mm across two planes and where the size of sample does allow tying or pinning down, this will help prevent contracture taking place. Where such contracture does occur it is usually extremely uneven resulting in stretching in part of the muscle fibre and condensation in other parts (Figure 1). This makes detailed analysis such as morphometry almost impossible. Where delay in penetration of fixative occurs, other artefacts such as hypoxic figures may be produced[17] and as such may present another major problem in immersion fixation. The addition of the fixative, particularly osmium tetroxide, frequently causes contraction and for this reason osmium tetroxide should be avoided for biopsy samples. The major technical difficulties and the very small sample sizes which can be processed in this way strongly reinforce the recommendation that this procedure should not be used for animal studies where alternatives are available.

Perfusion-fixation of the *in situ* Heart

Perfusion of a highly vascular tissue such as the heart is the easiest way to preserve good morphology. Whether the perfusion is carried out *in vivo* or *in vitro* there should be no difficulty in obtaining rapid penetration of fixative into the myocardium and hence retention of all constituents of the muscle. For *in vivo* studies, the usual route is by cannulation of one of the major vessels to the heart or by injection of fixative directly into the ventricular cavity. Cannulation of the main vessels generally gives excellent results and the reader is recommended to consult the paper by Tomanek

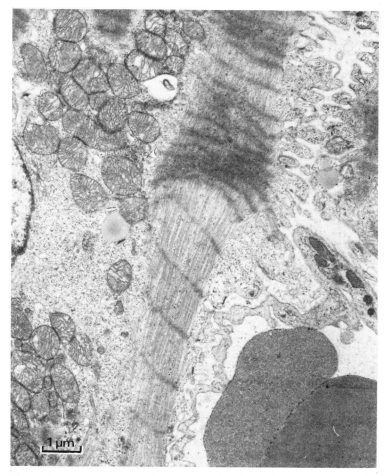

Figure 1. Immersion-fixed foetal guinea-pig heart showing contractures in the tissue. Magnification × 11,000. The author wishes to thank Miss D. Parry for this picture

and Karlsson[30] where good results were obtained using a pre-wash and fixation by means of a pressure head. In our own hands, the use of a pressure head was found to be unreliable in that vascular resistance developed in the tissue once fixation commenced and it proved difficult to maintain a steady flow. To overcome this problem, we prefer to use a peristaltic pump with the pressure of both the pre-wash and the first few minutes of fixation maintained at a level approximate to the normal *in vivo* peak systolic pressure of the animal. After this period, as the vascular resistance increases, the flow rate has to be reduced to maintain the pressure

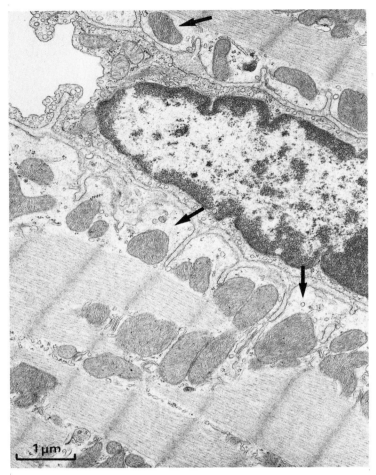

Figure 2. Perfused-fixed rat heart showing markedly contracted sarcomeres and loss of glycogen and ground material from the cytosol (arrows). Magnification × 15,000

at a steady rate.[6] In addition, the perfusate can be maintained at the desired temperature by incorporating a heater in the circuit for temperatures above room temperature. Excessive pressure can induce vesicle formation or even cell separation. The effect of extraction of cellular constituents by the fixative and contraction of the fibres can be seen in Figure 2.

The addition of $CaCl_2$ has been used in some studies[27,30] and appears to help preserve mitochondrial integrity but there are serious objections to including calcium ions in either pre-wash or fixative wherever the physiological effect of these ions is being studied. Excellent preservation can be obtained without the use of calcium salts and these should be avoided where possible.

Another common cause of patchy fixation is the formation of blood clots blocking the flow of the fixative. Adequate fixation may be achieved up to and around the clot but on the distal side the quality will fall off rapidly (Figures 3 and 4). Use of heparin (i.v. 1000 iu kg^{-1} body weight) shortly before cannulation of the vessels helps to eliminate this problem.

Monitoring the quality of overall fixation is best done by cutting 1 μm thick sections of the resin-embedded myocardium and examining them in

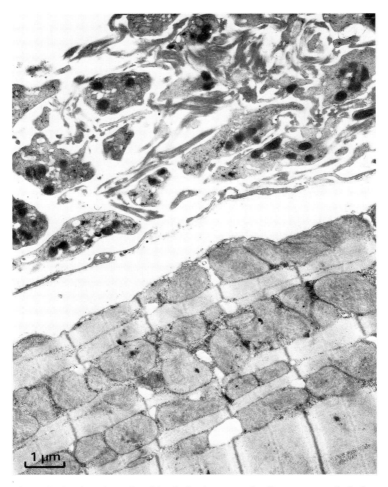

Figure 3. Section through a blood clot in a vessel adjacent to a relatively normal fibre. Magnification × 10,000

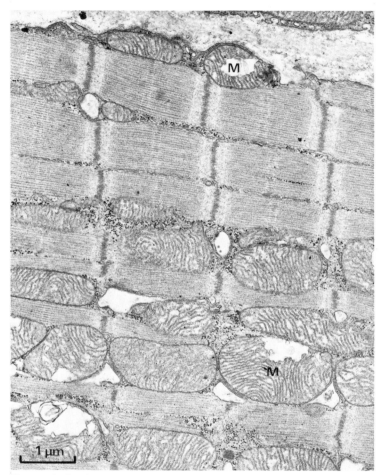

Figure 4. Cardiac muscle from the distal side of a blood clot. The mitochondria (M) show considerable signs of damage. Magnification ×15,000

the light microscope after staining with toluidine blue in borax.[31] This method avoids the time-consuming and often inadequate sampling obtained by thin sectioning and inspection in the electron microscope.

The same anaesthetic should be used throughout the experiments as these are generally highly polar agents and may produce tissue changes in their own right (such as that produced by ether on the lining endothelium of the lung where localized changes may take place). Many of the local anaesthetics can also produce morphological changes which may be indistinguishable from pathological changes.[24]

Perfusion-fixation of the Isolated Heart

A highly reproducible system for biochemical, pharmacological and morphological studies of cardiac muscle using the isolated heart has been utilized by Hearse et al.[19] At the end of the experimental period, the perfusion medium can readily be replaced by fixative which is added at constant pressure through the side arm using either a syringe or a peristaltic pump. Immediately after this preliminary fixation, the hearts can be either fixed further (after cutting the selected part of the myocardium into small pieces) or used directly for histochemical enzyme activity studies.

PARAMETERS AFFECTING THE QUALITY OF FIXATION

Choice of Fixative

Formaldehyde

It is perhaps of interest that one of the earliest and most widely used fixatives for routine histology, i.e. formalin, is still the best fixative for preserving HEA[11] and has not been superseded by other chemicals.

At least 40% of the histochemical enzyme activity of most enzymes is retained by using this fixative but used on its own for electron microscopy the preservation of the tissue is generally very poor (by comparison with glutaraldehyde). Improved results can be obtained using formaldehyde (prepared from paraformaldehyde immediately prior to use) but this is still a very inadequate method where fine detail is required. One of the reasons given for the inferior quality is that formaldehyde forms very unstable Schiff's bases with proteins and that during washing and dehydration these break down and a considerable amount of soluble material is extracted from the tissue.[7]

Glutaraldehyde

The introduction of glutaraldehyde/formaldehyde mixtures[22] allows the rapid penetration and primary fixation by formaldehyde to be stabilized by the bifunctional activity of glutaraldehyde. Very low concentrations of both constituents can be used and give excellent tissue preservation. Re-evaluation of the histochemical enzyme activity retained in cardiac tissue after fixation by this method would certainly have some merit.

Fixation by glutaraldehyde alone produces excellent results. Although extraction of certain cellular constituents may take place,[18] this extraction is much less than with other fixatives (see reference 21 for a full description of the ability of glutaraldehyde to retain different cellular components). The slow penetration of this fixative into tissues is largely overcome by perfusion-fixation and with the highly vascular cardiac muscle this penetration can be very rapid. Glaumann[16] showed that liver could be well fixed after a 3 minute perfusion period and at the end of this period 50–60% of the histochemical enzyme activity of three different types of phosphatases (acid phosphatase, glucose-6-phosphatase, and inosine-5'-diphosphatase) was retained. This high level of activity may have been partly due to perfusion of the substrate as well; thus allowing less restricted access to the cytosol.

Very low concentrations of glutaraldehyde have been tested for their effect on other enzymes[4,5] and the need for very brief periods of fixation has been emphasized by Hopwood.[21] Ellar et al.[14] found that low concentrations of glutaraldehyde would bind enzymes located in the plasma membrane more firmly to the membrane so that they could not be removed by washing, e.g. Ca^{2+}-dependent ATPase and NADH dehydrogenase, which would obviously be relevant to measurements of the cardiac muscle ATPases. However, this very enhancement of binding has been severely criticized by Hillman[20] who suggested that many enzymes may erroneously appear to be membrane bound due to this property of glutaraldehyde, whereas normally in the cell they may only be loosely associated with the membrane or even free in the cytoplasm. These criticisms must certainly be taken into account when discrepancies arise as to the original location of an enzyme in a cell when measured by biochemical or histochemical techniques.

Osmium tetroxide

Although this compound has been used frequently and successfully to fix cardiac muscle, it reduces enzymatic activity to zero and also interferes with energy dispersive X-ray microanalysis of elements such as phosphorus which have similar K-shell energy peaks to osmium and hence cannot easily be distinguished. Substitution of another fixative or cryo-techniques for preparing the tissue will eliminate this problem.

Osmolality of Buffer

Since it has been shown quite clearly in recent years that fixatives exert no effective osmotic pressure once they have penetrated the cell membrane, it is the osmolality of the buffer in which the fixatives are dissolved which must be balanced with that of the cell.[3] Fixatives such as glutaraldehyde and osmium tetroxide cause irreversible changes in the membranes such that they can no longer act as a semipermeable membrane.[3,25] The most commonly used buffers are phosphate and cacodylate and both seem to have their merits. It is important where primary fixation with glutaraldehyde is followed by fixation with osmium to use the same buffer throughout to avoid deposition of granular material within the tissue as has been shown for kidney and heart.[23] We confirmed this effect with the same tissues in our laboratories. When a buffer solution is used with an excessively high osmotic pressure, the tissue will shrink leaving wide interfibre spaces (Figure 5) and also the T tubules will widen to varying degrees (Figures 6 and 7). This widening of the T tubules has frequently been described as swelling of the tubular system whereas in fact the cell has *shrunk* leaving an extended

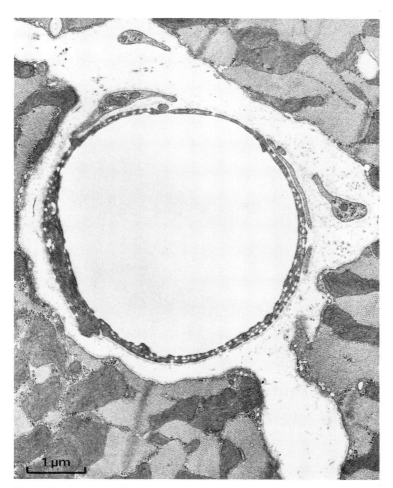

Figure 5. A myocardial blood vessel with widened intercellular spaces around it. A very hypertonic fixative has given rise to generalized tissue shrinkage. Magnification × 15,000

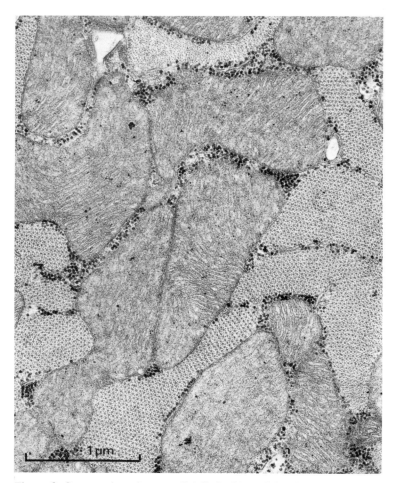

Figure 6. Cross-section of myocardial fibrils. Very slight distension of the T tubules (T) and condensation of the mitochondrial matrix. Magnification ×30,000

tubular space. This shrinkage is frequently found if sucrose is used to raise the osmotic pressure but a very slightly hypertonic solution enhances preservation of all other subcellular components and appears to reduce loss of glycogen (Figure 8). The structure of the intercalated disc is also very sensitive to osmotic pressure and again the use of a slightly hypertonic solution gives successful results (Figure 9). We have found that the use of a mixture of 2.5% glutaraldehyde with 4% formaldehyde made freshly from paraformaldehyde in 0.08 M-phosphate buffer pH 7.2 and used at 4 °C, gives excellent, highly reproducible fixation with minimal shrinkage of the tissue and opening of the T tubular system. This mixture also allows one to stain with excellent contrast when thick resin sections are stained with toluidine blue.

Davey[12] recommends the use of Ringers solution for frog muscle where fixation follows experimental procedures to reduce rapid changes in the ionic concentrations in the tissue on the introduction of the fixative and hence minimize chloride withdrawal contractures at the onset of fixation. This method of avoiding contractures could eliminate the use of injected KCl as used by Tomanek and Karlsson[30] to stop the heart beating although they did not in fact find any morphological change due to the KCl.

Figure 7. Marked dilation of the T-tubular system (T) due to use of 0.2 M sucrose in fixative. Magnification × 15,000

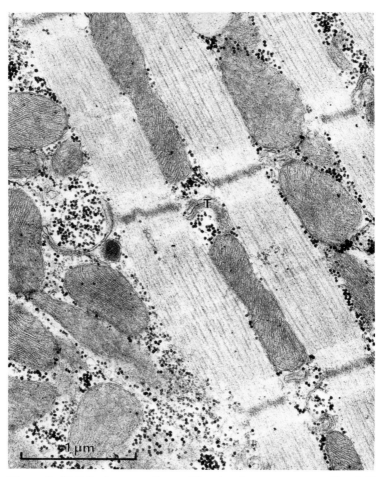

Figure 8. Well-fixed myocardium with relaxed sarcomeres, undilated T system (T) and excellent retention of glycogen. Magnification × 30,000

Control of pH

Adequate control of the pH of the solutions used for washing out the heart, perfusion-fixing or processing the tissue is of considerable importance. This applies particularly to glutaraldehyde where careful monitoring of the stock solution is required as impurities present may explain the variability found in enzyme electron histochemistry reported from different laboratories. A simple method for distillation of glutaraldehyde at atmospheric pressure has been described by Smith and Farquhar[28] or glutaraldehyde can be treated with charcoal before use[2] though the latter method is

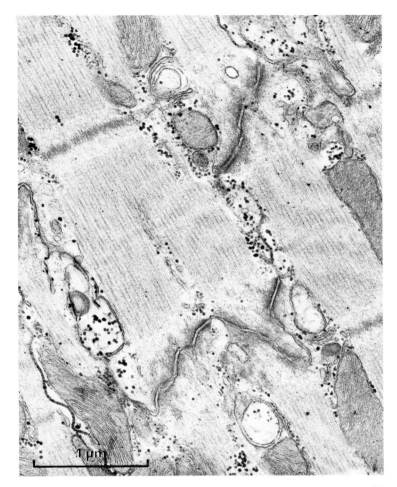

Figure 9. Intercalated disc fixed in slightly hypertonic fixative. Magnification × 30,000

not as satisfactory as distillation.[13] If glutaraldehyde is stored at alkaline pH (pH 8) it polymerizes rapidly.

Control of Temperature

Fixation at room temperature or at 34 °C gives excellent results with glutaraldehyde or combinations of aldehydes. This is almost certainly due to the increased speed of penetration of tissues by the warmer fixative giving a more uniform quality of tissue preservation. However, for osmium tetroxide, fixation in the cold (4 °C) is essential (this also applies to experiments where the muscle has been pre-perfused with cold solutions prior to fixation to minimize artefacts due to rapid temperature changes).

ANALYSIS OF RESULTS

Distinction between Fixation Artefacts and Pathological Changes

Knowledge of structures seen in cardiac tissue is vital when determining whether they are due to the handling of the tissue or pathological changes developed *in situ*. Table 1 shows some of the more common artefacts or non-experimentally induced changes which may be found in the tissue.

Table 1. The Commonly and Less Commonly Encountered Artefacts seen in Cardiac Muscle as a Consequence of Tissue Preservation

Description of artefact	Interpretation and action
1. Clear or empty spaces in or around the tissue	Delay between removal and fixation (post-mortem change). Tonicity of medium too low (Figure 10)
2. Myelin-like inclusions or large amorphous zones in mitochondria[15] (Figure 11)	This may be due to glutaraldehyde interacting with a defective or fragile membrane[9] which can be checked by using another fixative
3. Tissue shrinkage or swelling. Widening of T tubules (Figure 7)	Buffer may not be in balance with the tissue tonicity. This may cause change in pathological conditions
4. Z band distortion (Figure 12), necrotic foci	May be indigenous to the species being studied
5. Lipid extraction	Property of the fixative used[17]
6. Glycogen loss (Figure 1)	Increase tonicity of buffer
7. Intracristal changes in mitochondria (Figure 13)	Non-specific change due to alteration in external milieu which may be drug induced.[24] Seen in isolated mitochondria[8]
8. Loss of fibrillar or other proteinaceous materials	May be due to excessive exposure of the tissue to the fixative

Morphometric Analysis

Where gross changes have taken place in a tissue due to pathological changes or response to drug treatment, it is not difficult to devise a method to demonstrate the changes quantitatively. Unfortunately, very few published papers include such quantitative analysis so that valuable information may be lost. This is often due to inadequate care being taken in the preparation of the tissue or inclusion of insufficient controls to allow analysis to be done. For simple analytical techniques, the reader is recommended to consult the work by Bullock and Parsons[10] where either counting or a grid overlay can be used to demonstrate such changes. Where more subtle changes may have occurred, then more sophisticated techniques must be

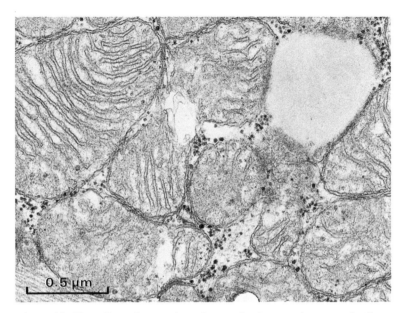

Figure 10. The effect of extraction of ground substance from muscle tissue incubated in phosphate buffer for 45 minutes. Magnification × 44,000

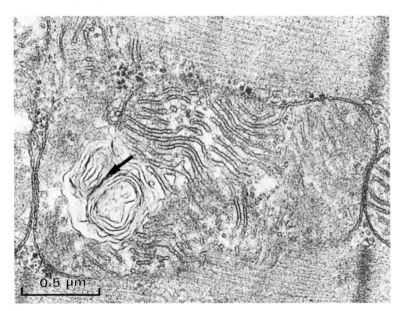

Figure 11. Glutaraldehyde-fixed skeletal muscle mitochondrion after steroid treatment showing focal necrosis (arrow). Magnification × 40,000

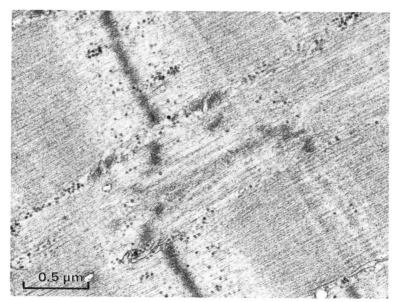

Figure 12. Z-Band distortion from normal muscle tissue. Magnification ×35,000

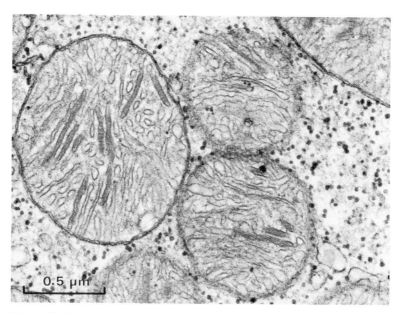

Figure 13. Intracristal alterations within mitochondria from diaphragm muscle incubated with caffeine (5 mg ml^{-1}) or alprenolol (2 mg ml^{-1}). Magnification ×45,000

used and the methods of analysis described by Weibel and Elias[32] where detailed grid analyses and a computer program are involved, give excellent results.

Whereas visual description of results may give useful information, for random sampling and genuine measurement of experimental changes, morphotometric techniques are essential.

RETENTION OF IONS WITHIN THE TISSUE

With increasing use of X-ray microanalysis to confirm findings shown by histochemical techniques, attention must be drawn to the fact that all the currently popular fixatives allow escape of ions due to the change in the permeability of cell membranes. The problems involved in retention of Ca^{2+} have been discussed thoroughly in the work of Yarom et al.[33] and the relevance of the distribution of zinc in damaged myocardium has been discussed by Peters et al.[26] Where it is necessary to carry out correlative work, the glutaraldehyde/urea embedding mixture described by Yarom et al.[33] has been used successfully in cardiac muscle.

The presence of intramitochondrial granules can be varied by incubating tissues in different media causing mitochondrial uptake of ions such as Ca^{2+}, Sr^{2+}, and Ba^{2+} all of which form electron-opaque granules which may then be extracted during the processing of the tissue.[29] These and other cytoplasmic granules may be best preserved by dry sectioning of frozen tissue.[1]

CONCLUDING COMMENTS

Good structural preservation of the myocardium depends on a number of factors. For biopsy samples, it is essential that they are removed rapidly and small pieces fixed by immersion to reduce penetration artefacts. Perfusion-fixation gives excellent, reproducible results in the animal model but attention must be paid to choice of fixative and buffer where histochemical measurements are to be made. The osmolality of the buffer, pH, and temperature of fixation can all markedly affect the end result and where X-ray microanalysis is to be performed on the tissue special preparative techniques must be used. Thick resin sections can give much useful information and morphometric analysis of the results should be included.

REFERENCES

1. Ali, S. Y. and Wisby, A. Mitochondrial granules of chondrocytes in cryosections of growth cartilage. *American Journal of Anatomy*, **144,** 243–8 (1975).
2. Anderson, P. J. Purification and quantitation of glutaraldehyde and its effects on several enzyme activities in skeletal muscle. *Journal of Histochemistry and Cytochemistry*, **15,** 652–61 (1967).

3. Arborgh, B., Bell, P., Brunk, V., and Collins, V. P. The osmotic effect of glutaraldehyde during fixation. A transmission electron microscopy, scanning electron microscopy and cytochemical study. *Journal of Ultrastructural Research*, **56,** 339–50 (1976).
4. Asano, M., Kurono, C., Wakabayashi, T., and Kimura, H. Stabilisation of configurational states and enzyme activities in subcellular fractions after fixation with extremely low concentrations of glutaraldehyde. *Histochemical Journal*, **8,** 113–20 (1976).
5. Boecking, A., Grossarth, C., and Deimling, O. von. Esterase X. Content and substrate patterns of esterase in fixed tissue. *Histochemie*, **37,** 265–73 (1973).
6. Bowes, D., Bullock, G. R., and Winsey, N. J. P. A method for fixing rabbit and rat hind limb skeletal muscle by perfusion. *Septieme Congres International de Microscopie Electronique*, Grenoble, 1970, pp. 397–8.
7. Bowes, J. H. and Cater, C. W. The reaction of glutaraldehyde with proteins and other biological materials. *Journal of Royal Microscopical Society*, **85,** 193–200 (1966).
8. Bullock, G., Carter, E. E., and White, A. M. The preparation of mitochondria from muscle without the use of a homogeniser. *FEBS Letters*, **8,** 109–11 (1970).
9. Bullock, G. R., Christian, R. A., Peters, R. F., and White, A. M. Rapid mitochondrial enlargement in muscle as a response to triamcinolone acetonide and its relationship to the ribosomal defect. *Biochemical Pharmacology*, **20,** 943–53 (1971).
10. Bullock, G. R. and Parsons, R. R. Photography in pharmacological research. In *Photographic Methods in Scientific Research*. Academic Press, London and New York, 1976.
11. Chayen, J., Bitensky, L., and Butcher, R. In *Practical Histochemistry*, John Wiley and Sons, 1973, pp. 234–46.
12. Davey, D. F. The effect of fixative tonicity on the myosin filament lattice volume of frog muscle fixed following exposure to normal or hypertonic Ringer. In *Fixation in Histochemistry* (Ed. Stoward, P. J.) Chapman and Hall, 1973, pp. 103–20.
13. Davies, K. J. and Garrett. J. R. Improved preservation of alkaline phosphatase in salivary glands of the cat. In *Fixation in Histochemistry* (Ed. Stoward, P. J.), Chapman and Hall, 1973, pp. 151–65.
14. Ellar, D. J., Munoz, E., and Salton, M. R. The effect of low concentrations of glutaraldehyde on *Micrococcus lysodeikticus* membranes: changes in the release of membrane associated enzymes and membrane structure. *Biochimicaet Biophysica Acta*, **225,** 140–50 (1971).
15. Franke W. W., Krien. S., and Brown, R. M., Jr. Simultaneous glutaraldehyde–osmium tetroxide fixation with postosmication. *Histochemie*, **19,** 162–4 (1969).
16. Glaumann, H. Ultrastructural demonstration of phosphatases by perfusion fixation followed by perfusion incubation of rat liver. *Histochemistry*, **44,** 169–78 (1975).
17. Hatt, P. Y. and Moravec, J. Acute hypoxia of the myocardium. Ultrastructural changes. *Cardiology*, **56,** 73–84 (1972).
18. Hayat, M. A. *Principles and Techniques of Electron Microscopy*, Vol. 1, Van Nostrand Reinhold Company, 1970, pp. 123–4.
19. Hearse, D. J., Humphrey, S. M., Feuvray, D., and de Leiris, J. A biochemical and ultrastructural study of the species variation in myocardial cell damage. *Journal of Molecular and Cellular Cardiology*, **8,** 759–78 (1976).
20. Hillman, H. *Certainty and Uncertainty in Biochemical Techniques*, Surrey University Press, 1972.
21. Hopwood, D. Theoretical and practical aspects of glutaraldehyde fixation. In *Fixation in Histochemistry* (Ed. Stoward, P. J.), Chapman and Hall, 1973, pp. 47–83.
22. Karnovsky, M. J. A formaldehyde–glutaraldehyde fixative of high osmolarity for use in electron microscopy. *Journal of Cellular Biology*, **27,** 177A (1965).
23. Kuthy, E. and Csapó, Z. Peculiar artefacts after fixation with glutaraldehyde and osmium tetroxide. *Journal of Microscopy*, **107,** 177–82 (1976).
24. Manchester, R. L., Bullock, G., and Roetzscher, V. M. Influence of methylxanthines and local anaesthetics on the metabolism of muscle and associated changes in mitochondrial morphology. *Chemical–Biological Interactions*, **6,** 273–96 (1973).
25. Penttila, A., Kalimo, H., and Trump, B. F. Influence of glutaraldehyde and/or osmium tetroxide on cell volume, ion content, mechanical stability and membrane permeability of Ehrlich ascites tumour cells. *Journal of Cellular Biology*, **63,** 197–214 (1974).

26. Peters, P. D., Yarom, R., Dormann, A., and Hall, T. X-Ray microanalysis of intracellular zinc: EMMA-4 Examinations of normal and injured muscle and myocardium. *Journal of Ultrastructural Research*, **57,** 121–31 (1976).
27. Smith, M. E. and Page, E. Morphometry of rat heart mitochondrial subcompartments and membranes: application to myocardial cell atrophy after hypophysectomy. *Journal of Ultrastructural Research*, **55,** 31–41 (1976).
28. Smith, R. E. and Farquhar, M. G. Lysosome function in the regulation of the secretory process in the cells of the anterior pituitary gland. *Journal of Cellular Biology*, **31,** 319–47 (1966).
29. Sutfin, L. V., Holtrop, M. E., and Ogilvie, R. E. Microanalysis of individual mitochondrial granules with diameters less than 1000 Angstroms. *Science*, **174,** 947–9 (1971).
30. Tomanek, R. J. and Karlsson, U. L. Myocardial ultrastructure of young and senescent rats. *Journal of Ultrastructural Research*, **42,** 201–20 (1973).
31. Trump, B. F., Smuckler, E. A., and Benditt, E. P. A method for staining epoxy sections for light microscopy. *Journal of Ultrastructural Research*, **5,** 343–8 (1961).
32. Weibel, E. R. In *Quantitative Methods in Morphology* (Eds. Weibel, E. R. and Elias, H.), Springer-Verlag, Berlin-Heidelberg, New York, 1967.
33. Yarom, R., Peters, P. D., and Hall, T. A. Effect of glutaraldehyde and urea embedding on intracellular ionic elements. X-Ray microanalysis of skeletal muscle and myocardium. *Journal of Ultrastructural Research*, **49,** 405–18 (1974).

CHAPTER 21

Metabolic manipulations: tissue damage and enzyme leakage

L. H. Opie and J. de Leiris

INTRODUCTION	481
GENERAL FEATURES OF THE CONTROL OF MYOCARDIAL ENERGY METABOLISM	481
SUBSTRATE SUPPLY AND ISCHAEMIC INJURY	483
EFFECTS OF EXOGENOUS GLUCOSE	485
In Anoxia	485
In Regional Ischaemia	487
EFFECT OF INSULIN	488
EFFECT OF GLUCOSE–INSULIN–POTASSIUM	490
EFFECT OF ACUTE DIABETES	490
EFFECT OF FREE FATTY ACIDS AND OF ALBUMIN	492
OTHER NON-GLUCOSE SUBSTRATES	492
ANTILIPOLYTIC AGENTS	493
METABOLIC MANIPULATIONS, GLYCOLYTIC FLUX, AND ISCHAEMIC DAMAGE	495
OTHER MANIPULATIONS	496
CONCLUDING COMMENTS	497
REFERENCES	498

INTRODUCTION

Recent evidence suggests that the rate of enzyme loss from the ischaemic heart cell can be modified by the exogenous substrate reaching the heart.[5,29-31,74] It is, therefore, appropriate briefly to review the substrate metabolism of first the normal heart, then that of the ischaemic heart, and to stress links between substrate supply and altered patterns of myocardial energy metabolism which may in part underlie the phenomenon of enzyme loss.[9]

GENERAL FEATURES OF THE CONTROL OF MYOCARDIAL ENERGY METABOLISM

Introduction of coronary sinus catheterization by Bing led to an appreciation of the fuels of the normal human myocardium[3] which relies principally

Table 1. Fuels for Oxidative Metabolism of Heart of Normal Man, of Patients with Pacing-induced Angina, of Normal Greyhound Dogs, and of Dog Heart during Regional Ischaemia

	Man			Greyhound dog[57]		
	fasting[47]	pacing-induced angina[20]	fasting	regional ischaemia		
				20 min	120 min	
Free fatty acids	60%	42%	50%	39%	35%
Glucose	28%	21%	43%	35%	58%
Lactate	11%	1%	11%	0%	0%
Total	99%	64%	104%	74%	93%

In each case the oxygen extraction ratio has been calculated as a percentage by assuming full oxidation of the extracted substrate.

on free fatty acids for its nutrition during the fasted state, but not during exercise nor after carbohydrate loading.[47] In patients fasting overnight (Table 1), the heart uses predominantly free fatty acids, then glucose, and then lactate.[47] More recent data confirm these figures but also assign a small role to triglyceride.[28] Myocardial free fatty acid uptake is partially dependent on the molar binding with plasma albumin. Glucose uptake depends on its circulating concentration, the availability of insulin, and the degree of heart work.[43,46,56] In isolated preparations, when both free fatty acid and glucose are present together in the perfusion fluid, fatty acid is used in preference to glucose.[43,47,72] In normal isolated hearts utilization of exogenous glucose is mainly limited by glucose transport, the rate of which depends upon the levels of insulin bound to the tissue, the presence or absence of anoxia, and the availability of fatty substrates.[43,46] The rate of fatty acid uptake and oxidation is controlled by the availability of exogenous fatty acid, the rate of acetyl CoA oxidation by mitochondria and the rate of acyl translocation across the inner mitochondrial membrane.[43]

Major modifications in myocardial metabolism occur during pathological conditions with oxygen deprivation. Anoxia markedly accelerates glucose and glycogen utilization.[43,46] In severely ischaemic tissue, evidence from both the whole heart ischaemic model of Neely's group[58] and from the infarcting dog myocardium[50] suggests an initial rapid acceleration of glycolysis, followed by inhibition at the level of the rate-controlling enzymes phosphofructokinase[26] and/or glyceraldehyde-3-phosphate dehydrogenase,[64] as a result of accumulation of hydrogen ions and lactate respectively. Intracellular acidosis could be harmful by activating lysosomes, by inhibition of interaction of calcium with the contractile proteins, and by increased binding of intracellular free calcium to the sarcoplasmic reticulum and mitochondria.[4,21,69] Therefore, metabolic modifications induced by either anoxia or ischaemia differ in several important ways.[50] In regional ischaemia, the effect of oxygen deprivation in increasing the glucose uptake is opposed by inhibition of glycolysis, and anaerobic production of ATP remains limited. In anoxia with maintained coronary perfusion, anaerobic production of ATP may contribute to cell survival.[10] In ischaemia an intracellular acidosis develops; in anoxia with maintained coronary perfusion the degree of intracellular acidosis is not severe.[50] Thus ischaemia causes a more dramatic metabolic situation in the heart than does anoxia, even though the collateral flow is still delivering oxygen and the predominant pattern of metabolism is oxidative.[50] Myocardial metabolism is disrupted and despite the persistence of oxidative metabolism (albeit at a reduced rate), cellular energy production is severely impaired by a variety of mechanisms, including accumulation of acyl CoA,[73] protons, lactate, and CO_2,[50] and the activity of metabolic cycles 'wasting' ATP.[50]

From studies with a variety of experimental models, it has been shown that following the onset of anoxia or ischaemia, myocardial ATP levels decline[11,50] and the cells enter a state of negative energy balance. The availability of adequate supply of ATP is important not only for contractile activity, but also for maintenance of cellular integrity. It has even been suggested that the supply of ATP is a major determinant of cellular survival and that once the level of myocardial ATP drops below a certain critical level, the heart would become irreversibly damaged and recovery would no longer be possible.[10,26]

SUBSTRATE SUPPLY AND ISCHAEMIC INJURY

Among the modifications observed in patients with acute myocardial infarction are increased circulating blood free fatty acid concentrations, glucose intolerance, and increased secretion of catecholamines.[49] Hypothetically, each of these changes might contribute to the outcome of acute myocardial ischaemia.

Kurien, Yates, and Oliver[27] stressed the relationship between circulating free fatty acids and post-occlusion arrhythmias. Others have suggested that it may not be the absolute level of free fatty acid that is of major importance, but rather the free fatty acid-to-albumin molar ratio.[49,86] Circulating free fatty acids are bound to circulating albumin by binding sites with various degrees of affinity for free fatty acids. Spector[79] indicated that free fatty acid uptake is dependent on the circulating concentration especially when the tight albumin-binding sites are exceeded, i.e. when the free fatty acid/albumin molar ratio is high. Thus the higher the free fatty acid/albumin molar ratio, the greater the possibility that the tight binding sites will be 'over-occupied' and that circulating free fatty acid molecules will be held only in a weak association with albumin and therefore, be even more readily taken up by the heart.[49]

Although controversial, experimental results have been presented concerning the relationship between blood free fatty acid levels and arrhythmias in acute developing infarction,[49] nevertheless recent evidence links increased ischaemic damage with fatty acid metabolism.[22,31,40,45] In isolated rat hearts, a high free fatty acid/albumin molar ratio in the perfusion fluid can produce arrhythmias in apparently normal, non-ischaemic hearts.[85,86] High concentrations of free fatty acids depress contractility of hypoxic rat papillary muscle.[15] Restriction of oxygen availability by reduction of perfusate oxygen tension to zero (anoxia) or by reduction of coronary flow (ischaemia) both reduces rates of oxidation and uptake of free fatty acids in isolated perfused hearts[43] and in dog heart.[50] There are increased cellular levels of long-chain acyl CoA and α-glycerol phosphate, thereby increasing conversion of fatty acids to tissue triglycerides, an accumulation of which is associated with pathological conditions of the heart.[47,71] In both anoxic and in ischaemic conditions, decreased uptake of oxygen leads to decreased delivery of oxygen to the mitochondria and decreased removal of acyl CoA formed from intracellular free fatty acids.[43] Acyl CoA also accumulates because of decreased activity of the carnitine-dependent transferase which normally transfers acyl CoA into the mitochondria. Substantial evidence shows that acyl CoA can inhibit the adenine nucleotide translocase system[59,73] thereby aggravating the cellular ATP deficit.[49]

Changes in carbohydrate metabolism are also observed in pathological conditions. Although glycolytic flux is inhibited in severely ischaemic cells,[50,64,65] the inhibition is less marked in cells subject to a milder degree of ischaemia[50] so that provision of glucose can enhance glycolytic flux in such cells;[5] enhanced glycolytic flux may provide glycolytic ATP which is postulated to have a special role in the maintenance of membrane integrity.[6] Such moderately ischaemic cells are probably located in the 'border zone' of developing acute myocardial infarction[14] which responds most to provision of glucose and insulin.[55] Hence agents promoting glycolytic flux could, according to these arguments, metabolically protect ischaemic cells, especially in the presence of moderate ischaemia when glycolytic flux is not severely inhibited (see later).

Conversely, the metabolism of free fatty acid appears to be harmful for the ischaemic myocardial cells. Hence interventions such as antilipolytic agents, including nicotinic acid analogues, clofibrate, or naturally occurring substances such as prostaglandin E_1 may also exert some beneficial effect by making less free fatty acid available for extraction by cardiac cells.[40]

These considerations are largely based on animal experiments and it is not known what the severity of reduction of blood flow is in patients with acute infarction. Presumably different cells have different severities of ischaemia, and hence, different responses of glycolytic flux. In patients with pacing-induced angina the contribution of lactate to the oxygen uptake ceases as lactate is produced instead of being taken up (Table 1). But the contribution of both glucose and of free fatty acids falls, indicating that an endogenous fuel, presumably glycogen, is being used. In dogs, the development of regional ischaemia is also accompanied by cessation of lactate uptake; initially (20 min post-ligation) contributions of glucose and FFA fall as in patients with angina, again presumably due to glycogenolysis.[57] But by 120 min post-ligation the contribution of glucose has increased to become the major fuel for the residual oxidative metabolism of those cells assessed in this way, i.e. only those cells which are supplied by the collateral arterial blood and which drain into the coronary vein sampled by the local catheterization technique to obtain predominantly ischaemic blood. Thus there is some correspondence between the trends in oxidative metabolism during angina in man and regional ischaemia in dogs. Problems and limitations relevant to the local vein sampling technique are detailed elsewhere.[50,57]

We now review data on experimental metabolic manipulations designed to protect myocardial cells against ischaemic or anoxic damage, the degree of protection being assessed by any significant reduction in enzyme release.

EFFECTS OF EXOGENOUS GLUCOSE

In Anoxia (Table 2)

In isolated arrested hearts perfused without exogenous glucose, severe anoxia leads to a rapid decline in the cellular content of glycogen, ATP, and CP, associated with a biphasic release of several enzymes.[8,10,11] Abrupt readmission of oxygen after 2 hours of anaerobiosis results in a massive acceleration of enzyme leakage[10,11] and extensive ultrastructural damage.[13] In some species (rat and guinea-pig) exogenous glucose consistently increases myocardial glycogen, ATP, and CP, reduces enzyme release and preserves myocardial ultrastructure.[13] However, in the mouse and the rabbit exogenous glucose appears to be far less effective in the protection of anoxic cardiac cells.[13] It is likely that glucose protects the anoxic well-perfused isolated heart through its ability to support glycolytic ATP production.[13] Intraspecies variability in glucose efficacy seems to be due to variations in maximum capacity for glycolytic flux between species. Inclusion of glucose in the anoxic perfusion medium of isolated hearts considerably delays the onset of lysosomal activation; such a beneficial effect is related to both metabolic and osmotic effects of the addition of glucose.[83]

Elevations in cardiac glycogen are associated with greater glycolytic reserve and improved mechanical resistance to anoxia, mainly due to enhanced glycogenolysis and anaerobic ATP production.[67]

Table 2. Effects of Glucose on Anoxia-induced Creatine Kinase Release from Isolated Perfused Hearts

References	Species	Conditions	Duration of (min)	Glucose concentration (mmol l.$^{-1}$)	Enzyme release expressed as	Per cent reduction of CK release in the presence of glucose
Hearse and Humphrey[11]	Rat	K$^+$-Arrested	360	11	Total released during 6 h	89%
Seabra-Gomes, Ganote, and Nayler[70]	Rat Rabbit	Spontaneously beating	150	10	Peak release in miu min^{-1} g^{-1} wet wt.	82% Delayed release, but no reduction
Hearse, Humphrey, and Garlick[12]	Rat Mouse Guinea-pig Rabbit	K$^+$-Arrested	400	11	Total released during 400 min	88% 29% 66% 50%
Hearse, Humphrey, Feuvray, and de Leiris[13]	Rat Mouse Guinea-pig Rabbit	K$^+$-Arrested	300	11	Total released during 300 min	96% 34% 63% 2 subgroups of rabbits either 97% or 23%
Welman and Peters[83]	Guinea-pig	Spontaneously beating	360	11	Release at 6 h anoxia in miu min^{-1}	43%
Nayler (this book)	Rabbit	Paced	60	11	Release at 60 min in miu min^{-1} g^{-1} wet wt.	49%

In Regional Ischaemia (Figures 1 and 2)

The therapeutic use of glucose in ischaemic heart disease has been widely discussed since the early years of the century (see reference 49). In totally ischaemic tissue with no collateral blood flow, it is difficult to see how exogenous glucose could contribute to the production of anaerobic energy. In severely ischaemic tissue, in which some glucose uptake still persists, the rate of glycolytic flux is determined by at least two different factors: the supply of substrate and the activity of the rate-limiting enzymes of glycolysis and especially of the phosphofructokinase and glyceraldehyde-3-phosphate dehydrogenase which can be markedly inhibited by accumulation of protons and lactate.[26,50,64] In 'border-zone' tissue both glucose and oxygen can reach mildly ischaemic myocardial cells by tne way of collateral flow, and glycolytic flux becomes the major source of energy.[50]

In dogs with experimental coronary artery ligation, the infusion of glucose for 24 hours after occlusion, did not modify heart rate, rhythm, arterial pressure, and haematocrit compared with ligated control dogs. However, the depletion of tissue creatine kinase activity was lessened.[37]

In the isolated rat heart with left coronary artery ligation, in which high rates of enzyme release are observed in the presence of free fatty acid as substrate (Figure 1), the addition of glucose leads to significant reduction of enzyme release, suggesting that glucose may be beneficial in reducing ischaemia-induced cellular damage (Figure 2). However, the mechanism involved is not yet clear, although this protective effect is accompanied by a marked reduction in the number of intramitochondrial dense bodies,[30] which are generally observed in the more severely damaged ischaemic cells.[19]

In the whole animal, administration of glucose (or glucose plus insulin) is accompanied by decreased circulating free fatty acid values, an effect which might have contributed to the beneficial effect of glucose and insulin in dogs and in baboons with developing myocardial infarction.[54,55] However, in the experiments of de Leiris et al.,[31] the beneficial effect of glucose developed in isolated hearts perfused with relatively constant circulating free fatty acid values. Opie and Bricknell have recently shown that fatty acid decreased the efficiency of work of the ligated isolated rat heart by over 20%[53] which is in excess of the degree of change which can be found in non-ligated hearts.[14] A change from a respiratory quotient of 3.15 (respiration of glucose alone) to 2.83 (palmitate alone) could only account for an increase of 10% in the myocardial oxygen uptake. The mechanisms involved may include intracellular accumulation of free fatty acid with consequent uncoupling of oxidative phosphorylation but direct experimental proof for this postulate is still lacking. Thus fatty acids 'wasted' and glucose 'conserved' the oxygen available to the ligated heart. Interventions decreasing the oxygen demand of

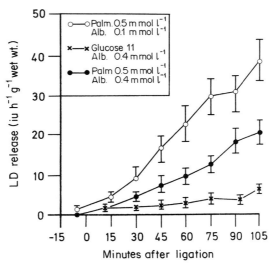

Figure 1. Rates of release of lactate dehydrogenase (LD) in isolated pumping rat heart with coronary artery ligation 15 min after the onset of recirculation perfusion. Note that: (1) palmitate-perfused hearts have higher rates of LD release than do glucose-perfused hearts; (2) an increased albumin concentration decreases lactate dehydrogenase release in palmitate-perfused hearts; and (3) lactate dehydrogenase release is approximately linear. Data based on de Leiris and Opie.[29]

ischaemic tissue are thought to limit infarct size. Other possible mechanisms of the beneficial effect of glucose during oxygen restriction are discussed elsewhere.[50]

EFFECT OF INSULIN (Figure 2)

In acute myocardial infarction, there is insulin resistance and a circulating insulin level inappropriately low in relation to the elevated blood sugar.[49,81] Hence, if glucose were to protect against ischaemia by crossing the cell membrane, it would seem logical also to provide insulin. But insulin may also have other effects, independent of glucose transport, such as inhibition of lysosomal activity.[84] Although it has been argued that insulin is able to enhance the rate of lactate accumulation in ischaemic tissue promoting intracellular acidosis,[65] nevertheless in isolated rat hearts made ischaemic by coronary artery ligation and perfused with free fatty acid as substrate, the

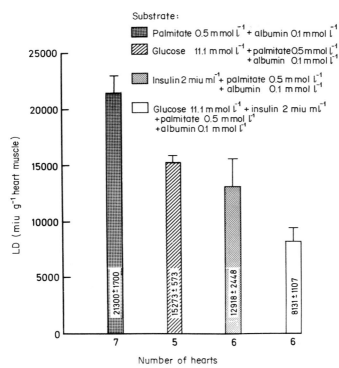

Figure 2. Effects on release of lactate dehydrogenase (LD) of addition of glucose, insulin or glucose + insulin to isolated working rat hearts perfused for 60 min after coronary artery ligation. Note reduction of enzyme release by addition of glucose, or insulin, or glucose + insulin. Based on de Leiris, Lubbe, and Opie.[31] p values: palmitate vs palmitate + glucose <0.025; palmitate vs palmitate + insulin <0.01; palmitate vs palmitate + glucose + insulin <0.0001; palmitate + glucose vs palmitate + glucose + insulin < 0.004; palmitate + insulin vs palmitate + glucose + insulin NS

addition of insulin decreased enzyme release (Figure 2) and diminished the incidence of intramitochondrial dense bodies.[30] More marked beneficial effects can be observed when glucose and insulin are added together to solutions containing free fatty acids (Figure 2). Thus in ischaemic isolated hearts, the detrimental effect of lipid substrates on enzyme release and ultrastructural damage could be reverted nearly to normal by the addition of insulin itself or of glucose and insulin.[29-31]

EFFECT OF GLUCOSE–INSULIN–POTASSIUM

Sodi-Pallares et al.[76] first advocated the use of glucose–insulin–potassium (GIK) infusions in ischaemic heart disease. Their studies suggested that provision of GIK could help to return potassium ions to the ischaemic cells and to 'repolarize' the cells. Others have argued that provision of glucose could exert a beneficial effect either by increasing glycolytic flux, or by reducing circulating free fatty acid levels, or by providing α-glycerophosphate for re-esterification of intracellular free fatty acids, or by an osmotic effect.[48–50] Finally, provision of insulin could enhance glucose transport or exert an inhibition on lysosomal enzyme activity.

The recent study of Maroko et al.[37] shows that glucose–insulin–potassium infusion is able to decrease infarct size following experimental coronary occlusion in dogs, and to maintain a higher creatine kinase activity in ischaemic cells, an effect that these authors attribute to an increase in anaerobic metabolism. However, the rate of production of glycolytic ATP in ischaemic tissue is very low in relation to the rate of aerobic production of ATP, hence it is unlikely that GIK acts by increased provision of energy by formation of anaerobic ATP.[49,50]

In the isolated working rat heart with coronary artery ligation, a low (3.0 mmol l.$^{-1}$) extracellular potassium has a marked negative chronotropic effect when hearts are perfused with fatty acid as substrate.[53] Increasing the potassium to 5.9 mmol l.$^{-1}$ and adding glucose plus insulin leads to mechanical performance values similar or superior to those found in hearts perfused with glucose alone while fatty acid-induced enzyme release is decreased towards values found with glucose alone.[53] Such effects develop in the face of constant fatty acid levels in the perfusion fluid and do not seem to be dependent on a decrease of circulating fatty acid concentration. Thus in these experiments, glucose–insulin–potassium gives a marked protective effect leading to enhanced mechanical performance and to decreased enzyme release.

Glucose–insulin–potassium also decreases the release of creatine kinase from the whole ischaemic rat heart model of Neely, reperfused after 10 minutes of ischaemia, compared with palmitate-perfused hearts.[82] However, the potassium concentration used was 10 mmol l.$^{-1}$ which is higher than is clinically useful.

It should be noted that the effects of GIK developed without any significant increase in cellular ATP or CP levels, but an effect on turnover of high-energy phosphate compounds cannot be excluded.

EFFECT OF ACUTE DIABETES (Figure 3)

As increased provision of glucose and insulin is beneficial to hearts with developing infarction, it may be supposed that the acute diabetic state should

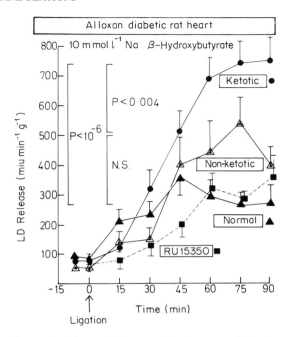

Figure 3. Release of lactate dehydrogenase (LD) from coronary ligated hearts perfused with DL-β-hydroxybutyrate (10 mmol l.$^{-1}$). Comparisons between normal and ketotic hearts were significant ($p < 10^{-6}$ for grouped values 15–90 min post-ligation) as were comparisons between non-ketotic and ketotic hearts ($p < 0.004$). Comparison between grouped data for hearts from non-ketotic diabetic rats and normal rats were not significant but at 75 min post-ligation values were lower in normal hearts ($p < 0.025$). RU 15350 = blood ketone lowering agent. Based on Sinclair-Smith and Opie[74]

predispose to ischaemic damage. Sinclair-Smith and Opie[74] have found that hearts from rats made ketotic by alloxan, release considerably more enzyme after coronary artery ligation than do hearts from non-ketotic, diabetic rats. The ketotic diabetic hearts have lower cardiac contents of ATP which could possibly explain the increased enzyme release.[9] Such enhanced release of enzyme from hearts from ketotic rats can be found during perfusions with glucose, β-hydroxybutyrate (10 mmol l.$^{-1}$) or free fatty acid (0.9 mmol l.$^{-1}$ palmitate, albumin 0.25 mmol l.$^{-1}$) but higher values of enzyme release are reached with ketone bodies or fatty acids. Pre-treatment of the rats injected with alloxan by an antilipolytic agent, tizoprolic acid (Roussel, RU 15350), leads to decreased circulating ketone bodies and decreased enzyme release.[74] Thus the ketotic state predisposes to myocardial ischaemic damage.

EFFECT OF FREE FATTY ACIDS AND OF ALBUMIN

Harmful effects of free fatty acids have been found in normal hearts perfused with oleate, linoleate, or octanoate at high fatty acid concentrations and/or high free fatty acid/albumin molar ratio.[49,86] Palmitate can be 'toxic' to either coronary ligated hearts, or to the whole ischaemic heart preparation.[29-31,82] High free fatty acid/albumin molar ratios enhance 'toxicity'. Enzyme release 1 hour after coronary artery ligation is consistently higher with high palmitate/albumin molar ratio (P/A = 5) than with normal ratios (P/A = 1) even with constant initial palmitate perfusate concentrations (Figure 1). However, with high palmitate/albumin molar ratios, increasing both palmitate and albumin concentrations leaves the rate of enzyme release unchanged.[29]

Lochner et al.[33] found that albumin added to the perfusate of the isolated anoxic rat heart had a beneficial effect on the function of isolated mitochondria. In dogs with coronary artery occlusion, a significant reduction of myocardial free fatty acid uptake and epicardial ST segment elevation has been reported by Oliver et al.[39] after albumin infusions.

With similar free fatty acid/albumin molar ratios (free fatty acid/albumin = 5) and similar free fatty acid concentrations (0.5 mmol l.$^{-1}$), linoleate and oleate cause less enzyme release from isolated working rat hearts after coronary artery ligation than does palmitate (de Leiris and Opie, unpublished data). The solubility of palmitate and of its calcium salt in water is much less than that of oleate or linoleate.[80] Precipitation of calcium soap of fatty acid in extracellular space may be one of the reasons accounting for the 'toxic' effect of free fatty acid substrates.[86]

OTHER NON-GLUCOSE SUBSTRATES (Figure 4)

Bricknell and Opie[5] have subjected hearts to mild global ischaemia by underperfusion and then induced exaggerated ischaemic damage by reperfusion. Hearts perfused with pyruvate, acetate, or free fatty acid all have increased enzyme release in the reperfusion period. This increase of enzyme release could be related to an increased intracellular accumulation of cyclic AMP, which could either be a reflection of increased ischaemic damage or could be viewed as an agent increasing cell permeability to enzymes.[18]

Two other non-glucose fuels, β-hydroxybutyrate and lactate also increase myocardial ischaemic damage as assessed by increased lactate dehydrogenase (LD) release.[53,74] Thus there may be a general rule that provision of

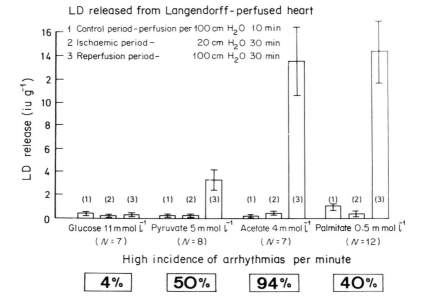

Figure 4. Effect of substrates on lactate dehydrogenase (LD) release and on arrhythmias in globally ischaemic rat heart.[6] The various substrates used were: glucose 11 mmol l.$^{-1}$, pyruvate 5 mmol l.$^{-1}$, acetate 4 mmol l.$^{-1}$, and palmitate 0.5 mmol l.$^{-1}$ (bound to albumin 0.1 mmol l.$^{-1}$). Note that in this model LD release is not increased in the ischaemic period but rises during the reperfusion period in pyruvate but especially in acetate and in palmitate hearts. The peak incidence of arrhythmias (per minute) in the reperfusion period is shown; the major difference lies between the very low incidence with glucose and the much higher incidence with non-glucose fuels. For full details see references 5 and 6. LD units indicate total release of LD for the duration of the whole control, ischaemic or reperfusion period

non-glucose fuels to the ischaemic myocardium can increase ischaemic damage.

ANTILIPOLYTIC AGENTS

In vivo, the release of free fatty acids (from adipose tissue and/or the heart) occurring during ischaemic conditions plays an important role in the evolvement of acute ischaemic injury of the myocardium. Inhibition of lipolysis may, therefore, confer some protection upon the myocardium. Thus inhibition of lipolysis with β-pyridyl carbinol was associated with less

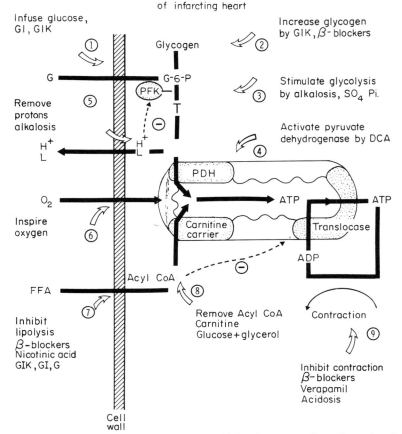

Figure 5. Substrate hypothesis for therapy of infarcting myocardium. Reproduced from Opie.[51] Nine possibilities for substrate-orientated therapy of the infarcting heart are shown. These possibilities are based on animal experimental data and require carefully controlled evaluation in patients in specialized centres. First infusions of glucose[37] or provision of glucose and insulin[31] or glucose–insulin–potassium (GIK) accelerate the uptake[55] and metabolism of glucose (arrow 1) by the whole heart (ischaemic plus non-ischaemic zones). Differential effects on ischaemic versus non-ischaemic zones are complex (see text and reference 50). Increased cardiac glycogen results from GIK[54,55] or β-blocking agents[62] (arrow 2). Glycolysis is stimulated by alkalosis in the well-perfused hypoxic heart[68] which may explain the success of Regan et al.[61] in using sodium carbonate in the infarcting dog heart (arrow 3). Activation of pyruvate dehydrogenase by dichloroacetate (DCA) also leads to decreased ischaemic damage[41] (arrow 4). Extracellular alkalosis removes extracellular protons and should thus remove intracellular protons and accelerate glycolysis (arrow 5). Increased provision of oxygen (arrow 6) decreases ischaemiac damage both in experimental models[38] and in man.[35] Lipolysis when inhibited by β-blockade,[58] nicotinic acid,[45] clofibrate,[40] or GIK[54,55] (or, presumably, glucose or insulin) leads to reduced circulating free fatty acids and to diminished ischaemic damage (arrow 7). Provision of carnitine (arrow 8) might act by removal of acyl CoA[52] although the role of carnitine is controversial. Glucose and glycerol and insulin[33] might conceivably act here. Ninthly, and very importantly (arrow 9), the rate of utilization of ATP can be decreased by inhibiting contractile activity as a result of β-blockers,[36,58] verapamil,[75] or acidosis.[2] A beneficial effect of extracellular acidosis in hypoxia must be assessed together with its harmful effects in inhibiting glycolysis in ischaemia; in practice, alkalinization improves developing dog infarcts[61]

creatine kinase depletion in the ischaemic area, indicating salvage of the ischaemic myocardium.[22] Similar protection has been described with other antilipolytic agents such as nicotinic acid, p-chlorophenoxyisobutyrate (clofibrate), or prostaglandin E_1.[40] In isolated rat hearts, nicotinic acid and clofibrate also inhibit catecholamine-induced lipolysis.[40] However, a cautionary note is introduced by the finding that the administration of the antilipolytic agent, 'Ronicol', to dogs with coronary artery thrombosis failed to improve tissue glycogen and potassium in the infarct zone 24 hours later when compared with control dogs.[7]

Inhibition of free fatty acid oxidation accompanied by a stimulation of glucose oxidation has been reported after treatment with sodium dichloroacetate in isolated perfused rat heart.[34] In dogs with coronary occlusion, Mjøs et al.[41] report that dichloroacetate reduces the degree of ST segment elevation while decreasing the uptake of fatty acid; the mechanism of dichloroacetate action is probably by activation of pyruvate dehydrogenase[34] (see Figure 5).

METABOLIC MANIPULATIONS, GLYCOLYTIC FLUX, AND ISCHAEMIC DAMAGE (Tables 3 and 4; Figure 5)

Most of the procedures discussed thus far can be related to their effects on glycolytic flux. Thus in general, agents increasing glycolytic flux decrease ischaemic injury (Table 3) except for increased heart work of which the major effect is the increased oxygen demand and, hence, increased ischaemic damage. Conversely, agents which decrease glycolytic flux generally increase myocardial ischaemic injury, except for decreased heart work (Table 4). It must be established whether there is any special role for increased glycolytic flux as argued by Bricknell and Opie[5,6] or whether these changes in glycolytic flux merely reflect changes in the relative rates of

Table 3. Relation between Increased Glycolytic Flux, Enzyme Release from the Heart, and Myocardial Ischaemic Injury

Factors increasing glycolytic flux	Effects on enzyme release	Effects on ischaemic injury other than enzyme release
Fed state	No data	No data
Insulin	Decreases fatty acid-induced enzyme release[31]	Decreased mitochondrial dense bodies[30]
Glucose, GIK	Decreases[29,31,53]	Decreases[30,37,54,55]
Antilipolytic agents	No data	Decrease[39–41,45]
Alkalosis	No data	Decreases[57]
Activation of pyruvate dehydrogenase	No data	Decreases[41]
Increased heart work	Increases[53]	Increases[36]

Based on Opie.[51]

Table 4. Relation between Decreased Glycolytic Flux and Myocardial Ischaemic Injury

Factors decreasing glycolytic flux	Effects on enzyme release	Effects on ischaemic injury other than enzyme release
Fasting	No data	No data
Hypoglycaemia	No data	Increases[32]
Diabetes (alloxan)	Increases[74]	No data
Alternate fuels:		
FFA	Increase[29,31]	Increase[22,30,40]
Lactate	Increases[53]	Increases[1]
β-Hydroxybutyrate	Increases[74]	No data
Pyruvate	Increases[a]	No data
Acetate	Increases[a]	No data
Acidosis	No data	Increases[1,61]
Decreased heart work	Decreases[53]	Decreases[36]

[a] Reperfusion release of enzyme measured, Bricknell and Opie[5] (see Figure 3). Based on Opie.[51]

utilization of glucose and of free fatty acids. The proposal has been made that survival of the ischaemic myocardium can be improved by enhancing the utilization of glucose relative to that of free fatty acid.[48] Most, but not all of the substrate data support this latter proposition. However, the possible effect of all non-glucose fuels in promoting ischaemic damage argue for a special protective role for glucose, perhaps by provision of glycolytic flux. Changes in glycolytic flux should be seen as only one aspect of the substrate hypothesis for the therapy of the infarcting heart (Figure 5).

OTHER MANIPULATIONS

Studies with methylprednisolone sodium succinate (MPSS) have shown that this compound is able to modify the effects of hypoxia in isolated perfused rat hearts, delaying the release of creatine kinase and succinate dehydrogenase, prolonging the phase of increased coronary flow, and decreasing the rate of depletion of the energy-rich phosphate stores.[42] Such an observation is consistent with those of Spath et al.[77] on cats with coronary occlusion, who reported that MPSS is able to limit infarct size and suggested that this beneficial effect is related to the ability of glucocorticoid to stabilize cardiac membranes, preventing the ischaemia-induced leakage of lysosomal enzymes. However, the effect of MPSS appears to be complex. Indeed Nayler and Seabra-Gomes have claimed that the sodium succinate component of the MPSS is metabolically effective by itself, slowing down the rate at which the high-energy phosphates are depleted and therefore, delaying the development of myocardial cellular damage.[42]

Other drugs like α-tocopherol or vitamin E are also able to preserve mitochondrial function during severe oxygen depletion and to exert a sparing effect on the rate at which the tissue stores of ATP are depleted (see Chapter 2).

After more than 60–120 minutes of coronary occlusion, reperfusion or restoration of arterial blood flow into the ischaemic tissue either does not occur or is greatly impeded. This phenomenon called 'no-reflow' is associated with extensive capillary damage and myocardial cell swelling.[24] Elevation of osmolality by mannitol results in improved function and increased coronary flow probably by preventing or reducing myocardial cell swelling.[25,60] Dihydroxyacetone and allopurinol have also been shown to improve reperfusion of ischaemic rat hearts by reducing myocardial cell swelling and tissue oedema.[8]

CONCLUDING COMMENTS

Metabolic manipulations of the substrate supply can afford some degree of protection to myocardial cells against ischaemic damage, either *in vivo*, or in isolated perfused preparations, preventing, delaying, or reducing intracellular cardiac enzyme release. The site of action of such metabolic manipulations might be on less severely ischaemic cells in the 'border zone'.

These manipulations may act by augmenting glycolytic flux (glucose, glucose + insulin, hypertonic glucose, dichloroacetate), by inhibition of lipolysis (β-pyridyl carbinol, clofibrate, nicotinic acid), by enhancing glucose transport to the ischaemic cell of a substrate which can be utilized in energy production (insulin), by decreasing the delivery of free fatty acids to the ischaemic cell (albumin infusion), by protecting lysosomal membranes (insulin, MPSS), or by reducing myocardial cell swelling (mannitol, dihydroxyacetone, allopurinol). In general, agents decreasing enzyme loss from the heart decrease myocardial ischaemic damage as assessed by other means (Tables 3 and 4), and agents increasing enzyme release increase ischaemic damage. Although enzyme loss from isolated hearts cannot always be equated with ischaemic damage, it is difficult to see how enzyme loss can be beneficial to the heart cell.

It would seem reasonable to conclude that reduction of enzyme loss by substrate therapy is possible and desirable, especially in the setting of ischaemia or developing infarction. Presumably metabolic agents should be able to decrease enzyme release in patients with acute myocardial infarction. It must, however, be stressed that the aim of this chapter has been to *outline* the possible metabolic manipulations which could influence enzyme release from the oxygen-deprived and especially the ischaemic heart.

Each proposal requires rigorous testing, firstly in a variety of animal preparations corresponding as closely as possible to the situation in developing myocardial infarction in man, and then in patients in carefully controlled studies in a coronary care unit. The field remains highly controversial but further support for metabolic manipulation (glucose promotion, lipid restriction) has come from the patient studies of Rogers et al.[63] who found a reduction of hospital mortality rate of acute myocardial infarction in a non-randomized study on the effects of glucose–insulin–potassium, and from Russell and Oliver[66] who used antilipolytic therapy to decrease ST segment elevation during the early phase of acute infarction in man. But Heng et al.[16] found that glucose–insulin–potassium infusions, when given to patients in doses that caused marked hyperglycaemia (30 mmol l.$^{-1}$) had an inotropic effect but did not decrease infarct size as judged by creatine kinase release curves; their data could be interpreted to show that glucose–insulin–potassium prevented the increase in infarct size which should have resulted from the inotropic effect, the latter presumably achieved by hyperosmolality. However, because glucose and insulin promote the clearance of creatine kinase from the plasma,[78] procedures other than creatine kinase clearance curves should be used to assess effects of glucose–insulin–potassium.

Arguments for testing mannitol and methylprednisolone in patients become more difficult to sustain in view of recent evidence that mannitol did not decrease infarct size following permanent coronary occlusion (most of the previous studies were on temporary occlusion), and may have hindered the development of collateral vessels,[17] while high-dose methylprednisolone impaired wound healing.[23]

REFERENCES

1. Armiger, L. C., Herdson, P. B., and Gavin, J. B. Mitochondrial changes in dog myocardium induced by lowered pH in vitro. Laboratory Investigation, **32**, 223–6, (1975).
2. Bing, O. H. L., Brooks, W. W., and Messer, J. V. Heart muscle inability following hypoxia: protective effect of acidosis. Science, **180**, 1297–8 (1973).
3. Bing, R. J. Cardiac metabolism. Physiological Reviews, **45**, 171–213 (1965).
4. Brachfeld, N. Maintenance of cell viability. Circulation Supplement IV, **40**, 202–19 (1969).
5. Bricknell, O. L. and Opie, L. H. Effects of various substrates on lactate dehydrogenase release and on arrhythmias in the isolated rat heart during underperfusion and reperfusion. Circulation Research, **43**, 102–15 (1978).
6. Bricknell, O. L. and Opie, L. H. Glycolytic ATP and its production during ischaemia in isolated Langendorff-perfused rat hearts. In Recent Advances in Studies on Cardiac Structure and Metabolism Vol 11, Heart Function and Metabolism (Eds. Kobayashi, T., Sano, T., and Dhaila, N. S.). University Park Press, Baltimore, 1978, pp. 509–19.
7. Dalby, A. J., Boulle, G. J., Lubbe, W. F., and Opie, L. H. Effect of two antilipolytic agents ('Ronicol' and tizoprolic acid) on development of myocardial infarction in closed-chest dogs. Unpublished data.
8. Fabiani, J. N. The no-reflow phenomenon following early reperfusion of myocardial infarction and its prevention by various drugs. Heart Bulletin, **7**, 134–42 (1976).
9. Gebhard, M. M., Denkhaus, H., Sakai, K., and Spieckermann, P. G. Energy metabolism and enzyme release. Journal of Molecular Medicine, **2**, 271–83, (1977).
10. Hearse, D. J. and Chain, E. B. The role of glucose in the survival and recovery of the anoxic isolated perfused rat heart. Biochemical Journal, **128**, 1125–33 (1972).
11. Hearse, D. J. and Humphrey, S. M. Enzyme release during myocardial anoxia: a study of metabolic protection. Journal of Molecular and Cellular Cardiology, **7**, 463–82 (1975).

12. Hearse, D. J., Humphrey, S. M., and Garlick, P. B. Species variation in myocardial anoxic enzyme release, glucose protection and reoxygenation damage. *Journal of Molecular and Cellular Cardinology*, **8**, 329–39 (1976).
13. Hearse, D. J., Humphrey, S. M., Feuvray, D., and de Leiris, J. A biochemical and ultrastructural study of the species variation in myocardial cell damage. *Journal of Molecular and Cellular Cardiology*, **8**, 759–78 (1976).
14. Hearse, D. J., Opie, L. H., Katzeff, I. E., Lubbe, W. F., Van Der Werff, T. J., Peisach, M., and Boulle, G. Characterization of the 'border zone' in acute regional ischemia in the dog. *American Journal of Cardiology*, **40**, 716–26 (1977).
15. Henderson, A. H., Most, A. S., Parmley, W. W., Gorlin, R., and Sonnenblick, E. H. Depression of myocardial contractility in rats by free fatty acids during hypoxia. *Circulation Research*, **26**, 439–49 (1970).
16. Heng, M. K., Norris, R. M., Singh, B. N., and Barratt-Boyes, C. Effects of glucose and glucose–insulin–potassium on haemodynamics and enzyme release after acute myocardial infarction. *British Heart Journal*, **39**, 748–57 (1977).
17. Hirzel, H. O. and Kirk, E. S. The effect of mannitol following permanent coronary occlusion. *Circulation*, **56**, 1006–15 (1977).
18. Horak, A. R., Podzuweit, T., and Opie, L. H. Adrenaline-induced release of enzymes from the myocardium; the role of beta-receptors, tissue cyclic AMP and high-energy phosphates. Unpublished data.
19. Jennings, R. B. and Ganote, C. E. Structural changes in myocardium during acute ischemia. *Circulation Research Supplement 3*, **34**, 156–72 (1974).
20. Kaijser, L., Carlson, L. A., Eklund, B., Nye, E. R., Rossner, S., and Wahlqvist, M. L. Substrate uptake by the ischaemic human heart during angina-induced atrial pacing. In *Effect of Acute Ischaemia on Myocardial Function* (Eds. Oliver M. F., Julian, D. G., and Donald, K. W.), Churchill Livingstone, Edinburgh, 1972, pp. 223–33.
21. Katz, A. M. and Hecht, H. H. The early 'pump' failure of the ischemic heart. *American Journal of Medicine*, **47**, 497–502 (1969).
22. Kjekshus, J. K. Effect of lipolytic and inotropic stimulation on myocardial ischemic injury. *Acta Medica Scandinavica Supplement*, **587**, 35–42 (1975).
23. Kloner, R. A., Fishbein, M. C., Lew, H., Maroko, P. R., and Braunwald, E. Mummification of the infarcted myocardium by high doses of corticosteroids. *Circulation*, **57**, 56–63 (1978).
24. Kloner, R. A., Ganote, C. E., and Jennings, R. B. The 'no-reflow' phenomenon after temporary coronary occlusion in the dog. *Journal of Clinical Investigation*, **54**, 1496–508 (1974).
25. Kloner, R. A., Reimer, K. A., Willerson, J. T., and Jennings, R. B. Reduction of infarct size with hyperosmolar mannitol. *Proceedings of the Society for Experimental Biology and Medicine*, **151**, 677–83 (1976).
26. Kübler, W. and Spieckermann, P. G. Regulation of glycolysis in the ischaemic and the anoxic myocardium. *Journal of Molecular and Cellular Cardiology*, **1**, 351–77 (1970).
27. Kurien, V. A., Yates, P. A., and Oliver, M. F. The role of free fatty acids in the production of ventricular arrhythmias after acute coronary artery occlusion. *European Journal of Clinical Investigation*, **1**, 225–41 (1971).
28. Lassers, B. W., Carlson, L. A., Kaijser, L., and Wahlqvist, M. The nature and control of myocardial substrate metabolism in healthy man. In *Effect of Acute Ischaemia on Myocardial Function* (Eds. Oliver, M. F., Julian, D. G., and Donald, K. W.), Churchill Livingstone, Edinburgh, 1972, p. 200.
29. de Leiris, J. and Opie, L. H. Beneficial effects of glucose, insulin and potassium and detrimental effects of free fatty acid on enzyme release and on mechanical performance of isolated rat heart with coronary artery ligation. *Cardiovascular Research*. In press, 1978.
30. de Leiris, J. and Feuvray, D. Ischaemia-induced damage in the working rat heart preparation: effect of perfusate substrate composition upon subendocardial ultrastructure of the ischaemic left ventricular wall. *Journal of Molecular and Cellular Cardiology*, **9**, 365–73 (1977).
31. de Leiris, J., Lubbe, W. F., and Opie, L. H. Effects of free fatty acid and glucose on

enzyme release in experimental myocardial infarction. *Nature (London)*, **253**, 746–7 (1975).
32. Libby, P., Maroko, P. R., and Braunwald, E. The effect of hypoglycaemia on myocardial ischemic injury during acute experimental coronary artery occlusion. *Circulation*, **51**, 621–6 (1975).
33. Lochner, A., Kotze, J. C. N., and Gevers, W. Mitochondrial oxidative phosphorylation in myocardial anoxia: effects of albumin. *Journal of Molecular and Cellular Cardiology*, **8**, 465–80 (1976).
34. McAllister, A., Allison, S. P., and Randle, P. J. Effects of dichloroacetate on the metabolism of glucose, pyruvate, acetate, 3-hydroxybutyrate and palmitate in rat diaphragm and heart muscle *in vitro* and on extraction of glucose, lactate, pyruvate and free fatty acids by dog heart *in vivo*. *Biochemical Journal*, **134**, 1067–81 (1973).
35. Madias, J. E., Madias, N. E., and Hood, W. B. Jr. Precordial ST mapping. 2) Effects of oxygen inhalation on ischemic injury in patients with acute infarction. *Circulation*, **53**, 411–7 (1976).
36. Maroko, P. R., Kjekshus, J. K., Sobel, B. E., Watanabe, T., Covell, J. W., Ross, J., Jr., and Braunwald, W. Factors influencing infarct size following experimental coronary artery occlusions. *Circulation*, **43**, 67–82 (1971).
37. Maroko, P. R., Libby, P., Sobel, B. E., Bloor, C. M., Sybers, H. D., Shell, W. E., Covell, J. W., and Braunwald, E. The effects of glucose–insulin–potassium infusion on myocardial infarction following experimental coronary artery occlusion. *Circulation*, **45**, 1160–75 (1972).
38. Maroko, P. R., Radvany, P., Braunwald, E., and Hale, S. L. Reduction of infarct size by oxygen inhalation following acute coronary occlusion. *Circulation*, **52**, 360–8 (1975).
39. Miller, N. E., Mjøs, O. D., and Oliver, M. F. Relationship of epicardial ST-segment elevation to the plasma free fatty acid to albumin ratio during coronary occlusion in dogs. *Clinical Science and Molecular Medicine*, **51**, 209–13 (1976).
40. Mjøs, O. D. Effect of reduction of myocardial free fatty acid metabolism relative to that of glucose on the ischemic injury during experimental coronary artery occlusion in dogs. *Acta Medica Scandinavica Supplement*, **587**, 29–34 (1976).
41. Mjøs, O. D., Miller, N. E., Riemersma, R. A., and Oliver, M. F. Effects of dichloroacetate on myocardial substrate extraction, epicardial ST-segment elevation and ventricular blood flow following coronary occlusion. *Cardiovascular Research*, **10**, 427–36 (1976).
42. Nayler, W. G. and Seabra-Gomes, R. Effect of methylprednisolone sodium succinate on hypoxic heart muscle. *Cardiovascular Research*, **10**, 349–58 (1976).
43. Neely, J. R. and Morgan, H. E. Relationship between carbohydrate and lipid metabolism and the energy balance of heart muscle. *Annual Review of Physiology*, **36**, 413–59 (1974).
44. Neely, J. R., Whitmer, K. M., and Mochizuki, S. Effects of mechanical activity and hormones on myocardial glucose and fatty acid utilization. *Circulation Research Supplement 1*, **38**, 210–2 (1976).
45. Oliver, M. F., Rowe, M. J., Luxton, M. R., Miller, N. E., and Nelson, J. M. Effect of reducing circulating free fatty acids on ventricular arrhythmias during myocardial infarction and on ST-segment depression during exercise-induced ischemia. *Circulation Supplement 1*, **53**, 210–2 (1976).
46. Opie, L. H. Metabolism of the heart in health and disease. Part I. *American Heart Journal*, **76**, 685–98 (1969).
47. Opie, L. H. Metabolism of the heart in health and disease. Part II. *American Heart Journal*, **77**, 100–22 (1969).
48. Opie, L. H. Metabolic response during impending myocardial infarction. I) Relevance of studies of glucose and fatty acid metabolism in animals. *Circulation*, **45**, 483–90 (1972).
49. Opie, L. H. Metabolism of free fatty acids, glucose and catecholamines in acute myocardial infarction. Relation to myocardial ischemia and infarct size. *American Journal of Cardiology*, **36**, 938–53 (1975).
50. Opie, L. H. Effects of regional ischaemia on metabolism of glucose and free fatty acids. Relative rates of aerobic and anaerobic energy production during myocardial infarction and comparison with effect of anoxia. *Circulation Research Supplement 1*, **38**, 52–68 (1976).
51. Opie, L. H. Metabolic heart disease, with special reference to carbohydrate metabolism in

health and disease. In *Myocardial Failure* (Ed. Riecker, G.), Springer-Verlag, Berlin, 1977.
52. Opie, L. H. Role of carnitine in cardiac fatty acid metabolism. Implications for myocardial ischaemia, arrhythmias, adriamycin toxicity and diphtheritic myocarditis. In press, *American Heart Journal*, 1978.
53. Opie, L. H. and Bricknell, O. L. Substrate-induced enzyme release and inefficiency of work in isolated rat hearts with coronary artery ligation. Effects of fatty acid, lactate, glucose and glucose-insulin-potassium. Submitted for publication, 1978.
54. Opie, L. H., Bruyneel, K., and Owen, P. Beneficial effects of glucose, potassium and insulin infusions on tissue metabolic changes within srst hours of myocardial infarction in the baboon. *Circulation*, **52**, 49–57 (1975).
55. Opie, L. H. and Owen, P. Effects of glucose–insulin–potassium infusions on arteriovenous differences of glucose and of free fatty acids and on tissue metabolic changes in dogs with developing myocardial infarction. *American Journal of Cardiology*, **38**, 310–21 (1976).
56. Opie, L. H., Owen, P., and Mansford, K. R. L. Effects of increased heart work on glycolysis and on adenine nucleotides in the perfused heart of normal and diabetic rats. *Biochemical Journal*, **124**, 475–90 (1971).
57. Opie, L. H., Owen, P., and Riemersma, R. A. Relative rates of oxidation of glucose and free fatty acids by ischemic and non-ischemic myocardium after coronary artery ligation in the dog. *European Journal of Clinical Investigation*, **3**, 419–35 (1973).
58. Opie, L. H. and Thomas, M. Propranolol and experimental myocardial infarction. Substrate effects. *Postgraduate Medical Journal Supplement 4*, **52**, 124–32 (1976).
59. Pande, S. V. and Blanchaer, M. C. Reversible inhibition of mitochondrial adenosine diphosphate phosphorylation by long chain acyl coenzyme A esters. *Journal of Biological Chemistry*, **246**, 402–11 (1971).
60. Powell, W. J., Dibona, D. R., Flores, J., and Leaf, A. The protective effect of hyperosmotic mannitol in myocardial ischemia and necrosis. *Circulation*, **54**, 603–15 (1976).
61. Regan, T. J., Effros, R. M., Haider, B., Oldewurthel, H. A., Ettinger, P. O., and Ahmed, S. S. Myocardial ischemia and cell acidosis: modification by alkali and the effects on ventricular function and cation composition. *American Journal of Cardiology*, **37**, 501–7 (1976).
62. Reimer, K. A., Rasmussen, M. M., and Jennings, R. B. Reduction by propranolol of myocardial necrosis following temporary coronary artery occlusion in dogs. *Circulation Research*, **33**, 353–63 (1973).
63. Rogers, W. J., Stanley, A. W., Breinig, J. B., Prather, J. W., McDaniel, H. G., Moraski, R. E., Mantle, J. A., Russell, R. O., and Rackley, C. E. Reduction of hospital mortality of acute myocardial infarction with glucose–insulin–potassium infusion. *American Heart Journal*, **92**, 441–54 (1976).
64. Rovetto, M. J., Lamberton, W. F., and Neely, J. R. Mechanisms of glycolytic inhibition in ischemic rat hearts. *Circulation Research*, **37**, 742–51 (1975).
65. Rovetto, M. J., Whitmer, J. T., and Neely, J. R. Comparison of effects of anoxia and whole-heart ischemia on carbohydrate utilization in isolated working rat hearts. *Circulation Research*, **32**, 699–711 (1973).
66. Russell, D. C. and Oliver, M. F. Effect of antilipolytic therapy on ST-segment elevation during myocardial ischemia in man. *British Heart Journal*, **40**, 117–23 (1978).
67. Scheuer, J. and Stezoski, S. W. Protective role of increased myocardial glycogen stores in cardiac anoxia in the rat. *Circulation Research*, **27**, 835–49 (1970).
68. Scheuer, J. and Stezoski, S. W. The effect of alkalosis on the mechanical and metabolic response of the rat heart to hypoxia. *Journal of Molecular and Cellular Cardiology*, **4**, 599–610 (1972).
69. Schwartz, A., Sordahl, L. A., Entman, M. L., Allen, J. C., Reddy, Y. S., Goldstein, M. A., Luchi, R. J., and Wyborny, L. E. Abnormal biochemistry in myocardial failure. *American Journal of Cardiology*, **32**, 407–22 (1972).
70. Seabra-Gomes, R., Ganote, C. E., and Nayler, W. G. Species variation in anoxic-induced damage of heart muscle. *Journal of Molecular and Cellular Cardiology*, **7**, 929–37 (1975).
71. Shipp, J. C., Menahan, L., and Crass, M. F. Heart triglycerides in health and disease. In *Recent advances in Studies on Cardiac Structure and Metabolism* (Ed. Dhalla, N. S.), Vol. 3, University Park Press, Baltimore, 1973, p. 179.

72. Shipp, J. C., Opie, L. H., and Challoner, D. Fatty acid and glucose metabolism in the perfused heart. *Nature (London)*, **189,** 1018–9 (1961).
73. Shug, A. L. and Shrago, E. A proposed mechanism for fatty acid effects on energy metabolism of the heart. *Journal of Laboratory and Clinical Medicine*, **81,** 214–7 (1973).
74. Sinclair-Smith, B. C. and Opie, L. H. Effect of diabetic ketosis on enzyme release from isolated perfused rat hearts with experimental myocardial infarction. *Journal of Molecular and Cellular Cardiology*, **10,** 221–34 (1978).
75. Smith, H. J., Singh, B. N., Nisbet, H. D., and Norris, R. M. Effects of verapamil on infarct size following experimental coronary occlusion. *Cardiovascular Research*, **9,** 569–78 (1975).
76. Sodi-Pallares, D., Testelli, M. R., Fishleder, B. L., Bisteni, A., Medrano, G. A., Friedland, C., and de Micheli, A. Effects of an intravenous infusion of a glucose–insulin–potassium solution on the electrocardiographic signs of myocardial infarction. A preliminary clinical report. *American Journal of Cardiology*, **9,** 166–81 (1962).
77. Spath, J. A., Lane, D. L., and Lefer, A. L. Protective action of methylprednisolone on the myocardium during experimental myocardial ischemia in the cat. *Circulation Research*, **35,** 44–51 (1974).
78. Spath, J. A., Ogletree, M. L., and Lefer, A. L. Lack of significant protective effect of augmented circulating glucose on the ischemic myocardium. *Canadian Journal of Physiology and Pharmacology*, **54,** 423–9 (1976).
79. Spector, A. Metabolism of free fatty acids. *Progress in Biochemistry and Pharmacology*, **6,** 130–76 (1971).
80. Spector, A., John, K., and Flechter, J. E. Binding of long chain fatty acids to bovine albumin. *Journal of Lipid Research*, **10,** 56–7 (1969).
81. Vetter, N. J., Strange, R. C., Adams, W., and Oliver, M. F. Initial metabolic and hormonal response to acute myocardial infarction. *Lancet*, **1,** 284–8 (1974).
82. Waldenström, A. and Hjalmarson, A. Factors modifying ischemic injury in the isolated rat heart. *Acta Medica Scandinavica*, **201,** 533–8 (1977).
83. Welman, E. and Peters, T. J. Enhanced lysosome fragility in the anoxic perfused guinea pig heart: effects of glucose and mannitol. *Journal of Molecular and Cellular Cardiology*, **9,** 101–20 (1977).
84. Wildenthal, K. Inhibition by insulin of cardiac cathepsin D activity. *Nature (London)*, **243,** 226–7 (1973).
85. Willebrands, A. F., Tasseron, S. J. A., Ter Welle, H. F., and Van Dam, R. Th. Effects of oleic acid and oxygen restriction on rhythm and contractility of the isolated rat heart; protective action of glucose. *Journal of Molecular and Cellular Cardiology*, **8,** 375–88 (1976).
86. Willebrands, A. F., Ter Welle, H. F., and Tasseron, S. J. A. The effect of a high molar free fatty acid to albumin ratio in the perfusion medium on rhythm and contractility of the isolated rat heart. *Journal of Molecular and Cellular Cardiology*, **5,** 259–73 (1973).

CHAPTER 22

Pharmacological protection of the hypoxic heart: enzymatic, biochemical, and ultrastructural studies in the isolated heart

W. G. Nayler and A. M. Slade

INTRODUCTION	503
SPECIES VARIATION IN HYPOXIA-INDUCED ENZYME LEAKAGE	504
THE METABOLIC AND ULTRASTRUCTURAL CONSEQUENCES OF CARDIAC HYPOXIA	505
Metabolism	505
Ultrastructure	510
PROTECTIVE PROCEDURES	511
EFFECT OF A MILD ACIDOSIS	513
β-ADRENOCEPTOR ANTAGONISTS	517
Ca^{2+}-ANTAGONIST DRUGS, E.G. VERAPAMIL	518
MEMBRANE-STABILIZING DRUGS	521
CONCLUDING COMMENTS	523
REFERENCES	525

INTRODUCTION

During the past decade many attempts have been made to elucidate the sequence of events that occur when, because of an inadequate or interrupted blood flow, the myocardium becomes ischaemic.[27] Ischaemia has two important consequences—a reduction in the supply of oxygen and of substrate, and a retention of the products of metabolism. The resultant functional, metabolic, and morphological changes are complex and, if the supply of oxygen and substrates is not rapidly restored, will result in irreversible damage to the myocardium leading to cell death and necrosis.[7,10] The appearance in the plasma of various enzymes, e.g. creatine kinase, succinate dehydrogenase, hydroxybutyrate dehydrogenase, aspartate aminotransferase, and other substances that are of intracellular origin, e.g. myoglobin[37,40] or inosine and hypoxanthine,[31] is commonly used to assess the severity and extent of the damage caused by ischaemic episodes.[35,36] In this chapter, however, emphasis will be placed not on the possible use of such marker profiles as an aid to diagnosis and prognosis but rather on their use in laboratory studies which are aimed at elucidating the sequence of events triggered by the initial insult and which, if left unchecked, will result in cell damage and infarction. One particular enzyme—creatine kinase (CK-ATP: creatine phosphotransferase EC 2.7.3.2) has been used more than any other as a marker enzyme[35,36] for this purpose and therefore will be used here. Our observations will be restricted to the effects of a reduced supply of oxygen, with or without a concommitant reduction in substrate supply. Only basic mechanisms are being considered, and the discussion will be centred primarily around results obtained from studies on isolated hearts in the author's laboratory. Before considering these results it may be wise to document the logic behind the selection of the particular model used—the isolated rabbit heart.

SPECIES VARIATION IN HYPOXIA-INDUCED ENZYME LEAKAGE

Irrespective of whether isolated hearts are perfused retrogradely as described by Langendorff[12] or as 'working' models such that the left ventricle expels fluid against a pressure load[25] different species show marked differences in their susceptibility to the combined effects of hypoxia and substrate lack, as evidenced by enzyme release studies.[8,9,34] Rat and mouse hearts are relatively resistant to hypoxia-induced change, whilst rabbit hearts[34] are extremely sensitive. The hypersensitivity of the rabbit heart relative to either the rat or guinea-pig is reflected in the results from the enzyme (CK) release studies summarized in Table 1.

Table 1. Species Variation in Release of CK from Isolated Langendorff Perfused Hearts during Substrate-free Hypoxic Perfusion

CK release (miu min^{-1} g^{-1} wet wt.). Mean ± SEM $n=6$				
Duration of perfusion (min)	15	30	45	60
Species: Rabbit	76 ± 4	102 ± 8	135 ± 16	220 ± 14
Rat	5 ± 1	14 ± 2	16 ± 6	12 ± 4
Guinea-pig	52 ± 4	68 ± 6	82 ± 5	76 ± 12

Perfusion at 37 °C using a modified substrate-free Krebs-Henseleit buffer solution. Hypoxia ($Po_2 < 6$ mmHg) introduced at zero time by gassing the perfusate with 95% N_2 + 5% CO_2 instead of 95% O_2 + 5% CO_2. Results are mean ±SEM of 6 separate experiments.[34]

Removal of substrate coincident with the introduction of hypoxic conditions causes (Table 2) an exacerbation of the hypoxia-induced enzyme release. Rabbit hearts that are perfused with substrate-free hypoxic buffer therefore provide a useful preparation in which to investigate how effectively certain drugs, e.g. steroids, β-adrenoceptor antagonists, Ca^{2+}

Table 2. CK Released from Isolated Langendorff Perfused Rabbit Hearts; Hypoxic Perfusion with and without Glucose Substrate

CK release (miu min^{-1} g^{-1} wet wt.). Mean ± SEM $n = 6$				
Duration of perfusion (min)	15	30	45	60
Aerobic perfusion	8 ± 2	10 ± 4	5 ± 3	9 ± 2
Hypoxia + glucose substrate	75 ± 5	78 ± 6	105 ± 4	108 ± 6
Hypoxic − substrate free	76 ± 4	102 ± 8	135 ± 16	220 ± 4
p		<0.05	<0.01	<0.001

Tests of significance calculated by Student's 't' test and related to the significance of the difference in CK release during hypoxic perfusion with and without glucose substrate. During substrate-free perfusion glucose was replaced by mannitol. Heart rate was kept constant (145 beats min^{-1}). Aerobic perfusion $Po_2 > 600$ mmHg; hypoxic perfusion $Po_2 < 6$ mmHg.[17]

antagonists, and physiological interventions protect the heart against the deleterious consequences of hypoxia. Whether the results from such studies can be applied to the ischaemic heart remains to be investigated. The exacerbation of the hypoxia-induced release of CK caused by substrate-lack (Table 2) contrasts with the failure (Figure 1) of substrate-lack to enhance enzyme leakage from hearts that are perfused with Ca^{2+}-free solutions.[17] It is perhaps worth noting that not all species of hearts (e.g. rat) leak CK enzyme during Ca^{2+}-free perfusion. It must also be appreciated that the pattern of CK release that occurs when Ca^{2+} is removed from the perfusion circuit (Figure 1) differs from that obtained during hypoxia (Figure 2). Many other conditions apart from the removal of substrate from the extracellular phase cause an exacerbation of hypoxia-induced enzyme leakage. These include (Figure 3) hyperthyroidism, tachycardia (Table 3), and various inotropic agents,[14] including digitalis, isoprenaline and a raised extracellular Ca^{2+}. These factors must, therefore, be carefully controlled in experiments which are designed to study the effects of hypoxia and ischaemia on heart muscle and in which enzyme release profiles are quantitated for purposes of comparison.

THE METABOLIC AND ULTRASTRUCTURAL CONSEQUENCES OF CARDIAC HYPOXIA

Although the effects of an inadequate supply of oxygen (with or without substrate lack) on the metabolism and fine ultrastructure of the heart are discussed in detail in other chapters of this book, it will be useful for us to consider them briefly here.

Metabolism

Studies from many laboratories[1,25,27,41] have shown that hypoxic perfusion of the mammalian heart results in a rapid deterioration of mechanical

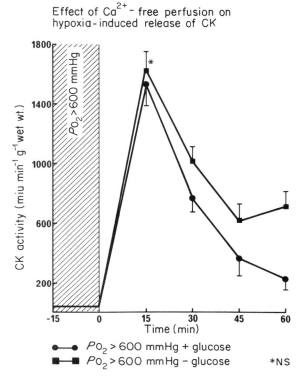

Figure 1. Release of CK from isolated Langendorff perfused rabbit hearts. At zero on the time scale Ca^{2+}-free buffer solution was introduced, with and without glucose substrate as indicated. Perfusion conditions aerobic, Ca^{2+}-free, 37 °C. Each point is mean ± SEM of six determinations. CK assay, see references 17 and 23

function, accompanied but not necessarily caused by[11] a depletion of the endogenous stores of adenosine triphosphate and creatine phosphate. The rapidity with which the tissue high-energy phosphate stores are depleted under these conditions is illustrated by the data summarized in Figure 4. These data were obtained from experiments[17] on isolated rabbit hearts perfused with hypoxic ($Po_2 < 6$ mmHg) substrate-free buffer but similar results have been obtained for other species.[16] This rapid decline in the tissue stores of ATP is accompanied, or preceded by, a similar precipitous decline in the creatine phosphate reserves. The real significance of such a decline in the high-energy phosphates of heart muscle is difficult to analyse, because of the strong possibility that these stores are compartmentalized, the different

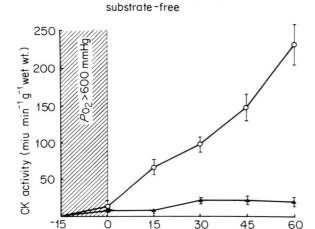

Figure 2. Release of CK from rabbit hearts perfused with substrate-free Krebs–Henseleit solution under either aerobic ($Po_2 > 600$ mmHg) or hypoxic ($Po_2 < 6$ mmHg) conditions. Each point is mean ± SEM of six determinations. CK assay, see references 17 and 23

compartments having different rates of turnover and supporting different functions within the cell. For example, recent data from Forrester and Williams[5] indicate that there is a specific, extremely labile, hypoxic-sensitive pool of ATP that is associated with cardiac cell membranes. Because of such compartmentalization there is a very real possibility that measurements of total tissue ATP may give a completely misleading idea as to the amount of ATP that is available to support a particular function within the cell, for example, to drive the Ca^{2+}-activated ATPase of the sarcoplasmic reticulum or the $Na^+ K^+$-activated ATPase of the cell membrane.

An hypoxia-induced reduction of the high-energy phosphate reserves in cardiac muscle, unlike that caused by the administration of excessive amounts of ouabain[24] cannot be explained in terms of an enhanced rate of energy utilization because it occurs at a time when cardiac work is diminished.[28] It can, however, be explained in terms of an altered mitochondrial function. Recent experiments[13,23] have shown that mitochondria that have been isolated from hypoxic-perfused and ischaemic heart muscle exhibit a marked alteration in respiratory activity—as indicated by reduced

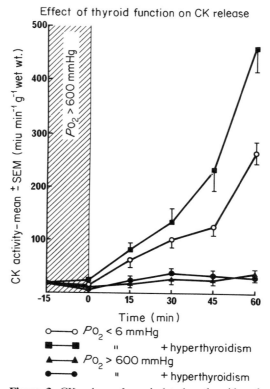

Figure 3. CK release from isolated euthyroid and hyperthyroid hearts perfused under either aerobic ($Po_2 > 600$ mmHg) or hypoxic ($Po_2 < 6$ mmHg) conditions. Hyperthyroidism was induced by daily injections of thyroxine.[21] Each point is mean ± SEM of six separate experiments

Table 3. Effect of Inotropic Agents on the Release of CK from Hypoxic Langendorff Perfused Rabbit Hearts

	CK release (miu min^{-1} g^{-1} wet wt.). Mean ± SEM $n=6$			
Duration of perfusion (min)	15	30	45	60
Spontaneously beating				
Hypoxic control	76 ± 4	102 ± 8	135 ± 16	220 ± 14
Hypoxia + digoxin (10^{-7} mol l^{-1})	116 ± 5	142 ± 9	178 ± 8	322 ± 21
Hypoxia + isoprenaline (10^{-8} mol l^{-1})	122 ± 6	176 ± 15	272 ± 9	386 ± 16
Paced preparations				
140 min^{-1}	72 ± 3	112 ± 7	138 ± 9	212 ± 8
220 min^{-1}	108 ± 7	173 ± 14	252 ± 8	312 ± 9

Perfusions at 37 °C using modified substrate-free Krebs–Henseleit buffer solution. Hypoxia ($Po_2 < 6$ mmHg) introduced by gassing the perfusate with 95% N_1 + 5% CO_2. n is number of separate perfusions.

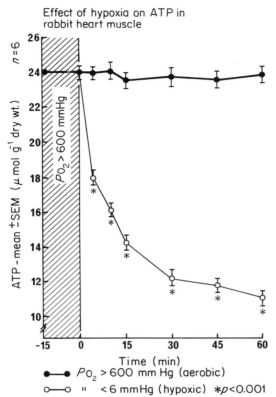

Figure 4. Effect of hypoxic substrate-free perfusion on the cardiac stores of ATP in isolated Langendorff perfused rabbit hearts. Each result is mean±SD of six experiments

Qo_2 (nanoatoms O_2 used $min^{-1} mg^{-1}$ mitochondrial protein) and RCI (respiratory control index—ratio of oxygen consumed in the presence of ADP to that taken up after phosphorylation to ATP). At the same time the mitochondria develop a marked avidity for Ca^{2+}.[23] These changes in mitochondrial function are illustrated by the data listed in Table 4.

This impairment of mitochondrial function, particularly when it is caused by a reduced supply of oxygen and of metabolic substrates to the affected muscle, inevitably results in a reduced rate of ATP synthesis and translocation. The cascade of events precipitated by the resultant decrease in the cytosolic availability of ATP can be best explained by recalling that cardiac muscle requires ATP not only as a substrate for its various ATPase enzymes (including the Na^+ K^+-activated ATPases of the plasmalemma, the myosin ATPase of the myofilaments, and the Ca^{2+}-activated ATPases of the

Table 4. Effect of Hypoxic Perfusion on Cardiac Mitochondrial Function

Perfusion	n	RCI	Q_{O_2}	Ca^{2+} uptake
Nil	6	16.6 ± 2.1	114.3 ± 3.6	22.4 ± 0.6
Aerobic-1 h ($P_{O_2} > 600$ mmHg)	6	15.8 ± 1.9	112.6 ± 2.8	21.3 ± 1.1
Hypoxic-1 h ($P_{O_2} < 6$ mmHg)	6	11.3 ± 0.8	46.8 ± 1.2	29.3 ± 2.1
Sig.	p	<0.001	<0.001	<0.001

RCI: ratio of oxygen consumed in the presence of ADP to that taken up after phosphorylation to ATP. Q_{O_2}: natoms O_2 consumed min^{-1} mg^{-1} protein. Ca^{2+} uptake: Ca^{2+} accumulated nmol min^{-1} mg^{-1} protein. When the hearts were perfused for 1 hour before the mitochondria were harvested, hypoxic conditions were obtained by gassing the perfusion buffer with 95% N_2 + 5% CO_2 instead of 95% O_2 + 5% CO_2. Perfusions were at 37 °C and the mitochondria were harvested by homogenization followed by differential centrifugation.[23] The reaction mixture used for the *respiratory studies* contained 125.0 mmol l^{-1} KCl; 12.5 mmol l^{-1} Tris–Hepes buffer; pH 7.3; 3.0 mmol l^{-1} Tris–glutamate; 3.0 mmol l^{-1} KH_2PO_4; 0.5 mmol l^{-1} EDTA; 2% dextran; 0.75 mg ml^{-1} mitochondrial protein; and 0.25 mmol l^{-1} ADP. The reaction mixture for the Ca^{2+}-accumulation studies contained 150.0 mmol l^{-1} sucrose; 75.0 mmol l^{-1} KCl; 2.0 mmol l^{-1} KH_2PO_4; 3.0 mmol l^{-1} Tris HCl (pH 7.2); 5.0 mmol l^{-1} $MgCl_2$; 1.7 mmol l^{-1} sodium succinate; 35.0 μ mol l^{-1} murexide; 3.0 mmol l^{-1} rotenone; 50 μ mol l^{-1} Ca^{2+}; and 250–300 mg ml^{-1} mitochondrial protein.[23] n refers to number of experiments.

sarcoplasmic reticulum, the mitochondria, and plasmalemma), but also for the maintenance of the fine ultrastructure and metabolism of the muscle.

Ultrastructure

Evidence supporting the hypothesis that the fine ultrastructure of the muscle is damaged during conditions of oxygen deprivation is provided by the electron micrographs shown in Figure 5(a–c). Figure 5(a) relates to the ultrastructure of rabbit heart muscle that has been perfusion-fixed[17] after 30 minutes' aerobic perfusion ($P_{O_2} > 600$ mmHg). Figures 5(b) and (c) relate to the fine ultrastructure of rabbit heart muscle that has been similarly perfusion-fixed for electron microscopy but after an equivalent period of perfusion under hypoxic ($P_{O_2} < 6$ mmHg) conditions. In the control, aerobic-perfused tissue (Figure 5a) the myofibrils remain in good array; there is no evidence of oedema and the mitochondria contain an orderly array of cristae. Figures 5(b) and (c) show evidence of cell damage typical of that caused by an inadequate supply of oxygen. There is damage to the cell membrane such that its selective permeability is destroyed: it becomes permeable even to quite large molecules (including CK, mol. wt. 80,000) and in some instances its continuity is destroyed (Figure 5b). Many of the mitochondria become swollen; some (Figure 5b and c) invaginate. Oedema is often present (Figure 5c) and glycogen disappears from the cytosol. Because a detailed description of the morphological consequences of myocardial ischaemia and hypoxia is given elsewhere in this book no more attention will be given to it here, except to emphasize the fact that it is because of this damage to the fine ultrastructure that intracellular components appear in the extracellular phase, and hence in the blood plasma.

PROTECTIVE PROCEDURES

Procedures which might be expected to protect the heart against or reduce the severity of the damage caused by inadequate oxygenation naturally include those which, because they limit the heart's demand for oxygen, tend to restore the balance between energy supply and demand. Pharmacological agents that belong to this category include those which reduce either:

(a) afterload, e.g. nitroglycerine;
(b) heart rate, e.g. β-adrenoceptor and Ca^{2+}-antagonists
(c) peak developed tension, e.g. Ca^{2+} antagonists and certain β-adrenoceptor antagonists.

There are, of course, other approaches that can be used, e.g. the glucose–insulin–potassium therapy[2] might be used to stimulate anaerobic glycolysis, or procedures might be instigated to increase arterial P_{O_2} or to improve coronary perfusion. Alternatively, a *mild* respiratory acidosis might be evoked, to reduce cardiac work. In the remainder of this chapter the discussion will be centered around results which suggest that the mitochondria can be protected, to some extent at least, against the deleterious effects of hypoxia. Possibly because mitochondrial function is protected these same procedures attenuate the release of CK.

Before discussing in detail the consequences, at the subcellular level, of adding these various protective agents to hypoxic heart muscle it may be useful to summarize our working schema, which is as follows:

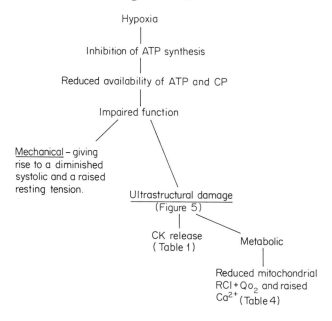

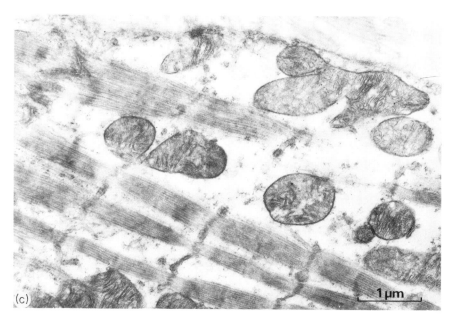

Figures 5(a–c). Electron micrographs of perfusion-fixed rabbit heart muscle[21] fixed after 30 minutes' perfusion under aerobic ($Po_2 < 600$ mmHg Figure 5a) or hypoxic ($Po_2 < 6$ mmHg Figures 5b and c) conditions. In Figure 5(a) note the normal appearance of the ultrastructure and the intact cell membrane. Magnification ×18,950. In Figure 5(b) note the damaged cell membrane and the invaginated mitochondria. Magnification ×18,500. In Figure 5(c) note the presence of oedema, of swollen mitochondria, and of damaged Z lines and myofilaments. Marker bars, 1 micron

EFFECT OF A MILD ACIDOSIS

When the oxygen supply to the heart is insufficient to provide enough ATP by oxidative phosphorylation to meet its energy requirements[26] the work output of the heart decreases until a new steady state is established. If there is a concomitant reduction in coronary flow the affected zone of muscle will retain the end products of its metabolism, with a resultant fall in intracellular pH.[42] Recently there has been considerable speculation concerning the magnitude of the fall in intracellular pH that occurs under these conditions and the consequences of the raised H^+ concentration.[32,41,42] Cardiac muscle responds rapidly to a respiratory acidosis–mimicked in laboratory studies by gassing the perfusion buffer with gas mixture containing varying proportions of CO_2[28] in either O_2 (for aerobic) or N_2 (for hypoxic perfusions). The mechanism of the altered myocardial contractility that occurs as the H^+ concentration increases is not fully understood.[3] Some

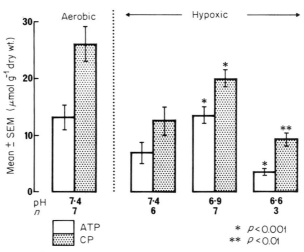

Figure 6. Effect of a mild 'respiratory' acidosis—induced by gassing the perfusate with varying mixtures of CO_2 in either N_2 or O_2 for hypoxic and aerobic conditions; and to provide pH values of 7.4, 6.9, and 6.6 respectively—on ATP and CP levels in rabbit heart muscle. The hearts were perfused by the Langendorff technique[21] at 37 °C for 1 hour before they were frozen and analysed for CP and ATP. Note that a mild acidosis (pH 6.9) had a protective effect on the high-energy phosphate stores of the hypoxic heart muscle

investigators believe that H^+ modifies the binding of Ca^{2+} to the sarcoplasmic reticulum;[15] others that it alters the affinity of the troponin–tropomyosin complex for Ca^{2+}.[6] Another possibility is that H^+ and Ca^{2+} compete for transport sites at the level of the cell membrane. Whatever its cause, an increase in H^+ is associated with reduction in cardiac work. When applied to the hypoxic heart, therefore, an increase in H^+ availability might, because it reduces the work done by the heart, provide some protection. The data summarized in Figures 6 and 7, and Table 5 show that a *mild* acidosis (pH reduced to 6.9, induced by changing the CO_2 content of the perfusate) does provide some degree of protection, evidenced by a reduced rate of CK release, a better preservation of the endogenous stores of ATP and CP (Figures 6 and 7) and the maintenance of mitochondrial respiratory and Ca^{2+}-accumulating activities at near normal levels. Reducing the pH even further, to 6.6 instead of 6.9 failed to provide any further protection. It should be noted, in passing, that the apparently different CK release patterns shown for hypoxic-perfused hearts at pH 7.4 (Figures 2 and 7) and

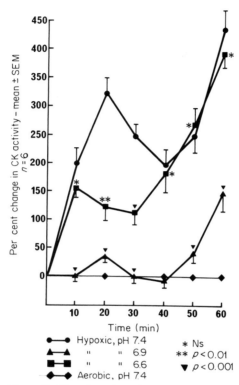

Figure 7. CK release from isolated Langendorff perfused rabbit hearts perfused under aerobic or hypoxic conditions, pH 7.4, 6.9, or 6.6. Changes in pH were induced by gassing the perfusate with gas mixtures containing varying proportions of CO_2 and N_2, per cent change refers to the per cent change in CK release relative to that detected during an initial 15 minutes' aerobic perfusion ($Po_2 > 600$ mmHg) at pH 7.4. n refers to the number of experiments and tests of significance related to the significance of the change in the hypoxic-induced CK release caused by changing the pH of the perfusate from 7.4 to 6.9 or 6.6

Table 5. Effect of a Mild Acidosis on the Respiratory and Ca^{2+}-accumulating Activity of Cardiac Mitochondria

Perfusion	Aerobic			Hypoxic		
pH	7.4	6.9	6.6	7.4	6.9	6.6
RCI	16.4±2.1	17.1±2.6	12.9±1.8	11.8±0.2	15.9±12.	12.1±1.3
Qo_2	112.0±2.2	118.4±1.6	86.2±3.1	52.4±1.6	116.5±3.1	38.2±1.2
Ca^{2+}	22.5±1.2	21.0±0.8	25.3±1.6	28.6±1.3	22.8±1.1	27.5±0.6

Each result is mean±SEM, 4 experiments. The rabbit hearts were perfused at 37 °C at the indicated pH, under either aerobic or hypoxic conditions. One hour later the mitochondria were isolated and their respiratory and Ca^{2+}-accumulating activity assayed. RCI: ratio O_2 consumed in the presence of ADP to that taken up after phosphorylation to ATP. Qo_2 (natoms O_2 used $min^{-1} mg^{-1}$ protein). Ca^{2+} accumulated (nmol $min^{-1} mg^{-1}$ protein). Reaction mixture as described in footnote to Table 4.

the different ATP concentrations (Figures 4 and 6) are due to a seasonal variation peculiar to the rabbit.

The results presented in Figures 6 and 7 and Table 5 were obtained during experiments in which isolated, electrically paced rabbit hearts were perfused under either aerobic or hypoxic conditions at pH of either 7.4, 6.9, or 6.6. The mitochondria were harvested after the required period of perfusion had elapsed and all batches of mitochondria were treated alike. The protective effect of the reduced pH (6.9) must, therefore, have resulted from the changed conditions of perfusion—perhaps a pH-induced reduction in cardiac (Figure 8) work during the early phases of the hypoxic episode is sufficient to maintain the endogenous stores of ATP above the critical levels needed for the maintenance of membrane integrity. Whatever its cause a *mild* reduction in pH may protect heart muscle against the deleterious effects of oxygen deprivation. By contrast, a severe acidosis is damaging.

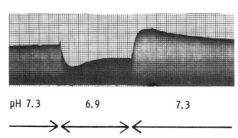

Figure 8. Mechanogram recorded from an isolated rabbit trabecular muscle, immersed in Krebs–Henseleit solution gassed with varying mixtures of O_2 and CO_2 to provide pH of 7.3 or 6.9 in the buffer. Stimulation rate, 60 min^{-1}. Temperature 37 °C. Tension recorded as an upward deflection. Note the decline in peak developed tension coincident with changing the pH of the perfusion buffer from 7.3 to 6.9

β-ADRENOCEPTOR ANTAGONISTS

Recent laboratory studies have shown that blockade of the cardiac β-adrenoceptors with racemate propranolol protects the heart against the deleterious effects of hypoxia[33] and ischaemia.[1,14,29,30,38,39] Figure 9 shows just how effective propranolol can be in reducing CK release from isolated hearts perfused with hypoxic Krebs–Henseleit buffer solution containing glucose substrate. In these particular studies[21] propranolol was added to the perfusion circuit at the start of the hypoxic episode but in other studies[22] the hearts of rabbits that had been pre-treated with propranolol for 3–5 days were used. The end result was the same, CK release had been attenuated and mitochondrial function maintained.[22]

Propranolol is a β-adrenoceptor antagonist lacking intrinsic sympathomimetic activity (ISA). Other β-adrenoceptor antagonists which have powerful agonist activity but which have, in addition, ISA—e.g. practolol, oxprenolol, acebutolol—are less effective than propranolol in preventing CK release and hence, presumably, in protecting the heart against the deleterious consequences of oxygen deprivation. An example of the failure of potent β-adrenoceptor antagonists to block CK release from hypoxic perfused heart muscle is shown by the data summarized in Figures 10 and 11, and similar results have been described in the literature for other β-antagonists with ISA activity, e.g. oxprenolol.[21] When comparing the data shown in Figures 9 and 10 it is interesting to note that the inclusion of glucose in the perfusion buffer (Figure 9) brought about approximately 50% reduction in enzyme release relative to that found in the absence (Figure 10) of substrate.

There is perhaps one additional point that needs to be emphasized here relating to the protective effect of propranolol shown in Figure 9, that is, that when rabbits have been pre-treated with propranolol and the hearts subsequently isolated and perfused under hypoxic ($P_{O_2} < 6$ mmHg) conditions the protection that is afforded by propranolol persists for 272 hours after the last dose of the drug has been given[22] and hence at a time when effective β-blockade has abated. This protective effect of propranolol extends, like that described for increased H^+, to the preservation of mitochondrial function.[23] Thus as is shown by the data presented in Table 6, mitochondria from hearts that had been perfused for 1 hour with hypoxic ($P_{O_2} < 6$ mmHg) propranolol-containing buffer solution had higher Q_{O_2} and lower Ca^{2+}-accumulating activities than did mitochondria that had been similarly isolated from hearts perfused for 1 hour with hypoxic ($P_{O_2} < 6$ mmHg) propranolol-free buffer. As in the acidosis series described in the preceding section of this chapter, therefore, we have a preservation of mitochondrial function induced by a procedure, in this case the introduction of propranolol, which reduces cardiac work[18] and attenuates CK release.[21]

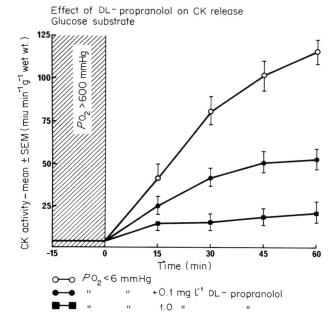

Figure 9. Effect of propranolol on the hypoxic ($Po_2 < 6$ mmHg)-induced release of CK from isolated perfused rabbit hearts. Mean ± SEM of six results is shown. Hypoxic perfusion, with and without propranolol was started at time zero on the time scale. The perfusate buffer contained glucose substrate

Ca^{2+}-ANTAGONIST DRUGS, e.g. VERAPAMIL

Another group of drugs that have been shown to protect hypoxic heart muscle is exemplified by the drug verapamil.[4,18] Verapamil is the forerunner of a new and exciting group of drugs that limit the entry of Ca^{2+} into the cell during the plateau phase of the action potential.[19] By reducing Ca^{2+} entry verapamil depresses cardiac function and, at the same time, limits myocardial oxygen consumption.[18] When added to isolated hearts perfused under hypoxic conditions verapamil reduces CK release (Figure 12). In Figure 12 it should be noted that 1 mg l^{-1} verapamil was less effective than 0.5 mg l^{-1} in preventing CK release. Although this has been a consistent finding for our *in vitro* studies no satisfactory explanation can be offered.

When added to hypoxia-perfused hearts verapamil also preserves mitochondrial function, so that (Figure 13) mitochondrial Qo_2 and Ca^{2+}-accumulating activity (Figure 14) are maintained at the levels expected for

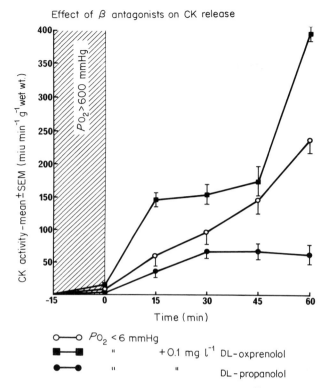

Figure 10. Effect of propranolol and oxprenolol[21] on the release of CK from hypoxic substrate-free perfused rabbit hearts. Each point is the mean of six experiments. The drugs were added to the buffer solution at time zero—at the start of the hypoxic episode. Note that oxprenolol caused an exacerbation of the CK release

aerobically perfused hearts. Although the results presented in Figures 13 and 14 were obtained from experiments in which verapamil was added to the isolated perfused hearts, similar results have been obtained when the rabbits have been pre-treated with the drug, prior to isolating the heart and exposing it to hypoxic conditions of perfusion.[17] Final proof that verapamil does protect heart muscle against the deleterious effects of hypoxic perfusion has come from ultrastructural studies showing near-normal ultrastructure (Figure 15(b)) even after 30–60 minutes of perfusion at a $Po_2 <$ 6 mmHg substrate free conditions. This preservation of the fine ultrastructure and the maintenance of near-normal mitochondrial activity even after prolonged periods of hypoxic perfusion in the presence of verapamil is

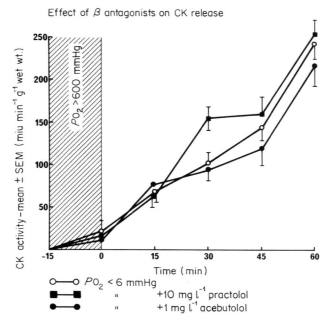

Figure 11. Effect of practolol and acebutolol[21] on CK release from hypoxic substrate-free perfused rabbit heart. Note that these β-antagonists failed to attenuate the release of CK. The drugs were added at zero time, coincident with the start of the hypoxic perfusion. Results mean±SEM, six experiments

Table 6. Effect of Propranolol (2 mg kg^{-1} subcutaneously) on the Effect of Hypoxic Perfusion on Cardiac Mitochondrial Function

Perfusion	RCI	Qo_2	Ca^{2+} uptake
Hypoxic control ($Po_2 < 6$ mmHg)	11.3±0.8	46.8±1.2	29.3±2.1
Hypoxic—propranolol ($Po_2 < 6$ mmHg)	16.8±1.2	116.4±3.1	22.1±0.5
Aerobic control ($Po_2 < 600$ mmHg)	15.8±1.9	112.6±2.8	21.3±1.1

Each result is mean±SEM, 6 experiments. RCI, Qo_2, and Ca^{2+}-accumulation as in Table 4. Hypoxic control received placebo injections of saline; hypoxic—propranolol—rabbits were pre-treated with 2.0 mg kg^{-1} propranolol for 5 days prior to sacrifice, 1 hour after the last dose of propranolol had been given. Duration of perfusion 60 minutes. Note, propranolol was not added to the isolated mitochondria.

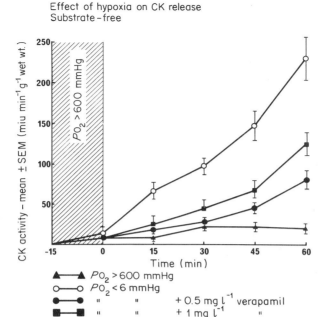

Figure 12. Effect of the Ca^{2+} antagonist verapamil on CK release from hypoxic substrate-free perfused rabbit hearts. Each point is mean±SEM, six determinations. Verapamil[17,18] was added at zero time, at the start of the hypoxic episode

accompanied[17] by the preservation of normal levels of ATP and CP. Since the drug does not enter the cardiac cell its protective effect must almost certainly result from its ability to reduce cardiac work.

MEMBRANE-STABILIZING DRUGS

Examples of drugs that fall into this group include methylprednisolone[16,20] and α-tocopherol or vitamin E. These drugs differ from those described above in that whilst they protect heart muscle against the damage caused by hypoxic perfusion (references 16, 20, and Nayler, unpublished data), and hence attenuate CK release (Figure 16), they do so without any attendant reduction in cardiac work. Nevertheless they, like the mild respiratory acidosis, the cardiac depressant β-antagonist—propranolol, and the Ca^{2+} antagonist—verapamil, all share one feature in common—they all act in such a way as to preserve mitochondrial function during conditions of severe O_2 depletion. Introduction of the glucose–insulin–potassium therapy has this

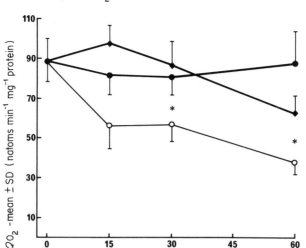

Figure 13. Oxygen utilization by cardiac mitochondria isolated from rabbit hearts perfused either aerobically, or under hypoxic substrate-free conditions ($Po_2 < 6$ mmHg) with and without verapamil. Note that the verapamil was added to the perfusion circuit and not to the mitochondria. Duration of perfusion is shown by the time axis. O_2 consumption was monitored with an O_2 electrode.[23] Each point is the mean ± SD of six experiments. Note the protection given by verapamil outlasted the isolation procedure

same end point[13]—that is, the preservation of mitochondrial function, resulting in the maintenance of the tissue stores of ATP. In retrospect it seems likely that this maintenance of the tissue stores of ATP above a critical level provides the vital clue as to why certain drugs are protective—because if sufficient energy remains available to maintain the activity of the various Ca^{2+}-sequestering systems the Ca^{2+} content of the cytosol will not rise. The various phospholipases that are Ca^{2+} activated will, therefore, remain dormant so that the fine ultrastructure of the lipid-containing membranes will remain intact. At the same time sufficient ATP will remain available as substrate for the Na^+/K^+-ATPase enzyme, an enzyme whose activity is required if the correct distribution of Na^+ and K^+ is to be

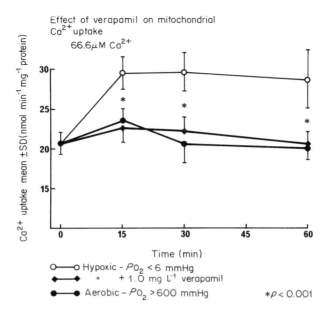

Figure 14. The Ca^{2+}-accumulating activity of cardiac mitochondria isolated from hearts perfused under either aerobic or hypoxic conditions, with and without verapamil. Ca^{2+} uptake was monitored by dual-beam spectrometry[23]

maintained. The maintenance of this correct distribution of Na^+ and K^+ is of vital importance if the cells are to remain excitable, and if oedema is not to occur.

CONCLUDING COMMENTS

Pharmacological and physiological interventions can be used to protect the heart against the deleterious effects of hypoxia. These interventions include a mild acidosis, and the addition of certain drugs, including propranolol, verapamil, methylprednisolone, and other membrane stabilizers. The end result of the addition of these drugs and lowering the pH is the preservation of mitochondrial function. At the same time enzyme release is attenuated. The results described here have been obtained from studies using isolated hearts, which necessarily lack neuronal control. When applying such techniques to intact animals it is, of course (see next chapter) important to consider the consequences of such intervention on the whole animal.

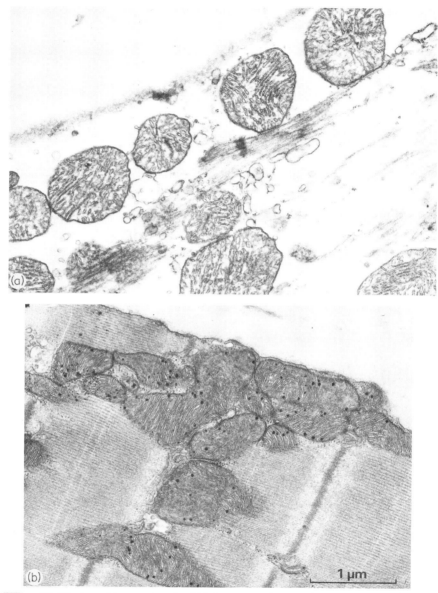

Figure 15. Electron micrographs of rabbit heart muscle perfused for 60 minutes under hypoxic conditions ($P_{O_2} < 6$ mmHg) without (a) and with (b) 0.5 mg l^{-1} verapamil added to the perfusion buffer. Note that in (a) the fine ultrastructure of the muscle is damaged, the mitochondria are disorganized, and there is oedema. In (b) the fine ultrastructure has been retained, the cell membrane remains intact, and the myofilaments are in good array. Magnification (a) × 18,500; (b) × 22,800

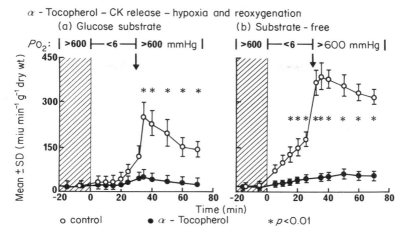

Figure 16. Effect of α-tocopherol on CK release from isolated rabbit hearts perfused under aerobic or hypoxic conditions as shown, with and without substrate. α-Tocopherol was at zero on the time scale, at the start of the hypoxic perfusion. At arrow O_2 was reintroduced. α-Tocopherol was infused at a rate of 23.2 nmol min^{-1} to provide a concentration similar to that found in blood

REFERENCES

1. Brachfeld, N. Metabolic evaluation of agents designed to protect the ischaemic myocardium and to reduce infarct size. *American Journal of Cardiology*, **37,** 528–32 (1976).
2. Calva, E., Mujica, A., Bisteni, A., and Sodi-Pallares, D. Oxidative phosphorylation in cardiac infarct: effect of glucose–KCl–insulin solution. *American Journal of Physiology*, **209,** 371–5 (1965).
3. Cingolani, H. E., Mattiazzi, A. R., Blesa, E. S., and Gonzales, N. C. Contractility in isolated mammalian heart muscle after acid–base changes. *Circulation Research*, **26,** 269–78 (1970).
4. Fleckenstein, A. Specific inhibitors and promotors of calcium action in the excitation–contraction coupling of heart muscle and their role in the prevention or production of myocardial lesions. In *Calcium and the Heart* (Eds. Harris, P. and Opie, L. H.), Academic Press, London, 1971, pp. 135–88.
5. Forrester, T. and Williams, C. A. Release of adenosine triphosphate from isolated adult heart cells in response to hypoxia. *Journal of Physiology*, **268,** 371–90 (1977).
6. Fuchs, F., Reddy, Y., and Briggs, F. M. The interaction of cations with the calcium-binding site of troponin. *Biochimica et Biophysica Acta*, **22,** 407–9 (1970).
7. Ganote, C. E., Seabra-Gomes, R., Nayler, W. G., and Jennings, R. B. Irreversible myocardial injury in anoxic perfused rat hearts. *American Journal of Pathology*, **80,** 419–40 (1975).
8. Hearse, D. J., Humphrey, S. M., and Garlick, P. B. Species variation in myocardial anoxic enzyme release, glucose protection and re-oxygenation damage. *Journal of Molecular and Cellular Cardiology*, **8,** 329–39 (1976).
9. Hearse, D. J., Humphrey, S. M., Feuvray, D., and de Leiris, J. A. Biochemical and ultrastructural study of species variation in myocardial cell damage. *Journal of Molecular and Cellular Cardiology*, **8,** 759–78 (1976).

10. Jennings, R. B., Sommers, H. M., Smyth, G. A., Flack, H. A., and Linn, H. Myocardial necrosis induced by temporary occlusion of a coronary artery in the dog. *Archives of Pathology*, **70**, 68–78 (1960).
11. Katz, A. M. and Hecht, H. H. The early 'pump' failure of the ischaemic heart. *American Journal of Medicine*, **47**, 497–502 (1969).
12. Langendorff, O. Untersuchungen am überlebendem Saügetierherzen. *Pflügers Archives*, **61**, 291–332 (1895).
13. Lochner, A., Kotze, J. C. N., and Gevers, W. Mitochondrial oxidative phosphorylation in myocardial ischaemia: Effects of glucose and insulin on anoxic hearts perfused at low pressure. *Journal of Molecular and Cellular Cardiology*, **8**, 575–84 (1976).
14. Maroko, P. R. and Braunwald, E. Effects of metabolic and pharmacologic interventions on myocardial infarct size following coronary occlusion. *Circulation Supplement 1*, **53**, 162–7 (1976).
15. Nakamuru, Y. and Schwartz, A. Possible control of intracellular calcium metabolism G[H$^+$]; sarcoplasmic reticulum of skeletal and cardiac muscle. *Biochemical and Biophysical Research Communications*, **41**, 830–6 (1970).
16. Nayler, W. G. and Seabra-Gomes, R. The effect of methylprednisolone on hypoxic heart muscle. *Cardiovascular Research*, **10**, 349–58 (1976).
17. Nayler, W. G., Grau, A., and Slade, A. A protective effect of verapamil on hypoxic heart muscle. *Cardiovascular Research*, **10**, 650–62 (1976).
18. Nayler, W. G. and Szeto, J. Effect of verapamil on contractility, oxygen utilization and calcium exchangeability in mammalian heart muscle. *Cardiovascular Research*, **6**, 120–8 (1972).
19. Nayler, W. G. and Krikler, D. Verapamil and the myocardium. *Postgraduate Medical Journal*, **50**, 441–6 (1974).
20. Nayler, W. G., Yepez, C., Grau, A., and Slade, A. The protective effect of methylprednisolone sodium succinate on the ultrastructure and resting tension of hypoxic heart muscle. *Cardiovascular Research*, **12**, 91–8 (1977).
21. Nayler, W. G., Grau, A., and Yepez, C. β-Adrenoceptor antagonists and the release of creatine phosphokinase from hypoxic heart muscle. *Cardiovascular Research*, **11**, 344–52 (1977).
22. Nayler, W. G., Yepez, C., and Fassold, E. Experimental studies on the effect of β-adrenoceptor antagonists on hypoxic heart muscle. In *Beta-Blockade* George-Thieme, (Ed.) Stuttgart: Verlag, p. 32–48 (1977).
23. Nayler, W. G., Fassold, E., and Yepez, C. The pharmacological protection of mitochondrial function in hypoxic heart muscle: Effect of verapamil, propranolol and methylprednisolone. *Cardiovascular Research*, **12**, 152–61 (1978).
24. Nayler, W. G. and Williams, A. Some morphological and biochemical aspects of the sarcoplasmic reticulum. *European Journal of Cardiology*, **7**, Supplement 35–50 (1978).
25. Neely, J. R., Liebermeister, H., Battersby, D. J., and Morgan, H. E. Effect of pressure development on oxygen consumption by the isolated rat heart. *American Journal of Physiology*, **212**, 804–14 (1967).
26. Neely, J. R. and Morgan, H. E. The relationship between carbohydrate and lipid metabolism and the energy balance of heart muscle. *Annual Review of Physiology*, **36**, 413–59 (1973).
27. Opie, L. H. Effects of anoxia and regional ischaemia on metabolism of glucose and fatty acids. *Circulation Research Supplement 1*, **38**, 52–76 (1976).
28. Poole-Wilson, P. A. and Langer, G. A. Effect of pH on ionic exchange and function in rat and rabbit myocardium. *American Journal of Physiology*, **229**, 570–81 (1975).
29. Reimer, K. A., Rasmussen, M., and Jennings, R. B. Reduction by propranolol of myocardial necrosis following temporary coronary artery occlusion in dogs. *Circulation Research*, **33**, 353–63 (1973).
30. Reimer, K. A., Rasmussen, M. M., and Jennings, R. B. On the nature of protection by propranolol against myocardial necrosis after temporary coronary occlusion in dogs. *American Journal of Cardiology*, **37**, 520–7 (1976).
31. Remmie, W. J., de Jong, W., and Verdouw, P. D. Effects of pacing-induced myocardial ischaemia on hypoxanthine efflux from the human heart. *American Journal of Cardiology*, **40**, 55–62 (1977).

32. Schaer, H. Decrease in ionized calcium by bicarbonate in physiological solutions. *Pflügers Archives*, 347, 249–54 (1974).
33. Sakai, K. and Spieckermann, P. G. Effects of reserpine and propranolol on anoxia-induced enzyme release from the isolated perfused guinea-pig heart. *Naunyn-Schmiedeberg's Archiv für Experimentelle Pathologie und Pharmakologie*, **291**, 123–30 (1975).
34. Seabra-Gomes, R., Ganote, C. E., and Nayler, W. G. Species variation in anoxic-induced damage of heart muscle. *Journal of Molecular and Cellular Cardiology*, **12**, 929–38 (1975).
35. Sobel, B. E., Bresnahan, G. F., Shell, W. E., and Yoder, R. D. Estimation of infarct size in man and its relation to prognosis. *Circulation*, **46**, 640–8 (1972).
36. Sobel, B. E. Roberts, R., and Carson, K. B. Considerations in the use of biochemical markers in ischaemic injury. *Circulation Research Supplement 1*, **38**, 99–108 (1976).
37. Stone, M. J., Willerson, J. T., Gomez-Sanchez, C. E., and Waterman, M. R. Radioimmunoassay of myoglobin in human serum. Results in patients with acute myocardial infarction. *Journal of Clinical Investigation*, **56**, 1334–9 (1975).
38. Theroux, P., Franklin, D., Ross, J., Jr., and Kemper, W. S. Regional myocardial function during acute coronary occlusion and its modification by pharmacologic agents in the dog. *Circulation Research*, **35**, 896–908 (1974).
39. Theroux, P., Ross, J., Jr., Franklin, S., Kemper, W. S., and Sasayama, S. Regional myocardial function in the conscious dog during acute coronary occlusion and responses to morphine, propranolol, nitroglycerine and lidocaine. *Circulation*, **53**, 302–13 (1976).
40. Volk, P. Myoglobin in the serum after myocardial infarction. *Munchener Medizinische Wochenschrift*, **115**, 2122–8 (1973).
41. Williamson, J. R., Safer, B., Rich, T., Schaffer, S., and Kobayashi, K. Effects of acidosis on myocardial contractility and metabolism. In *Experimental and Clinical Aspects on Preservation of the Ischaemic Myocardium*, (Eds. Hjalmarson, A. and Werko, L.), Lindgren and Soner; Molndal, Sweden; 1975, pp. 95–112.
42. Williamson, J. R., Schaffer, S. W., Ford, C., and Safer, B. Contribution of tissue acidosis to ischemic injury in the perfused rat heart. *Circulation Supplement 1*, **53**, 3–14 (1976).

CHAPTER 23

Pharmacological limitation of infarct size: enzymatic, electrocardiographic, and morphological studies in the experimental animal and man

P. R. Maroko, D. Maclean, L. G. T. Ribeiro, and E. Braunwald

INTRODUCTION	529
METHODS FOR EVALUATING CHANGES IN MYOCARDIAL DAMAGE IN EXPERIMENTAL ANIMALS	530
MECHANISM OF ACTION OF VARIOUS INTERVENTIONS	536
EFFECT OF INTERVENTIONS ON THE HEALING PROCESS	544
THE INFLUENCE OF THE TIME INTERVAL BETWEEN CORONARY ARTERY OCCLUSION AND THE APPLICATION OF AN INTERVENTION ON SALVAGE OF ISCHAEMIC MYOCARDIUM	546
CLINICAL OBSERVATIONS	548
CONCLUDING REMARKS	553
REFERENCES	553

INTRODUCTION

The conventional treatment of patients with acute myocardial infarction is directed principally at the complications, i.e. arrhythmias and myocardial pump failure. Drugs often successfully suppress the arrhythmias, but the management of severe cardiac failure, as manifested by cardiogenic shock, or pulmonary oedema remains a major therapeutic challenge. A different approach is to attempt to influence the natural evolution of the impending myocardial necrosis by interrupting the process while it is still in a reversible phase, thus reducing the size of the eventual infarction.[56,57,63,69] The rationale for this approach is that such a reduction in the extent of necrosis will leave a larger mass of viable, contracting cells so that the frequency and severity of these complications of myocardial infarction, particularly myocardial pump failure, would be expected to be reduced.

In order to test this hypothesis and to evaluate the possible applicability of this therapeutic approach to patients, our investigations have been carried out in five stages:

(i) Determination of whether the extent of necrosis is determined principally by the site of occlusion and anatomic factors such as the degree and location of collaterals, or whether infarct size following coronary artery occlusion can indeed be altered by interventions.

(ii) Identification of the physiologic, pharmacologic, and metabolic interventions that can alter infarct size following experimental coronary artery occlusion (Table 1).[1,4–6,9,10,12,14,15,18–20,23,26–28,36,42,44,45,47–50,52,54,56–69,73–75,78,79,81,84,87,92–96,98–104,107]

(iii) Determination of the time interval following coronary occlusion during which ischaemic cells retain their viability. During this part of this investigation, our attention naturally focused on those cells which, although jeopardized, were not yet irreversibly damaged and were therefore still capable of recovery, rather than on those cells that were already irreversibly injured.

(iv) Determination of the effect of drugs on the process of healing (i.e. on fibrosis). Special attention was paid to possible impairment of scar formation.

(v) Development of atraumatic techniques for measuring alterations in myocardial ischaemic injury that are applicable to patients and the demonstration in patients with acute myocardial infarction that some or all of the interventions that reduce experimental myocardial necrosis are effective.

METHODS FOR EVALUATING CHANGES IN MYOCARDIAL DAMAGE IN EXPERIMENTAL ANIMALS

To test the hypothesis that acute myocardial ischaemic injury can be modified by interventions applied following coronary occlusion, a technique was developed in which acute myocardial ischaemic injury in dogs was indirectly quantified by determining electrographic ST segment elevation in multiple epicardial sites during repeated 20 minute coronary artery occlusions.[56,57] This method overcame several difficulties inherent in other techniques which had previously been used for assessing changes in the extent of acute myocardial damage. Since all sequential electrographic recordings are made at the same epicardial sites on the same heart, the effect of the considerable variations in the distribution of the coronary arteries and in the extent of the intercoronary collaterals among different animals is eliminated, and each animal can serve as its own control. An alteration in the extent and magnitude of ST segment elevation during one of these repeated occlusions is taken as an index of a change in acute myocardial ischaemic injury. In

Table 1. Interventions That Modify Myocardial Injury Following Coronary Occlusion

Interventions That Reduce Myocardial Injury

By decreasing myocardial oxygen demand
    Propranolol[20,56,57,59,98]
    Practolol[49,74]
    Barbituates
    Cardiac glycoside in the failing heart[64,102]
    Counterpulsation
        Intraaortic balloon[44,47,58]
        External counterpulsation[36]
    By decreasing afterload in hypertensive individuals—'Arfonad'[94]
    By inhibition of lipolysis—β-pyridyl carbinol[42,73]
By increasing myocardial oxygen supply
    Directly
        Coronary artery reperfusion[5,18,60,96]
        Elevating arterial P_{O_2}[1,54,67]
    Through collateral vessels
        Elevation of coronary perfusion pressure by methoxamine, neosynephrine, or norepinephrine[56,57,59,81]
        Intraaortic balloon counterpulsation[44,47,58]
        External counterpulsation[36]
    Nitroglycerine[4,6,9,12,15,19,28,95]
        'Reverse coronary steal' or favourable redistribution of regional myocardial blood flow[10,84]
    By increasing plasma osmolality
        Mannitol[75,104]
        Hypertonic glucose[62]
By augmenting anaerobic metabolism (presumed)
    Glucose–insulin–potassium[62,100]
    Hypertonic glucose[62]
By enhancing transport to the ischaemic zone of substrate utilized in energy production (presumed)—hyaluronidase[26,27,52,61,66,68]
By protecting against autolytic and heterolytic processes (presumed)
    Glucocorticoids[48,99]
    Cobra venom factor[65]
    Aprotinin[14,23]
    Non-steroidal antiinflammatory agents—'Ibuprofen'

Interventions That Increase Myocardial Injury

By increasing myocardial oxygen requirements
    Isoproterenol[42,45,56,57,59,92,103]
    Glucagon[45,57]
    Ouabain[57,101]
    Bretylium tosylate[57]
    Tachycardia[57,81,93]
By decreasing myocardial oxygen supply
    Directly
        Hypoxaemia[78]
        Anaemia[107]
    Through collateral vessels
        Reducing coronary perfusion pressure (haemorrhage)[56,57,59,81]
        By 'coronary steal' or unfavourable redistribution of regional myocardial blood flow[9,79]
By decreasing substrate availability—hypoglycaemia[50]

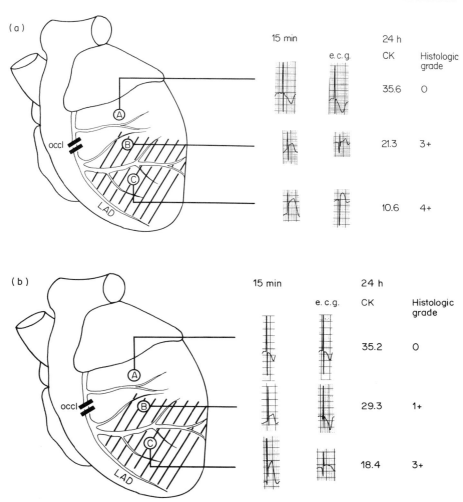

Figure 1. (a and b), On the left side is a schematic representation of the heart and its arteries. The left anterior descending (LAD) was occluded at its midportion (occl). The shaded area represents the zone of ST segment elevation 15 minutes after occlusion. On the right-hand side are examples of epicardial electrograms, myocardial CK values (in iu mg^{-1} protein), and histologic grades, from (a) a control dog and from (b) a dog-treated with hyaluronidase. In the control dog (a), site A (from non-ischaemic myocardium) exhibited no ST segment elevation at 15 minutes (ST_{15m}). At 24 hours it had no changes in QRS configuration and normal CK activity, and it appeared hitologically normal. Site B (border zone) showed moderate ST_{15m}, while at 24 hours there was a significant Q wave and partial loss of R wave voltage. The CK activity was moderately depressed, and the histologic section was graded 3+ (51–75%) necrosis. Site C (centre of the ischaemic zone) had marked ST_{15m} and at 24 hours is demonstrated a total loss of R wave with a QS complex. The myocardial CK activity was greatly depressed, and the histologic section was graded 4+ (>75%) necrosis. [Reproduced, with permission, from

order to verify the value of this technique in reflecting acute myocardial ischaemic damage and to confirm its ability to predict necrosis, the correlation between acute epicardial ST segment elevation 15 minutes after occlusion and the ultimate fate of the myocardium beneath the recording electrode was studied[2] (Figure 1(a)). The latter was assessed by the microscopic appearance of the subjacent myocardium 24 hours after the occlusion; in addition, the creatine kinase (CK) activity of the myocardium was measured in these same sites in order to quantify this damage further.

In these experiments we demonstrated that epicardial sites without markedly abnormal ST segments (elevations of 0–2 mV) had normal myocardial CK activity, while those sites with ST segment elevations exceeding 2 mV showed a distinct depression of myocardial CK activity. Thus in one series of dogs, sites with ST segment elevation at 15 minutes of 0–2 mV showed 24 hours later CK activity of 33.1 ± 0.9 iu mg^{-1} protein ($n = 49$); sites with ST segment elevations of 3–4 mV, CK of 16.3 ± 0.7 ($n = 37$); and sites with ST segment elevations of 5–7 mV, CK of 11.8 ± 0.5 ($n = 33$). There was an inverse relationship between the ST segment elevation at 15 minutes after the coronary artery occlusion and the CK activity 24 hours later. Thus the ST segment elevation at each epicardial site 15 minutes after the occlusion can be used to predict the biochemical integrity of the underlying myocardium 24 hours later, at least as reflected by this enzymatic parameter. Similarly, those sites with normal ST segments 15 minutes post-occlusion had a normal histologic appearance 24 hours later, as determined by haematoxylin–eosin stain, as well as staining for glycogen and for fat droplets, whereas the sites at which the ST segment elevation exceeded 2 mV showed histologic signs of early myocardial infarction 24 hours later. In addition, electron microscopic changes paralleled the changes seen on light microscopy.[57,100] The correlations between ST segment changes and myocardial metabolism have been examined in several studies. When global ischaemia of the left ventricle was produced, ST segment changes occurred almost simultaneously with the first biochemical indices of ischaemia, i.e. reduction of myocardial lactate extraction, and efflux of K$^+$ from the heart.[90]

Figure 1. (*Continued*)

Circulation, **54**, 591–8 (1976)]. In the hyaluronidase-treated dog (Figure 1b), site A (from non-ischaemic myocardium), as in the control dog, exhibited no ST_{15m} and at 24 hours it had no changes in QRS configuration and normal CK activity, and it appeared normal histologically. Site B (border zone) showed moderated ST_{15m}, but unlike in the control dog, at 24 hours there was little change in QRS configuration, CK activity was better preserved, and the histologic section was graded only 1+ (up to 25%) necrosis. Site C (centre of the ischaemic zone) had marked ST_{15m}, but again, unlike the control dog, at 24 hours it demonstrated only a small Q wave and only partial loss of R wave voltage. The CK activity was also better preserved than in the control dog and the histologic section was graded only 3+ (51–75%) necrosis

In a correlation of epicardial ST segment changes with metabolic alterations in the underlying myocardium following occlusion of the anterior descending coronary artery in dogs, Karlsson found that biopsies of the myocardium subjacent to sites with epicardial ST segment elevations showed lactate accumulation, as well as depletion of ATP and creatine phosphate,[38] reflecting anaerobic myocardial metabolism. Sayen et al.[89] using polarographic measurements of intramyocardial oxygen tension, found that ST segment elevations in the epicardial electrocardiogram promptly followed reduction of oxygen tension below 65% of control. More recently, Angell et al.[1] compared the magnitude of ST segment elevations in surface electrograms with the intramyocardial oxygen tension in the subjacent tissue recorded by means of platinum–iridium electrodes. The ST map correlated closely with the oxygen tension as the latter was varied by altering coronary perfusion pressure.[1] Also, Khuri and associates[39] varied coronary blood flow and recorded myocardial P_{O_2} and P_{CO_2} using a mass spectrometer. When regional ischaemia was produced, epicardial ST segment elevations correlated with changes in myocardial gas tensions. However, intramyocardial ST segments proved to be more sensitive than those recorded from the epicardium.[39] More recently, Hearse et al.[24] used a biopsy instrument with multiple drill heads to retrieve simultaneously multiple individual transmural tissue samples from the left ventricular wall of the dog heart. After coronary artery occlusion they measured: (i) metabolic changes (adenosine triphosphate, creatine phosphate, glycogen, lactate, potassium, sodium, water); (ii) electrocardiographic ST changes; and (iii) blood flow distribution (microspheres). Adenosine triphosphate, creatine phosphate and lactate values in the non-ischaemic tissue were essentially constant until 2–3 mm from the edge of visible cyanosis. The high-energy phosphate content of the tissue then decreased sharply across a zone 8–15 mm wide that spanned the visible edge. Across this zone lactate content increased sharply as did ST segment elevation, and coronary flow decreased to approximately 20 per cent of the control value. Multiple cross-correlation studies revealed that changes in the tissue content of adenosine triphosphate, creatine phosphate, and lactate were actually reflected by ST segment changes, and further, that all of these variables were directly related to the degree of ischaemia, as indicated by the reduction in coronary flow.

These techniques have provided interesting and important results, but their value is limited by the fact that ST segment elevation is not specific for myocardial ischaemia.[72] Changes in local electrolyte concentrations, particularly of K^+, both in the ischaemic area and in the normal tissue, can markedly alter the height of the ST segment elevation.[77] Pericarditis and intraventricular conduction defects,[70,72] as well as several commonly used cardiac drugs, including the cardiac glycosides, are also known to alter the height of the ST segment elevation. Although these limitations can be

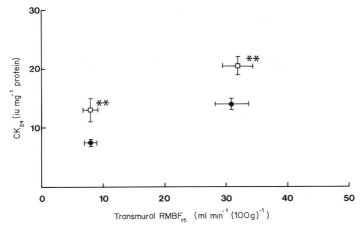

Figure 2. Comparison between transmural creatine kinase activity 24 hours after occlusion (CK_{24}) in control (○) and hyaluronidase-treated (□) dogs in severely (RMBF 0–15 ml min^{-1} (100 g)$^{-1}$) and moderately (16–50 ml min^{-1} (100 g)$^{-1}$) ischaemic sites. Note sparing of CK by hyaluronidase treatment. Horizontal and vertical bars represent ±SEM.
$** = p < 0.01$

overcome in experimental investigations, all of these considerations prompted us to search for a technique independent of changes in the ST segment, which could assess the efficacy of an intervention in reducing acute ischaemic myocardial damage.[85] Accordingly, it was demonstrated that the relationship between regional myocardial blood flow 15 minutes after coronary occlusion measured by the radioactive microsphere technique and myocardial CK activity at the same sites 24 hours later[25,41] could also be used to assess the effect of an intervention on myocardial salvage and that interventions changed this relationship[85] (Figure 2). Evidence for sparing of CK by an intervention, when analysed both by the ST and by the flow techniques, provides useful confirmation of both techniques. More recently, in order to avoid completely all indirect methods of measuring changes in infarct size and myocardial salvage, the rat model of coronary artery occlusion[13,37,52,91] has been utilized to quantify infarctions directly by serial histologic sections and by total CK activity of the left ventricle.[52,53] Using this technique, it has been possible to: (i) measure the effect of an intervention in reducing myocardial infarction size when the necrotic process is at its peak (2 days after occlusion); (ii) determine the long-term effect of an intervention in reducing infarct size and salvaging myocardium by analysis of the hearts 21 days after the occlusion (when the process of infarction is complete); and (iii) determine the effect of an intervention on scar formation, including its possible interference with the healing process.

Using this technique, a standard size of infarct was produced in the left ventricles of 250–300 g male Sprague–Dawley rats by occluding the left coronary artery 1–2 mm from its origin.[52] In order to assess the effect of an intervention on expected infarct size, 4 groups of rats were studied; sham-operated, intervention-treated sham-operated, rats with an occlusion but no intervention, and intervention-treated occluded rats. The animals were killed either 48 hours after occlusion or 21 days after occlusion. At the time of sacrifice, 10 per cent carbon black in saline (particle size, 300 Å) was injected either intravenously or directly into the left coronary ostium. If the anterior surface of the left ventricle was not discoloured by the carbon, a complete coronary artery occlusion was deemed to be present. The excised heart either had its left ventricle homogenized for enzymatic study or it was placed in 10 per cent phosphate-buffered formalin prior to histologic processing and examination.

For the enzymatic studies, the CK activity of the homogenized whole left ventricle (free wall plus interventricular septum) was measured as described by Kjekshus and Sobel.[40] In order to permit calculation of 'enzymatic' infarct size, the minimum CK activity in the infarcted centre 48 hours after occlusion was also determined and found to be 2.6 iu mg^{-1} protein. For the histologic studies the left ventricles were sectioned into four slices (2–2.5 mm thick) from the apex to the base of the heart in a plane parallel to the atrioventricular groove. Paraffin-embedded sections (5 μm in thickness) were prepared from each slice, stained with haematoxylin and eosin, projected onto a screen, and planimetered to determine the cross-sectional area of the left ventricle and of the infarcted myocardium.[53] The fraction of the left ventricle that was infarcted was calculated as a mean of this value in each of the four slices, then expressed as a percentage. The thickness of the transmurally infarcted portion of the left ventricle was also measured.

MECHANISM OF ACTION OF VARIOUS INTERVENTIONS

In studies begun in 1968, it was shown that the area of ischaemic injury, as reflected by the epicardial ST segment elevations, can be either increased or reduced by changing the balance between oxygen supply and demand.[56,57] Thus when the oxygen supply was altered by changing systemic arterial pressure and, therefore, coronary collateral flow to the ischaemic area, the extent of acute myocardial injury was found to be inversely proportional to these changes in arterial pressure. Furthermore, the changes in myocardial oxygen requirements were important determinants of the final extent of cell necrosis; interventions which increase oxygen requirements, such as isoproterenol, augmented myocardial injury, whereas interventions that decrease myocardial oxygen requirements, such as propranolol, reduced damage, as assessed both by epicardial ST segment elevation and by myocardial CK activity. It is important to note that, in general, changes in collateral blood flow appeared to override the changes in oxygen requirements. For example, the increase in systemic arterial pressure on the one hand, increases the afterload and thereby augments myocardial oxygen consumption which presumably has a detrimental effect on cell damage, whereas on the other, this increase in arterial pressure augments the perfusion pressure of the coronary arteries and thereby increases the coronary collateral flow to the ischaemic zone thus favourably influencing the damaged cells. However, the net result of these two opposite actions is favourable since the effect of an increase in collateral blood flow to the ischaemic zone overrides the detrimental effect of the increase in afterload. This indicates the important role played by collateral flow in determining the final extent of necrosis.[57]

Subsequent experimental studies using electrocardiographic, enzymatic and histologic indices of tissue damage confirmed this initial hypothesis. Thus interventions which increased myocardial oxygen requirements increased myocardial damage. These included stimuli that exert positive inotropic or chronotropic actions such as isoproterenol, glucagon, digitalis in the non-failing heart, bretylium tosylate, and pacing-induced tachycardia, all of which augment myocardial oxygen needs.[57] Reductions in oxygen delivery to the ischaemic myocardium also resulted in extensions of the injury; these reductions could be produced either by hypoxaemia[78] or by reducing collateral blood flow to the ischaemic zone by lowering the perfusion pressure by means of haemorrhagic hypotension,[57,59] or because of an unfavourable redistribution of regional myocardial blood flow such as may occur when coronary vasodilators, such as minoxidil[79] and nitroprusside,[9] which act on the resistance vessels are administered (Figure 3).

The recognition that several vasodilators may be detrimental in the setting of acute myocardial infarction is potentially of considerable clinical importance. The possibility that this class of vasodilators augments blood flow to the non-ischaemic zone, where this increase is not necessary, and reduces collateral perfusion of the ischaemic zones, where collateral flow is of paramount importance, warrants careful determination of the precise effects of each vasodilator on coronary flow.

In contrast, interventions which increased myocardial oxygen delivery or decreased myocardial oxygen needs reduced the extent of tissue death. Included in this category are:

(i) Elevation of systemic arterial pressure (which probably acts by increasing coronary perfusion pressure and collateral flow to the ischaemic area);[56,57,59,81]

(ii) Coronary vasodilators, such as nitroglycerine, which act on the conductance vessels to increase collateral flow;[4,6,9,12,15,19,28,95]

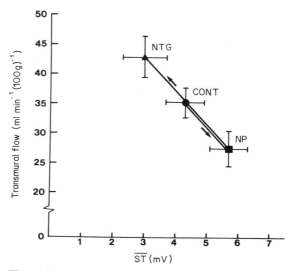

Figure 3. Simultaneous changes in average ST segment ($\overline{ST}$) and transmural flow in dogs. Flow decreased with nitroprusside (NP), causing elevation in $\overline{ST}$. After nitroglycerine (TNG), flow increased and $\overline{ST}$ decreased. [Reproduced, with permission, from *Circulation* **54**, 766–73 (1976).]

(iii) Coronary vasoconstrictors, such as methoxamine, which probably act on the coronary resistance vessels, thereby favourably redistributing regional myocardial flow;[10] the inhalation of 100% oxygen also acts by a similar mechanism,[84] i.e. it induces constriction of coronary vessels in the non-ischaemic zone but increases the collateral coronary blood flow to the ischaemic zone (Figure 4).

It has been suggested that vasodilators which act on the proximal conductance vessels are beneficial,[4,6,9,12,15,19,28,95] while those which act on the peripheral resistance vessels are detrimental[9,79,106] because they produce a redistribution of flow away from the ischaemic zone. Accordingly, it was postulated that coronary constrictors that act on resistance vessels will be beneficial while constrictors that act on conductance vessels will be harmful. Therefore, an effort was made to ascertain whether redistribution of blood flow to the ischaemic zone can be induced by constriction of coronary vessels in the normal myocardium. The constant infusion of methoxamine, while the systemic pressure was maintained constant, resulted in a reduction of regional myocardial blood flow to the normal areas. This reduction in flow can be ascribed both to the lowering of myocardial oxygen consumption and to the presumed constrictive action of the drug on the coronary arterioles. In contrast, blood flow to the ischaemic areas increased. The failure of methoxamine to constrict the arteries in the ischaemic areas may be explained either by a reduced amount of the drug reaching the underperfused zones or, more probably, by the more powerful 'metabolic' vasodilation induced by the ischaemia itself. It has been suggested that the collateral vessels have their origin at the level of the arterioles.[17] When the pressure in large arteries is kept constant despite infusion of vasoconstrictors, arteriolar pressure rises.[8] This increase in perfusion pressure at the origin of the collateral vessels may therefore be responsible for the redistribution of coronary blood flow and it was accompanied by a significant reduction in acute myocardial ischaemic injury as reflected by the change in the average ST segment elevation.

The possible multifaceted effects of methoxamine on regional myocardial blood flow in the normal areas and the consequent redistribution of coronary flow to the ischaemic zone should be analysed closely since they may help to explain the salutary effects of other agents which have similar general effects. Methoxamine, as a result of α-receptor stimulation, causes

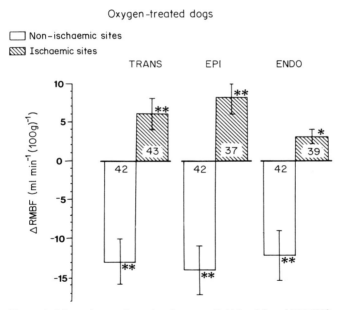

Figure 4. Mean changes in regional myocardial blood flow (ΔRMBF) in dogs (due to 100% oxygen breathing) in the non-ischaemic sites (plain columns) and in the ischaemic sites (shaded columns). Left-hand pair of columns depicts changes in transmural flow (TRANS), central pair of columns depicts changes in epicardial flow (EPI), and right-hand pair of columns depicts changes in endocardial flow (ENDO). Numbers in parentheses = number of sites. Bars represent $\pm$ SEM. $* = p < 0.05$ and $** = p < 0.01$. Note that with 100% oxygen breathing blood flow was redistributed from the non-ischaemic to the ischaemic myocardium

coronary constriction. However, a reduction in myocardial oxygen consumption will also reduce coronary flow in the normal myocardium. In the haemodynamic experiments performed in this study, heart rate, pre-load, and afterload did not change following the drug administration; however, left ventricular dP/dt decreased by 20%, presumably resulting in lower myocardial oxygen consumption. This fall in dP/dt can be ascribed to this β-blocking effect of methoxamine.[30] Thus at least two factors could have contributed to the coronary vasoconstriction: the direct effects of α-receptor stimulation, and the indirect effects of β-blockade induces reduction in myocardial oxygen consumption caused by decrease in contractility.

Other interventions which may decrease injury are those that act by reducing oxygen needs and include the β-adrenergic blockers, such as propranolol,[56,57,59,98] practolol, oxprenolol, and timolol,[49,74] the barbituates, such as pentobarbital, bradycardia, and digitalis in the failing heart.[64,102] Some interventions may act both by decreasing myocardial oxygen needs and by increasing perfusion pressure, as in the case with intraaortic balloon counterpulsation.[44,47,58] Of special interest is the action of digitalis, which exhibits opposite effects depending on the inotropic state.[64,101,102] Thus in the non-failing heart, digitalis increased damage, while in the same animal when failure was induced pharmacologically, digitalis reduced injury. It is postulated that in the absence of failure the stimulation of contractility increases oxygen needs, and consequently ischaemic injury, while in the latter the reduction in energy requirements resulting from the digitalis-induced decrease in left ventricular size overrides the energy cost of an increase in contractility.

Other studies have shown that many other factors which probably do not act through changing myocardial oxygen balance can also significantly alter infarct size. The infusion of hypertonic glucose, with or without insulin and potassium, reduced the extent of myocardial infarction 24 hours after occlusion.[62] These conclusions were based on the observation that there was less myocardial creatine kinase depression and less extensive histologic damage in the treated dogs than in the untreated dogs at sites with similar epicardial ST segment elevations. The protection of the ischaemic myocardium by glucose–insulin–potassium was also evident from the better preservation of the ultrastructure in the ischaemic zone. The mechanism of action of glucose is debatable, but it may act by increasing substrate availability for glycolysis and/or lowering free fatty acid levels that may themselves be deleterious. In contrast, hypoglycaemia induced by insulin administration increased myocardial damage (Figure 5).[50] Its mode of action is most probably related to a reduction in available substate rather than the accompanying adrenergic discharge, since β-adrenergic blockade with propranolol did not prevent this increase in damage.

Hyaluronidase, an enzyme which depolymerizes mucopolysaccharides, is

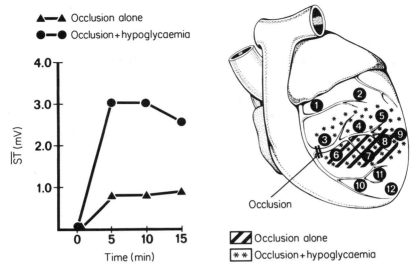

Figure 5. Example of an experiment showing increased ST segment elevation during hypoglycaemia. (Right panel) The diagram depicts the anterior surface of the heart showing location of the arbitrarily selected sites. The striped area represents the area of sites with abnormal ST segments 15 minutes after occlusion alone. The starred area shows the substantially greater area of ST segment abnormality 15 minutes after the second occlusion which was carried out during hypoglycaemia. (Left panel) Time course of average ST segment elevation ($\overline{ST}$). Triangles denote data from occlusion alone, circles show increased ST segment elevation during second occlusion in the same dog after insulin treatment. [Reproduced, with permission, from *Circulation*, **51**, 621–6 (1975).]

also very effective in reducing the acute ST segment elevation, and it reduced necrosis 24 hours after occlusion, as analysed by myocardial CK depression, histologic appearance, and epicardial ST segment changes[26,27,61] (Figure 1(b)). Moreover, following left coronary artery occlusion in the rat, infarct size in hyaluronidase-treated animals is substantially smaller than in non-treated rats, when analysed either by serial histologic sections or by total left ventricular CK activity.[52] This is evident both after 2 and after 21 days, thus demonstrating that not only is the necrosis less after 2 days but also that the final size of the scar after the process of infarction is complete is likewise substantially smaller. The smaller scar demonstrates that hyaluronidase administration does not simply delay the necrotic process but that it actually reduces the extent of necrosis and so results in a larger amount of normally contracting myocardium (Figure 6).

The beneficial results obtained with this drug during the first 48 hours exceeded those produced by the other interventions tested in the rat model. By the CK method there was approximately 50% more residual myocardium in the hyaluronidase-treated than in control occluded rats at the peak

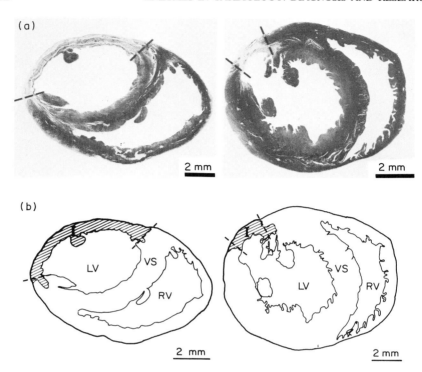

Figure 6. (a) Histologic sections and (b) diagrams of transverse slices of hearts of rats killed 21 days after occlusion of the left main coronary artery without (left side) and with (right side) hyaluronidase treatment: LV, left ventricle; RV, right ventricle; VS, ventricular septum; and I, infarcted myocardium. The borders of the infarctions are shown by the interrupted lines and their areas by the shaded portions of the diagrams. Note that in the rat with an occlusion alone the infarction involved 54.1 per cent of the endocardial circumference of the left ventricle, whereas in the hyaluronidase-treated rat it involved only 20.1 per cent. [Reproduced, with permission, from *Science*, **194**, 199–200, (1976).]

of the necrotic process. After the infarct had healed the volume of preserved myocardium in the treated rats was still significantly greater than in the untreated (control) rats with coronary occlusion although less than at 48 hours, so that much of the early benefit derived from hyaluronidase administration appears to persist. Furthermore, hyaluronidase need be given only over the first 24 hours after occlusion to achieve its long-term effects.

The mechanism of action of hyaluronidase is still not fully understood. Hyaluronidase has been shown to penetrate the ischaemic zone, depolymerizing the mucopolysaccharides even in the centre of the infarction.[61] This action may facilitate either the transport of nutrients to, or the washout of potentially harmful metabolites from, the ischaemic zone. The electron microscopic observation that glycogen is abundant even in the centre of the ischaemic zone in hyaluronidase-treated rats may support the postulate that the transport of nutrients such as glucose is facilitated.[43] In addition, it was found that hyaluronidase prevents the fall in collateral flow to the ischaemic zone that normally occurs between 15 minutes and 6 hours after coronary artery occlusion.[3] This may be explained by a reduction in oedema and therefore in compression of the microvasculature. Alternatively, the prevention of the fall in collateral flow may be an effect, rather than a cause of the reduced size of the infarct.

Another promising group of substances that are effective in reducing necrosis is drugs which influence the inflammatory process. As the necrotic process evolves, there is usually an increase in capillary permeability, in chemotaxis and in phagocytosis, each of which plays a role in the transformation of the injured tissue to a scar. In the case of myocardial infarction, it may be advantageous to arrest this process of killing the ischaemic cells, since many of them may not be irreversibly damaged at the time that an intervention can be applied and the increase in collateral circulation may still salvage them. Thus several substances that have widely differing pharmacologic actions may decrease necrosis by reducing inflammatory reactions. These include glucocorticoids,[48,99] cobra venom (CVF),[65] aprotinin[14,23] and ibuprofen. Glucocorticoids have multiple and complex actions and the mechanism by which these agents benefit ischaemic myocardium remains to be elucidated. A number of mechanisms of action may be proposed. These include:

(i) stabilizing myocardial cell membranes and so preventing or delaying the release of lysosomal enzymes;[16,99]
(ii) stabilizing the phagocytic vacuoles and thereby reducing the heterolytic activity of infiltrating inflammatory cells;[48]
(iii) inhibiting the generation of prostaglandins and thromboxanes;[21]
(iv) exerting other anti-inflammatory properties;
(v) increasing collateral blood flow to the ischaemic myocardium.[71]

It is interesting, however, that CVF, which has a specific action, i.e. the cleavage and therefore the reduction of the C3 component of the complement system, also decreases infarct size. It is postulated that this reduction in necrosis is caused by a reduction in capillary permeability, prevention of the normal release of chemotactic factor by the C5 component, and of the non-specific injury to the cells normally induced by C7 through C9 components. Aprotinin, by inhibiting the kallikrein–kinin system, may similarly reduce the effects of the released kinins that can increase cell permeability. Ibuprofen suppresses prostaglandin-mediated components of inflammation. The potential advantage of this group of anti-inflammatory agents, as compared to interventions such as propranolol which alter myocardial oxygen consumption, is that while the latter may induce undesirable haemodynamic side effects, the former simply change the natural evolution of the infarct and adverse haemodynamic side effects are not expected.

Other interventions which may have clinical potential for decreasing myocardial necrosis include hyperosmolar substances, such as mannitol,[75,104] that may act by reducing cell swelling and thus facilitating collateral blood flow, and interventions which decrease free fatty acid levels such as β-pyridyl carbinol.[42,73]

EFFECT OF INTERVENTIONS ON THE HEALING PROCESS

The healing process is distinct from that of necrosis of the myocardial fibres. It was considered possible that drugs that have a protective effect on the ischaemic myocardium may have a detrimental effect on scar formation. In the rat model, the thickness of the scars was examined 3 weeks after the occlusions. Hyaluronidase, reserpine, cobra venom factor, ibuprofen, hydrocortisone, and a single dose of methylprednisolone did not thin the scar. However, when multiple doses of methylprednisolone were administered during the 24 hours after coronary artery occlusion the scars were excessively thinned[53] (Figure 7). At the time of sacrifice 21 days after coronary occlusion these abnormally thin scars had already developed into prominent ventricular aneurysms. In these methylprednisolone-treated rats, the distension of the scar 3 weeks after occlusion may give the false impression that myocardial necrosis has been more extensive than is actually the case, there simply being greater representation of the scar as a per cent of the left ventricular surface. This situation represents expansion of the infarct rather than extension or reinfarction due to a true increase in necrosis.[29] Whatever the mechanism of this excessive thinning, the hazard of the formation of large ventricular aneurysms is clear. Interestingly, in man, multiple doses of methylprednisolone following acute myocardial infarction have been reported possibly to increase the incidence of ventricular rupture[86] although, so far, this has not been substantiated. It is possible that for a more accurate definition of scar formation biochemical methods of quantification will need to be used. This approach was attempted by Leon et al.[46] who measured hydroxyproline showing that dimethylsulphoxide (DMSO) interferes with the healing process of experimental myocardial infarctions.

The present controversy about the role glucocorticoids should play in the treatment of myocardial infarction may be related, as least in part, to their opposite effects on necrosis and healing. When administered during the first 24 hours following a skin incision, glucocorticoids interfere with wound healing.[88] Although the rat is particularly sensitive to other metabolic effects

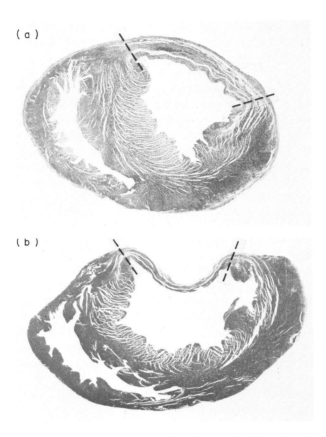

Figure 7. Scar thinning resulting from administration of 50 mg kg^{-1} of methylprednisolone 5 minutes, 3, 6, and 24 hours after occlusion (MP×4). A representative histologic section from a non-treated occluded rat at 21 days post-occlusion is shown in (a) and one from an (MP×4)-treated rat in (b). Note that for a scar of comparable size, that from the (MP×4)-treated rat is abnormally thin (Masson's trichrome, ×3)

of steroids, it is no more sensitive to steroid-induced interference with healing than are other species.[82] Furthermore, in the dog pharmacologic doses of methylprednisolone adversely affect the early healing phase of acute myocardial infarction,[22] and in one patient receiving prolonged hydrocortisone therapy following acute myocardial infarction, the development of a ventricular aneurysm was associated with delayed myocardial healing.[7]

THE INFLUENCE OF THE TIME INTERVAL BETWEEN CORONARY ARTERY OCCLUSION AND THE APPLICATION OF AN INTERVENTION ON SALVAGE OF ISCHAEMIC MYOCARDIUM

It has been shown that myocardial necrosis first occurs 20–25 minutes after occlusion and that many myocardial cells die during the first 60 minutes of severe ischaemia.[31,33] More recently, Reimer et al.[83] have shown in the dog that papillary muscle necrosis increases progressively as the duration of circumflex artery occlusion is extended. However, even as late as 6 hours after occlusion, there is a zone of ischaemic but still viable myocardium which is potentially salvageable. Certain interventions have a significant salutary effect on the ischaemic myocardium even when they are administered several hours after coronary artery occlusion.[48,57–59,62,69]

Although the irreversibility of acute myocardial ischaemic injury has been studied extensively, the events which characterize the transition from reversible to irreversible injury remain clouded. The mitochondria demonstrate striking anatomic and metabolic changes as the cells which contain them enter an irreversible phase of injury.[32,34] It is not clear whether such mitochondria are capable of recovery if the ischaemia is relieved. The loss of cell volume regulation and increased sarcolemmal permeability have been found to correlate more closely with irreversible injury than has mitochondrial failure.[35] Therefore, the primary event leading to irreversibility may be a sarcolemmal defect which allows excess calcium and water to enter the injured cells.

In order to define the period in which the ischaemic myocardium is still amenable to treatment, hyaluronidase was administered at several different time intervals following coronary artery occlusion in the dog—20 minutes, 3, 6, and 9 hours.[27] The extent of myocardial necrosis was measured biochemically (myocardial CK activity), histologically, and electrocardiographically (changes in QRS configuration 24 hours after coronary artery occlusion).[26] All three methods of measuring necrosis were complementary, and the results of each correlated well with one another. By correlating ST segment elevation shortly after occlusion (15 minutes) with these indices of necrosis at 24 hours, it was demonstrated that when hyaluronidase was administered after occlusion its beneficial effects declined progressively as the interval between the coronary artery occlusion and the administration of the drug lengthened (Figure 8). However, even when it was administered 6 hours after occlusion, a significant beneficial effect was still demonstrable, although it was less than that observed when the drug was given 20 minutes after occlusion. In contrast, hyaluronidase administered 9 hours after occlusion had no discernible beneficial effect. It remains possible, however, that a more potent agent or one with a mechanism of action different from that of

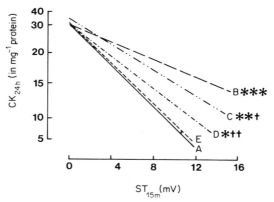

Figure 8. The relationship between ST segment elevation 15 minutes after coronary artery occlusion (ST_{15m}) and log CK values of specimens obtained from the same sites 24 hours later. Group A (occlusion alone) (———): log CK = $(-0.064 \pm 0.007)ST_{15m} + (1.49 \pm 0.02)$; 14 dogs, $r = -0.81 \pm 0.03$. Group B (hyaluronidase given 20 minutes after occlusion) (— — —): log CK = $(0.025 \pm 0.003)ST_{15m} + (1.48 \pm 0.02)$; 12 dogs, $r = -0.72 \pm 0.04$. Group C (hyaluronidase given 3 hours after occlusion) (·········): log CK = $(-0.037 \pm 0.005)ST_{15m} + (1.53 \pm 0.01)$; 8 dogs, $r = -0.85 \pm 0.02$. Group D (hyaluronidase given 6 hours after occlusion) (-·-·-·-): log CK = $(-0.044 \pm 0.003)ST_{15m} + (1.49 \pm 0.01)$; 8 dogs, $r = -0.78 \pm 0.03$. Group E (hyaluronidase given 9 hours after occlusion) (- - - - -): log CK = $(-0.060 \pm 0.006)ST_{15m} + (1.50 \pm 0.02)$; 6 dogs, $r = -0.86 \pm 0.06$. Note that for any ST_{15m}, hyaluronidase given 20 minutes, 3 hours, or 6 hours after occlusion results in significantly greater myocardial CK activity; in contrast, hyaluronidase administered 9 hours after occlusion has no such effect ($* = p < 0.025$, $** = p < 0.025$, $*** = p < 0.0005$ in comparison to control; $† = p < 0.025$, $†† = p < 0.0005$ in comparison to hyaluronidase at 20 minutes). [Reproduced, with permission, from *Circulation Research*, **41**, 26–31 (1977).]

hyaluronidase, may still cause significant myocardial salvage when started at this later time.

The extrapolation of the results of this study to patients must be done cautiously, since there are obvious differences between the animal model and acute myocardial infarction in man. However, since the histopathologic evolution of infarction in the dog is similar to that in man,[55] this investigation supports the hope that those patients reaching the hospital within a few hours of the onset of symptoms may benefit from interventions designed to limit ischaemic injury.

CLINICAL OBSERVATIONS

One of the most formidable barriers to the clinical application of the information which has been obtained in the laboratory is the lack of a suitable technique for assessing the efficacy, or lack thereof, of these interventions. The ideal technique for assessing the effectiveness of interventions designed to protect injured but potentially salvageable myocardium in patients would be:

(i) safe and non-invasive;
(ii) capable of predicting the extent of necrosis to be expected if no interventions were employed;
(iii) capable of assessing the extent of necrosis that actually develops;
(iv) capable of providing the data in items (ii) and (iii) accurately and in quantitative terms, i.e. in grams;
(v) effective if applied immediately upon the patient's admission, so that the intervention under study can be promptly applied, since delay in treatment may be expected to reduce the population of injured cells that are salvageable;
(vi) relatively simple, easy to apply, and inexpensive, so that its use will not be limited to specialized centres;
(vii) applicable to all patients with acute myocardial infarction.

Items (ii) and (iii) are of particular importance, since they would allow each patient to be used as his own control.

At present, myocardial ischaemic injury in the clinical setting can be assessed by the following methods:

(i) radionuclide scintigraphy by gamma or positron imaging of the myocardium;
(ii) release of enzymes from the injured myocardium (CK disappearance curves);
(iii) praecordial ST segment and QRS mapping.

The application of the radionuclide techniques has great potential, but their use for monitoring changes in the extent of an ischaemic zone awaits the development of high-resolution cameras and the availability of an agent which would specifically identify injured cells and that could be used serially. The CK disappearance curves offer the possibility of predicting 'infarct size' following at least 7 hours of sampling (i.e. at least 10 hours from the onset of pain), but they are therefore of less value in studies carried out in the early hours of ischaemia when the population of reversibly injured cells is maximal.

Although the use of praecordial ST segment elevation in patients does not permit the quantitative evaluation of the size of an infarction, the technique is useful for determining whether the myocardial ischaemic injury is increasing or decreasing.[59] This is based on the observation that praecordial electrocardiograms are semidirect measurements of the epicardial events,[80,105] and that interventions which change the extent and magnitude of ST segment elevations induce parallel alterations in ST segment elevations on the canine praecordium and epicardium.[72] Experimentally, epicardial ST segment elevation correlates well with changes in myocardial P_{O_2},[1,89] in coronary blood flow,[41,90] in myocardial high-energy phosphate and lactate concentrations,[38] in myocardial CK activity,[57] and, most importantly, it accurately predicts the areas of necrosis, as defined by light and electron microscopy.[62,100]

The clinical relationship between praecordial ST segment elevation and the reversible damage to myocardial cells is best illustrated by patients with 'variant' angina as described by Prinzmetal.[76] In these patients, there is a temporary occlusion of a major coronary artery, most often as a result of spasm.[51] The onset of the ischaemic injury to the myocardium is characterized by the triad: pain, ST segment elevation, and ventricular arrhythmias. The release of coronary spasm and subsequent abolition of ischaemic injury is reflected by cessation of pain and ventricular irritability and abolition of the ST segment elevation. However, in patients in whom the coronary flow fails to return, the ischaemic injury progresses and the ST segments remain elevated until a frank infarction develops.

The praecordial ST segment mapping technique has been successfully used in man to show that several interventions reduce praecordial ST segment evation more rapidly than expected; this implies that these interventions effectively reduce myocardial injury while it is still in the reversible phase. These interventions include propranolol[20,59] (Figure 9), hyaluronidase,[68] nitroglycerine[9,12,15] and the inhalation of high concentrations of oxygen.[54] ST segment elevation can, however, result from other causes such as pericarditis, epicardial trauma, changes in ionic concentrations (i.e. hyperkalaemia), and normal early repolarization. Moreover, ST segments will not reflect myocardial injury on the epicardium when focal block, as defined by widened QRS complexes, occurs.[72] In this instance, the wide QRS complexes, cause a displacement of the ST segment in the opposite direction, as explained by the gradient theory.[2] This consideration may account for the absence of ST segment elevation in the centre of an ischaemic zone, as reported by Cohen and Kirk[11] and by Smith et al.[97] This phenomenon of lower ST segment elevation in the centre of an ischaemic zone is not observed in dogs without widened QRS complexes,[72,80] nor in praecordial electrocardiograms recorded in patients.

More recently, praecordial QRS complex mapping was added to ST

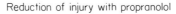

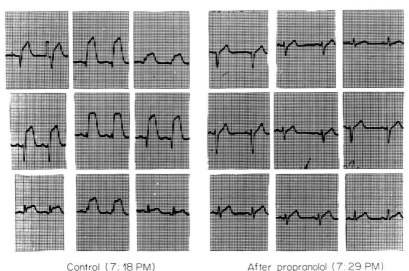

Figure 9. The effect of propranolol on the reduction in ST segment elevation in an average patient. The nine leads depicted are from the V_1, V_3, and V_5 positions and the corresponding sites one intercostal space above and below the standard positions. Note the marked reduction in ST segment elevations after propranolol administration

segment mapping. By this technique, a fall in R waves and the appearance of new Q waves are analysed during the first week following the onset of an infarct in praecordial leads that initially (i.e. up to 8 hours after onset of pain) showed ST segment elevations. The rationale of this approach therefore is to use the praecordial ST segment recorded as soon as possible after the onset of the clinical event as a predictor of the ultimate fate of the tissue, in a manner analogous to the epicardial ST segment in the experimental animal.[26] This praecordial ST segment may then be compared to the changes in the QRS complex that occur subsequently, such as the development or deepening of Q waves and the reduction of R waves (Figure 10); these changes in the QRS complex could be employed for the assessment of myocardial ischaemic injury in a manner analogous to the alterations in CK activity or histologic appearance of the myocardium subjacent to the epicardial electrode in the experimental animal.

Thus the simultaneous analysis of alterations of ST segments and QRS complexes, while not capable of expressing the mass of infarcted myocardium in quantitative terms, would appear to be:

(i) capable of predicting the surface representation of the extent of necrosis to be expected when much of the myocardial injury is still in a reversible phase, i.e. at a time of ST segment elevation, prior to the development of changes in the QRS complex;

(ii) capable of assessing the surface representation of the extent of necrosis that actually develops, as reflected by the change in the QRS complexes;

(iii) capable of being applied immediately, so as not to delay therapy;

(iv) safe and atraumatic;

(v) simple to apply, easy to interpret, and inexpensive.

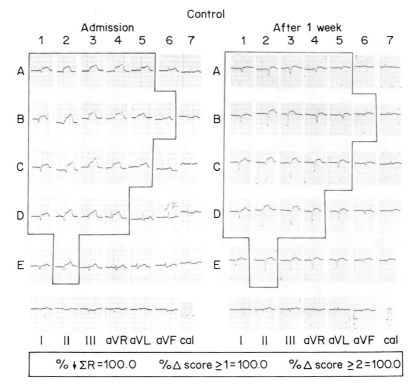

Figure 10. An example of the 35 praecordial leads and the 6 classical electrocardiographic leads in a control patient with acute myocardial infarction of the anterior left ventricular wall on admission (left panel) and 1 week later (right panel). The leads showing signs of acute ischaemia at the time of admission are enclosed by the boundary line in each of the two panels. Note that the R wave was completely lost over this time period in all leads where an R wave was present initially; all sites on the initial electrocardiogram with ST segment elevation ≥ 0.15 mV and an initial score of 0, 2, or 3 showed a rise in score of 1 or more (% Δ score $\geq 1 = 100.0$); all sites on the initial electrocardiogram with ST segment elevation ≥ 0.15 mV and an initial score of 0 or 2 showed a rise in score by 2 or more (% Δ score $\geq 2 = 100.0$)

Using this technique the effects of hyaluronidase have been assessed in a multicentre study.[66] We randomized 91 patients with anterior infarction to control (45 patients) or to hyaluronidase-treatment (46 patients) groups. A 35-lead praecordial electrocardiogram was recorded on admission and 7 days later. Hyaluronidase was administered intravenously after the first electrocardiogram and every 6 hours for 48 hours.

The sum of R wave voltages of vulnerable sites fell more in the control group than in the hyaluronidase group (70.9 ± 3.6 per cent {± 1 SEM} vs 54.2 ± 5.0 per cent $p < 0.01$). Q waves appeared in 59.3 ± 4.9 per cent of the vulnerable sites in control vs 46.4 ± 4.9 per cent in hyaluronidase-treated patients ($p < 0.05$) (Figure 11). Thus the findings in this study demonstrate that hyaluronidase, when administered within the first 8 hours after the

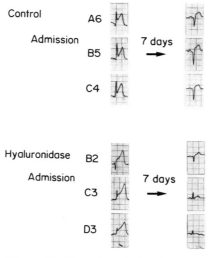

Figure 11. Examples of the electrocardiographic evolution in 7 days. On the left are three leads from admission tracings and on the right those from the same sites one week later. The three top leads are taken from a patient in the control group. Note that after 7 days all complexes exhibited a QS pattern (i.e. 'score' of 4). The three bottom leads are from a patient treated with hyaluronidase. All leads show only moderate declines in the R waves. Each letter and number pair identifies the position of the electrode in the 35-lead mat. The letter identifies the horizontal row, and the number the vertical one. [Reproduced, with permission, from *New England Journal of Medicine*, **296**, 898–903 (1977).]

onset of the clinical event, reduces the extent of myocardial necrosis in patients with acute myocardial infarction. Although the differences in these electrocardiographic changes between the two groups were approximately 20 per cent, it should be noted that this type of analysis indicates a directional change in the extent of myocardial infarction, but does not provide quantitative information.

At present, it is not possible to identify the specific intervention likely to be most effective in reducing infarct size. Indeed, there may not be one single best treatment; rather, it seems that in the future patients will be carefully but rapidly subdivided and categorized according to their clinical, electrocardiographic, and haemodynamic states, and the intervention given tailored appropriately. For example, in hypertensive patients, reduction of afterload may be the most effective intervention; in patients without any evidence of myocardial depression, cardiospecific β-adrenergic blockade may be the treatment of choice; in normotensive or hypotensive patients with pump failure, circulatory support may be in order. Moreover, all patients with acute infarction, regardless of their haemodynamic state, may benefit from the administration of an anti-inflammatory agent, such as hydrocortisone, methylprednisolone, ibuprofen, or aprotinin, or from a drug such as hyaluronidase.

CONCLUDING REMARKS

Acute myocardial infarction is the most common cause of death in the western world. The possibility of reducing the mortality associated with this syndrome and of improving cardiac function by salvaging much of the acutely ischaemic myocardium in those who survive, raises new hopes for the future. It seems desirable, therefore, that an intensive research effort should now be directed toward ascertaining the definitive clinical value of therapies aimed at the protection of the acutely ischaemic myocardium.

REFERENCES

1. Angell, C. S., Lakatta, E. G., Weisfeldt, M. L., and Shock, N. W. Relationship of intra-myocardial oxygen tension and epicardial ST segment changes following acute coronary artery ligation: effects of coronary perfusion pressure. *Cardiovascular Research*, **9**, 12–8 (1975).
2. Ashman, R. The normal human ventriculargradient. IV. The relationship between the magnitudes A_{QRS} and G, the deviations of the RS-T segment. *American Heart Journal*, **26**, 495–510 (1943).
3. Askenazi, J., Hillis, L. D., Diaz, P. E., Davis, M. A., Braunwald, E., and Maroko, P. R. The effects of hyaluronidase on coronary blood flow following coronary artery occlusion in the dog. *Circulation Research*, **40**, 566–71 (1977).
4. Bleifeld, W., Wende, W., Bussmann, W. D., and Meyer, J. Influence of nitroglycerin on the size of the experimental myocardial infarction. *Nauyn-Schmeidebergs Archiv Pharmacology*, **277**, 387–400 (1973).

5. Bloor, C. M. and White, F. C. Coronary artery reperfusion: effects of occlusion duration on reactive hyperemia responses. *Basic Research in Cardiology*, **70,** 148–58 (1975).
6. Borer, J. S., Redwood, D. R., Levitt, B., Cagin, N., Bianchi, C., Vallin, H., and Epstein, S. T. Reduction in myocardial ischemia with nitroglycerin or nitroglycerin plus phenylephrine administered during acute myocardial infarction. *New England Journal of Medicine*, **293,** 1008–12 (1975).
7. Bulkley, B. H. and Roberts, W. C. Steroid therapy during acute myocardial infarction. A cause of delaying healing and of ventricular aneurysm. *American Journal of Medicine*, **56,** 244–50 (1974).
8. Burton, A. C. *Physiology and Biophysics of the Circulation; An Introductory Test*, Year Book Medical Publishers, 1965, pp. 86–9.
9. Chiariello, M., Gold, H. K., Leinbach, R. C., Davis, M. A., and Maroko, P. R. Comparison of the effects of nitroprusside and nitroglycerin on ischemic injury during acute myocardial infarction. *Circulation*, **54,** 766–73 (1976).
10. Chiariello, M., Ribeiro, L. G. T., Davis, M. A., and Maroko, P. R. Reverse Coronary Steal induced by coronary vasoconstriction following coronary artery occlusion in dogs. *Circulation*, **56,** 809–15 (1977).
11. Cohen, W. V. and Kirk, E. S. Reduction of epicardial ST segment elevation following increased myocardial ischemia: Experimental and theoretical demonstration. *Clinical Research*, **22,** 269A (1974).
12. Come, P. C., Flaherty, J. T., Baird, M. G., Rouleau, J. R., Weisfeldt, M. L., Greene, H. L., Becker, L., and Pitt, B. Reversal by phenylephrine of the beneficial effects of intraveneous nitroglycerin in pateints with acute myocardial infarction. *New England Journal of Medicine*, **293,** 1003–7 (1975).
13. Deloche, A., Fontaliran, F., Fabiani, J. N., Pennecot, G., Carpentier, A., and Dubost, C. Etude experimentale de la revascularisation chirurgicale precoce de l'infarctus du myocarde. *Annales Chirurgie Thoracique et Cardio-Vasculaire*, **11,** 89–105 (1972).
14. Diaz, P. E., Fishbein, M. C., Davis, M. A., Askenazi, J., and Maroko, P. R. Effect of the kallikrein inhibitor aprotinin on myocardial ischemic injury after coronary occlusion in the dog. *American Journal of Cardiology*, **40,** 541–9 (1977).
15. Flaherty, J. T., Reid, P. R., Kelly, D. T., Taylor, D. R., Weisfeldt, M. L., and Pitt, B. Intraveneous nitroglycerin in acute myocardial infarction. *Circulation*, **51,** 132–9 (1975).
16. Fox, A. C., Hoffstein, S., and Weissman, G. Lysosomal mechanisms in production of tissue damage during myocardial ischemia and the effects of treatment with steroids. *American Heart Journal*, **91,** 394–7 (1976).
17. Fulton, W. F. M. *The Coronary Arteries; Arteriography, Microanatomy and Pathogenesis of Obliterative Coronary Artery Disease*, Charles C. Thomas, Springfield, 1965, pp. 98–101.
18. Ginks, W. R., Sybers, H. D., Maroko, P. R., Covell, J. W., Sobel, B. E., and Ross, J., Jr. Coronary artery reperfusion. II. Reduction of myocardial infarct size at one week after the coronary occlusion. *Journal of Clinical Investigation*, **51,** 2717–23 (1972).
19. Gold, H. K., Leinbach, R. C., and Sanders, C. A. Use of sublingual nitroglycerin in congestive failure following acute myocardial infarction. *Circulation*, **46,** 839–45 (1972).
20. Gold, H. K., Leinbach, R. C., and Maroko, P. R. Reduction of signs of ischemic injury during acute myocardial infarction by intraveneous propranolol. *American Journal of Cardiology*, **38,** 689–95 (1976).
21. Goldstein, I. M., Malmsten, C. L., Kaplan, H. B., Jindahl, H., Samuelsson, B., and Weissmann, G. Thromboxane generation by stimulated human granulocytes: inhibition by glucocorticoids and superoxide dismutase. *Clinical Research*, **25,** 518A (1977).
22. Green, R. M., Cohen, J., and DeWeese, J. A. Short-term use of corticosteroids after experimental myocardial infarction: effects of ventricular function and infarct healing. *Circulation Supplement III*, **50,** 111–3 (1974).
23. Hartmann, J. R., Robinson, J. A., and Bunnar, R. M. Chemotactic activity in the coronary sinus after experimental myocardial infarction: effects of pharmacologic interventions on ischemic injury. *American Journal of Cardiology*, **40,** 550–5 (1977).
24. Hearse, D. J., Opie, L. H., Katzeff, I. E., Lubbe, W. F., Van der Werff, T. J., Reisach, M., and Boulle, B. Characterization of the 'Border Zone' in acute regional ischemia in the dog. *American Journal of Cardiology*, **40,** 716–26 (1977).

25. Heng, M. K., Singh, B. N., Norris, R. M., John, M. B., and Elliott, R. Relationship between epicardial ST segment elevation and myocardial ischemic damage after experimental coronary artery occlusion in dogs. *Journal of Clinical Investigation*, **58,** 1317-26 (1976).
26. Hillis, L. D., Askenazi, J., Braunwald, E., Radvany, P., Muller, J. E., Fishbein, M. C., and Maroko, P. R. Use of changes in the epicardial QRS complex to assess interventions which modify the extent of myocardial necrosis following coronary artery occlusion. *Circulation*, **54,** 591-8 (1976).
27. Hillis, L. D., Fishbein, M. C., Braunwald, E., and Maroko, P. R. The influence of the time interval between coronary artery occlusion and the administration of hyaluronidase on salvage of ischemic myocardium in dogs. *Circulation Research*, **41,** 26-31 (1977).
28. Hirshfeld, J. W., Jr., Borer, J. S., Goldstein, R. E., Barrett, M. J., and Epstein, S. E. Reduction in severity and extent of myocardial infarction when nitroglycerine and methoxamine are administered during coronary occlusion. *Circulation*, **49,** 291-7 (1974).
29. Hutchins, G. M. and Bulkley, B. H. Expansion versus extension: Two different complications of acute myocardial infarction. *American Journal of Cardiology*, **39,** 323 (1977).
30. Imai, S., Shigel, T., and Hashimoto, K. Cardiac action of methoxamine with special reference to its antagonistic action to epinephrine. *Circulation Research*, **9,** 552 (1961).
31. Jennings, R. B. Early phase of myocardial ischemic injury and infarction. *American Journal of Cardiology*, **24,** 753-65 (1969).
32. Jennings, R. B., Herdson, R. B., and Sommers, H. M. Structural and function abnormalities in mitochondria isolated from ischemic dog myocardium. *Laboratory Investigation*, **20,** 548-57 (1969).
33. Jennings, R. B., Sommers, H. M., Herdson, R. B., and Kaltenbach, J. P. Ischemic injury of myocardium. Annals of the New York Academy of Sciences, **156,** 61-78 (1969).
34. Jennings, R. B. and Ganote, C. E. Mitochondrial structure and function in acute myocardial ischemic injury. *Circulation Research Supplement 1*, **38,** 1-80-89 (1976).
35. Jennings, R. B. Relationship of acute ischemia to functional defects and irreversibility. *Circulation Supplement 1*, **53,** 1-26-29 (1976).
36. Johansen, K. H., DeLaria, G. A., and Bernstein, E. F. Effect of external counterpulsation in reduction of myocardial ischemia following coronary artery occlusion. *Transactions American Society for Artificial Internal Organs*, **19,** 419-23 (1973).
37. Johns, T. N. P. and Olson, B. J. Experimental myocardial infarction. 1. Method of coronary occlusion in small animals. *Annals of Surgery*, **140,** 675-82 (1954).
38. Karlsson, J., Templeton, G. H., and Willerson, J. T. Relationship between epicardial ST segment changes and myocardial metabolism during acute coronary insufficiency. *Circulation Research*, **32,** 725-30 (1973).
39. Khuri, S. F., Flaherty, J. T., O'Riordan, J. B., Pitt, B., Brawley, R. L., Donahoo, J. W., and Gott, V. L. Changes in intramyocardial ST segment voltage and gas tension with regional ischemia in the dog. *Circulation Research*, **37,** 455-64 (1975).
40. Kjekshus, J. K. and Sobel, B. E. Depressed myocardial creatine phosphokinase activity following experimental myocardial infarction in rabbit. *Circulation Research*, **27,** 403-14 (1970).
41. Kjekshus, J. K., Maroko, P. R., and Sobel, B. E., Distributions of myocardial injury and its relation to epicardial ST segment changes after coronary artery occlusion in the dog. *Cardiovascular Research*, **6,** 490-9 (1972).
42. Kjekshus, J. K. and Mjos, O. D. Effect of inhibition of lipolysis on infarct size after experimental coronary artery occlusion. *Journal of Clinical Investigation*, **52,** 1770-8 (1973).
43. Kloner, R. A., Fishbein, M. C., Maclean, D., Braunwald, E., and Maroko, P. R. Effect of hyaluronidase on myocardial ultrastructure following coronary artery occlusion in the rat. *American Journal of Cardiology*, **40,** 43-9 (1977).
44. Leinbach, R. C., Gold, H. K., Buckley, M. J., Austen, G. W., and Sanders, C. A. Reduction of myocardial injury during acute infarction by early application of intraaortic balloon pumping and propranolol. *Circulation Supplement IV*, **47** and **48**, IV-100 (1973).
45. Lekven, J., Kjekshus, J. K., and Mjos, O. D. Effects of glucagon and isoproterenol on

severity of acute myocardial ischemic injury. *Scandinavian Journal of Clinical and Laboratory Investigation,* **30,** 129–37 (1973).
46. Leon, A. S., White, F. C., Bloor, C. M., and Saviano, M. A. Reduced myocardial fibrosis after dimethylsulfoxide (DMSO) treatment of isoproterenol-induced myocardial necrosis in rats. *American Journal of the Medical Sciences,* **261,** 41–5 (1971).
47. Levine, I. D., Maroko, P. R., and Bernstein, E. F. Comparison of intraaortic balloon pumping and left ventricular decompression on myocardial ischemic injury after experimental coronary artery occlusion. *Surgical Forum; Clinical Congress of the American College of Surgeons,* **22,** 149–50 (1971).
48. Libby, P., Maroko, P. R., Bloor, C. M., Sobel, B. E., and Braunwald, E. Reduction of experimental myocardial infarct size by corticosteroid administration. *Journal of Clinical Investigation,* **52,** 599–607 (1973).
49. Libby, P., Maroko, P. R., Covell, J. W., Malloch, C. I., Ross, J., Jr., and Braunwald, E., Effect of practolol on the extent of myocardial ischemic injury after experimental coronary occlusion and its effects on ventricular function in the normal and ischemic heart. *Cardiovascular Research,* **7,** 167–73 (1973).
50. Libby, P., Maroko, P. R., and Braunwald, E. The effect of hypoglycemia on myocardial ischemic injury during acute experimental coronary artery occlusion. *Circulation,* **51,** 621–6 (1975).
51. MacAlpin, P. N., Kattus, A. K., and Alvaro, A. B. Angina pectoris at rest with preservation of exercise capacity. Prinzmetal's variant angina. *Circulation,* **47,** 946–58 (1973).
52. Maclean, D., Fishbein, M. C., Maroko, P. R., and Braunwald, E. Hyaluronidase induced reductions in myocardial infarct size. Direct quantification of infarction following coronary artery occlusion in the rat. *Science,* **194,** 199–200 (1976).
53. Maclean, D., Fishbein, M. C., Braunwald, E., and Maroko, P. R. Long-term preservation of ischemic myocardium after experimental coronary artery occlusion. *Journal of Clinical Investigation,* **61,** 541–51 (1978).
54. Madias, J. E., Madias, N. E., and Hood, W. B., Jr. Precordial ST segment mapping. 2. Effects of oxygen inhalation on ischemic injury in patients with acute myocardial infarction. *Circulation,* **53,** 411–7 (1976).
55. Mallory, G. K., White, P. D., and Salcedo-Salgar, J. The speed of healing of myocardial infarction. A study of the pathologic anatomy in seventy-two cases. *American Heart Journal,* **18,** 647–71 (1939).
56. Maroko, P. R., Braunwald, E., Covell, J. W., and Ross, J., Jr. Factors influencing the severity of myocardial ischemic following experimental coronary occlusion. *Circulation Supplement III,* **39** and **40,** III-140 (1969).
57. Maroko, P. R., Kjekshus, J. K., Sobel, B. E., Watanabe,T., Covell, J. W., Ross, J., Jr., and Braunwald, E. Factors influencing infarct size following coronary artery occlusions. *Circulation,* **43,** 67–82 (1971).
58. Maroko, P. R., Bernstein, E. F., Libby, P., DeLaria, G. A., Covell, J. W., Ross, J., Jr., and Braunwald, E. Effects of intraaortic balloon counterpulsation on the severity of myocardial ischemic injury following acute coronary occlusion. Counterpulsation and myocardial injury. *Circulation,* **45,** 1150–9 (1972).
59. Maroko, P. R., Libby, P., Covell, J. W., Sobel, B. E., Ross, J., Jr., and Braunwald, E. Precordial ST segment elevation mapping; An atraumatic method for assessing alterations in the extent of myocardial ischemic injury. The effects of pharmacologic and hemodynamic interventions. *American Journal of Cardiology,* **29,** 223–30 (1972).
60. Maroko, P. R., Libby, P., Ginks, W. R., Bloor, C. M., Shell, W. E., Sobel, B. E., and Ross, J., Jr. Coronary artery reperfusion. 1. Early effects on local myocardial function and the extent of myocardial necrosis. *Journal of Clinical Investigation,* **51,** 2710–6 (1972).
61. Maroko, P. R., Libby, P., Bloor, C. M., Sobel, B. E., and Braunwald, E. Reduction by hyaluronidase of myocardial necrosis following coronary artery occlusion. *Circulation,* **46,** 430–7 (1972).
62. Maroko, P. R., Libby, P., Sobel, B. E., Bloor, C. M., Sybers, H. D., Shell, W. E., Covell, J. W., and Braunwald, E. Effect of glucose–insulin–potassium infusion on myocardial

infarction following experimental coronary artery occlusion. *Circulation*, **45,** 1160–75 (1972).
63. Maroko, P. R. and Braunwald, E. Modification of myocardial infarction size after coronary occlusion. *Annals of Internal Medicine*, **79,** 720–33 (1973).
64. Maroko, P. R., Braunwald, E., and Ross, J., Jr. The metabolic costs of positive inotropic agents. In *Myocardial Infarction* (Eds. Corday and Swan HKC), Williams and Wilkins, Baltimore, 1973, pp. 244–50.
65. Maroko, P. R. and Carpenter, C. B. Reduction in infarct size following acute coronary occlusion by the administration of cobra venom factor. *Clinical Research*, **22,** 298A (1974).
66. Maroko, P. R., Hillis, L. D., Muller, J. E., Tavazzi, L., Heyndrickx, G. R., Ray, M., Chiariello, M., Distante, A., Askenazi, J., Salerno, J., Carpentier, J., Reshetnaya, N. I., Radvany, P., Libby, P., Raabe, D. S., Chazov, E. I., Bobba, P., and Braunwald, E. Favorable effects of hyaluronidase on electrocardiographic evidence of necrosis in patients with acute myocardial infarction. *New England Journal of Medicine*, **296,** 898–903 (1977).
67. Maroko, P. R., Radvany, P., Braunwald, E., and Hale, S. L. Reduction of infarct size by oxygen inhalation following acute coronary occlusion. *Circulation*, **52,** 360–8 (1975).
68. Maroko, P. R., Davidson, D. M., Libby, P., Hagan, A. D., and Braunwald, E. Effects of hyaluronidase administration on myocardial ischemic injury in acute infarction. A preliminary study in 24 patients. *Annals of Internal Medicine*, **82,** 516–20 (1975).
69. Maroko, P. R., and Braunwald, E. Effects of metabolic and pharmacologic interventions on myocardial infarct size following coronary occlusion. *Circulation Supplement 1*, **53,** 1-162–8 (1976).
70. Massie, E. and Walsh, T. J. *Clinical Vectorcardiography and Electrocardiography*, Yearbook Medical Publishers, Chicago, 1969, pp. 70–91.
71. Masters, T. N., Harbold, N. B., Hall, D. G., Jackson, R. D., Muller, D. C., Daugherty, H. K., and Robicsek, R. Beneficial metabolic effects of methylprednisolone sodium succinate in acute myocardial ischemia. *American Journal of Cardiology*, **37,** 557–63 (1976).
72. Muller, J. E., Maroko, P. R., and Braunwald, E. Evaluation of precordial electrocardiographic mapping as a means of assessing changes in myocardial ischemic injury. *Circulation*, **52,** 16–27 (1975).
73. Oliver, M. F. The influence of myocardial metabolism on ischemic damage. *Circulation Supplement I*, **53,** 1-168–70 (1976).
74. Pelides, L. J., Reid, D. W., Thomas, M., and Shillingford, J. P. Inhibition by beta-blockade of the ST segment elevation after acute myocardial infarction in man. *Cardiovascular Research*, **6,** 295–302 (1972).
75. Powell, W. J., Jr., DiBona, D. R., Flores, J., Frega, N., and Leaf, A. Effects of hyperosmotic mannitol in reducing ischemic cell swelling and minimizing myocardial necrosis. *Circulation Supplement 1*, **53,** 1-45–9 (1976).
76. Prinzmetal, M., Kennamer, R., Merless, R., Wada, T., and Bor, N. Angina pectoris. 1. Variant form of angina pectoris; preliminary report. *American Journal of Medicine*, **27,** 375–88 (1959).
77. Prinzmetal, M., Toyoshima, H., Ekmecki, A., Mizuno, Y., and Nagaya, T. Myocardial ischemia; nature of ischemic electrocardiographic patterns in mammalian ventricles as determined by intracellular electrographic and metabolic changes. *American Journal of Cardiology*, **8,** 493–503 (1961).
78. Radvany, P., Maroko, P. R., and Braunwald, E. Effects of hypoxemia on the extent of myocardial necrosis after experimental coronary occlusion. *American Journal of Cardiology*, **35,** 795–800 (1975).
79. Radvany, P., Davis, M. A., Muller, J. E., and Maroko, P. R. The effect of minoxidil on coronary collateral flow and acute myocardial injury following experimental coronary artery occlusion. *Cardiovascular Research*, **12,** 120–26 (1978).
80. Rakita, L., Borduas, J. L., Rothman, S., and Prinzmetal, M. Studies on the mechanism of ventricular activity. XII. Early changes in the ST-T segment and QRS complex following acute coronary artery occlusion; experimental study and clinical applications. *American Heart Journal*, **48,** 351–72 (1954).
81. Redwood, D. R., Smith, E. R., and Epstein, S. E. Coronary artery occlusion in the

conscious dog: effects of alterations in heart rate and arterial pressure on the degree of myocardial ischemia. *Circulation*, **46,** 323–32 (1972).
82. Rehder, E., and Enquist, I. F. Species differences in response to cortisone in wounded animals. *Archives of Surgery*, **94,** 74–8 (1967).
83. Reimer, K. A., Loew, J. E., Rasmussen, M. M., and Jennings, R. B. The wave front phenomenon of ischemic cell death. 1. Myocardial infarct size vs duration of coronary occlusion in dogs. *Circulation*, **56,** 786–94 (1977).
84. Ribeiro, L. G. T., Louie, E. K., Davis, M. A., and Maroko, P. R. Beneficial effect of 100% O_2 breathing on redistribution of regional myocardial blood flow. *Clinical Research*, **25,** 248A (1977).
85. Ribeiro, L. G. T., Hillis, L. D., Fishbein, M. C., Davis, M. A., Maroko, P. R., and Braunwald, E. A new technique for demonstrating the efficacy of interventions designed to limit infarct size following coronary occlusion: beneficial effect of hyaluronidase. *Clinical Research*, **25,** 248A (1977).
86. Roberts, R., deMello, V., and Sobel, B. E. Deleterious effects of methylprednisolone in patients with myocardial infarction. *Circulation Supplement 1*, **53,** 1-204–6 (1976).
87. Rogers, W. J., Russell, R. O., Jr., McDaniel, H. G., and Rackley, C. E. Acute effects of glucose–insulin–potassium infusion on myocardial substrates, coronary blood flow and oxygen consumption in man. *American Journal of Cardiology*, **40,** 421–8 (1977).
88. Sandberg, N. Time relationship between administration of cortisone and wound healing in rats. *Acta Chirurgica Scandinavica*, **127,** 446–55 (1964).
89. Sayen, J. J., Pierce, G., Katcher, A. H., and Sheldon, W. F. Correlation of intramyocardial electrocardiograms with polarographic oxygen and contractility in the nonischemic and regionally ischemic left ventricle. *Circulation Research*, **9,** 1268–79 (1961).
90. Scheur, J. and Brachfeld, N. Coronary insufficiency: relations between hemodynamic, electrical, and biochemical parameters. *Circulation Research*, **18,** 178–89 (1966).
91. Selye, J., Bajusz, E., Grassos, S., and Mendell, P. Simple techniques for the surgical occlusion of coronary vessels in the rat. *Angiology*, **11,** 398–407 (1960).
92. Shell, W. E., Kjekshus, J. K., and Sobel, B. E. Quantitative assessment of the extent of myocardial infarction in the conscious dog by means of analysis of serial changes in serum creatine phosphokinase activity. *Journal of Clinical Investigation*, **50,** 2614–25 (1971).
93. Shell, W. E. and Sobel, B. E. Deleterious effects of increased heart rate on infarct size in the conscious dog. *American Journal of Cardiology*, **31,** 474–9 (1973).
94. Shell, W. E. and Sobel, B. E. Protection of jeopardized ischemic myocardium by reduction of ventricular afterload. *New England Journal of Medicine*, **291,** 481–6 (1974).
95. Smith, E. R., Redwood, D. R., McCarron, W. E., and Epstein, S. E. Coronary artery occlusion in the conscious dog. Effects of alterations in arterial pressure produced by nitroglycerin, hemorrhage, end alpha-adrenergic agonists on the degree of myocardial ischemia. *Circulation*, **47,** 51–7 (1973).
96. Smith, G. T., Soeter, J. R., Haston, H. H., and McNamara, J. J. Coronary reperfusion in primates. Serial electrocardiographic and histologic assessment. *Journal of Clinical Investigation*, **54,** 1420–7 (1974).
97. Smith, H. J., Singh, B. N., Norris, R. M., John, M. B., and Hurley, P. J. Changes in myocardial blood flow and S-T segment elevation following coronary artery occlusion in dogs. *Circulation Research*, **36,** 697–705 (1975).
98. Sommers, H. M. and Jennings, R. B. Ventricular fibrillation and myocardial necrosis after transient ischemia. *Archives of Internal Medicine*, **129,** 780–9 (1972).
99. Spath, J. A., Jr., Lane, D. L., and Lefer, A. M. Protective action of methylprednisolone on the myocardium during experimental myocardial ischemia in the cat. *Circulation Research*, **35,** 44–51 (1974).
100. Sybers, H. D., Maroko, P. R., Ashraf, M., Libby, P., and Braunwald, E. The effect of glucose–insulin–potassium on cardiac ultrastructure following acute experimental coronary occlusion. *American Journal of Pathology*, **70,** 401–20 (1973).
101. Varankov, Y., Shell, W., Smirnov, V., Gukovsky, D., and Dhazov, E. I. Augmentation of serum CPK activity by digitalis in patients with acute myocardial infarction. *Circulation*, **55,** 719–27 (1977).
102. Watanabe, T., Covell, J. W., Maroko, P. R., Braunwald, E., and Ross, J., Jr. Effects of

increased arterial pressure and positive inotropic agents on the severity of myocardial ischemia in the acutely depressed heart. *American Journal of Cardiology*, **30,** 371–7 (1972).
103. Watanabe, T., Shintani, F., Fu, L., Fujil, J., Watanabe, H., and Kato, K. Influence of inotropic alteration on the severity of myocardial ischemia after experimental coronary occlusion. *Japanese Heart Journal*, **13,** 222–30 (1972).
104. Willerson, J. T., Powell, W. J., Jr., Guiney, T. E., Stark, J. J., Sanders, C. A., and Leaf, A. Improvement in myocardial function and coronary blood flow in ischemic myocardium after mannitol. *Journal of Clinical Investigation*, **51,** 2989–98 (1972).
105. Wilson, F. N., Johnston, F. D., Rosenbaum, F. F., Erlanger, H., Knossman, C. E., Hecht, H., Cotrim, N., deOliveira, R. M., Scarsi, R., and Barker, P. S. The precordial electrocardiogram. *American Heart Journal*, **27,** 19–85 (1944).
106. Winbury, M. M., Howe, B. B., and Hefner, M. A. Effects of nitrates and other coronary dilators on large and small coronary vessels: an hypothesis for the mechanism of action of nitrates. *Journal of Pharmacology and Experimental Therapeutics*, **168,** 70 (1960).
107. Yoshikawa, H., Powell, J. W., Jr., Bland, J. H. L., and Lowenstein, E. Effect of acute anemia on experimental myocardial ischemia. *American Journal of Cardiology*, **32,** 670–8 (1973).

CHAPTER 24

Metabolism, enzyme release, and cell death: possibilities for future investigation
L. H. Opie

From the clinical point of view, enzyme release from the heart is virtually synonymous with cell necrosis and constitutes a diagnostic sign used by cardiologists all over the world. As yet there has been little fundamental work about the molecular mechanisms involved. Is large-scale enzyme release directly related to cell necrosis? Are there common metabolic events occurring at about the same time, giving rise to both events? If so, identification of those metabolic changes would be imperative. Not only would elucidation of the 'enzyme release-necrotic' factor be theoretically desirable, but if it be accepted that cell necrosis in developing acute myocardial infarction should be limited, then knowledge of the 'enzyme release-necrotic' factor must have practical importance.

Observations reported in this book make a direct link between enzyme release and necrosis unlikely, because enzyme release can be found in isolated hearts soon after coronary artery ligation (see Chapter 21) when necrosis cannot be diagnosed pathologically. But, what is necrosis? Strictly speaking, it is a pathological diagnosis, fully evident only in those cells which are completely dead, and not readily detectable at about 6-8 hours post-occlusion when enzyme release is already evident (for references, see Rose et al.[24]) Thus more probably, the relationship between the metabolic events underlying enzyme release and those underlying necrosis would only be indirect and it seems that the probable sequence of events is: first, metabolic changes, then enzyme release, and then necrosis. Thus the metabolic changes are more likely to be linked to the events leading up to necrosis, i.e. 'pre-necrosis'. It may also be that the association between any given metabolic change and enzyme release is also indirect, at least in our present state of knowledge. In that case, the search for the missing "enzyme release-pre-necrotic factor' could be reduced to a more pragmatic question: are there any metabolic events closely associated with the development of cell necrosis which could also regulate cell membrane activity? Calcium may be such an agent. Calcium accumulates in cells damaged by ischaemia and especially after reperfusion. The uptake of calcium by mitochondria is 'energy wasting' and must, therefore, be occurring before total energy depletion, i.e. in the pre-necrotic phase. Once taken up, calcium can be recovered in the mitochondrial fraction and such calcium can contribute to the development of ischaemic contracture.[12] Calcium also helps to maintain normal cell wall integrity, possibly by participating in calcium–carbohydrate couplings.[9] In the sinus node, perfusion with chelating agents partially ruptures the cell junctions leaving only the gap junctions intact.[13] Thus it is at least conceivable that alterations in calcium ion movements or calcium gradients could both damage the cell membrane and promote irreversible injury. More direct evidence for this point of view has recently been provided by Burton et al.[4] They used an ionic lanthanum probe technique to monitor changes in the cat papillary muscle exposed to hypoxia. Lanthanum is a trivalent ion with calcium-resembling properties and is not normally visualized within the heart cell. With progressive hypoxia lanthanum moves from an extracellular to an intracellular location and after 2–3 hours of hypoxia there is both irreversible ultrastructural damage and deposition of lanthanum at intracellular sites of high calcium affinity such as the I band and the inner mitochondrial membrane and the adjacent matrix.

These observations of course beg the question, because the primary event which must now be explained is increased membrane permeability during hypoxia. But what exactly is meant by 'increased membrane permeability'? Michell and Coleman[17] have reviewed relevant aspects of the structure of the cell membrane. The turnover time of the membrane components is so short, in some instances every few hours, that although the membrane is a long-lived structure, its individual components are continuously being replaced. Energy may be required for several different processes, such as biosynthesis, biodegradation, and possibly to influence the 'fluidity' of the membrane, as argued by Gebhard et al.[10] When there is energy depletion, the net effect is increased cell permeability.

In ischaemia, the complexity of the concurrent metabolic processes[19] makes it difficult to conclude that there could be firm links between the extent of cellular depletion of ATP and enzyme release. However, in a number of simpler models, ATP depletion can be related to enzyme release. In rat lymphocytes damaged by phospholipase A, the addition of exogenous ATP decreases the leakage of lactate dehydrogenase.[28] The extent of fall of tissue ATP in the perfused anoxic dog heart correlates very well with the amount of enzyme release.[10] However, in regional ischaemia (developing myocardial infarction), substantial ATP depletion could occur in the first 120 minutes at a time when the tissue content of creatine kinase had not decreased,[8] and before major release of creatine kinase into the circulation.[26] It may, however, be that a subcompartment of ATP, readily accessible to the cell membrane, has a special role in the maintenance of cell integrity, as argued by Michell and Coleman[17] and by Bricknell and Opie.[2] Whether it be the total ATP or a subcompartment of ATP, two correlations are possible. First, that ATP depletion advances until a certain critical level is reached which irreversibly damages the cell, as argued by Kübler and Spieckermann[15] and Hearse and Chain.[11] Alternatively, ATP depletion may act in a graded way, so that the lower the ATP content, the greater the membrane leakiness.[10]

It has also been argued that ATP generated by glycolysis has a special role in the maintenance of cell membrane integrity but evidence for this point of view is largely electrophysiological (for references see Bricknell and Opie[2]) and not directly applicable to the problem of enzyme release. However, Bricknell and Opie[3] were able to relate the extent of post-ischaemic enzyme release to metabolic changes found in the ischaemic period. During ischaemia (produced by underperfusion in an isolated Langendorff perfused heart) glucose and pyruvate were equally able to maintain cellular contents of high-energy phosphate compounds, but only with glucose was there production of glycolytic ATP. During reperfusion, enzyme release and the development of arrhythmias were much higher in pyruvate than in glucose-perfused hearts. Thus the possibility was raised that glycolytically produced ATP could in some manner protect against the development of those membrane changes associated with enzyme release.

Similarly, in a perfused heart model with coronary ligation, the rate of release of lactate dehydrogenase could be inversely related to the rate of production of glycolytic ATP.[20] In those experiments, the substrate is varied from free fatty acid alone (high rates of enzyme release, low rates of glycolytic flux) to glucose alone (very low rates of enzyme release, highest rates of glycolytic flux) with combinations producing intermediate values. However, it is likely that in severely ischaemic tissue, the rate of glucose uptake is limited[18,19] and hence, these relations may hold only for mildly or moderately ischaemic tissue and not for severely ischaemic tissue. Thus severe ischaemia by limiting glycolytic ATP, could be promoting membrane damage according to this point of view.

Two major questions are provoked. First, in exactly which way is the postulated ATP pool in relation to the membrane able to maintain membrane integrity? Is it by provision of energy to maintain lysosomal membranes, and hence to reduce membrane biodegradation? Could ATP in some way directly stabilize the membrane and minimize the 'fluidity'? Or is it that ATP is required for the activity of ionic pumps, which in turn maintain the balance of Ca^{2+}, as proposed by Michell and Coleman?[17]

Secondly, how can more direct evidence be obtained for the postulated 'membrane ATP'? It should be noted that general arguments against the partitioning of ATP in the cytoplasm have been presented by Altschuld and Brierley.[1]

If ATP depletion acts essentially to allow increased ingress of Ca^{2+} into the cell, then there should be circumstances in which other agents could promote the same ingress of Ca^{2+} with the same result, even though ATP does not decrease. Isoproterenol and other catecholamines may provide the model required to satisfy this requirement. Isoproterenol can cause increased permeability to proteins within 10 minutes,[23] yet in certain models catecholamines can cause increased loss of enzymes from the heart without major depletion of ATP and phosphocreatine.[27] Possibly accumulation of cyclic AMP may be the mechanism whereby more Ca^{2+} is admitted, as postulated by Bricknell and Opie.[3] The role of calcium influx is supported by the prevention of isoproterenol-induced cardiac necrosis when transmembrane Ca^{2+} influx is reduced by calcium-antagonistic drugs including magnesium salts.[14] Possibly increased Ca^{2+} influx could explain the interesting phenomenon of cyclic AMP-induced cardiac necrosis.[16] Thus depletion of ATP may only be one of several mechanisms operating to increase cell membrane damage, and several different mechanisms may be involved. Firm data on the molecular interactions between ATP, Ca^{2+}, cyclic AMP, and the cell membrane would appear to be sorely needed.

Can the proposed roles of calcium and of glycolytic ATP be reconciled? The connecting link may lie in the cardiac glycogen. The severity of glycogen depletion is closely related to the development of cardiac necrosis[22,24] and to the severity of depression of blood flow[21] after coronary ligation in the dog. But in a fascinating series of papers, Schwartz and his group[5-7] have shown the existence of a glycogenolytic–sarcoplasmic reticulum complex in which glycogenolysis takes place even more efficiently than in optimal *in vitro* conditions. Although generalized glycogen depletion must be distinguished from depletion of glycogen in the glycogenolytic–sarcoplasmic reticulum complex, yet both types of depletion release phosphorylase[5,25] and the process is irreversible.[5] It may be that this glycogenolytic–sarcoplasmic reticulum complex could be required for provision of energy for calcium uptake by the sarcoplasmic reticulum; conversely, in the presence of glycogen depletion, cellular overload with calcium could be more likely to occur. Thus another basic observation would be to establish the relation, if any, between the integrity of the glycogenolytic–sarcoplasmic reticulum complex, abnormalities of calcium metabolism, and enzyme release from the heart.

Does glycogen depletion represent the much sought after 'point of no return'? If so, that would explain why an increased cardiac glycogen, induced by experimental therapeutic procedures such as glucose–insulin–

potassium infusions, propranolol, or hyaluronidase, is associated with lessened cell necrosis, and by implication with lessened enzyme release. However, both glycogen depletion and enzyme release are graded events and one of the major contributions of some of the animal studies reported in this book is to show that cell damage, as represented by enzyme release, is not necessarily an all-or-nothing phenomenon. Various degrees of cell damage may be reflected by various degrees of enzyme release. Thus we must know whether Ca^{2+} entry, depletion of glycogen and of ATP, are phenomena which reach critical end-points or not. If they are all graded, and if enzyme release is also graded, then it seems likely that the interrelations are going to be extremely complex.

In conclusion, the exact links between the metabolic derangements resulting from continued regional ischaemia and enzyme release remain to be defined. It would be a great step forward in our therapeutic considerations if a single event or group of events could be incriminated both in increasing membrane permeability (and causing enzyme release) and in causing cell necrosis (or, more exactly, pre-necrosis). Among the possible candidates for this key position are:

(i) abnormalities of calcium metabolism;
(ii) decreased production of glycolytic ATP;
(iii) accumulation of cyclic AMP;
(iv) glycogen depletion.

The newly described glycogenolytic–sarcoplasmic reticulum complex may link some of these events. It is in the definition of the critical metabolic event(s) linking enzyme release, membrane permeability, and eventual cell necrosis in developing myocardial infarction that the greatest challenge now lies.

REFERENCES

1. Altschuld, R. A. and Brierley, G. P. Interaction between the creatine kinase of heart mitochondria and oxidative phosphorylation. *Journal of Molecular and Cellular Cardiology*, **9,** 875–96 (1977).
2. Bricknell, O. L. and Opie, L. H. Glycolytic ATP and its production during ischaemia in isolated Langendorff-perfused rat hearts. In *Recent Advances in Studies on Cardiac Structure and Metabolism*, Vol. 11, *Heart Function and Metabolism* (Eds. Kobayashi, T. Sano, T., and Dhalla, N. S.) University Park Press, Baltimore, 1978, pp. 509–19.
3. Bricknell, O. L. and Opie, L. H. Effects of substrates on tissue metabolic changes in the isolated rat heart during underperfusion and on release of lactate dehydrogenase and arrhythmias during reperfusion. *Circulation Research*, **43,** 102–15 (1978).
4. Burton, K. P., Hagler, H. K., Templeton, G. H., Willerson, J. T., and Buja, L. M. Lanthanum probe studies of cellular pathophysiology induced by hyposia in isolated cardiac muscle. *Journal of Clinical Investigation*, **60,** 1289–302 (1977).
5. Entman, M. L., Bornet, E. P., Van Winkle, W. B., Goldstein, M. A., and Schwartz, A. Association of glycogenolysis with cardiac sarcoplasmic reticulum. II. Effect of glycogen

depletion, deoxycholate solubilization and cardiac ischemia: evidence for a phosphorylase kinase membrane complex. *Journal of Molecular and Cellular Cardiology*, **9**, 515–28 (1977).
6. Entman, M. L., Goldstein, M. A., and Schwartz, A. The cardiac sarcoplasmic reticulum–glycogenolytic complex, an internal beta adrenergic receptor. *Life Sciences*, **19**, 1623–30 (1976).
7. Entman, M. L., Kaniike, K., Goldstein, M. A., Nelson, T. E., Bornet, E. P., Futch, T. W., and Schwartz, A. Association of glycogenolysis with cardiac sarcoplasmic reticulum. *Journal of Biological Chemistry*, **251**, 3140–6 (1976).
8. Font, B., Vial, C., and Goldschmidt, D. Enzyme levels and metabolite release in the ischaemic myocardium. *Journal of Molecular Medicine*, **2**, 291–7 (1977).
9. Frank, J. S., Langer, G. A., Nudd, L. M., and Seraydarian, K. The myocardial cell surface, its histochemistry and the effect of sialic acid and calcium removal on its structure and cellular ionic exchange. *Circulation Research*, **41**, 702–14 (1977).
10. Gebhard, M. M., Denkhaus, H., Sakai, K., and Spieckermann, P. G. Energy metabolism and enzyme release. *Journal of Molecular Medicine*, **2**, 271–83 (1977).
11. Hearse, D. J. and Chain, E. B. The role of glucose in the survival and 'recovery' of the anoxic isolated perfused rat heart. *Biochemical Journal*, **128**, 1125–33 (1972).
12. Henry, P. D., Shuchler, R., Davis, J., Weiss, E. S., and Sobel, B. E. Myocardial contracture and accumulation of mitochondrial calcium in ischaemic rabbit heart. *American Journal of Physiology*, **233**, H677–H684, (1977).
13. James, T. N. Selective experimental chelation of calcium in the sinus node. *Journal of Molecular and Cellular Cardiology*, **6**, 493–504 (1974).
14. Janke, J., Jaedicke, W., and Fleckenstein, A. Prevention of isoproterenol-induced cardiac necrosis by reduction of transmembrane Ca influx with the use of K and Mg salts or of Ca-antagonistic inhibitors of excitation–contraction coupling. *Pflügers Archiv ges Physiologie*, **319**, R8 (1970).
15. Kübler, W. and Spieckermann, P. G., Regulation of glycolysis in the ischaemic and anoxic myocardium. *Journal of Molecular and Cellular Cardiology*, **1**, 351–77 (1970).
16. Martona, P. A. The role of cyclic AMP in isoprenaline-induced cardiac necroses in the rat. *Journal of Pharmacy and Pharmacology*, **23**, 200–3 (1971).
17. Michell, R. H. and Coleman, R. Structure and permeability of normal and damaged membranes. In *Enzymes in Cardiology: Diagnosis and Research* (Eds. Hearse, D. J. and de Leiris, J.) John Wiley, London, 1979.
18. Neely, J. R., Whitmer, J. T., and Rovetto, M. J. Effect of coronary blood flow on glycolytic flux and intracellular pH in isolated rat hearts. *Circulation Research*, **37**, 733–41 (1975).
19. Opie, L. H. Effects of anoxia and regional ischemia on metabolism of glucose and fatty acids. Relative rates of aerobic and anaerobic energy production during first 6 hours of experimental myocardial infarction. *Circulation Research Supplment* 1, **38**, 152–68 (1976).
20. Opie, L. H. and Bricknell, O. L. Role of glycolytic flux in effect of glucose in decreasing fatty-acid-induced release of lactate dehydrogenase from isolated coronary ligated rat heart. *Cardiovascular Research*. Submitted for publication (1978).
21. Opie, L. H., Owen, P., and Lubbe, W. F. Estimated glycolytic flux rate in infarcting heart. In *Recent Advances in Studies on Cardiac Structure and Metabolism*, Vol. 7, *Biochemistry and Pharmacology of Myocardial Hypertrophy, Hypoxia and Infarction* (Eds. p., Harris, Bing, R. J., and Fleckenstein, A.), University Park Press, Baltimore, 1976, pp. 249–55.
22. Reimer, K. A., Rasmussen, M. M., and Jennings, R. B. Reduction by propranolol of myocardial necrosis following temporary coronary artery occlusion in dogs. *Circulation Research*, **33**, 353–63 (1973).
23. Rona, G., Boutet, M., and Hüttner, I. Membrane permeability alterations as manifestation of early cardiac muscle cell injury. In *Recent Advances in Studies on Cardiac Structure and Metabolism*, Vol. 6, *Pathophysiology and Morphology of Myocardial Cell Alteration* (Eds. Fleckenstein, A. and Rona, G.), University Park Press, Baltimore, 1975, pp. 439–51.
24. Rose, A. G., Opie, L. H., and Bricknell, O. L. Early experimental myocardial infarction. Evaluation of histologic criteria and comparison with biochemical and electrocardiographic measurements. *Archives of Pathology and Laboratory Medicine*, **100**, 516–21 (1976).
25. Schultze, W., Krause, E. G., and Wollenberger, A. On the fate of glycogen phosphorylase

in the ischemic and infarcting myocardium. *Journal of Molecular and Cellular Cardiology*, **2,** 241–51 (1971).
26. Shell, W. E., Kjekshus, J. K., and Sobel, B. E. Quantitative assessment of the extent of myocardial infarction in the conscious dog by means of analysis of serial changes in serum creatine phosphokinase activity. *Journal of Clinical Investigation*, **50,** 2614–25 (1971).
27. Waldenström, A., Hjalmarson, A., and Thornell, L. T. A possible role of noradrenaline in the development of myocardial infarction. An experimental study in the isolated rat heart. *American Heart Journal*, **95,** 43–51 (1978).
28. Wilkinson, J. H. and Robinson, J. M. Effect of ATP on release of intracellular enzymes from damaged cells. *Nature*, **249,** 662–3 (1974).

Index

Acetate
  metabolism in ischaemia 492
Acetyl cysteine
  creatine kinase activation 210
Acidosis
  ischaemia 31
Activation of enzymes 166, 170
Activation of enzymes
  CK assay 210
  dilution of CK 210
  metal ions 171
  separation of CK isoenzymes 214
  sulphydryl groups 171
Active centre
  substrate specificity 151
Alanine amino transferase
  myocardial infarction 230
  plasma half-life 128
  tissue distribution 117, 118, 119
Alcohol intoxication
  enzyme leakage 239
Aldolase isoenzymes
  structural sub-units 136
Alkaline phosphatase
  isoenzymes 136
Allelozymes 136
Allosteric enzymes
  activators and inhibitors 158
  control of enzyme activity 157
  kinetics 155
Angina
  enzyme changes 231
  substrate utilization 482
Anoxia
  acidosis and enzymes leakage 514, 515
  albumin effects 492
  beta-blocker protection 517
  calcium paradox 407
  cellular changes 5
  definition 2
  enzyme leakage 421, 504, 507
  fatty acid effects 492
  glucose and enzyme leakage 448, 486
  glucose effects 485
  high energy phosphate changes 509
  mitochondrial function 509, 510
  mitochondrial function with acidosis 516
  rat model 419
  species differences, enzyme leakage and ultrastructure 450
  ultrastructural changes 426, 446, 505, 510
  verapamil protection 518, 519
Anoxia models, *see* Models of anoxia
Antibody preparation
  radioimmunoassay of CK 249
Antilipolytic agents
  ischaemia 493
Arrhenius equation
  temperature and enzyme activity 175
Arrhythmias
  enzyme leakage 233
Artefacts
  fixation and ultrastructural studies 474
Aspartate amino transferase
  assay 184, 208
  changes during uncomplicated myocardial infarction 223, 224
  half-life in plasma 128
  in various disease states 208
  isoenzymes 141, 208
  kinetics 164
  mitochondrial form in myocardial infarction 229
  multiple forms 136
  sample collection and storage 209
  stability after death 343
  tissue distribution 117, 118, 119
  variation between individuals 341
Automated enzyme assays
  practical considerations 181–206

Beer–Lambert law 177
Beta blockers
  enzyme leakage 517, 519, 520
  infarct size reduction 370, 540
Blanking
  enzyme assays 179
Border zone 2, 532, 534
Border zone
  collateral flow 29
  contractility gradients 31
  gradients 29
  regional ischaemia 487
  transmural progression of injury 47

Calcium
  energy metabolism 74
  enzyme leakage 399
  enzyme leakage and cell death 562
  membrane damage 72
  reperfusion 46
  transport mechanisms 74
  uptake by isolated mitochondria 523
Calcium antagonists
  enzyme leakage 370, 449, 518
Calcium-free perfusion
  enzyme leakage 506
Calcium paradox 400
Calcium paradox
  anoxia 407
  clinical relevance 412
  contractile effects 401
  duration of calcium depletion 427, 430
  electrical effects 401
  energy metabolism 404, 406
  enzyme leakage 402, 403, 409, 410, 423
  factors influencing 404
  mechanism 439, 440
  mitochondria, possible role 411
  rat heart model 420
  temperature effect 433
  tissue damage and enzyme leakage 403, 417
  ultrastructural changes 425, 454
Capillary
  anatomy 82
  pores 82
Carbohydrate
  metabolism in ischaemia and anoxia 6

Cardiac failure
  enzyme leakage 234
Catalytic differences
  separation of isoenzymes 194
Catecholamine
  enzyme leakage and tissue damage 88, 449, 508
  induced ischaemia 385
  myocardial necrosis 400
  release in ischaemia 33
Catheterization
  enzyme leakage 236
Cell death
  critical mechanisms 565
  metabolic determinants 561
  progression in ischaemia 28
  progression with time 22, 27, 48, 50, 294
  relation to enzyme leakage 260
  transmural progression in ischaemia 47
Cell swelling
  ischaemia 10
  membrane injury 69, 70
  reperfusion 44
Cerebrovascular injury
  enzyme leakage 238
Clearance
  coronary vessels 13, 81, 86
  enzymes 100
  enzymes, factors influencing 109, 110
  enzymes from blood 14, 53, 98
  enzymes from blood, differential rates 99
  enzymes from blood, factors influencing 99, 284
  enzymes from tissue 53
  enzymes, HBD 347
  enzymes, mechanisms 110
  enzymes in myocardial infarction 89
  enzymes, variability between individuals 108, 109
  enzymes via lymphatics 13, 81, 98
  enzymes, via lymphatics and coronaries 86, 91
  ions from blood 14
  metabolites from blood 14
Clinical utility
  infarct size reduction 548
Coenzymes 172

Collateral flow
  border zone 29
  ischaemia 5, 24, 25
Compartmentation
  enzymes in fluid pools 100
  enzymes at a sub-cellular level 121
Control
  enzyme activity, allosteric mechanisms 157
  myocardial metabolism 481
Cooperativity
  enzyme kinetics 156
Competitive inhibition
  enzyme kinetics 166, 167
Coronary artery ligation 380, 381, 382, 383
Coronary artery occlusion
  relevance to myocardial infarction 391
Coronary flow
  clearance of enzymes 81
  clearance of enzymes during myocardial infarction 89
  distribution 23
  distribution in ischaemia 24, 25, 26, 27, 28, 538
  during cardiac cycle 23
  gradients in ischaemia 28
  infarct size 537
  measurement by fluorescence 26
  measurement by microspheres 25
  reduction in ischaemia 25
  regulation 23
  transmural gradients 23
  transmural gradients in ischaemia 25
  transport of leaked materials 13
Coronary fluid
  enzyme content under normal conditions 87
Coronary insufficiency
  enzyme changes 231
Coronary vessels
  distribution in dog 24
  functional anatomy 23, 82
  regulation of blood flow 23
Creatine kinase
  assay 184, 210
  distribution in populations 205
  distribution volume for infarct sizing 272
  during uncomplicated myocardial infarction 233, 234
  fractional disappearance rates 101, 270
  gram equivalents 263, 363
  half-life in plasma 128
  inactivation 273, 362
  inactivation in lymph 277
  in various disease states 209
  sample collection and storage 211
  stability after death 343
  time-activity in plasma of injected enzyme 101
  tissue distribution 117, 118, 119
  variability between individuals 341
Creatine kinase isoenzymes 139, 209
Creatine kinase isoenzymes
  antibodies 249
  assay 135
  assay, activation methods 214
  assay, catalytic difference methods 195
  assay, electrophoretic methods 190, 212
  assay, immunochemical methods 195, 213
  assay, ion exchange methods 192, 213
  assay, practical considerations 212, 213, 214
  assay, radioimmunoassay methods 247, 248
  during cardiac surgery 237
  during myocardial infarction 278
  during uncomplicated myocardial infarction 223, 224, 253
  fractional disappearance rates 279
  histological injury 261
  infarct size estimation 277, 363
  mitochondrial form 141
  radioimmunoassay, sensitivity and specificity 251, 252
  species differences in tissue content 279
  specificity 282
  specificity and sensitivity in diagnosis 226, 227, 279
  storage 211
  structural sub-units 134, 140
  sub-units changes 141
  tissue distribution 140, 226, 280, 283
Creatine kinase leakage
  calcium paradox 403, 409, 410

Creatine kinase leakage (*contd.*)
  correlation of time course with infarct size 329
  correlation to electrocardiographic changes 532
  correlation with clinical indices of severity of infarction 365
  correlation with metabolic markers 365
  hyaluronidase protection 547
  ST-segment changes 533
Cultured foetal hearts
  models of anoxia and ischaemia 390
Cyclic AMP
  enzyme leakage and cell death 564
  ischaemia 33
Cytoplasm
  marker enzymes 121

Diabetes
  ischaemic damage 490
Diagnostic enzymology
  basic principles 116, 129, 130, 201
  choice of enzymes 129, 241
  complicating factors 129, 236
  differentiation at cellular level 120
  differentiation at organ level 118
  differentiation at sub-cellular level 120
  during complicated myocardial infarction 232
  during uncomplicated myocardial infarction 221.
  electrocardiographic correlation 240, 241
  multiple enzyme profiles, importance 119
  practical considerations 204
  specificity and sensitivity 226, 227
  specificity and sensitivity of CK radioimmunoassay 251, 252
Dichloroacetate 495
Digoxin
  enzyme leakage 508
Dilution activation
  CK activity 210
Direct current countershock
  CK leakage 278
  enzyme leakage 233
  infarct sizing 278
Distribution of enzymes
  between fluid compartments 100, 102, 103, 104
  between fluid compartments for infarct sizing 273
  between vascular and extravascular compartments 104, 105
  factors influencing 109, 110, 274
  one-compartment model 359
  one- and two-compartment models for CK 274
  physiologic models 276
  population distribution 204
  variability between individuals 108, 109, 342
Dog heart ischaemia 380–389
Double reciprocal plots
  competitive and non-competitive inhibitors 168
Drug overdose
  enzyme leakage 238

Electrocardiography
  correlation with enzymatic infarct size 323
  correlation with enzyme leakage 240, 241, 535, 549, 550, 551, 552
  estimation of infarct size 357, 530, 531, 549, 550, 551, 552
  glucose protection 541
Electrophoretic separation
  CK isoenzymes 212
  isoenzymes 189
  LD isoenzymes 138, 221
Energy metabolism
  anoxia 5
Enzyme
  active centre 151
  activity 15
  clearance 15, 97, 98
  differential clearance from blood 99
  distribution and clearance, individual variability 108, 109
  distribution and elimination 97, 103, 104
  distribution at sub-cellular level 121
  distribution, factors influencing 122
  fractional disappearance rates 100
  half-life in plasma of diagnostically important enzymes 128
  inactivation 97
  inactivation in lymph 111

isoenzymes 133
normal plasma activities 125
removal by reticuloendothelial system 111
selection for diagnostic use 129
specificity 15, 202, 203, 226, 227, 228
time activity curves in plasma 101
time course after myocardial infarction 202
tissue distribution profile 119
tissue specificity 15
value for detection of tissue injury 116
vascular and extravascular pools 102
Enzyme activity
activation of enzymes 170
allosteric control 157
allosteric factors 158
coenzymes and prosthetic groups 172
competitive inhibition 166, 167
consecutive reactions 158, 159, 160
double reciprocal plots 153
effect of enzyme concentration 147
effect of pH 173
effect of substrate concentration 151, 152, 153
effect of temperature 174, 175
enzyme-substrate complex 149, 150
fixed time assays 149, 150
inhibition and activation 165
inhibition, irreversible 166
inhibition, mixed 169
inhibition, non-competitive 167
inhibition, reaction products 169
inhibition, reversible 166, 167
inhibition, uncompetitive 168, 169
initial rates of reaction 148
international units 182
K_m 153
K_m and V_{max} 154
kinetic measurements 149
lag phase 147
measurement of reaction progress 147
measurement of substrates or products? 147
metal ion participation 171
myocardial infarction 224
phases of reaction 147, 148

pH optimum 174
ping-pong bi-bi mechanism 163, 164
rectilinear phase 147
reference values 204
reversible and irreversible reactions 158, 159
specificity and sensitivity 146
standards 183
substrate specificity 151
temperature optimum 174, 175
two substrate reaction 160, 163, 164
units of activity 182, 341
Enzyme assay
AST assay 184, 209
automated procedures 181
blanking 179
choice of enzyme 204
choice of temperature 176
clinical requirements 201
CK assay 184, 210
CK assay, practical considerations 210
CK isoenzyme assay 212, 213, 214
CK radioimmunoassay 247
fluorimetric methods 180
international recommendations 206
isoenzyme assays 187
isoenzyme assays, practical considerations 190
LD assay 184, 215
kits 207
methods for measuring chemical changes 176
optimized assays 185
pH and buffering 174
practical considerations 206
reproducibility 186
sensitivity 201
sensitivity and specificity 202, 203
specificity 201
standardization and quality control 183
standardization and quality control, practical considerations 186
temperature 174, 175
Enzyme clearance
influence of drugs 110
influence of haemodynamics 109, 110
influence of metabolism 110
mechanisms 110

Enzyme clearance (*contd.*)
  two-compartment model 106, 107
Enzyme content
  of coronary and lymphatic fluid after heart work 88
  of coronary and lymphatic fluid after myocardial infarction 90
  of coronary and lymphatic fluid under normal conditions 87
Enzyme distribution
  determination of fractional disappearance rates 102
  fluid compartments 99
  two-compartment model 106, 107
Enzyme inactivation
  lymph 98
  mechanisms 110
Enzyme kinetics
  activation and inhibition of allosteric enzymes 158
  activation of enzymes 170
  allosteric enzymes 155, 157
  catalytic differences for the separation of isoenzymes 194
  consecutive reactions 158, 159
  cooperativity 156
  double reciprocal plots 153, 161, 162
  effect of enzyme concentration 147
  effect of substrate concentration 151, 152, 153
  enzyme-substrate complex 149, 150
  equilibria and steady-states 153
  Hill plots 156
  inhibition and activation 165
  inhibition by reaction products 169
  inhibition, competitive 166, 167
  inhibition, mixed 169
  inhibition, non-competitive 167
  inhibition, reversible 166, 167
  inhibition, uncompetitive 168, 169
  K_m 153
  K_m and V_{max} 161, 162
  Michaelis–Menten 154
  ping-pong bi-bi mechanism 163, 164
  sigmoidal relationships 155
  substrate concentration and reaction velocity 163, 164
  two substrate reactions 155, 160
  V_{max} 152, 153
Enzyme leakage

acetate effects 493
beta-blocker effects 372
calcium effects 399, 441, 562
catecholamine effects 88, 400
clearance via coronaries 86
correlation between tissue depletion and serum enzyme levels 343, 362, 364
correlation of infarct size and serum enzyme time course 329
correlation with electrocardiographic changes 240, 241, 260, 323, 532, 535, 549, 550, 551, 552
correlation with infarct size, influencing factors 340
correlation with other indices of injury 534
correlation with ultrastructural changes 445
cyclic AMP effects 564
diabetic effects 491
differential time courses 126, 225
differential tissue distribution patterns 117
during accidental trauma 237
  during acidosis and anoxia 514, 515
during acute coronary insufficiency 231
during alcohol intoxication 239
during angina 231
during anoxia 421, 422, 447, 504, 507
during arrhythmias 233
during calcium-free perfusions 506
during calcium paradox 402, 409, 410, 422, 423, 430
during cardiac catheterization 236
during cardiac failure 234
during cardiac massage 233
during cardiac surgery 237
during cell turnover 126
during cerebrovascular injury 238
during coma 237
during complicated myocardial infarction 232
during direct current countershock 233, 278
during dissecting aneurysm of aorta 234
during drug overdose 238

during exercise 239
during infarct extension 232
during intramuscular injections 238
during irreversible injury 51
during ischaemia 52, 422, 423
during myocarditis 234
during non-cardiac surgery 236
during pericarditis 234
during pulmonary embolism and infarction 235
during reinfarction 232
during reoxygenation 54, 410, 421, 422, 428, 429, 447
during reperfusion 54, 410, 422, 423
during stepwise calcium repletion 432
during stepwise reoxygenation 431
during tachycardia 234
energetics 441
factors influencing 53, 126, 233
fatty acid : albumin ratio effects 492
fatty acid effects 453
glycogen effects 564
glycolytic effects 495, 564
glucose effects 453
glucose effects in anoxia 448, 486
glucose effects in regional ischaemia 487
glucose–insulin–potassium effects 490
HBD clearance 346
heart work effects 88
high energy phosphate effects 562
hyaluronidase effects 541, 547
hyperthyroidism effects 508
hypothermia effects 238
inotropic effects 508
insulin effects in ischaemia 489
intrinsic sympathomimetic activity effects 517
irreversible damage marker 259
local denaturation 344
lymphatic and coronary clearance during myocardial infarction 89
lymphatic clearance 86, 394
lymphatic clearance during myocardial infarction 89
lymphatic transit time 92
mechanisms of normal leakage 64
membrane mechanisms 75
metabolic basis 561

models for anoxia and ischaemia 393, 394
necrosis 561
prediction by log-normal function 297
prediction by non-linear solutions 297
prediction methods currently used 298
prognosis 224
proportion appearing in blood 97
pyruvate effects 493
rat heart model 403
reduction with therapeutic agents 369, 370, 371
relation between plasma and tissue enzyme profiles 127, 263
relation to cell death 294
relation to infarct location 322
relation to ischaemic damage 10, 41
relation to reversible and irreversible damage 11, 360
sampling times for enzyme assays 225
species variation 450, 504
steroid effects 521
substrate effects 488, 495
temperature effects 435, 436, 437, 438, 442
time course 53, 54
time course and peak activity during uncomplicated myocardial infarction 224
tocopherol effects 521, 525
transient membrane lesions 75
verapamil effects 518, 519
viable cells 261
Enzyme profiles
  during uncomplicated myocardial infarction 222, 223, 224
Enzyme specificity
  basic principles 117
Enzyme substrate complex 149, 150, 161, 162
Enzyme tissue distribution
  disease effects 124
  physiological factors affecting 124
  species differences 123
Exercise
  enzyme leakage 239
Exocytosis
  of proteins 65, 67

Fatty acids
  antilipolytic agents 493
  fatty acid:albumin binding ratios 483
  metabolism in ischaemia and anoxia 7, 483
  tissue damaging effect 484, 492
  tissue damaging effect in ischaemia 453
Fixation methods
  anticoagulation 465
  calcium content of fixative effects 464
  cell swelling and shrinking 469, 470, 471
  distinction between artefacts and pathological changes 474
  factors influencing quality of fixation 467
  formaldehyde 467
  for ultrastructural studies 462
  glutaraldehyde 467
  immersion fixation 462
  ion retention for X-ray microanalysis 477
  osmium tetroxide 468
  osmolality effects 468, 469, 470, 471
  perfusion fixation 463
  pH effects 472
  pressure of fixation 463
  temperature effects 464, 473
Fluorimetry
  measurement of enzyme activity 180
Fractional disappearance rates
  CK isoenzymes 279
  complications and calculations 102, 103, 104, 105, 360
  differences between CK and HBD 347
  differences between endogenous and exogenous CK 359
  enzymes from blood 100
  factors influencing 109, 110, 270, 347, 348
  HBD 347
  in infarct sizing 263, 269, 361
  two-compartment and one-compartment models 107
  variability between individuals 108, 109

Gamma glutamyl transferase
  in myocardial infarction 230
Genetic enzyme variants
  allelozymes 136
Genetic factors
  in enzyme tissue distribution or activity 124
Glucose
  metabolism in ischaemia 483
  protection in anoxia 448, 485, 505
  protection in ischaemia 452, 487, 540
Glucose–insulin–potassium
  protection in ischaemia 490
  reduction of infarct size 370
Glutamate dehydrogenase
  tissue distribution 117
Glycogen
  enzyme leakage effects 564
Glycolytic activity
  enzyme leakage and cell death 495, 563, 564
Glycolytic flux
  and tissue protection 495
  control in ischaemia and anoxia 6
Gram equivalents
  CK 363

Half-life in plasma
  diagnostically important enzymes 128
Healing process 544
Heart work
  and ischaemic injury 496
Heat inactivation
  separation of LD isoenzymes 217
Heat stable lactate dehydrogenase 217
High energy phosphates
  compartmentation 563
  during anoxia 509
  during anoxia and acidosis 514
  during calcium paradox 405, 406
  during irreversible injury 36
  during ischaemia 30, 33, 483
  enzyme leakage 562
  membrane integrity 69
  relation to cell death and tissue necrosis 34
  relation to reversibility of damage 563
High pressure liquid chromatography

separation of isoenzymes 192
Hill plots
  enzyme kinetics 156
Histochemistry
  assessment of infarct size 536, 542
Hyaluronidase
  infarct size reduction 541
  ischaemic damage 454
  mechanism of action 542
  time of intervention 546
Hydroxybutyrate dehydrogenase
  during cardiac surgery 346
  during myocardial infarction 346
  infarct size estimation 339
  LD isoenzyme separation 218
  stability after death 343
  variability between individuals 341
Hypothermia
  enzyme leakage 238
Hypoxia
  definition 2
Hypoxia models, see Models of anoxia

Immunochemical methods
  separation of CK isoenzymes 248
  separation of isoenzymes 195
  separation of isoenzymes, practical considerations 196
Inactivation of enzymes
  blood 99
  lymph 98, 111
  mechanisms 110
Inactivation methods
  separation of isoenzymes 193
Individualized fractional disappearance rates
  CK 269, 360
Infarct size
  beta-blocker effects 372, 540
  congestive heart failure 327
  coronary flow effects 537
  electrical instability 266, 365, 366
  extension 531
  glucose effects 540
  hyaluronidase effects 541
  membrane stabilizer effects 543
  mortality and morbidity 266
  progression with time 48, 50, 51
  reduction by therapeutic agents 281, 315, 369, 370, 371, 531, 537
  steroid effects 544

timing of intervention 546
ventricular compliance 266
ventricular failure 324, 325, 326, 328
ventricular function 266, 366
Infarct size estimation 257
Infarct size estimation
  assumptions 259, 261, 263, 359, 360, 361, 362
  AST 345
  baseline corrections 266
  basic principles 261
  body weight 363
  choice of enzyme 339
  clinical follow-up study 334
  clinical studies 266, 267, 279, 319
  clinical utility 335, 356, 371
  CK isoenzyme advantages 363
  CK isoenzyme disadvantages 350
  CK isoenzymes 277
  CK versus CK isoenzymes 265
  correlation between different enzyme markers 345
  correlation with electrocardiographic assessments 323
  correlation with histochemical assessments 319, 367
  correlation with morphological assessments 264
  correlation with serum CK time course 329, 330, 331, 332
  direct current countershock effects 278
  distribution volumes for CK 272
  electrocardiographic methods 357
  enzyme leakage and irreversible injury 259
  enzyme time activity curves 261
  equations 287, 288, 289, 320
  errors, artefacts and complications 266, 268, 280, 340, 350, 360, 362
  experimental dog studies 265
  extra-cardiac enzyme 280
  factors influencing enzymatic estimations 268, 339, 340
  factors influencing enzyme clearance 284
  factors influencing enzyme fractional disappearance rates 271
  fractional disappearance rates 269, 270, 271

CK versus CK isoenzymes (contd.)
  fractional disappearance rates for CK 263
  fractional disappearance rates, errors 360
  fractional disappearance rates, individualized 361
  future developments 282
  gram equivalents 363
  HBD utility 345, 349, 350
  histochemical methods 536
  individualized fractional disappearance rates 361
  limitations 281
  location of infarct 321, 322
  lymphatic system 277
  metabolic methods 358, 367
  model 320
  mortality 333
  non-enzymatic methods 258, 292, 356
  one-compartment model 358
  one-compartment model, practical considerations 359
  parameters of the enzymatic model 269
  physiologic models 275, 276
  possibilities, limitations, and future developments 281, 355
  practical considerations 281
  problem with CK isoenzymes 350
  prognosis 267, 365
  reperfusion effects 275, 340
  sampling techniques 268
  sub-groups of patients 330, 331, 332
  systematic and non-systematic errors 348
  technetium 99-m 368
  tissue enzyme depletion in relation to plasma levels 273
  units 363
  variations of enzyme content between individuals 341
  validation of model 266
Infarct size index 363
Infarct size prediction 291
Infarct size prediction
  assumptions 295
  baseline corrections 303
  baseline determinants 304
  basic principles 295
  biologic algorithms for curve projection 300
  choice of initial parameter estimates 310
  computer algorithms 304
  correlation between observed and predicted infarct size 311
  critical determinants 298
  current methods 298, 304, 305, 306
  delay in prediction time 310
  effect of interventions 308
  errors of projection 301
  factors influencing 314
  imprecision of early data 311
  initial CK value 303, 305
  initial enzyme elevation, definition 310
  initial estimates 298, 299, 310
  large and small infarcts 299
  limitation of parameter range 310
  log-normal function for curve projection 296
  non-linear solutions 297
  parameter limitation in projection of enzyme leakage 302, 303
  physiologic and empiric models 296, 312
  possibilities and limitations 293, 310
  projection of enzyme leakage 296
  reduction of prediction time 314
  time 313
  validation 305
  validation studies in dogs 306
  validation studies in patients 306, 308, 309
  validation studies in patients using matched controls 307
  zero time definition 303, 305
Inhibition of enzyme activity
  by reaction products 169
Inhibition of enzymes
  reversible and irreversible 165
Insulin
  effect upon ischaemic damage 488
International recommendations
  for enzyme assay conditions 207
International units
  of enzyme activity 182
Interstitium
  anatomy and composition 83
  fluidity 83

Intramuscular injections
  enzyme leakage 238
Intrinsic sympathomimetic activity
  beta-blockers and enzyme leakage 517
Iodination
  procedures for radioimmunoassay 250
Ion exchange methods
  separation of CK isoenzymes 213
  separation of isoenzymes 191
Ion pumps
  cell membrane 63, 69
Ion transport
  across cell membranes 64
  membrane defects 69
Ionic changes
  during reperfusion 43, 45
  calcium during reperfusion 46
Irreversible enzyme inhibition
  characteristics and kinetics 165, 166
Irreversible enzyme reactions 158
Irreversible injury 10
Irreversible injury
  enzyme leakage 51, 259
  final events 51
  loss of membrane resealing ability 71
  metabolic changes in ischaemia 27, 36
  progression with time 48
  transmural progression 47
  ultrastructural changes in ischaemia 37, 38, 39, 40
Ischaemia
  acetate metabolism 493
  albumin effects 492
  anaerobic glycolysis 29
  antilipolytic agents 493
  border zone 2, 29
  carbohydrate metabolism 484
  catecholamine release 33
  cellular changes 3, 4, 27
  collateral flow 25
  contracture 8
  coronary flow redistribution 25
  coronary flow, regional 539
  definition 1, 29
  diabetes 490
  difference to anoxia 5
  differential susceptibility of cell types 3
  electrophysiological changes 5, 7
  energy metabolism 29, 30
  enzyme leakage 52, 423
  factors influencing 2
  fatty acid effects 484, 492
  fatty acid metabolism 8
  final events of cell injury 51
  glucose effects 487
  glucose–insulin–potassium effects 490
  glycolytic flux, importance 495
  gradients of coronary flow 28
  insulin effects 488
  ionic changes 8, 10, 73
  ion leakage 13
  macromolecule leakage 10
  membrane changes 10
  membrane damage 72
  metabolic changes 5, 6, 7, 32, 483, 487
  metabolic changes during irreversible phase 36
  metabolic changes during reversible phase 28
  metabolite leakage 10
  oxidation reduction changes 28
  pH changes 31
  progression with time 3, 22
  pyruvate metabolism 493
  rat model 419
  reduction of tissue damage and infarct size 281, 531
  regional coronary flow 539
  regional contractility 31
  relation to enzyme leakage 294
  reperfusion 27
  reperfusion during irreversible phase 40
  reperfusion during reversible phase 35
  substrate metabolism 482, 483, 494
  time course of enzyme leakage 51
  time course of injury 294
  ultrastructural changes 8, 9, 451, 452
  ultrastructural changes during irreversible phase 37, 38, 39, 40
  ultrastructural changes during reversible phase 35
  vascular changes 8, 25, 46
Ischaemia models, *see* Models of ischaemia

Ischaemia versus anoxia
  definitions 1, 29
  in experimental models 391
Isoelectric focusing methods
  separation of isoenzymes 192
Isoenzymes
  aldolase structural sub-units 136
  alkaline phosphatase 136
  assay 187
  AST 141
  CK 139
  CK isoenzyme assay 212
  CK structural sub-units 134
  CK sub-units changes 141
  definition 133
  diagnostic sensitivity and specificity 226, 227
  genetic and structural principles 187, 188
  identification and measurement 187
  LD 138
  LD structural sub-units 134
  LD sub-unit changes 139
  separation by catalytic difference methods 194
  separation by electrophoretic methods 189
  separation by high pressure liquid chromatography methods 192
  separation by immunochemical methods 195
  separation by ion exchange methods 191
  separation by isoelectric focusing methods 192
  separation by selective inactivation methods 193
  structural basis 134
  structural differences 188
Isolated mitochondria
  tissue damage 509

Katals
  units of enzyme activity 183
Kinetics
  distribution and clearance of enzymes 100, 102

Lactate dehydrogenase
  assay, practical considerations 216
  assay procedure 184, 215
  during uncomplicated myocardial infarction 223, 224
  half-life in plasma 128
  isoenzymes 138, 187
  isoenzyme sub-units 138
  reversibility of reaction 215
  tissue distribution 117, 118, 138
Lactate dehydrogenase isoenzymes
  assay, practical considerations 214, 215, 218, 219
  during myocardial infarction 139
  sample collection and storage 220
  sensitivity and specificity for diagnosis 228
  separation 135, 138
  separation by catalytic difference methods 194
  separation by electrophoretic methods 189, 220
  separation by heat inactivation methods 217
  separation by hydroxybutyrate reaction 218
  separation by ion exchange chromatography methods 191
  separation by selective inactivation methods 193
  separation by stability in urea 218
  sub-unit composition 134
  sub-unit changes 139
Lag phase
  in CK assay 210, 211
  in consecutive reactions 159
  in measurement of enzyme activity 147
Leakage of cellular constituents
  clearance routes 13
  controlling mechanisms 11
  enzymes and macromolecules 116
  enzymes in ischaemia 10, 52
  ions 11, 116
  ions in ischaemia 10
  macromolecules 11
  macromolecules in ischaemia 10
  metabolites 11, 116
  metabolites in ischaemia 10
  relation to ischaemic damage 12
  time course of leakage 11, 12, 14
Location of infarct
  relation to infarct size 321
Log-normal function

infarct size prediction 296
infarct size prediction, complications 310
Lymphatic system
  clearance of enzymes 81, 91, 276, 394
  clearance of enzymes during myocardial infarction 89
  clearance of leaked materials 13
  inactivation of CK in cardiac lymph 273, 277, 362
  inactivation of enzymes in cardiac lymph 98, 111
  flow rate of lymph 91, 277
  functional anatomy 82, 84, 85
  normal enzyme content of cardiac lymph 86, 87
  pores 85
  transit time of cardiac lymph 92
Lysosomal enzymes
  ischaemia 7
  marker enzymes 122
  membrane damage 62, 72

Malate dehydrogenase
  tissue distribution 117
Markers
  of ischaemic damage 15
Membrane
  biosynthesis 62
  capillary membrane transport and permeability 82, 91
  cellular and sub-cellular 59
  damage 68
  damage and calcium 73
  damage and enzyme leakage during irreversible injury 51
  damage and enzyme leakage 563
  damage due to cell swelling 70
  damage transient 75
  degradation 62
  energy dependence 62
  enzyme leakage 63
  enzyme leakage mechanisms 75
  exocytosis of proteins 65
  extrinsic proteins 61
  intrinsic proteins 61
  ion pumps 63
  ion transport 63, 64
  ion transport in ischaemia 69
  leakiness 11, 68

lipid bilayer 60
lipoproteins 60, 61
lysosomal breakdown of membranes 62
macromolecule transport 63, 67
metabolite transport 63
normal transport of macromolecules 65
normal transport of proteins 64
permeability changes 68
permeability characteristics 59
protease damage 72
reoxygenation induced changes 45
repair 62
reperfusion induced changes 45
resealing 70
resealing, effects of pH, calcium, and energy 71
selective permeability 63
sieving 70
signal peptides for protein transport 66
slits 70
stabilization and infarct size 543
structure 60, 61
structure, intrinsic and extrinsic proteins 61
temperature effects 442
transient damage 75
turnover 60, 62
Membrane stabilizing drugs
  enzyme leakage 521
Metabolism
  control 481
  of the ischaemic and anoxic myocardium 5, 494
Methoxamine
  and infarct size 539
Michaelis-Menten
  the effect of substrate concentration on rate of reaction 141, 152, 153
Microsome
  marker enzymes 122
Microspheres
  coronary flow measurement 25
Mitochondria
  anoxia 510
  anoxia, function 509
  anoxia with acidosis, function 516
  AST 208
  calcium uptake 523

Mitochondria (contd.)
  CK isoenzyme 141
  ischaemia 8
  marker enzymes for mitochondria 121
  membrane effects 522
  reperfusion effects 44
  role in tissue damage and enzyme leakage 411
Models of anoxia and hypoxia 389, 390
Models of anoxia and hypoxia
  choice of animal 390
  cultured hearts 390
  enzyme leakage, measurement 392
Models of anoxia and reoxygenation 419
Models of the calcium paradox 420
Models of enzyme leakage 379
Models of ischaemia 21, 380, 389, 419
Models of ischaemia
  catecholamine induced 385
  chemically induced 384
  choice of animal 390
  closed-chest techniques 383, 384, 385
  cultured hearts 390
  enzyme leakage measurement 392
  ligation of circumflex coronary artery 381, 382
  ligation of left anterior descending coronary artery 380, 381, 382, 383
  microsphere induced 384
  open-chest regional ischaemia 380, 381, 382, 383
  papillary muscle 382
  progressive occlusion of coronary artery 383
  rat heart model 536
  regional ischaemia 380–385, 387, 388, 389
  species: dog, rat, mouse, baboon, rabbit, pig, guinea pig 380–389
  whole heart ischaemia 385–389
Models of ischaemia and reperfusion 419
Models of myocardial infarction 21
Models of myocardial infarction
  dog heart 293
Models of myocardial tissue damage 379

Molar absorptivity
  nicotinamide adenine dinucleotides 180
Molecular weight
  as a determinant of transport routes 92
  of diagnostically important enzymes 128
Morphometric analysis
  for ultrastructural studies 474
Mucopolysaccharides
  and molecular transport 83
  of interstitium 83
Myocardial infarction
  AST profiles 229
  choice of diagnostic enzymes 241
  CK isoenzyme profiles 253
  correlation between CK leakage and other indices of damage 365
  criteria for diagnosis 201
  detection and quantification 258
  during cardiac surgery 237
  enzymes in cardiac lymph 89
  enzyme profiles in complicated cases 232
  enzyme profiles in serum 223
  enzyme profiles in uncomplicated cases 221
  enzyme time course 202
  extension 232
  infarct location 321
  infarct size, see Infarct size estimation and Prediction
  infarct size estimation, clinical study 319
  infarct size prediction, model 293
  insulin levels 488
  models, see Models of myocardial infarction; Models of ischaemia; Models of anoxia
  non-myocardial enzyme changes 230
  reinfarction 232
  sampling times for enzyme assays 224, 225
  substrate metabolism 484, 494

Necrosis
  and enzyme leakage 561
  and healing process 545
  histochemical quantitation 52
  progression with time 50

INDEX
583

Nicotinamide adenine dinucleotides
  absorption spectra 178
  linked assay for AST 208
  linked assay for CK 210
  linked assay for LD 215
Non-competitive inhibition
  of enzyme activity 167
Normal ranges 204

One-compartment model
  CK distribution 274
  infarct size estimation 358
  infarct size prediction 312
  practical considerations 359
Optimization
  of enzyme assays 185, 207
Ornithine carbamoyl transferase
  in myocardial infarction 230
Osmolality
  of fixatives for ultrastructural studies 468
Oxygen delivery
  and infarct size 537
  and tissue damage 531
Oxygen paradox
  definition 440
  mechanism 440

Papillary muscle
  anoxic model 389
  ischaemic model 382
Permeability
  of capillary endothelium 91
pH
  enzyme activity 173
  fixatives for ultrastructural studies 472
  ischaemic changes 31
  lysosomes and membrane damage 72
  protective effect of mild acidosis 513
  protons and contractile failure 31
Physiologic models
  infarct size estimation and prediction 275, 276, 296, 312
Pig heart
  ischaemic model 381–389
Plasma enzymes
  normal activities 125
  relation to tissue enzyme profiles 127
  roles and sources 125

Prediction of infarct size *see* Infarct size prediction
Product inhibition
  of enzyme activity 169
Prognosis
  and enzyme leakage 224
  and infarct size 267
Prosthetic groups 172
Protection of the anoxic myocardium 505
Protection of the anoxic myocardium
  acidosis 513
  beta-blockers 517
  future possibilities 561
  glucose 448, 485
  membrane stabilizing drugs 521
  reserpine 449
  therapeutic agents 511
  verapamil 449, 518, 519
Protection of the ischaemic myocardium 292
Protection of the ischaemic myocardium
  anti inflammatory agents 543
  basic principles 530, 531
  basic requirements 548
  beta-blockers 372, 540
  clinical utility 548
  future possibilities 561
  glucose 452, 487, 540
  glucose–insulin–potassium 490
  glycolytic flux 495
  healing process 544
  hyaluronidase 455, 541, 547
  infarct size reduction 281, 315, 369, 370, 371
  insulin 489
  interventions 531
  mechanisms 536, 537
  membrane stabilizers 543
  metabolic manipulations 497
  osmolality 496
  steroids 496
  substrate metabolism 494
  timing of interventions 546
Protons
  and contractile failure 31
  in ischaemia 6, 31
Pulmonary embolism and infarction
  enzyme leakage 235
Pyruvate
  LD assay 215
  metabolism in ischaemia 492

Quality control
  in enzyme assays 183
Quantification of infarct size 257
  see also Infarct size estimation and Prediction

Rabbit heart
  anoxic model 389
  calcium paradox 402
  enzyme leakage 402, 418
  ischaemic model 382–389
  perfusion 506
Radioimmunoassay
  antibody preparation 249
  basic requirements 248
  CK iodination procedures 250
  CK isoenzyme determination 196, 247
  CK isoenzymes, clinical study 253
  sensitivity and specificity 146
  technical considerations 254
Reference values 204
Regional ischaemia
  glucose effects 487
  models 380–389
  substrate utilization 482
Reoxygenation
  duration of anoxia 427
  enzyme leakage 54, 410, 417, 421, 447
  membrane changes 45
  models 419
  stepwise 431
  temperature effects 433
  tissue damage and enzyme leakage 417
  ultrastructural changes 426, 447
Reperfusion
  calcium changes 46
  cell swelling 44, 45
  during irreversible phase of injury 40, 41, 42, 43
  during reversible phase of injury 35
  enzyme leakage 54, 410, 417, 423
  extension of tissue damage 55
  ionic changes 43, 45
  membrane changes 45
  mitochondrial changes 44

  reversibility of damage 27
  tissue damage and enzyme leakage 417
  ultrastructural changes 40, 41, 42, 43, 451, 452
Resealing of membranes
  energy, pH, calcium 71
  reversibility of damage 71
Reserpine
  protection in anoxia 449
Reticuloendothelial system
  enzyme removal 111
Reversible enzyme inhibition 165, 166, 167
Reversible enzyme reactions 158
Reversible damage
  metabolic changes in ischaemia 28
  ultrastructural changes in ischaemia 35
Reversibility of damage
  in ischaemia 27
  in relation to enzyme leakage 10
  in relation to high energy phosphates 563
  in relation to loss of membrane resealing 71
  in relation to metabolism 10
  timing of interventions 546
  transition of reversible to irreversible damage 27

Salvage
  of ischaemic myocardium 281
  of ischaemic myocardium, reduction of infarct size 369, 370, 371
Secretory proteins
  synthesis 66
Sectioning methods
  for ultrastructural studies 465
Sensitivity 226, 227, 228
Sensitivity
  CK isoenzymes 279
  definition 202, 203
  radioimmunoassay for CK isoenzymes 251, 252
SI units of enzyme activity
  katals 183
Signal peptides
  for membrane protein transport 67
Species differences
  in anoxia-induced damage 450
  in enzyme leakage 504
  in tissue content of enzymes 279, 341
  in tissue distribution of enzymes 123

Specificity   226, 227, 228
Specificity
  CK isoenzymes   278, 282
  definition   202, 203
  radioimmunoassay for CK isoenzymes
    251, 252
Spectrophotometry
  absorption and transmission   178
  Beer–Lambert law   177
  blanking   179
  measurement of enzyme activity   177
  molar absorptivity   177
  practical considerations   178, 179
  spectrophotometer design   179
Stability
  cardiac enzymes after death   343
Standards
  for enzyme assays   183
Steroids
  enzyme leakage   521
  infarct size   544
Sub-cellular distribution profiles
  enzymes of diagnostic importance
    121
Substrate metabolism
  enzyme leakage and tissue damage
    453
  in angina   482
  in diabetes   491
  in infarcting myocardium   494
  in ischaemia   482, 483
  in normal tissue   482
Sulphydryl groups
  activation of CK   210
  activation of enzymes   170
Surgery
  enzyme leakage   236

Tachycardia
  enzyme leakage   234
Temperature
  effect on calcium paradox   404, 434, 435, 436, 437, 438
  effect on enzyme activity   174, 175
  effect on enzyme inactivation   175
  effect on enzyme leakage   435, 436, 437, 438
  effect on reoxygenation   434, 435, 436, 437, 438
  enzyme assays, practical considerations   207

LD isoenzyme inactivation   217
Thyroid state
  enzyme leakage   508
Time course
  for enzyme profiles during uncomplicated myocardial infarction   222
  for ischaemic injury   294
Tissue damage
  catecholamines   400
  methods of assessment   530
Tissue distribution of enzymes
  CK isoenzymes   140, 280, 283
  combined enzymes   119
  diagnostically important enzymes   117, 118, 119
  differential patterns   117
  factors influencing   122
  LD isoenzymes   138
Tissue enzyme profiles
  relation to plasma profiles   127
Tocopherol
  enzyme leakage   521, 525
Transport
  across capillary membranes   82
  across cell membranes, macromolecules   63, 65
  across cell membranes, proteins   67
  across interstititium   84
  between vascular and extravascular compartments   103, 104
Two-compartment model
  CK distribution   274
  enzyme distribution and clearance   106, 107
  infarct size prediction   312
  versus one-compartment model   107

Ultrastructural changes
  distinction between artefacts and pathological changes   474
  during anoxia   426, 446, 505, 510
  during calcium paradox   424, 425, 454
  during irreversible injury   37, 38, 39, 40
  during ischaemia   8, 9, 35, 49, 451, 452
  during reoxygenation   426, 447
  during reperfusion   451, 452
  during reperfusion in irreversible phase   40, 41, 42, 43

Ultrastructural changes (contd.)
  during reperfusion, mitochondrial changes 44
  during reversible ischaemia 35
Ultrastructural methods
  artefacts 461
  artefacts of fixation 462
  fixation 420, 462
  fixation, see also Fixation methods
  morphometric analysis 474
  perfusion fixation 462
  sectioning 466
  X-ray microanalysis and ion retention 477
Ultrastructure
  artefacts 474
  cell swelling and shrinking 469, 470, 471
  correlation with enzyme leakage 445, 455
  distinction between artefacts and pathological changes 474
  effect of anaesthetics 466
  enzyme leakage 445, 455
  glucose protection during anoxia 448
  glucose protection during ischaemia 452
  normal tissue 36, 424
  species differences in anoxia 450
  verapamil protection during anoxia 519
Uncompetitive inhibition

enzyme activity 168, 169
Urea-stable
  assay for LD isoenzymes 218
Utility of enzymes for detection of injury 116

Variability of enzymes
  content between individuals 341
  content in different areas of the same tissue 341
  distribution between individuals 342
  factors influencing 205
Vascular exchange
  CK 274
Vasoconstrictors
  and infarct size 538
Vasodilators
  and infarct size 537
Verapamil
  enzyme leakage effects 518, 519, 521
  infarct size reduction 370
  mitochondrial function effects 522
  protection in anoxia 449

Work
  effect upon enzyme levels in coronary and lymphatic fluid 88

X-ray microanalysis
  ion retention for ultrastructural studies 477

THE LIBRARY
UNIVERSITY OF CALIFORNIA
San Francisco
THIS BOOK IS DUE ON THE LAST DATE STAMPED BELOW

Books not returned on time are subject to fines according to the Library Lending Code. A renewal may be made on certain materials. For details consult Lending Code.

14 DAY	14 DAY	14 DAY
FEB 20 1980	MAY 6 1981	NOV 1 3 1981
RETURNED	RETURNED	RETURNED
FEB 25 1980	MAY 4 1981	NOV 9 1981
14 DAY		14 DAY
NOV 2 1980	14 DAY	MAR 2 2 1982
RETURNED	JUN 9 1981	RETURNED
NOV - 4 1980	RETURNED	
	JUN 7 1981	MAR 1 5 1982
14 DAY	14 DAY	14 DAY
APR 7 1981	JUL 12 1981	SEP 10 1982
RETURNED		RETURNED
APR 7 1981	RETURNED	SEP 1 0 1982
	JUL 12 1981	

Series 4128